Ionizing radiation and life

Ionizing radiation and life

An introduction to radiation biology and
biological radiotracer methods

VICTOR ARENA, Sc.D.

With 271 illustrations

The C. V. Mosby Company

Saint Louis 1971

To the late Professor G. B. Alfano,
who stimulated my interest in the biologic sciences during
my college years

Preface

The purpose of this text is to present, in a semielementary form, an integrated and systematic treatment of the biologic effects of ionizing radiations and the basic principles of radiotracer methods as applied to the life sciences.

The idea of this book was born from my experience in teaching radiation biology and radiotracer methods to undergraduate and graduate biology students. I noticed that it was impossible for me to suggest only one text to my students for this course. Indeed, excellent books are available, but none of them, to my knowledge, covers both radiation biology and radiotracer methodology in an integrated manner. And yet, courses of this type are being introduced into an increasing number of colleges and universities.

This book was designed and written for undergraduate junior and senior students majoring in biology, for graduate biology students who have never taken a course in radiation biology and radiotracer methods, for radiology residents as a preliminary introduction to the more specialized courses of diagnostic radiology and radiation therapy, and for medical students oriented toward nuclear medicine. It may also be useful to specialists in other fields of the life sciences who need some general knowledge of radiation biology and radiotracer methods.

I have assumed the reader to have the following minimum background: general, inorganic, and organic chemistry; general biology; botany; zoology; genetics; and elementary anatomy and physiology. Whenever necessary, background information in more specialized areas is presented in this text. I am aware that many users of this book may not be familiar with mathematics beyond elementary algebra and trigonometry; so this is the only mathematical knowledge assumed here. As much as possible, equations are derived with simple algebraic manipulations. More advanced procedures are carefully avoided. I am also aware that some readers may not have taken any physics courses at the college level or may need some review of fundamental physical concepts. Chapter 1 is an attempt to fill this vacuum. The material reviewed in this chapter is the minimum essential background for the understanding of many aspects of atomic, nuclear, and radiation physics which will be covered in later chapters and for the intelligent use of nuclear instrumentation. Readers who are already familiar with the basic concepts of physics may ignore Chapter 1 altogether.

The primary intent of the book is suggested by a key word in the subtitle — "introduction." The term is not used merely to signify that this book is an elementary treatment of radiation biology and radiotracer methods; the book is also designed to provide the student with some preliminary information that is essential as a tool for a thorough understanding of radiobiologic problems and for the intelligent use of radiation-generating and -detecting equipment. This is the premise for the organization of the whole subject matter and for the emphasis given to the content of some chapters.

The chapter on the electromagnetic spectrum presents an integrated survey of the nature and physical properties of radiations that are relevant not only to radiation biology but also to other branches of the biologic sciences. An X-ray unit is perhaps the most common radiation source used in radiobiologic experimentation. It is not conceivable that a student who is being introduced to radiation biology and who is likely to use an X-ray machine should be left with only scant and incomplete information about the generation and properties of X-rays and how to use an X-ray machine. Therefore, a chapter is exclusively devoted to these topics.

The main purpose of the chapter on nuclear fission is to serve as an introduction to the nuclear reactor and the atomic bomb. A nuclear reactor is the most intense source of neutron radiation available at present. To fully appreciate how neutrons are used for production of artificial radioisotopes, for irradiation of living material, or for activation analysis studies, the student must know what a reactor is and how it works. The knowledge of the phenomenon of fission is also essential for a comprehension of the biologic and medical aspects of criticality accidents. Information on atomic explosions aids the reader in understanding the biologic effects of acute exposure to mixed neutron-gamma radiation and the biologic implications of radioactive fallout.

Radiation dosimetry is another topic whose knowledge is preliminary to radiation biology. For that reason, it has been covered in considerable detail.

In the chapters dealing more directly with radiation biology and with the use of radioisotopes in biologic and medical research, classic investigation and experiments have been adequately emphasized. However, the interested reader is also entitled to a taste of current research; thus, contemporary investigations and experiments in areas promising future exciting developments have been mentioned, with proper stress on their preliminary nature.

The format of each chapter is designed to make the text readable. The subject matter has been divided into sections and subsections with appropriate headings and subheadings. An outline is presented at the beginning of each chapter so that the reader may have a quick glimpse of what he is about to read. Sections or paragraphs containing less important, auxiliary, or collateral information are printed in reduced type. The bibliography at the end of each chapter includes all the references cited in the text. A classified and annotated list of general references and periodicals is presented in Appendix I.

Attention is also called to Appendix III "Main characteristics of radionuclides of biologic importance," which, I believe, is a rather novel feature. This appendix contains important "personal data" of radionuclides commonly encountered in radiation biology and radiotracer methodology. The student will find all the relevant information about a radionuclide on one page, without being compelled to consult many other sources.

This text is complemented by a laboratory manual, *Ionizing Radiation and Life: Laboratory Experiences*, also published by The C. V. Mosby Company. Several references to the manual are found as footnotes in this book.

I am indebted to many persons who have contributed to the preparation of this volume. Dr. George J. Highland, Associate Professor of Physics, assisted in the preparation of Chapter 1. Virginia May, a former student of mine, critically reviewed the whole manuscript from a student's point of view. William Burns, Susan Latsko, and Virginia May typed most of the manuscript. My appreciation goes also to those investigators, organizations, and manufacturers of equipment who have kindly supplied photographs and other illustrative material or have been most gracious in granting permission to quote published material.

It should not be surprising that in a textbook of this nature, written by one author, certain topics are treated with more detail than others. Radiation biology and radiation physics comprise so broad a subject area that one person cannot claim full competence in all aspects. I have made an honest attempt to provide comprehensive coverage, but nearly every book reflects the biases, strengths, and weaknesses of its author; this is no exception.

This text may contain errors of commission, for which I assume full responsibility. It definitely contains sins of omission, for which I accept only part of the blame. Much important material had to be left out in order to keep the size and production costs within the limits imposed by the nature and scope of the book. I shall be delighted to receive constructive criticism and suggestions, particularly from scientists who teach introductory courses of radiation biology and who have taken the time to read this preface, as well as the whole book.

Victor Arena

Contents

Ionizing radiation and life

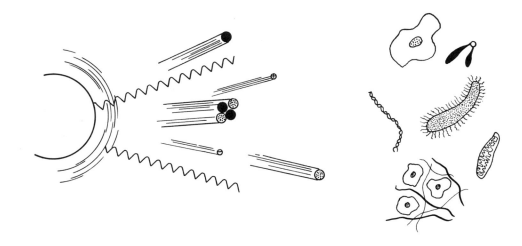

Introduction

One essential characteristic of life is the exchange of matter and energy between organisms and their environment. Through metabolism, a living organism transforms a variety of inorganic and organic molecules taken from the environment into its own protoplasmic compounds and then releases the the products of degradation back into the environment. Because of the metabolic transformations, the output material is chemically different from the input material.

But life involves also energy conversions. An organism absorbs energy from its environment in one form or another. Through metabolism the energy absorbed is transformed into different forms. The light energy absorbed by a photosynthetic organism is converted into chemical potential energy stored in the carbon-carbon bonds of complex organic molecules. The degradation and oxidation of these molecules (cellular respiration) releases the stored energy in forms that may be quite different from the input energy. An organism can be viewed, therefore, as a transformer of matter and energy.

Radiation is one form of energy that has always been available in nature; it has also played an important role in biologic processes and in the evolution of life on this planet.

Radiation is energy that can be transferred from one body to another across empty space, either in association with electromagnetic waves or with subatomic particles traveling at high speed. The energy associated with electromagnetic radiations is known as photon energy; the energy associated with particulate radiations is the kinetic energy of the particles.

The types of radiations of interest to biologists may be summarized as follows:

Nonionizing — photon energy from 10^{-2} to 10^2 eV*
   Infrared
   Visible
   Ultraviolet
Ionizing — photon or particle energy from a few keV to hundreds of MeV
   Electromagnetic   Particulate
     X-rays   Alpha particles
     Gamma rays   Beta particles
                    Neutrons
                    Protons
                    Deuterons, etc.

*For the meaning of eV, keV, and MeV, see p. 19.

Infrared, visible, and ultraviolet radiations are segments of the electromagnetic spectrum; they are classified as nonionizing because their photon energy is too small to remove orbital electrons from atoms or molecules (ionization). They may have, however, sufficient energy to produce excitation, that is, to raise atoms or molecules to higher energy levels and therefore to induce chemical and biologic changes. The extremely important role played by certain wavelengths of visible light in photosynthesis is too well known to be stressed here. Certain wavelengths of ultraviolet radiation can convert provitamin D into the active vitamin; others are strongly absorbed by nucleic acids and proteins and lethal for bacteria. Excessive absorption of ultraviolet radiation is known to cause skin burns and even carcinoma of the skin.

The energy associated with ionizing radiations is more than sufficient to overcome the binding energy of orbital electrons and thus to ionize atoms and molecules. X-rays and gamma rays are segments of the electromagnetic spectrum, like infrared, visible light, and ultraviolet radiation. Other ionizing radiations consist of high-speed particles emitted through radioactive decay processes or by certain nuclear reactions.

In a very broad sense, *radiation biology* (or radiobiology) could be understood as the study of interactions of living organisms with any type of radiation. The classic three-volume comprehensive review called *Radiation Biology*, edited by A. Hollaender and published in 1954, is a treatise concerned with the effects of both ionizing and nonionizing radiations on living organisms. More recently, however, the objective of radiation biology has been limited to the investigation in living systems of ionizing radiations only. The effects produced in organisms by visible and ultraviolet radiation are the subject matter of another science known as *photobiology*.

Specifically, the object of radiation biology is the description of the gross observable phenomena that occur when ionizing radiations interact with biologic systems and the elucidation of the basic biologic mechanisms and principles that bring about the observed response. In addition, radiation biology may have an ancillary value for other biologic sciences because radiation may be used as an experimental means to stimulate a biologic response that can clarify a biologic structure or process. Ionizing radiation may also be used as a tool in biomedical research, in medical diagnosis, and for therapeutic purposes.

Table 1 summarizes the subject matter of radiation biology and its relationships with other sciences. The basic problems are centered around the interactions of ionizing radiation with living systems. When an organism is exposed to radiation, the first changes are induced in the molecules that constitute its protoplasm (water, proteins, nucleic acids, enzymes, etc.). Thus the first injury is produced at molecular level. The elucidation of these radiation-induced chemical changes is provided by *radiation chemistry*, which is specifically concerned with the interaction of ionizing radiations with nonliving chemical systems. In organisms, the molecular injury triggers changes (usually harmful) that are observable in intracellular structures, in the cell as a whole, in the organism, and also in the biologic balance of an ecosystem. Genetic effects (gene mutations and chromosomal aberrations) are examples of damage induced at subcellular level. Irradiation of a complex organism (as a mammal) may elicit a prompt response (radiation burns, epilation, radiation sickness, death) or a delayed response (cancer, leukemia, sterility).

Table 1. Radiation biology: its subject matter and its relationships with other sciences

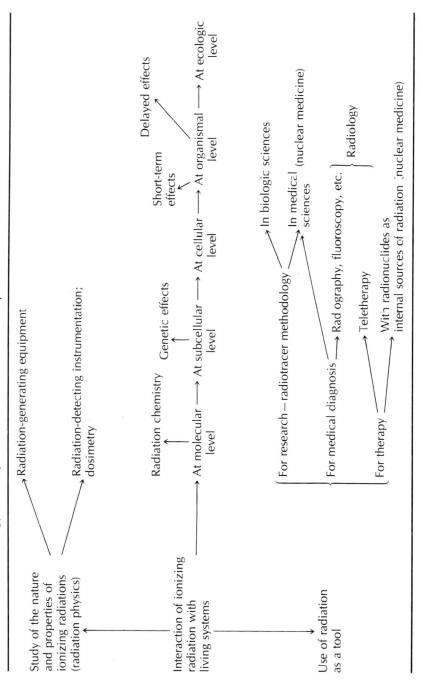

X-rays and the ionizing radiation emitted by radioisotopes of certain chemical elements may be used as tools in biologic research and in medical diagnosis and therapy. In biologic research radioisotopes are mainly used as tracers; *radiotracer methodology* is a natural offshoot of radiation biology. In the medical sciences, radiography, fluoroscopy, and teletherapy are examples of diagnostic and therapeutic applications of external sources of ionizing radiations; these applications are the concern of a medical science known as *radiology*. The use of radioisotopes introduced into the human body for diagnostic or therapeutic purposes is the domain of *nuclear medicine*. It is obvious then that radiation biology is fundamental for the study of radiotracer methodology, radiology, and nuclear medicine.

On the other hand, the fact should be also obvious that the study of radiation biology cannot be seriously undertaken without a thorough understanding of the nature and properties of ionizing radiation. This study is the object of *radiation physics*. The knowledge of man-made devices (X-ray tubes, particle accelerators, nuclear reactors) for the generation of ionizing radiation is also essential for any type of radiobiologic research. Likewise, the knowledge of the instruments designed to detect radiation or to measure the radiation dose delivered to, or absorbed by, an organism is basic in radiation biology, radiotracer methodology, and nuclear medicine.

Radiation biology grew out of atomic, nuclear, and radiation physics and tried to keep pace with the rapid progress of those sciences. For all practical purposes, one may think of atomic physics as having been born during the last few years of the nineteenth century. At that time, history-making discoveries succeeded one after the other quickly and unveiled the intimate complex structure of the atom, which had been previously viewed as the simplest "indivisible" particle of matter. X-rays were discovered by Roentgen in 1895, radioactivity was discovered in 1896 by Becquerel, and the electron by Thomson in 1897. The following year Pierre and Marie Curie discovered the radioactive elements radium and polonium in uranium ore extracted from the Joachimsthal mines. The discovery of radioactivity revealed the existence of theretofore unsuspected ionizing radiations that are spontaneously emitted by some chemical elements. The modes and laws of radioactivity were clarified by the pioneering work of Rutherford and his associates through a long series of experiments, which also led to the first acceptable model of the atomic structure.

Beginning in 1900, Planck and Einstein developed the quantum theory of electromagnetic radiation, which was a significant and substantial improvement of Maxwell's electromagnetic theory. The interactions of ionizing radiations with matter were investigated intensively during the second and third decades of this century. Notable was the discovery of the Compton effect in 1922; this interaction between X- or gamma rays and matter threw more light on the nature and properties of high-energy photons.

The experimental evidence given in 1932 by Chadwick in favor of the existence of the neutron confirmed what had already been suspected before: the atomic nucleus is made of positively charged particles (protons) and particles with no electric charge. Rutherford's atomic model was perfected by Bohr with his remarkable contributions to the knowledge of the extranuclear orbital electrons. That stable elements can be made artificially radioactive was shown in 1934 by F. Joliot and I. Joliot-Curie when they obtained radioactive phosphorus by bombarding aluminum with alpha particles. The discovery of artificial radioactivity led to the search of suitable nuclear reactions to obtain radioactive isotopes of other elements. Today one can prepare radioisotopes of practically every element.

Radiation biology was practically and informally born within months after the announcement of the discovery of X-rays. Biologists and physicians did not take long to notice that these strange and invisible radiations were capable of producing in organisms effects that were not known to be caused by other types of radiations. In 1896 X-ray–induced loss of hair was reported for the first time by J. Daniels of Vanderbilt University. The same year an X-ray–induced dermatitis was diagnosed in a Chicago vacuum-tube manufacturer and experimenter. In 1896 L. Freund, a Viennese physician, used X-rays to remove a hairy birthmark from one of his patients, and D. Walsh pointed out that a whole-body exposure of man to X-rays may be followed by a "radiation sickness." The first textbook of radiology was published in the United States by Francis H. Williams in 1901 with the title *Roentgen Rays in Medicine and Surgery*. During the same year, the discoverer of natural radioactivity, Becquerel, noticed a skin burn on his body caused by a small vial of radium that he had inadvertently carried in a pocket of his vest.

The first casual observations of radiation damage stimulated biologists and physicians to experiment with laboratory animals. The sterilizing effect of X-rays in rodents was first reported by Albers-Schönberg in 1903. In 1911 Morgan and his co-workers exposed fruit flies to the radiation emanating from radium in an attempt to increase the frequency of spontaneous mutations. The results of those experiments were rather inconclusive because of the inappropriate methods used. Several years later (1927), using better methods, H. J. Muller could prove that indeed X-rays have a mutagenic effect in *Drosophila*. In 1921 Holthusen discovered the oxygen effect with irradiation of *Ascaris* eggs.

The destructive effects of X-rays on cells and tissues were exploited very early in the attempt to destroy tumors and cancerous growths. Radiation therapy was thus born out of some preliminary radiobiologic observations. The first radiation therapists observed with interest and surprise that X-rays may kill cancerous cells with no apparent damage to the surrounding healthy tissues. In 1906 the French biologists Bergonié and Tribondeau echoed this discovery by formulating a law that still bears their names and is one of the most important contributions in the history of radiation biology (see p. 330). Bergonié and Tribondeau were the first scientists to recognize formally that X-rays may discriminate between different types of cells.

The need for measuring the radiation dose delivered to patients treated with radiation therapy was recognized at the very beginning of this century. In 1908 P. Villard suggested the roentgen as a unit of exposure dose, although its definition was very vague and unsatisfactory. The roentgen was accurately and officially defined in 1937 by the International Commission on Radiologic Units (see p. 215).

In the 1920s the American chemist H. Fricke made substantial contributions to the radiation chemistry of aqueous solutions. The new knowledge was helpful to radiation biologists for a better understanding of the effects of ionizing radiation on molecules of biologic importance. Fricke was also the inventor of the ferrous sulfate dosimeter, which is still today the most reliable chemical dosimeter.

In the 1930s the target theory was first suggested by Crowther, Timoféeff-Ressovsky, and Delbrück and then developed and expanded by the British radiobiologist D. E. Lea in his classic book *Actions of Radiations on Living Cells*. The theory was a useful model for the explanation and interpretation of radiation-induced effects in certain chemical and simple biologic systems.

Radiobiologic experimentation and radiology made significant progress when the existing artifical sources of ionizing radiations were improved and when new sources were invented. In 1913 Coolidge invented an X-ray tube that is still known by his name and that completely replaced the old unreliable "gas-filled tube." More penetrating X-rays of higher energy could be produced later with the Van de Graaff generator. In the early 1930s, mainly through the efforts of the American physicist E. O. Lawrence and his associates, the first linear accelerators of charged particles were built in Berkeley, California; they were soon followed by the first cyclotrons. Within the following decades increasingly more powerful accelerators ("atom smashers") were built in America and Europe. The particle accelerators are of enormous importance to high-energy physicists in their search for elementary particles; but they have been also very useful for the production of many artificial radioisotopes. Until 1930 only the few natural radioactive elements were available for biologic tracer studies. Radioisotopes of light and biologically important elements became available in fairly large amounts only after the development of the cyclotron. G. Hevesy may be considered the pioneer of the radiotracer method; his first experiments (1923) were done with a natural radioisotope of lead that was absorbed by the root system of the broad bean plant *(Vicia faba)*. Hevesy laid out also the basic rules and principles for radiotracer work. In 1936 sodium 22 was the first cyclotron-produced artificial radioisotope to become available in considerable quantity for use as tracer and in radiation therapy. Shortly after, phosphorus 32 could also be produced and used for the same purposes. Tritium was discovered in 1939 when deuterium was bombarded with accelerated deuterium nuclei (deuterons) in a cyclotron. The following year, carbon 14 was discovered by Ruben and Kamen. The tremendous impact of tritium and carbon 14 on biologic research is stressed in Chapter 10.

Until the 1940s the only radiation-detecting and -measuring devices were photographic emulsions, spinthariscopes, crude electroscopes, and ionization chambers. The cloud chamber was invented by C. T. R. Wilson in 1911; Geiger counters were invented by Geiger, one of Rutherford's disciples.

Radiation biology received a sudden new impetus at the dawn of the atomic age, when man discovered that tremendous amounts of atomic energy can be unleashed through the processes of nuclear fission and fusion. The birth of the atomic age took place quite rapidly between the years 1939 and 1945. The announcement of the discovery of nuclear fission of uranium was made in 1939 at the end of a series of observations and experiments for which the credit goes to several nuclear physicists. Because of the possible military applications of the newly discovered atomic energy, further research on nuclear fission was carried on under military secrecy and entrusted to the so-called Manhattan Engineering District Project. In 1942 E. Fermi achieved the first self-sustaining fission chain reaction in the first nuclear reactor ever built. The first nuclear weapon was exploded on July 16, 1945, in the desert of Alamagordo, New Mexico. One month later, two nuclear bombs were detonated over the Japanese cities of Hiroshima and Nagasaki for hostile purposes. For several years after the end of World War II nuclear explosions continued to take place for testing purposes. In the meantime, nuclear reactors were built in increasing number for the peaceful utilization of atomic energy. Nuclear reactors were used as intense sources of neutron radiation for the production of artificial radioisotopes and later for the production of electric energy. In 1946 the control of atomic energy was transferred from military hands to the Atomic Energy Commission (AEC), a civilian agency established by the Atomic Energy Act.

With the advent of the atomic age, radiation biology assumed a new and more important role in the family of the physical and natural sciences. New and intense sources of ionizing radiations, hardly dreamed of before, had suddenly become available for peaceful and military purposes. Increasing numbers of human individuals were exposed or likely to be exposed to the new sources of radiation, and yet the health hazards possibly deriving from such exposures were largely unknown. Radiation biologists were called upon and entrusted with the responsibility of intensifying the investigation of the biologic effects of ionizing radiations. The Atomic Bomb Casualty Commission was established immediately after the end of the war to investigate the possible radiation injuries suffered by the Japanese atomic bomb survivors and by their offspring. The Atomic Energy Act authorized and directed the AEC to arrange research and develop activities related to the utilization of special nuclear and radioactive material for medical, biologic, and agricultural applications and to the protection of health and the promotion of safety during research and production activities. Intensive research programs were sponsored to study the possible radiation damage deriving from low-level exposures and especially from radioactive fallout produced by atmospheric nuclear testing.

Nuclear reactors could produce much larger quantities of radioisotopes and at a lower cost than could be produced with a cyclotron. Radioisotopes became, therefore, easily available for biologic research; the radiotracer method became increasingly more popular and useful for the solution of a large number of problems in practically every biologic science. New radiation-detecting and dosimetric instruments and techniques were rapidly developed (solid and liquid scintillation counters, semiconductor detectors, thermoluminescent dosimetry, improved radioautographic techniques, etc.) Radioisotopes quickly found their way in the medical sciences as diagnostic and therapeutic tools; nuclear medicine came out of infancy after the advent of the atomic age.

From this sketchy and incomplete historical excursus one can plainly see that the birth and development of radiation biology have been very exciting indeed. The progress of radiation biology has been fast because life scientists could not afford to lag behind their colleagues working in the rapidly advancing fields of nuclear and radiation physics, lest mankind itself becomes a victim of the scientific advances in the knowledge of the atom.

The interdisciplinary nature of radiation biology is also quite evident. Radiation research is interesting, among other reasons, because it has performed the miracle of pulling together scientists with diverse backgrounds, who otherwise had no other reason to meet and communicate. The annual meetings of the Radiation Research Society bring together physicists, chemists, biochemists, anatomists, microbiologists, botanists, and geneticists, even though these scientists may look at radiation from different viewpoints. The meetings are healthy because they create a climate that discourages overspecialization, scientific myopia, isolationism, and intellectual sterility. One may note with interest that some outstanding scientists with a training in nuclear physics have been so fascinated by the challenges posed by radiobiologic problems that they have decided to spend the rest of their lives among the ranks of radiation biologists. On the other hand, the scientist formally trained in radiation biology is compelled to keep constantly abreast of the progress made in radiation physics and chemistry.

The study of radiation biology is not only fascinating but also useful to any educated person living in this atomic age. In a democracy in which the people are ultimately responsible for the direction of domestic and foreign policies, everyone should have some knowledge of the facts our policies are designed to cope with. Nuclear energy, in one form or another, is the subject of some of the most vital policy decisions today. Atmospheric fallout, nuclear warheads, nuclear power, civil defense problems, and nuclear test bans are all exposing today's young generation to a wide panorama of nuclear terminology. Tomorrow's high school science teachers will be called to educate their pupils in these problems. No matter what specialty they will choose, tomorrow's graduate biology students will be very likely confronted with research problems that can be easily approached with radiotracer techniques.

I earnestly hope that this book will help the diligent reader in obtaining an introductory insight into the fascinating fields of radiation biology and radiotracer methodology and that his interest will be stimulated toward further inquiry into more specialized literature.

CHAPTER 1

Review of some fundamental concepts of physics

This chapter is not intended to be a summary of physics or chemistry. Only a few basic concepts that, in my opinion, are essential to the understanding of the nature, properties, and behavior of electromagnetic and particulate ionizing radiations are presented and treated here in a very elementary form. Other such basic concepts are introduced in later chapters whenever they are needed.

The mathematics involved has been kept within reasonable limits and can be easily grasped by any student with a fair command of elementary algebra, geometry, and trigonometry.

Before proceeding further you are invited to assimilate and master the fundamental concepts, laws, equations, definitions, and units presented in this chapter; the help of standard textbooks of college physics, some of which are suggested in the general references, may be useful.

BASIC UNITS OF MEASUREMENT—THE CGS AND MKS SYSTEMS

Most of the entities of the physical world (mass, energy, work, force, displacement, velocity, heat, temperature, etc.) have a quantitative nature, and as such they can be measured.

Some of these quantities can be thoroughly described if their magnitude is measured, because they have only magnitude. If one speaks of a temperature of 50° F, the quantity is thus described sufficiently. This also holds true for mass, energy, and heat. Such quantities are called *scalars*.

On the other hand, some physical quantities can be thoroughly described only when the magnitude as well as the direction are given. Giving the magnitude of a force is not sufficient; one must also indicate the direction along which the force is being exerted. The same is true for displacement and velocity. Such quantities are called *vectors*.

Whenever the magnitude of a quantity is to be measured, one must choose an arbitrary but appropriate unit, which must be defined. The measurement of a quantity will consist then in comparing its magnitude with that of the unit previously chosen.

All physical units are so interrelated that usually in the definition of a unit another simpler unit appears, whose definition has already been given. If one goes from the definition of complex units to those of simpler ones, eventually he will arrive at three basic units constituting the foundations of all others—the units of length, mass, and time.

The basic unit for the measurement of length is the *meter* (m), which is the length of a prototype platinum-iridium bar kept in the Archives of Sèvres (France). More exactly, the meter is defined as a length equal to 1,650,763.73 wavelengths of an orange-red line present in the emission spectrum of krypton 86. Some submultiples and multiples of the meter are well known: the millimeter (1 mm = 1/1000 m), the centimeter (1 cm = 1/100 m), and the kilometer (1 km = 1000 m).

The concept of mass is not an easy one to define; its correct and complete definition is given later in this chapter. At this point it suffices to say that mass can be thought of as the quantity of matter contained in a body. The basic unit for the measurement of mass is the *gram* (g), which is defined as 1/1000 of the mass of a prototype platinum-iridium cylinder kept at Sèvres. It is approximately equal to the mass of 1 cm^3 of distilled water at the temperature of 4°C. Some submultiples and multiples of the gram are well known, such as the milligram (1 mg = 1/100 g), the microgram (1 μg = one millionth of a gram) and the kilogram (1 kg = 1000 g).

For measuring the mass of molecules, atoms, and subatomic particles, the gram would be very inconvenient because those particles have a mass much smaller than 1 gram. Physicists and chemists have agreed to use the *atomic mass unit* (amu), recently renamed *dalton*, for the measurement of molecular, atomic, and subatomic masses. The amu is defined as 1/12 of the mass of one atom of the predominant isotope found in natural carbon. The amu is equal to about 1.6603×10^{-24} g.

The basic unit for the measurement of time is the *second* (s or sec), which is defined as 1/31,556,925.9747 of the year 1900. This length of time is approximately equal to 1/86,400 of the mean solar day.

Some of the units used for the measurement of other physical quantities are based on the centimeter, gram, and second (the cgs system); others are based on the meter, kilogram, and second (the mks system). Examples of both types of units will be encountered in the following sections. At the present time, the units based on the mks system are more widely used because they have some distinct advantages over the units based on the cgs system.

Recently the International Committee on Weights and Measures has recommended the use of the following prefixes to denote the various multiples and submultiples of a physical unit:

PREFIX*	SYMBOL	MULTIPLIES THE UNIT BY
tĕra	T	10^{12}
gĭga	G	10^{9}
mĕga	M	10^{6}
kĭlo	k	10^{3}
hĕcto	h	10^{2}
dĕca	da	10^{1}
dĕci	d	10^{-1}
centi	c	10^{-2}
milli	m	10^{-3}
mīcro	μ	10^{-6}
năno	n	10^{-9}
pĭco	p	10^{-12}
femto	f	10^{-15}
atto	a	10^{-18}

After these recommendations, some familiar units should be renamed. For instance a micron (μ) should be called micrometer (μm); and a millimicron (mμ) should be a nanometer (nm). An angstron unit (A) is equal to 100 picometers (= 100 pm).

DYNAMICS

A body changing its position in space with respect to an arbitrarily chosen frame of reference is said to be in motion. Dynamics is that branch of study dealing with the motion of bodies under the action of forces. It is subdivided into the following two parts: *kinematics*, which refers to the study of motion without reference to the causes that may generate it, and *kinetics*, which relates the forces acting upon a body to its resulting motion.

Kinematics
Constant velocity

The simplest type of motion is considered first. A body that undergoes equal displacements during (any) equal intervals of time while progressing along a straight line is said

---

*The signs ¯ and ˘ indicate a long or short vowel, respectively.

to be moving with *constant velocity*. Since velocity is a vector, there are the following two quantities that completely define constant velocity: (constant) *speed* and (constant) *direction of motion*. The (constant) speed is defined by the following equation:

$$s = \frac{d}{t} \tag{1-1}$$

in which d is the total displacement of the body and t is the time it took to traverse that distance. From equation 1-1 we derive

$$d = st \tag{1-2}$$

and

$$t = \frac{d}{s} \tag{1-3}$$

One should note that for this type of motion the numerical value of the speed of a body as obtained from equation 1-1 is identical with the numerical value of its instantaneous speed at any instant (since the speed it possesses at any instant is unchanging).

The direction and sense of motion can be specified by a directed line segment, such as an arrow. Geometrically, we may therefore represent constant velocity by an arrow chosen to indicate the actual direction and sense of motion of the body, with the arrow's length chosen to be numerically equal (on a scaled basis) to the value of the body's speed given by equation 1-1.

A trivial instance of constant velocity exists whenever a body is at rest. Although the body's direction and sense of motion become nondefinable or at best ambiguous, we nevertheless maintain that it possesses constant velocity with constant speed numerically equal to zero.

A word of caution is the logical consequence of the concepts developed above. Constant speed is not synonymous with constant velocity. A body that possesses the same instantaneous speed at every instant but that does not travel in a straight line does not possess constant velocity. An automobile made to negotiate a turn at a constant speed, as indicated by its speedometer, is not traveling with constant velocity because its direction of motion changes from instant to instant.

Acceleration

The concept of nonconstant velocity can be developed by first introducing the concept of *instantaneous speed*. The instantaneous speed of a body at some instant is the time rate of change of position of the body at that instant. If the body is traveling at a constant speed, then its instantaneous speed is the same at every instant and is given by equation 1-1, regardless of whether it is traveling in a straight path or along a curved path. For example, suppose a cyclist is traveling at a constant speed around a perfectly circular track having a radius of 50 meters. If it takes him 100 seconds to make the complete trip, then his instantaneous speed is given by equation 1-1 and is equal to the total distance traversed, $2\pi50$ meters, divided by the time it took to make the trip, 100 seconds, or approximately 3.14 m/sec. This is precisely the value his speedometer would have shown while he was in motion.

For the more general case, when the instantaneous speed changes from one instant to another, equation 1-1 does not apply and one must then employ the methods of calculus to obtain the body's instantaneous speed at any instant of time. Since such methods are beyond the scope of this review, we will make the following clarification instead. The instantaneous speed of a body in motion at a given instant is the speed the body would have if, from that instant on, it traveled at constant speed. Consequently, if a body travels at a constant speed (as the cyclist of the example above), its instantaneous speed does not change with time, whereas if a body does not travel at constant speed, its instantaneous speed changes with time.

The *instantaneous velocity* of a body at some instant is defined by specifying both the body's instantaneous speed and its instantaneous direction (that is, its intended direction of travel at that instant). Consequently, if a body travels at constant speed but not on a straight path, its instantaneous velocity is changing because its direction of motion is changing at every instant. Since the instantaneous velocity is changing, the body is not traveling at constant velocity. Generally, we may say that a body is not traveling at constant velocity when its instantaneous velocity is changing from one instant to the next, that is, when either its instantaneous speed or instantaneous direction, or both, are changing in time.

A body traveling at nonconstant velocity is said to be accelerating. Therefore, *acceleration* is a change of velocity with time. Since the cyclist of the above example was changing direction at every instant, he was accelerating, although he was moving with a constant speed. The moon, in orbiting around the earth, is continuously changing direction and therefore is accelerating. The same is true for the electrons orbiting around an atomic nucleus. We shall consider only one special case involving accelerated motion. A body is said to undergo *constant acceleration* if it is moving in a straight path and its instantaneous speed changes at a constant rate. The instantaneous speed may either be increasing at a constant rate, in which case it is undergoing a *positive acceleration*, or it may be decreasing at a constant rate, in which case it is undergoing a *negative acceleration*. The rate at which the instantaneous speed is changing is called the instantaneous (constant) acceleration. Suppose that a body moving in a straight line has an instantaneous speed of 10 cm/sec at the end of the first second, 20 cm/sec at the end of the second second, 30 cm/sec at the end of the third second etc. The instantaneous speed is increasing at a constant rate of 10 cm/sec/sec or 10 cm/sec². This value is the constant acceleration of the body and is a positive acceleration. If we assume that at time 0 (seconds) the body was at rest, then at time t its instantaneous speed s is given by

$$s = at \tag{1-4}$$

where a represents the body's constant acceleration.

Also easily shown is that the distance d traversed by the body in a time t is given by

$$d = \frac{1}{2}at^2 \tag{1-5}$$

An example of almost uniformly accelerated motion is that of a body falling freely in a vertical direction in a vacuum toward the surface of the earth. The acceleration in this case is about 980 cm/sec². By using equation 1-5, one can see that the distance covered by a body falling in these conditions is 490 cm at the first second, 1960 cm at 2 seconds, 4410 cm at 3 seconds, etc.

Kinetics
Newton's first law of motion—the concept of force

The scientific foundations of kinetics were laid down by the physicists Galilei and Newton. Newton summarized his findings by stating three general laws of motion.

The first law states that a body will maintain a state of rest or constant velocity, unless acted upon by a net force. A net force will change the state of motion of a body; that is, a net force is required not only to start in motion a body that is at rest, but also to stop or to change its instantaneous speed or its instantaneous direction. However, in order to maintain a body in a state of rest or in a state of constant velocity, no net force is required. If a motorist turns his motor off while traveling on a level road, the automobile will slow down and eventually come to a stop, but only because there are forces acting on his vehicle (friction in the moving parts, friction between the tires and the pavement, air resistance, etc.). If a body is released above the surface of the earth, it will fall down toward the center of the earth, because there is a net force acting upon the body (the force of gravity). But if the same body is left on a table, it will remain at rest because the net force acting upon the body is now zero (the downward force of gravity acting upon the body is balanced by the upward force exerted by the table), and according to Newton's first law of motion a body at rest remains at rest as long as no net force acts upon it.

The second law of motion—linear momentum

What happens when a nonzero net force acts upon a body? According to the first law, the body no longer maintains a constant state of motion, that is, it no longer either remains at rest or moves with constant velocity. If it is no longer at rest, it must be accelerating; that is, either its instantaneous speed or its instantaneous direction, or both, change. What connection exists between the net force applied to the body and its resulting acceleration? If we let F represent the net force and a the magnitude of the body's acceleration, we find that the ratio F/a is independent of the value of F and the corresponding value of a. That is, suppose we apply a series of net forces, $F_1, F_2, F_3, \ldots$ and determine the body's corresponding accelerations $a_1, a_2, a_3, \ldots$. If the ratios $F_1/a_1, F_2/a_2, F_3/a_3, \ldots$ are compared, one finds that

$$F_1/a_1 = F_2/a_2 = F_3/a_3 = \ldots = m \qquad (1\text{-}6)$$

such that m is a constant for the same body and a measure of some property of the body itself.

If we now use another body and apply the same procedure, we will obtain a similar result, except that perhaps the ratio denoted by m will have a different value. This lends further credence to the interpretation that m is a measure of some property of the specific body being observed. In the cgs and mks systems of units, m represents the *mass* of the body. From equation 1-6 we derive

$$m = \frac{F}{a} \qquad (1\text{-}7)$$

or

$$F = ma \qquad (1\text{-}8)$$

or

$$a = \frac{F}{m} \qquad (1\text{-}9)$$

The first form defines mass in terms of the applied net force and the body's resulting acceleration. The second form can be used to determine the magnitude of the net force, given both the mass and the acceleration of the body. The third form permits us to determine the magnitude of the acceleration, given the net force applied and the mass of the body.

Suppose that a net force of a constant magnitude is applied consecutively to two bodies, one having a larger mass than the other. Then according to the third form, the body of lesser mass will have the greater acceleration, in agreement with our intuitive notion of mass.

The unit of force in the cgs system is called the *dyne* and is defined, by using the second form $(F = ma)$, as that force which when applied to a gram of matter will cause it to accelerate at 1 cm/sec/sec. Therefore, 1 dyne = 1 gram $\times$ 1 cm/sec². In the mks system the unit of force is the *newton* and is defined as that force which when applied to a kilogram of matter will cause it to accelerate at 1 meter/sec/sec. Therefore, 1 newton = 1 kilogram $\times$ 1 m/sec². It is easy to verify that 1 newton = 10^5 dynes.

In summary we can make the following statement. The mass of a body multiplied by (the magnitude of) its acceleration equals (the magnitude of) the net force acting on it, and the direction of the acceleration is the same as that of the applied net force. The content of this statement is often referred to as Newton's second law of motion.

In what follows we will assume that the applied net force is constant in magnitude and direction. Therefore the body will move with uniform (constant) acceleration.

The product Ft of the net force and the time in which the force is being applied is called the *impulse of the force*. Since $F = ma$ and $at = s$, it follows that

$$Ft = mat = ms \qquad (1\text{-}10)$$

The product ms is called the magnitude of the *linear momentum* of a body in motion.

The third law of motion — conservation of momentum

The third law of motion applies to interactions between bodies. "Whenever one body exerts a force upon a second body, the second body exerts an equal and opposite force on the first."

Whenever some agent exerts a force upon a body, there always exists an equal and opposite force of this body on the agent. If for example, a man pulls on a rope with a force of 1 newton, the rope pulls back on the man's hand with a force of 1 newton. A train pulls back the locomotive with a force that is exactly as great as the force that the locomotive exerts forward on the train. When a man presses with his thumb on a table, the table presses back equally on his thumb. The sun pulls on the earth and the earth pulls on the sun with equal and opposite forces. Indeed, when a body falls, not only the earth is pulling the body towards its center, but also the body is pulling the earth towards its center. Newton stated this law in a more general form: "To every action there is an equal and contrary reaction." Here the term "action" is used to imply force.

The recoil of a firing rifle is another important example of this law, and jet propulsion is a spectacular application of the same law.

Consider a freely suspended rifle, cocked and ready to fire. When the trigger is pulled, a force F is exerted on the bullet. By Newton's third law, the bullet exerts an equal and opposite force f on the gun, so that $F = -f$. By Newton's second law, it must be

$$F = \frac{MS}{t} \quad \text{and} \quad f = \frac{ms}{t}$$

where M and m are the masses of the gun and bullet respectively and S and s are the speeds of the gun and bullet respectively. Since the forces F and f act for exactly the same time, then

$$MS = -ms$$

which simply means that the momenta of the gun and bullet are identical in magnitude but opposite in direction. The algebraic sum of the two momenta therefore is zero, just as it was before the trigger was pulled. The total momentum of the system (rifle + bullet) has not changed.

As a consequence of these considerations based on both the second and third law, we derive a principle of the uppermost importance in physics; the principle of *conservation of momentum*. It can be stated as follows: "If two or more bodies interact, the total momentum after the interaction is equal to the momentum before the interaction, provided that there are no external forces acting on the bodies." Or: "The total momentum of any system of bodies is unchanged by any actions that occur between members of the system."

A quantity like momentum, which is conserved and cannot be created or destroyed, is of particular interest in physics. Only a few such quantities are known; mass and

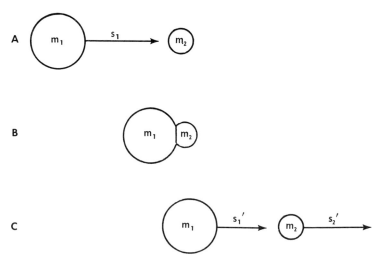

Fig. 1-1. Conservation of momentum in the collision of billiard balls. **A,** Before collisi·
B, During collision. **C,** After collision. m_1 and m_2, Masses. s_1 and s_2, Speeds.

energy are other examples, and we shall see that there are conservation laws that apply to these quantities as well.

Because of the importance of the principle of conservation of momentum, other examples will be given to illustrate its applications. Conservation of momentum can be seen, for instance, in the collision of billiard balls. Fig. 1-1 shows the situation before, during, and after the collision of two billiard balls. Their masses are m_1 and m_2; their instantaneous speeds before collision are s_1 and s_2 ($= 0$) respectively, and after collision s_1' and s_2'. The total momentum before collision is $m_1 s_1 + m_2 s_2$ and after collision is $m_1 s_1' + m_2 s_2'$. Because of conservation of momentum, it must be

$$m_1 s_1 + m_2 s_2 = m_1 s_1' + m_2 s_2'.$$

Notice that during collision, the interaction between the two balls is reciprocal; the shape of both balls is distorted because, by the third law, each ball exerts a force on the other.

When a battleship fires a broadside volley, the shells are given a large momentum in one direction. But the ship recoils with an equal momentum in the opposite direction. Also, consider one aircraft chasing another at about the same speed. If the pursuing plane opens fire, the bullets acquire forward momentum and the plane loses an equal momentum, which means that it will slow down. On the contrary, when the pursued craft opens fire, the bullets have momentum toward the rear and the aircraft gains speed.

Later we will see how the law of conservation of momentum applies to the interaction between ionizing radiations and matter, and specifically to the collisions between particulate radiations and the atoms of the matter through which they travel.

FUNDAMENTAL FORCES

The complex forces that we perceive at both the atomic and the macroscopic levels can be conceived as arising from a combination of three fundamental forces. They are the gravitational, the electric, and the magnetic forces. Nuclear forces (another type of fundamental force, see Chapter 2) are short-range forces acting over a range of the order of 10^{-13} cm and thus do not directly influence either the atomic or the macroscopic level.

What sources are responsible for the existence of these fundamental forces? On what do these forces depend? We may answer these questions by considering the classical concept of field.

A *field* may be thought of as some kind of "disturbance existing in space." The strength of the disturbance at any point in space depends on the distance of a point in space from the source creating the disturbance and on the strength of the particular source.

Gravitational field and gravitational force

The fact that material bodies left free in air do not remain there but start moving toward the earth has been known for centuries to men. The causes of this phenomenon remained mysterious until the great mind of Isaac Newton discovered that a force is responsible for the falling of the bodies. This is the force of *gravity*. However, Newton also found that this force is not peculiar to the earth alone but is only a particular case of a type of attractive force that is exerted between all material bodies existing in the universe. These forces are known as *gravitational forces* and the law that describes the way they operate is called the *law of universal gravitation*. A gravitational field is generated by the presence of matter, and every particle of matter in the universe generates a gravitational field. The strength of the gravitational field at a chosen point in space is inversely proportional to the square of the distance between the particle and the chosen point. It is also found to be directly proportional to the mass of the particle. If we place a second particle at the chosen point of the field generated by the first particle, it will experience a force directed toward the first particle, along the line joining the centers of the particles. The magnitude of this force is equal to the product of the gravitational field at the point where the second particle is located and the quantity of mass placed there. Likewise, with similar considerations, one can see that the first particle will also experience a force directed toward the second particle. The magnitude of this force is equal to the product of the gravitational field generated by the second particle at the point where the first particle is located and the quantity of mass placed there. Thus each particle experiences a (gravitational) force directed toward the other. The forces are equal in magnitude but oppositely directed; this fact is in accordance with Newton's third law.

From the above considerations one may easily derive an equation that gives the magnitude of this force of attraction as follows:

$$F = G\frac{M \times m}{d^2} \qquad (1\text{-}11)$$

where F is the magnitude of the gravitational force between two bodies of masses M and m, and d is the distance between them. G is the gravitational constant that was determined experimentally and found to be equal to 6.67×10^{-8} in the cgs system.

Equation 1-11 is then the mathematical expression of the law of universal gravitation; it states that the gravitational force of attraction between any two material bodies existing in the universe (from galaxies to subatomic particles) is directly proportional to the masses of the bodies and inversely proportional to the square of their distance.

Electric field and electric force

All of us at one time or another have encountered electrical phenomena. The curious "defiance of gravity" by the bits of paper attracted by the "charged" comb and the electric shock we receive on a winter day on touching a door knob after having brushed our feet on a rug are some common examples of our experience with electric phenomena.

The elementary particles of nature (electron, proton, etc.), which collectively define the atoms, are endowed not only with a property we have termed "mass", but some have in addition a property called *charge*. These particles generate not only gravitational fields via their mass, but also electric fields via their charge. The electric field represents a "disturbance of space" of a different nature than that of the gravitational field, insofar as the electric field interacts with charge to produce an *electric force*. The strength of the electric field is directly proportional to the quantity of charge generating the field, and at any chosen point it is inversely proportional to the square of the distance of the point from the charged particle or body generating the field. Another charged particle placed at some point in an electric field will experience an electric force quite different from the gravitational force. In fact, unlike mass, there are two kinds of charges, *positive* and *negative*. Particles of opposite type of charge always attract each other, whereas particles of the same type of charge always repel each other. As in the case of gravitational forces, each of two charged particles placed at a distance d from each other experiences an electric force (of attraction or repulsion) that is directly proportional to the magnitude of the two charges and inversely proportional to the square of their distance. Therefore the magnitude of the electric force is given by the following equation:

$$F = K\frac{Q \times q}{d^2} \qquad \text{(Coulomb's law)} \qquad (1\text{-}12)$$

in which Q and q are the quantities of charge of the two particles and K is a proportionality constant.

Experiments have shown that all elementary charged particles carry the same quantity of charge regardless of whether it is positive or negative. This basic quantity of charge is called the *electronic charge*. Thus any charged object, such as a comb, carries an integral number of electronic charges.

The electronic charge turns out (from a practical point of view) to be too small a quantity to be used as unit of charge. Instead a unit of charge that is often adopted in practical applications is the *coulomb*, which is defined as the charge on each of two spheres experiencing an electric force (of attraction or repulsion) of 8.98×10^9 newtons, when their distance is 1 meter. The coulomb is made up of approximately 6×10^{18} electronic charges of the same sign (positive or negative).

Another smaller unit of charge, often used in practical applications, is the *electrostatic unit* (esu), or *statcoulomb*, which is defined as the charge of each of two bodies that experience an electric force of 1 dyne in a vacuum, when their distance is 1 cm. One coulomb equals 3×10^9 esu.

To compare the magnitude of gravitational and electric forces, let us consider one specific example. With equation 1-12 it is possible to calculate* the electric force of repulsion between two identical elementary charged particles separated by a distance of 5.3×10^{-11} m (one atomic radius). This electric force turns out to be equal to 8.2×10^{-8} newtons. If we assume that these elementary particles are protons, then the gravitational force of attraction between them (as given by equation 1-11) is equal to 6.9×10^{-43} newtons. If we compare the electric repulsion with the gravitational attraction between the two particles, we obtain the following:

$$\frac{F\ electric}{F\ gravitational} \approx 1.2 \times 10^{35}$$

which in this particular instance shows how much greater the electric repulsion is than the gravitational attraction. In general, the electric forces between atomic or subatomic systems are much greater than the gravitational forces.

Magnetic field and magnetic force

A magnetic field is generated by a moving charged object (including, of course, moving elementary charged particles). The strength of the source of the field is directly proportional to the quantity of charge and to the instantaneous speed of the moving object. The strength of the field at an arbitrarily chosen point is inversely proportional to the square of the distance between the moving object and that point and directly proportional to the strength of the source. The magnetic field also depends on the (instantaneous) angle made between the instantaneous direction of motion and the line joining the particle to the point. A magnetic field will interact, for instance, with charged objects (including elementary charged particles) that are in motion; such interactions will produce *magnetic forces* of attraction or repulsion.

The magnetic relations are in general more complex than those that refer to gravitational or electrical interactions.

ENERGY
Concept of work

In a physical sense, *work* is done by a force whenever there is a movement of a body or particle with or against this force. In this sense, therefore, no work is being done when a man tries to push a wall, in spite of the very intense force he may exert upon it, because there is no movement involved.

In the situation illustrated by Fig. 1-2, we assume that the body B can be displaced only along the line ab and that the direction of the force F applied to B is not identical to ab, but forms an angle θ with it. Practically this situation might be that of a railroad car being pulled by a tractor that moves in a direction at an angle with the railroad tracks. In this case, not all of the magnitude of the force is being used effectively to move B along ab, but only a component of it. In dynamics this can be expressed geometrically by saying that the force F has two components—BC, which actually moves the body, and BD, which is ineffective.

With trigonometrical considerations, one can show that if the distance traveled by the body is d, then the work done is

$$W = d \times F \times \cos \theta \qquad (1\text{-}13)$$

If the force is exerted in the same direction of the displacement, then $\theta = 0$ and $\cos \theta = 1$. Equation 1-13 is thus reduced to the simpler form

$$W = d \times F \qquad (1\text{-}14)$$

---

*The interested reader is invited to verify these calculations, keeping in mind that the electronic charge equals 1.602×10^{-19} coulombs and that the mass of a proton equals 1.7×10^{-27} kg.

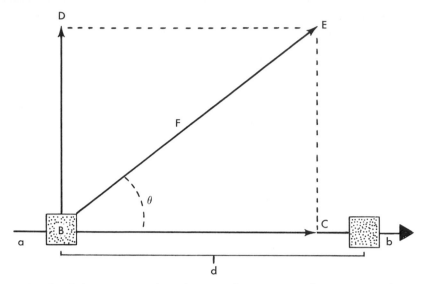

Fig. 1-2. Graph of the concept of work. **B,** Body. **F,** Force. **ab,** Direction of movement. **d,** Displacement of body. **BE,** Direction of force.

In this specific case work can be expressed as the product of the force applied to the body by the displacement of the same body.

Assuming that the magnitude of the force remains constant, the motion of B will be uniformly accelerated. Since $F = ma$ and $d = \dfrac{at^2}{2}$, by substituting these quantities in equation 1-14, we have

$$W = ma \times \frac{at^2}{2} = \frac{ma^2 t^2}{2}$$

But in the uniformly accelerated motion $at = s$. Therefore,

$$W = \frac{ms^2}{2} \tag{1-15}$$

which measures work as a function of the mass of the body and its velocity at any instant during its displacement.

General concept of energy

A particular entity concerning us at this point is *energy*. Probably because the concept of energy is so familiar to us, we have difficulty in defining it in a satisfactory manner. Energy is an abstract concept and is best defined mathematically. However, as an approximate definition, we can say that the energy of a body is a measure of the capacity or ability of the body to perform work. Energy can appear in one of many different forms, each of which requires a different mathematical formulation. Furthermore the principle of conservation of energy states that energy may be transferred from one physical system to another or transformed from one form into another, but it can neither be created nor destroyed. Thus, according to this principle, the numerical sum of all forms of energy in the universe is unchanged. If the total energy in the universe is measured at some instant today and measured again a thousand years hence, we should obtain the same value. The distribution and forms of energy may change with the passing of time, but the total energy remains constant in value. If a physical system (a body, a molecule, or an atom) gains energy in some form, another physical system must lose exactly the same amount of energy.

Forms of energy
Kinetic energy

Kinetic energy is that form of energy associated with the motion of a body, a molecule, or an atom, etc. If the physical system is vibrating, it possesses vibrational (kinetic) energy; if it is rotating, it possesses rotational (kinetic) energy; and if it is translating, it possesses translational (kinetic) energy.

If the instantaneous speed of the system is much less than the speed of light, its translational kinetic energy is approximately equal to $\frac{1}{2}ms^2$ (see equation 1-15), which shows that the translational kinetic energy is directly proportional to the mass of the system and to the square of its speed.

The same expression $\frac{1}{2}ms^2$ is used to define two units of energy. In the cgs system the unit of energy is the *erg*, which is defined as the energy possessed by a body of mass of 2 g, moving with the velocity of 1 cm/sec. In the mks system, the unit is the *joule*, which is defined as the energy possessed by a body of mass of 2 kg moving with the speed of 1 m/sec. One can easily verify that 1 joule = 10^7 ergs.

For the measurement of energy in processes involving atomic and subatomic systems (such as the kinetic energy of molecules, the kinetic energy of subatomic particles in motion, etc.), the erg is a unit too large to be employed conveniently. For these processes another unit of energy is used, the *electron volt* (eV), which is defined as the energy an electron gains in falling through a difference of electric potential of 1 volt (see definition of volt, p. 20). Widely used in atomic and nuclear physics are the multiples kilo electron volt (keV) = 10^3 eV and mega electron volt (MeV) = 10^6 eV. One eV is equal to 1.60×10^{-12} erg, and conversely 1 erg = 6.24×10^{11} eV = 6.24×10^5 MeV.

For a better appreciation of the magnitude of the electron volt, the following list gives approximate energy values (in eV or multiples thereof) associated with some molecular, atomic, or nuclear processes or structures:

PROCESS OR STRUCTURE	APPROXIMATE ENERGY
Kinetic energy of a molecule at room temperature	0.02 eV
Strength of a chemical bond	1-4 eV
Activation energy of a chemical reaction	0-4 eV
Energy of a photon of visible light	1.5-3 eV
Binding energies of atomic outer electrons	5-20 eV
Binding energies of atomic inner electrons	15 to 150,000 eV
Kinetic energy of a particle at 1,000,000°C	130 eV
Binding energy of deuterium nucleus	About 2 MeV
Energy of an alpha particle	4 MeV to 9 MeV
Energy of a gamma-ray photon	0.5 MeV to 5 MeV
Kinetic energy of some cosmic-ray particles	20,000 MeV

Potential energy

Potential energy is a form of stored or reserved energy. Potential energy is associated with every pair of physical entities that interact with each other by means of either gravitational, magnetic, or electric forces. For example, the gravitational forces acting between two bodies give rise to a gravitational potential energy that is associated with both bodies and depends on their masses and separation; electric forces give rise to an electric potential energy that depends on the magnitude and separation of the interacting charges. Attractive forces give rise to potential energies that increase with increasing separation of the physical entities. For example, as this book is raised above a table, the gravitational potential energy associated with the book-earth system increases. Repulsive forces give rise to potential energies that decrease with increasing separation or, conversely, increase with decreasing separation. For example, the closer we force two charged bodies of like sign together, the larger their (mutual) electric potential energy becomes.

The total potential energy of a system of interacting bodies equals the sum of all the various forms of potential energy associated with every pair of bodies within the system. The sum of all the kinetic and potential energies of a system of bodies is called the *mechanical energy* of the system.

Thermal energy

The molecules of a body are not only continually in random motion, but also exert (molecular) forces upon one another. These forces are primarily electric in nature and depend on the type of molecules interacting, as well as on their relative molecular separations. Thus molecular motion gives rise to what may be termed the *molecular kinetic energy* of a body, and the molecular forces give rise to what may be termed the *molecular potential energy* of the body. The temperature of a body depends on the magnitude of its molecular kinetic energy. The sum of the body's molecular kinetic and potential energies is called the *thermal energy* of the body.

Chemical energy

The electrons surrounding the nuclei of atoms are constantly in motion and subject to interatomic forces. However, these nuclear-electron systems can possess only discrete values of energy *(atomic energy levels)*. Whenever a chemical reaction takes place so that the products of the reaction are at a lower energy level than the original reactants, the difference between these energy levels is the *chemical energy* released by the reaction (exoergic or exergonic reaction).

Electric energy

Just as the presence of mass in a gravitational field will give rise to a gravitational potential energy that is a function of position, so too the presence of charge in an electric field will give rise to an electric potential energy that is a function of position.

If an electric field is applied within a material, any electric charges present within the material that are free to move will start flowing in a direction that depends on their sign ($+$ or $-$). This flow of charge within a material is called an *electric current*. There are several factors that determine the rate of this flow. The strength of the applied electric field is obviously a factor. Another is the degree of impedance that a material will offer to the flow of charge. This factor, due primarily to collisions between the flowing charges and the atoms composing the material, is called the *electric resistance* of the material. The resistance depends on the nature and the dimensions of the material. Materials that by their nature offer very little resistance to the electric current (such as metals) are called *conductors*; those which offer high resistance are called *insulators* (glass, porcelain), and those which offer an intermediate degree of resistance are called *semiconductors* (silicon, germanium). More specifically, a metallic wire of uniform cross section offers a resistance to the electric current, which is obviously a function of its length, its cross-sectional area and its nature. The characteristic that reflects the electric nature of the wire material is called the *resistivity*. Among the conductors, silver and copper have the lowest resistivity.

Since an applied electric field exists throughout a conductor, the electric potential energy at one end of the conductor will differ from that at the other end. Consequently, as the charged particles travel from one end of the conductor to the other, the kinetic energy they gain is transferred to the atoms of the conductor as they collide with them. The atoms vibrate more vigorously and the conductor therefore becomes warmer. Simply stated, electrical energy is transformed into (additional) thermal energy of the conductor. The quantity of electrical energy that is transformed into thermal energy is proportional to the quantity of charge that passes through the conductor. The proportionality factor is called the *potential difference* (across the conductor) and represents the quantity of electrical energy transformed into thermal energy for every coulomb that passes through the conductor. Specifically it is equal to the number of joules of electrical energy transformed into thermal energy per coulomb of charge transported through the conductor. The dimensional representation of the potential difference is the joule per coulomb. The joule per coulomb is commonly referred to as the *volt* (V), so that

$$1 \text{ volt} = \frac{1 \text{ joule}}{1 \text{ coulomb}}$$

Since potential difference is commonly measured in volts, it is also called "voltage."

The unit of electric current (the unit used to measure its rate) is the *ampere* (a), which is defined as the passage through a conductor of 1 coulomb/sec, so that

$$1 \text{ ampere} = \frac{1 \text{ coulomb}}{1 \text{ second}}$$

Only three parameters, the electric current, the resistance of the conductor, and the potential difference across the conductor are needed to describe the phenomenon of charge flow in conductors. Their relationship is expressed by Ohm's law, which states that the electric current I (the rate of flow) is directly proportional to the potential difference across the conductor, V, and inversely proportional to the conductor's resistance, R, or expressed as follows:

$$I = \frac{V}{R} \qquad \text{(Ohm's law)} \qquad (1\text{-}16)$$

Equation 1-16 may also be written $IR = V$ $\qquad\qquad\qquad\qquad$ (1-17)

or $\qquad\qquad\qquad\qquad\qquad$ $R = \frac{V}{I}$ $\qquad\qquad\qquad\qquad\qquad$ (1-18)

The last form can be used to define the unit of electric resistance, the *ohm*, which is the resistance of a conductor that requires a potential difference of 1 volt to cause a current of 1 ampere, or:

$$1 \text{ ohm} = \frac{1 \text{ volt}}{1 \text{ ampere}}$$

Other forms of energy (radiant, nuclear, etc.) are discussed in later chapters.

Power

In the definition of work given previously the time element was not involved. The same amount of work is done in lifting a weight to the height of 1 meter, regardless of whether it is done in 1 second or in 1 minute. However, in many instances the time rate at which work is done is an important quantity, and it is called *power*, which can be defined with the following expression:

$$\text{Power} = \frac{\text{Work}}{\text{Time}} = \frac{W}{t}$$

An automobile and a horse can do probably the same amount of work, such as hauling a load from sea level to an altitude of 100 m, but since they can do it in different times, they are said to have different powers. In the mks system the unit of power is the *watt*, which is defined as the work of 1 joule per sec,* or:

$$1 \text{ watt} = \frac{1 \text{ joule}}{1 \text{ second}}$$

Multiples of the watt are the kilowatt (10^3 watts) and the megawatt (10^6 watts).

MASS-ENERGY RELATIONSHIPS

Newton's laws of motion and other principles of the so-called classical mechanics are satisfactory and valid when we consider relatively large masses (larger than a molecule) and speeds much smaller than the speed of light. When we try to apply the same laws to events involving the very small masses of subatomic particles traveling at speeds approaching the speed of light, those laws are no longer valid. New laws and equations that apply to large and small masses as well as to moderate and high speeds must be found.

Most of these laws and equations have been well established during the present century, especially through the history-making work of Albert Einstein, the father of the theory of relativity. It is not within the scope of this summary to review the experiments,

---

*Although the joule has been considered above as a unit of kinetic energy, it can also be used as a unit of work. As such, it is defined as the work done by the force of 1 newton when the body on which the force is exerted moves a distance of 1 m in the direction of the force.

the considerations, and the arguments that led to the discovery of relativity at the beginning of this century. However, some corollaries emerging from the theory and finding applications in radiation physics are briefly summarized here.

1. It is necessary to abandon the classical concept of mass as an invariable property of a body that remains constant under all circumstances. On the contrary, the mass of a body increases with its speed, and the increase is appreciable if the speed is close to the speed of light. The mass a body has when it is at rest is called the *rest mass*, m_0, and the mass it has when it is in motion is called the *relativistic mass*, m. The two masses are correlated by the equation:

$$m = \frac{m_0}{\sqrt{1 - \dfrac{s^2}{c^2}}} \tag{1-19}$$

where s is the speed of the body and c is the speed of light (3×10^{10} cm/sec). A careful inspection of equation 1-19 shows the following:

a. When $s = 0$ (the body is at rest), $m = m_0$.
b. When s is very small compared to c, the value of m is slightly different from m_0.
c. If $s = c$, then the relativistic mass would be infinite.
d. If equation 1-19 is represented graphically by plotting mass as function of s (Fig. 1-3), one can see that m increases slowly at moderate speeds and very rapidly as the speed approaches the speed of light. If, for instance, $s = 0.99c$, one may easily verify that the relativistic mass is about 7.07 times the rest mass. It is extremely difficult to impress to a macroscopic body a speed comparable to the speed of light. But some subatomic particles emitted by atomic nuclei in radioactive decay can reach speeds as high as $0.9c$. Likewise, particles have been accelerated to similar speeds with powerful particle-accelerators (synchrotrons, cyclotrons), and their

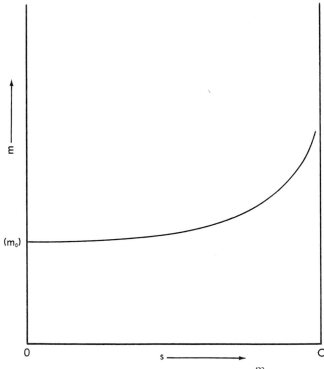

Fig. 1-3. Graph of the equation $m = \dfrac{m_0}{\sqrt{1 - \dfrac{s^2}{c^2}}}$

relativistic masses and speeds have been measured experimentally. Table 1-1 gives some of the pertinent data.

The fact that the mass of a body increases with its speed explains why an object cannot be accelerated to a speed greater than the speed of light. Let us assume that one can apply to the object a net force of constant magnitude and direction for any length of time. According to Newton's second law of motion, the object should then maintain a constant acceleration as long as its mass remains unchanged. However, its mass does not remain unchanged, but increases with speed. Therefore, with the same constant force, the acceleration decreases (recall that $a = F/m$). A smaller acceleration implies that the speed of the object will increase less rapidly, and the faster the particle moves, the greater its mass becomes. The greater its mass the smaller its acceleration; the smaller its acceleration the less rapidly will its speed increase. In reality, this results in making its attainment of the speed of light in a finite time impossible. In fact, any net force, constant or otherwise, would have to be applied for an indefinite period of time in order for an object to acquire the speed of light. Furthermore, an object traveling at the speed of light would possess an infinite mass and thus it could not possibly be accelerated. Consequently, it could not be made to acquire a speed greater than that of light. The speed of light represents an upper limit to the speed any object can assume.

2. Earlier in this chapter the fact was stated that in classical mechanics the momentum of a body in motion is the product of its mass multiplied by its speed (ms). For a particle moving at high speed the momentum is given more precisely by its relativistic mass multiplied by its speed, that is:

$$Momentum = ms = \frac{m_0}{\sqrt{1 - \frac{s^2}{c^2}}} \times s \tag{1-20}$$

which expresses the momentum of the particle as a function of its rest mass and its speed.

3. In the course of the development of his theory of relativity, Einstein made the startling discovery of the intimate relationship that exists between mass and energy. Mass and energy (contrary to the former views of classical physics) may be regarded as different manifestations of some fundamental property of matter. Furthermore, energy can be "converted" into mass, and mass into energy.

The equivalence of mass and energy is expressed by the well-known equation:

$$E = mc^2 \tag{1-21}$$

wherein c is the speed of light in cm/sec, m is the mass of a body or particle in grams, and E is the energy in ergs associated with the mass m.

If m is given in amu and E in MeV, then equation 1-21 becomes

$$E = 931 \times m \tag{1-22}$$

In fact, since 1 amu $= 1.6603 \times 10^{-24}$ g,

$$E = 1.6603 \times 10^{-24} \text{ g} \times m \times 9 \times 10^{20} \text{ cm}^2/\text{sec}^2 = 14.92 \times 10^{-4} \text{ erg} \times m$$

Since 1 erg $= 6.24 \times 10^5$ MeV, it follows that

$$E = 6.24 \times 10^5 \times 14.92 \times 10^{-4} \times m = 931 \times m \text{ MeV}$$

Table 1-1*

Speed of electrons (s) (cm/sec)	Ratio s/c	Ratio m/m₀ (experimental)
1.8998×10^{10}	0.6337	1.298
2.0868×10^{10}	0.6961	1.404
2.2470×10^{10}	0.7496	1.507

*Results of the experiments conducted in 1939 by M. M. Rogers, A. W. McReynolds, and F. T. Rogers.

In equation 1-21, m denotes the relativistic mass of a body or particle. If the body or particle is at rest, then the equation becomes

$$E_{rest} = m_0 c^2 \tag{1-23}$$

wherein m_0 is the rest mass and E_{rest} is the *rest mass energy*, that is, the amount of energy associated with the mass of the body or particle when it is at rest. The fact is obvious then that the total E of a body in motion is greater than E_{rest}, since m is always greater than m_0.

If in equation 1-21 we substitute m with its value as given by equation 1-19, we obtain:

$$E = \frac{m_0 c^2}{\sqrt{1 - \dfrac{s^2}{c^2}}} \tag{1-24}$$

which gives the total energy of a body in motion as a function of its rest mass and its speed.

The total energy of a body in motion, according to the preceding considerations, is then equal to its rest mass energy (E_{rest}) plus its kinetic energy (KE):

$$E = E_{rest} + KE$$

or (see equations 1-23 and 1-24):

$$\frac{m_0 c^2}{\sqrt{1 - \dfrac{s^2}{c^2}}} = m_0 c^2 + KE$$

and therefore:

$$KE = \frac{m_0 c^2}{\sqrt{1 - \dfrac{s^2}{c^2}}} - m_0 c^2 = m_0 c^2 \left(\frac{1}{\sqrt{1 - \dfrac{s^2}{c^2}}} - 1 \right) \tag{1-25}$$

which expresses the kinetic energy of a body as a function of its rest mass and its speed. Equation 1-25 is valid for any value of s from 0 to c. However, one can show with a few considerations that if s is small compared to c (so that $s/c < 0.7$), the same equation becomes:

$$KE \approx \frac{1}{2} m s^2$$

which is the classical equation of the kinetic energy! Once more one can see that many laws and equations of the classical mechanics are approximately valid if they are limited to situations where moderate speeds are involved.

If $s = c$, then according to equation 1-19 the mass of the body becomes infinite, and according to equation 1-24 its energy becomes infinite. Therefore, when we try to accelerate a particle in a particle accelerator, its energy becomes very large as its speed approaches the speed of light. Its speed cannot be made equal to the speed of light without giving it an infinite amount of energy. Obviously, an accelerator cannot impress an infinite amount of energy to a particle, but modern accelerators can accelerate particles to speeds very close to the speed of light (up to $0.99 \ldots \times c$). The particles will have therefore enormous energies (of several hundred MeV).

From all the preceding considerations, it follows that according to the relativistic views of the physical world, the two classical principles of conservation of mass and energy should be unified into one conservation principle. In other words, in any process or change, what is actually conserved is the combined mass and energy of a system.

REFERENCES

Cable, E. J., et al. 1969. The physical sciences. 5th ed. Prentice-Hall, Inc., Englewood Cliffs, N. J.
Krauscopf, K., and A. Beiser. 1966. Fundamentals of physical science. Ed. 5. McGraw-Hill Book Co., New York.
White, H. H. 1966. Modern college physics. 5th ed. D. Van Nostrand Co., Inc., Princeton, N. J.

The atom

HISTORICAL BACKGROUND

"An atom is the smallest particle of an element that can enter into chemical combination." This statement is one of several definitions that can be given for the atom and is satisfactory for the chemist because in chemical reactions the atom really behaves as an indivisible elementary particle. It was consistent with the general understanding scientists had about the nature and properties of the atom practically throughout the nineteenth century.

At the turn of that century, mainly through the work of Lavoisier and Dalton, the fact had been definitely established that atoms can be of different kinds. This point of view was in contrast with what Democritus had thought several centuries before. The number of atomic species was thought to be the same as the number of chemical elements. Nevertheless, the concept of the atom as the elementary and indivisible particle of matter, devoid of any internal structure, was left unshaken simply because all the facts known to physicists and chemists could be explained with this theory. However, a few scientists did have some doubts about the indivisibility and lack of structure of the atom. For instance, the historian of chemistry R. Angus Smith wrote in 1876: "In using the word atom, chemists seem to think that they bind themselves to a theory of indivisibility. This is a mistake. The word atom means that which is not divided, as easily as it may mean that which cannot be divided, and indeed the former is the preferable meaning."

As the nineteenth century was coming to a close, some important discoveries and contributions were made in rapid succession—the study of electric discharges in gases, the discovery of the cathode rays (1870s), X-rays by Roentgen (1895), radioactivity by Becquerel (1896), the electron by Thomson (1897), the classical studies on alpha-particle scattering by Rutherford, Bohr's studies on the hydrogen spectrum, and much later the discovery of the neutron (whose existence had already been postulated before) by Chadwick (1932). These discoveries and studies shook the widespread belief in the indivisibility and elementary nature of the atom and laid the foundations for the modern theory of atomic structure.

Shortly after the discovery of the electron and the calculation of its mass and electric charge, a few physicists suggested models to represent the possible structure of the atom. Perhaps the model that has enjoyed more fame, although it turned out to be incorrect in the light of later discoveries, is Thomson's "plum-pudding" model (Fig. 2-1).

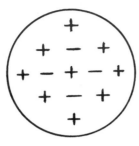

Fig. 2-1. Thomson's atomic model.

Thomson conceived the atom as a "sphere of positive electrification" embedded with electrons arranged in concentric layers or shells. The existence of a positive electrification had to be postulated to explain why an atom is electrically neutral, while it is made of negatively charged electrons. The total positive charge of the "sphere" must be equal to the sum total of the negative charges of the electrons.

Thomson's theory is important at least in one respect: it called attention to the electron as one of the elementary particles in the structure of an atom, and to the possibility that the atom might consist of an arrangement of positive and negative electric charges. However, the theory soon collided against several obstacles and was found to be inconsistent with some well-established facts. In particular, the uniform distribution of the electric charges and the tacitly implied compactness of the particles that make up an atom, with no possibility of empty spaces within it, were in contrast with some experimental observations made later by E. Rutherford and his co-workers.

RUTHERFORD'S NUCLEAR ATOM

During the first decade of this century, shortly after the discovery of radioactivity, Rutherford was investigating the behavior of the alpha particles emitted by radium and polonium. The fact was already known at that time that these particles are positively charged with a mass four times as large as the mass of the hydrogen atom (see the discussion on alpha particles, p. 89).

Results of experiments on alpha-particle scattering (Fig. 2-2) led Rutherford to assume that the atom contains a central electric charge with a small volume but with a relatively large mass. He further assumed that the central charge (later called nucleus) was positive and that the negative charges (electrons) were located outside of the nucleus.

One can see that Rutherford's views contained the germs of a new theory that envisions the atom as a miniature solar system, with its positive charge and most of its mass not uniformly distributed but concentrated in a relatively small central nucleus. The size of the whole atom is much larger than the size of the nucleus*; the extranuclear volume is filled with a limited number of electrons and with empty space. With so much empty space available within the volume of an atom, one is not surprised that very small, fast-moving particles, like beta particles and neutrons, may pass through matter of considerable thickness.

With regard to the nature and structure of the nucleus, many experiments, and in particular Moseley's studies of the characteristic X-rays (see Chapter 4),

---

*Today we know that, on the average, the diameter of the whole atom is about 10^4 times that of the nucleus. The average radius of the atomic nucleus is about 10^{-12} cm.

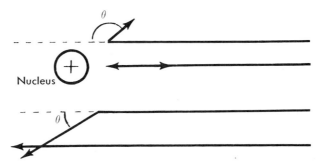

Fig. 2-2. Scattering of alpha particles. Rutherford noticed that a collimated beam of alpha particles passing through a thin sheet of metal would leave a rather diffuse image on a photographic plate, an indication of a deflection from their course. Further observations showed that some particles were actually deflected with a considerable angle and that some were even bouncing back toward the source. The diagram shows how these observations are in agreement with Rutherford's atomic model. The particles that are deflected very little, or not at all, are those that happen to travel through the empty space of an atom, or of several atoms, at a great distance from the nucleus. In these conditions, the coulombic force of electrostatic repulsion between the particle and the nucleus is very weak, and the particles proceed on their straight path. But those particles that happen to pass closer to the nucleus are deflected with an angle θ, which is roughly inversely proportional to the distance from the nucleus. If the distance is very short and the nuclear mass and charge are very large (as in elements of high atomic number), the alpha particle may be even turned back ($\theta = 180°$).

indicated that the number of positive charges in the atomic nucleus of an element must be equal to the ordinal number of the same element in the periodic system. This number had been previously called the *atomic number*. As a matter of fact, it is a common practice today to define the atomic number of an element as the number of positive charges carried by its atomic nucleus. To explain the electric neutrality of the atom, it was concluded that the atomic number of an atom must also be equal to the number of its extranuclear electrons.

In 1914 Rutherford reached the conclusion that the positive nuclear charges were carried by protons. These particles, which were found to be identical to the nuclei of the hydrogen atoms, had been discovered many years before as "rays" emitted by some modified cathode-ray tubes. Since the positive charge of the proton is equal to the negative charge of the electron but its mass is about 2000 times as large, it was apparent why most of the atomic mass is concentrated in the nucleus.

One problem to solve was the significant difference for most elements between the atomic number and the atomic weight (recall that atomic weights are expressed in atomic weight units, that the atomic weight unit is almost identical in magnitude to the amu, and that the mass of a proton is only slightly greater than 1 amu). For many elements the atomic weight is twice the atomic number or even larger. After other hypotheses were discarded, it seemed plausible that a nucleus is made not only of protons but also of some other unidentified, electrically neutral particles. These particles were called *neutrons* and were discovered and identified later in 1932 by Chadwick, as already mentioned above. Therefore, atomic weight = total weight of nuclear protons + total weight of nuclear neutrons + total weight of extranuclear electons.

THE EXTRANUCLEAR ELECTRONS

With his model, all Rutherford could establish about the extranuclear electrons was that they must be animated by some sort of rotatory movement around the nucleus. In fact, if electrons are to remain at a considerable distance from the nucleus, without being pulled onto it by its strong force of electrostatic attraction, one must postulate that they are in rotatory motion around the nucleus. Consequently, according to this conception the electrons would revolve around the nucleus in definite orbits, like the planets around the sun, with velocities that decrease as their distances from the nucleus increase.

The arrangement and number of electronic orbits in the atoms of different chemical elements and the behavior of the orbital electrons during the processes of absorption and emission of energy by the atom were definitely established by the Danish physicist Niels Bohr.

Bohr's work began with his attempt to find a satisfactory explanation of the optical spectrum of hydrogen. Such a spectrum is obtained when the light emitted by incandescent hydrogen is analyzed with a spectrograph. The fact had been observed many years before that the H spectrum consists of a few discrete bright lines of different colors, each line corresponding to a definite wavelength.

Through a long series of considerations and mathematical treatments, which cannot be discussed here, Bohr concluded that the single electron of the H atom can revolve around the nucleus only in one or another of a series of discrete concentric orbits. Bohr also derived equations to calculate the radii of these orbits. When the electron revolves in the innermost orbit, the atom is said to be at its ground-energy level. When the atom absorbs energy, the electron may be lifted to an outer orbit; the atom is thus raised to a higher energy level and is said to be "excited." If the electron jumps back to one of the inner orbits or to the innermost orbit, energy is emitted; in this case it is in the form of light.

Bohr extended his theory to all other elements, for which he introduced the concept of electronic shells, which can be envisioned as concentric groups of orbits. Beginning with the innermost shells, we call them K, L, M, N, O, etc. The maximum number of electrons that can be accommodated in each shell varies according to some laws discovered by Bohr and others; this number increases from the K-shell to the outer shells.

Within each shell electrons are grouped in subshells. The number of subshells for each shell also increases from the K-shell to the outer shells. Each electron, or electronic energy level, of any atom can be univocally identified with four *quantum numbers* (principal, orbital, magnetic, spin) (for further details, see Chapter 2 references), which are like the "Cartesian coordinates" of the electron. No two electrons of the same atom may have all four quantum numbers identical (Pauli exclusion principle).

The distribution of the orbital electrons in shells and subshells is shown in Table 2-1.

One should bear in mind that the numbers in the table are the maximum numbers of electrons that can possibly be accommodated within each shell and subshell. The reader is invited to consult textbooks of general chemistry for a review of the actual distribution of the electrons for each chemical element. One can find in the atoms of some elements, unfilled or incomplete inner shells or subshells. There is a relationship between the distribution of the orbital electrons and the periodic chemical properties of the elements.

Table 2-1. Distribution of orbital electrons in shells and subshells

Shell	Number of subshells	Maximum number of electrons per subshell	Total number of electrons in each shell
K	1	2	2
L	2	2,6	8
M	3	2,6,10	18
N	4	2,6,10,14	32
O	5	2,6,10,14,18	50

Table 2-2. Electron binding energies, in keV, of three elements

Element	Atomic number	K-shell	L-shell	M-shell
Hydrogen	1	0.013	–	–
Molybdenum	42	20.00	2.6	0.5
Tungsten	74	69.51	11.5	2.8

The electrons of the outermost shell are referred to as *valence electrons* because they alone are involved in chemical combinations and in the formation of electrovalent or covalent bonds between atoms.

In radiation physics, the concept of *electron binding energy* is of considerable importance. Consider the electrostatic attractive force exerted by the nucleus of an atom on a K-electron and on an L-electron. Evidently, since the K-electron is closer to the nucleus than the L-electron, the former is subject to a more intense attraction to the nucleus and therefore is more tightly bound in the atom than the latter. Consequently, to dislodge a K-electron from its orbit, one requires an expenditure of more energy. The energy required to remove an electron from an atom is called the *binding energy* (E_b) of that electron. For the same element, the binding energies decrease from the inner to the outer shells. Within the same shell they also decrease from the innermost to the outermost subshell.

The E_b of electrons of the same shell for atoms of different elements increase with the atomic number. The reason is that the positive nuclear charge increases with the atomic number and consequently the electrostatic attraction on the orbital electrons also increases.

To illustrate these variations, some examples of E_b values are reported in Table 2-2 in keV.* The values for the L- and M-shells are averages of the values for different subshells.

One can see that the E_b values vary from a few eV to several keV, depending on the shell and on the atomic number of the element. When one or more electrons are removed from an atom, the atom becomes a positive ion, and the process is called *ionization*. Several causes of ionization are known to the chemist and to the physicist. Ionization as it occurs in the solutions of electro-

---

*Complete tables can be found in the *Handbook of Chemistry and Physics.*

lytes involves removal of peripheral electrons from certain atoms and their transfer to the outermost shells of other atoms, which become negative ions; very small amounts of energy (a few eV) are needed for this type of ionization.

On the other hand, the large amounts of energy carried by various types of ionizing radiations are required to dislodge inner electrons from atoms of relatively high atomic number. Several examples of this type of ionization can be found in the following chapters. It is important that when an electron is removed from an inner shell, its vacancy will be filled by an electron of an outer shell; the consequence of this "filling in" is radiation (that is, emission) of energy.

A note of caution should be sounded at this point. There are many reasons to suspect that when we speak of electrons revolving around the nucleus in definite orbits with measurable radii, of electrons jumping from one orbit to another, etc., we are using a language that might not reflect the physical reality correctly. The practice followed in many elementary books of using diagrams to represent atoms consisting of a central nucleus with a number of electronic orbits oriented in various directions in space, is justifiable only if these diagrams are regarded as merely symbolic and are not interpreted too literally. As mentioned on pp. 43 to 45, electronic shells and subshells become more meaningful if they are considered as different atomic energy levels. Likewise, since prediction of the position of an electron in its orbit is impossible (Heisenberg's uncertainty principle), some physicists think of the electron as spread out into a cloud of electricity; the density of the cloud represents the statistical probability of finding an electron at any given point within the atom.

At this point, it is important for the reader to become aware of the following basic facts about atomic structure:

1. The atom, as conceived by Rutherford and Bohr, is made of a positively charged nucleus surrounded by negative electrons orbiting around it. In a neutral atom, the electric charge of the nucleus is equal to the total electric charge of the orbiting electrons.

2. The "emptiness of matter" is portrayed by the fact that the average radius of the nucleus is about 10^{-12} cm and that of the whole atom is about 10^{-8} cm; the nuclear radius is therefore $1/10^4$ of the atomic radius. If the nucleus were the size of a Ping-Pong ball, the electrons would be revolving around it at an average distance of about 200 yards!

3. The mass of an atom is almost all concentrated in the nucleus. The nuclear density is estimated at 10^8 tons/cm^3! Consequently, 1 cm^3 of packed atomic nuclei would weigh 10^8 tons. If all the empty space within the human body could be reduced to zero, it would probably be smaller than the head of a pin and yet have the original weight.

ATOMIC NUCLEUS, NUCLIDES, AND NUCLEAR FORCES
Nuclear structure

The atomic nucleus is made of at least two elementary particles—the *proton* and the *neutron*. Both are also called *nucleons*.

Some information on nucleons and negative electrons is given in Table 2-3.

Table 2-3.

Name of particle	Symbol	Electric charge	Mass
Negative electron	e^-	-1 $(4.8 \times 10^{-10}$ esu$)$	0.000548 amu
Proton	p	$+1$ $(4.8 \times 10^{-10}$ esu$)$	1.007276 amu
Neutron	n	0	1.008665 amu

Some composite particles are incidentally mentioned below because they may constitute ionizing radiation when traveling at high speed:

1. The *deuteron* (1p + 1n) is nothing but the nucleus of deuterium. In some particle accelerators (as cyclotrons), the emerging beam may be made of deuterons.

2. The *helion* (2p + 2n) is the nucleus of natural helium. When this particle is emitted by a radioactive nucleus, it is called an alpha particle.

In both the deuteron and the helion, the configuration of the nucleons seems to be particularly stable.

The following definitions are of great importance for this discussion:

1. The *atomic number* (Z) is the number of protons present in a nucleus.

2. The *mass number* (A) is the number of nucleons that make up a nucleus.

3. The *neutron number* (N) is the number of neutrons present in a nucleus.

The following relationships are derived from the above definitions:

$$A = Z + N \qquad\qquad Z = A - N \qquad\qquad N = A - Z$$

Nuclides

A *nuclide* is an atomic species univocally defined by:

1. An atomic number
2. A mass number
3. A nuclear energy level*

Consequently:

1. If two atoms have the same Z, the same A, and the same nuclear energy level, they are atoms of the same nuclide.

2. If two atoms have one or more of these characteristics different, they are atoms of different nuclides.

In accordance with a recommendation of the International Union of Pure and Applied Chemistry, the following notation should be used for the identification of a nuclide:

$$_Z^A X_N^{(valence)} \qquad\qquad\qquad \text{Example:}\quad _{20}^{45}Ca_{25}^{++}$$

(X is the symbol of the chemical element to which the nuclide belongs.)

For most applications it is sufficient to indicate only the mass number on the upper left side of the chemical symbol such as ^{45}Ca and ^{60}Co. The old notations like Ca^{45} and Co^{60} are no longer permitted. Notations like Ca-45 are tolerated. Notations like calcium 45 are permitted.

The following terms are quite often used to indicate certain types of relationships between two or more nuclides:

1. Two or more nuclides with the same atomic number but different mass numbers are called *isotopes*; they belong to the same chemical element; for example:

$$^{14}_{6}C \qquad \text{and} \qquad ^{12}_{6}C$$

Many chemical elements, as found in nature, are mixtures of two or more

---

*As electronic energy levels represent different states of the orbital electrons, nuclear energy levels apply to different possible energy states of the nucleus; in regard to amount of energy, also nuclei may be in a ground state or in an excited state. More about nuclear energy levels is discussed in later chapters.

isotopes. Their percentages (natural abundances) are constant regardless of the origin of the element.

2. Two or more nuclides with the same mass number but different atomic numbers are called *isobars*; they belong to different chemical elements; for example:

$$^{14}_{6}C \quad \text{and} \quad ^{14}_{7}N$$

3. Two or more nuclides with the same neutron number but different atomic numbers are called *isotones*; they belong to different chemical elements; for example:

$$^{39}_{19}K_{20} \quad \text{and} \quad ^{40}_{20}Ca_{20}$$

4. Two or more nuclides with the same atomic number and mass number but different nuclear energy levels are called *isomers*; they belong to the same chemical element; for example:

$$^{110m}Ag \quad \text{and} \quad ^{110}Ag \quad (m = \text{metastable})$$

The *atomic mass* of a nuclide (not to be confused with the mass number) is the mass of an atom of the nuclide expressed in amu. The atomic mass of a nuclide usually is not identical to the atomic weight of the element to which it belongs (see definition of atomic weight and atomic weight unit in standard textbooks of general chemistry).

At the present time (1971) about 1500 nuclides are known. Their Z number ranges from 1 to 103; their A number ranges from 1 to 258. About 330 of them are naturally occurring, and the others are artificially produced by nuclear fission, nuclear reactions, and various other methods.

Nuclides are also classified as follows:

1. *Stable* (about 265) are those nuclides that cannot transform themselves into other nuclides without the addition of energy from outside. None of them has a Z number above 83 (= Bi).

2. *Unstable (radioactive nuclides, radionuclides)* are those nuclides capable of transforming themselves spontaneously into other nuclides by changing their nuclear configuration or nuclear energy level. This transformation is called *radioactivity*. Sixty-five of the radionuclides occur in nature but most of them are artificially produced.

All nuclides with $Z > 83$ are unstable. Among the natural radionuclides with $Z \leq 83$, the following are noted: ^{14}C, ^{40}K, ^{87}Rb, ^{142}Ce, and some isotopes of Sm.

The reader is strongly advised to familiarize himself with a chart of the nuclides (the chart published by the Knolls Atomic Power Laboratory is suggested). Charts of this type contain most of the information about nuclides usually needed in radiation biology and radionuclide methodology.

Nuclear forces

The problem of what kind of forces are responsible for tightly holding together the nucleons in an atomic nucleus is one of the most intriguing, elusive, and fascinating in nuclear physics, yet it is far from being solved. In the meantime, let us begin with the following statement: "The atomic mass of an atom is always less than the sum of the masses of its individual nucleons and orbital electrons." The difference is known as *mass defect*. For example, consider the following:

The atomic mass of $_1^2$H is 2.01410 amu
Yet,

mass of proton = 1.007276 amu
mass of neutron = 1.008665 amu
mass of electron = 0.000548 amu
_____
total   = 2.016489 amu

Mass difference = 0.002389 ($\approx$ 0.0024)

The explanation of this startling fact has been provided by the theory of relativity (see Chapter 1) and in particular by Einstein's equation $E = mc^2$. According to the theory, mass can be converted into energy and vice versa. The above equation expresses the quantitative aspects of these conversions. Therefore the explanation of the mass defect is that when a proton and a neutron come together to form a deuterium nucleus, part of their mass is converted into energy, which is released. This energy is called the *nuclear binding energy* (E_b).

The nuclear E_b is, in effect, the amount of energy released in the formation of a nucleus, and conversely to break up a nucleus into its component particles, an amount of energy equivalent to the nuclear E_b must be supplied.

One can easily calculate the E_b of the deuterium nucleus. In fact the annihilation of 1 amu results in the release of 931.2 MeV of energy (see equation 1-22). Therefore the E_b of the deuterium nucleus is 0.0024 amu (= mass defect)$\times$ 931.2 MeV/amu = 2.23 MeV. For nuclides of higher mass, the nuclear binding energy is much greater. For bismuth ($Z = 63$, $N = 126$, atomic mass = 208.98), the mass defect is 1.76 amu and therefore $E_b \approx$ 1640 MeV. (One can verify these figures by the same method of calculation used above for the deuterium nucleus).

If energy must be provided to pull apart the nucleons of a nucleus, then they must be held together by some type of attractive forces. These forces must be very strong if more than 2 MeV are required to break apart the deuterium nucleus (compare with the electron binding energies that range from a few eV to about 100 keV).

At least two types of forces well known to classical physics should be operative between the nucleons within the atomic nucleus:

1. *Electrostatic forces.* The positively charged protons of a nucleus repel one another with a force that is inversely proportional to the square of their distance (Coulomb's law). Consequently, with a large number of neutrons in a nucleus, the distances between protons increase and the intensity of the electrostatic repulsion between any two protons decreases. Alone, the electrostatic forces would have a disruptive effect on the nucleus.

2. *Gravitational forces.* Because of the law of universal gravitation (see p. 15), all nucleons attract one another with an intensity expressed mathematically by Newton's equation. However, gravitational forces cannot account for the very high values of the nuclear binding energies. Furthermore, one can show that the electrostatic forces of repulsion are much more intense than the gravitational forces of attraction (see p. 17).

It is thus necessary to think of the existence of other forces of attraction in the nucleus. These forces must be strong enough to account for the high values of the nuclear binding energies and to overcome the electrostatic forces of repulsion. These forces, still very poorly understood, are called *nuclear forces*.

It is not possible yet (as it is for gravitational and electrostatic forces) to express the behavior of the nuclear forces with a mathematical equation. However, the current impression is that they are short-range forces; that is, they are significantly strong only when the distances are very small (such as those between nucleons). This fact is probably true because their intensity may be inversely proportional to the fourth or fifth power of the distance.

Nuclear forces are also referred to as "exchange forces" because they would be exerted through the exchange between nucleons of very short-lived, ephemeral particles called "mesons." The study of these exchanges has led some nuclear physicists to believe that the neutron and the proton are but two different states of essentially the same particle and that the change of state is reversible. Mesons are referred to sometimes as "nuclear glue." Free mesons are components of cosmic radiation (see Chapter 6), and their mass is intermediate between the mass of the proton and that of the electron.

For additional information on nuclear structure, see the suggested references.

REFERENCES

Fermi, E. 1950. Nuclear physics. University of Chicago Press, Chicago.

Glasstone, S. 1958. Sourcebook on atomic energy. D. Van Nostrand Co., Princeton, N.J. Chapters 1 and 4.

Lapp, R., and H. Andrews. 1954. Nuclear radiation physics. Prentice-Hall, Inc., Englewood Cliffs, N.J. Chapter 2.

Mayer, M. G. March 1951. The structure of the nucleus. Sci. Amer. **184:**22.

Romer, A. 1960. The restless atom. Doubleday & Co., Inc., Garden City, N.Y.

CHAPTER 3

The electromagnetic spectrum

Concept and properties of waves
The nature of electromagnetic radiations
Planck's quantum theory
General survey of the electromagnetic spectrum
Excitation and electronic transitions
Particle properties of photons

CONCEPT AND PROPERTIES OF WAVES

Some of the ionizing radiations of interest to the radiation biologist have wave properties (X-rays, gamma rays). These radiations are not essentially different in nature from more familiar kinds of radiations like light and radio waves. An analogy will illustrate the nature and the fundamental laws governing the behavior of these radiations. Let us consider a tank containing still water. If a stone is dropped vertically on the surface of the water, a series of concentric ripples or waves will appear around the point where the stone falls. These waves can be seen moving outward. One can create a persistent series of waves by partially immersing a vertically vibrating rod in the tank; waves will be seen originating from the point of disturbance as long as the rod is oscillating. Fig. 3-1 represents a vertical section of this wave system. At the point d where the disturbance is occurring, the water molecules are oscillating in a vertical direction. This disturbance is not confined to point d but is transmitted to the surrounding molecules, farther and farther away from point d. Consequently, even water molecules that are not directly disturbed by the vibrating rod will be subject to a vertical oscillatory motion. The wave system consists then of a series of crests and troughs. Looking at such a wave system, one has the impression that water is moving horizontally away from the point of disturbance. However, a floating cork proves that this is only a fallacious impression; the cork simply oscillates up and down because it is following the identical motion of the water molecules. The only horizontal "motion" is the propagation of the disturbance itself; it is also proper to say that in this case waves propagate horizontally.

These water waves are an example of *transversal waves*, because the oscillations of the water molecules are perpendicular to the direction of wave propaga-

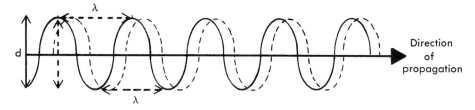

Fig. 3-1. Transversal wave system. **d,** Origin of disturbance. **λ,** Wavelength.

tion. The distance between two successive crests (or troughs) is called *wavelength* (λ) and depends on several factors related with the cause of the disturbance. The speed (*c*) of wave propagation is another relevant parameter and is a property of the medium, depending on its nature. The *frequency* (ν) of the wave system is the number of oscillations (or cycles) of a water molecule per unit of time. Speed of propagation, wavelength, and frequency of a wave system are correlated by the equation

$$c = \lambda \nu \qquad (3\text{-}1)$$

From equation 3-1 we derive $$\lambda = c/\nu \qquad (3\text{-}2)$$

and $$\nu = c/\lambda \qquad (3\text{-}3)$$

These equations are valid for any type of wave system.

THE NATURE OF ELECTROMAGNETIC RADIATIONS

One of the most interesting controversies in the history of science was that concerning the nature of light. Newton, who contributed so much to the knowledge of several properties of light, believed that this type of radiation has a corpuscular nature; accordingly, a beam of light would consist of a shower of corpuscles or particles projected from the source. This theory was thought to be consistent with such facts as the propagation of light in straight lines, formation of shadows, reflection, and other optical effects. Although a few scientists (notably C. Huygens) expressed doubts about this view during Newton's time, the alternative theory of light as a form of wave or vibratory motion was not widely accepted until the beginning of the nineteenth century, when the optical effects of interference and diffraction were further analyzed and found to be incompatible with the corpuscular theory. The speed of light had been measured earlier with good approximation and was found to be 3×10^{10} cm/sec* in a vacuum and somewhat less in other media.

In view of the fact that light can travel in empty space (for example, from the sun and stars to the earth), the real nature of the light waves without a material medium that is set in vibration was difficult to imagine. In 1864 James C. Maxwell, with mathematical methods, showed that certain electric disturbances can be propagated through the surrounding space; this propagation consists of the periodic variation in intensity of the electric and magnetic fields created by the electric disturbance. For instance, if an electron moves back and forth in a metallic wire with a certain frequency, the electric field surrounding the wire (and the magnetic field associated with it) will undergo variations in intensity with the same frequency. These periodical variations will be propagated farther and farther away from the wire, as in wave motion. This wave system is called *electromagnetic radiation* because it consists of an electric and a magnetic wave. More specifically, Maxwell predicted that the two waves are at right angles with each other and with the direction of propagation (Fig. 3-2). The electromagnetic waves are therefore transversal waves; because electric and magnetic fields are not bound to matter, they can travel in a vacuum; their speed of propagation, which is equal to the speed of light, is independent of their wavelength and cause of electric disturbance. Maxwell arrived at the conclusion that light itself is a form of electromagnetic radiation. Maxwell's genius did not stop at this point. Although light was the only form of electromagnetic radiation known to him, this great theoretical scientist predicted that other similar radiations should exist and might be discovered, not different from light in their nature and speed of propagation, but only in their wavelengths and frequencies. In fact, shortly afterwards it was shown that the previously discovered infrared and ultraviolet radiations are also electromagnetic radiations; their wavelengths are longer and shorter respectively than the wavelengths of light. Radiations of wavelengths longer than infrared were discovered by Heinrich R. Hertz (radio waves). The discovery of X-rays (1895) and of

---

*About 186,000 miles/sec.

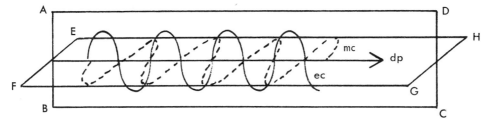

Fig. 3-2. Electromagnetic wave system. **ABCD,** Plane of vibration of the electric component (**ec**). **EFGH,** Plane of vibration of the magnetic component (**mc**). **dp,** Direction of propagation. The two waves travel together in phase. Both fields reach their maximum and zero values at the same time. When one reverses, the other does too.

gamma radiation emanating from some radioactive materials and of their electromagnetic nature meant the discovery of electromagnetic radiations of extremely short wavelengths (less than 0.1 A). Today we know of a continuous, uninterrupted gamut of electromagnetic radiations, natural and man-made, with wavelengths ranging from a few miles (radio waves) to small fractions of an angstrom unit (some components of cosmic rays); the series is known as the *electromagnetic spectrum.*

PLANCK'S QUANTUM THEORY

In the analogy of the water waves, energy is transferred by the waves from the point of origin to other points of the water surface. Likewise, in regard to electromagnetic radiations, energy is transferred by the electromagnetic waves through space from the source to other bodies. This form of energy is called *radiant energy.* Consequently, one can say that the earth receives light energy from the sun, that the antenna of a radio transmitter radiates radio energy, etc.

The electromagnetic theory, as developed by Maxwell, was sufficient to explain all the properties and behavior of electromagnetic radiations known until the end of the nineteenth century. Later, some phenomena were discovered that were not easy to explain without some corrections of and additions to, the theory. One such phenomenon is the photoelectric effect.

The photoelectric effect can be easily studied with a simple instrument called a "photoelectric cell" (Fig. 3-3), which consists basically of an evacuated glass bulb with a metallic sheet attached to its internal surface and connected with the negative pole of an electric battery. An anode is connected to the positive pole. There is a galvanometer in the circuit to detect and measure any electric current that might flow through the circuit itself.

If light strikes the cathode, it may eject electrons from the surface of the metal (photoelectric effect). If this phenomenon happens, the ejected electrons will be attracted by the anode and the galvanometer will indicate the presence of an electric current through the circuit and its intensity, which depends on the number of electrons ejected per unit of time. The intensity of this current would seem to be exclusively proportional to the intensity of light falling on the cathode, that is, on whether the light is dim or bright, but it turns out that a very bright red light might be unable to eject any electron from the metallic surface (no current detected by the galvanometer), whereas a dim violet light might eject at least a few electrons. This observation is not consistent with the theory of electromagnetic radiation as formulated by Maxwell. It seems that what is important in producing a photoelectric effect is not the intensity of light, but its color, that is, its wavelength and frequency (see below and Table 3-1). Violet light has a shorter wavelength and higher frequency than red light. It was thought that violet light must be, regardless of its intensity, somehow "more energetic" than red light if it can eject electrons from the cathode of a photoelectric cell.

Table 3-1. Chart of the electromagnetic spectrum

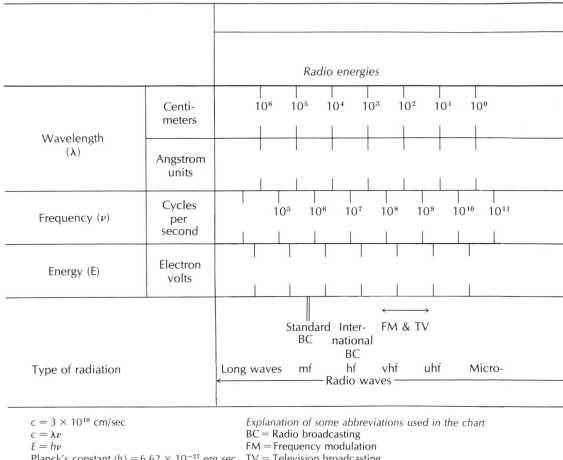

			Radio energies
Wavelength (λ)	Centi-meters	10^6 10^5 10^4 10^3 10^2 10^1 10^0	
	Angstrom units		
Frequency (ν)	Cycles per second	10^5 10^6 10^7 10^8 10^9 10^{10} 10^{11}	
Energy (E)	Electron volts		
Type of radiation		Standard Inter- FM & TV BC national BC Long waves mf hf vhf uhf Micro- ← Radio waves →	

$c = 3 \times 10^{10}$ cm/sec
$c = \lambda\nu$
$E = h\nu$
Planck's constant $(h) = 6.62 \times 10^{-27}$ erg sec

Explanation of some abbreviations used in the chart
BC = Radio broadcasting
FM = Frequency modulation
TV = Television broadcasting
mf = Medium frequency
hf = High frequency
vhf = Very high frequency
uhf = Ultrahigh frequency

These considerations and other related facts convinced Max Planck and Albert Einstein at the beginning of the twentieth century that electromagnetic radiation is not emitted as a continuous flow of energy, but rather as a shower of discontinuous waves, each wave carrying a small package of energy. Each discrete wave is called a *photon* and the package of energy associated with it is referred to as a *quantum* of energy. The consequence of this point of view is that a source may emit only an integral number of photons (1, 2, 3, etc.) and not intermediate or fractional amounts ($\frac{1}{2}$, $1\frac{1}{2}$, etc.). In terms of radiant energy emitted, this fact means that a source emits energy, in the form of radiation, in integral multiples of a definite amount or quantum. Likewise, a body can absorb radiant energy only in integral multiples of a quantum.

Max Planck found that the quantum of energy carried by a single photon (hereafter referred to as *photon energy*) is directly proportional to the frequency of the electromagnetic radiation in question, according to the equation

$$E = h\nu \qquad \text{(Planck's equation)} \qquad (3\text{-}4)$$

The electromagnetic spectrum

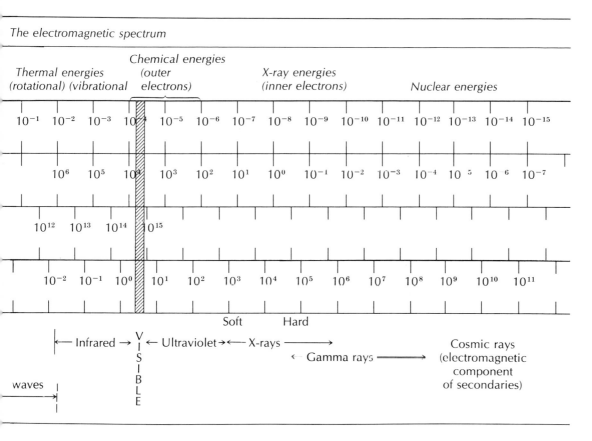

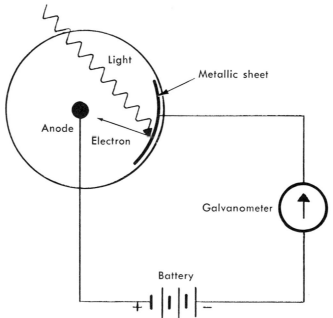

Fig. 3-3. Simple photoelectric cell.

wherein E is photon energy in ergs, and h is a constant (6.62×10^{-27} erg sec), called Planck's constant. This constant is one of the most fundamental in nature, although its significance is not completely understood.

According to Planck's equation, the photon energy increases with frequency and therefore from radio waves to gamma rays, from red light to violet light.* Returning to the photoelectric effect illustrated above, one sees that even one photon of violet light has sufficient energy to pull an electron from the cathode, whereas of the many photons of a bright red light, not a single one has enough energy to eject a single electron and the number of photons striking the cathode will not change this situation (see p. 100 for further discussion on the photoelectric effect).

Combining equation 3-4 with 3-3, one has

$$E = h\frac{c}{\lambda} \tag{3-5}$$

which simply states that photon energy is inversely proportional to wavelength.

Since the magnitude of photon energies is very small, one may more conveniently express them in electron volts (or its multiples) than in ergs.

Let us recall that

$$1 \text{ erg} = 6.24 \times 10^{11} \text{ eV}$$

Therefore

$$E(\text{eV}) = h\frac{c \times 6.24 \times 10^{11}}{\lambda(\text{cm})} = \frac{6.62 \times 10^{-27} \times 3 \times 10^{10} \times 6.24 \times 10^{11}}{\lambda(\text{cm})} = \frac{1.24 \times 10^{-4} \times 10^{8}}{\lambda(\text{A})} = \frac{1.24 \times 10^{4}}{\lambda(\text{A})}$$

And since $1 \text{ MeV} = 10^{6} \text{ eV}$,

$$E(\text{MeV}) = \frac{0.0124}{\lambda(\text{A})} \tag{3-7}$$

EXAMPLES:
1. The photon energy in electron volts of monochromatic violet light with a wavelength of 4000 A is given by

$$E = \frac{1.24 \times 10^{4}}{4 \times 10^{3}} = 3.1 \text{ eV}$$

2. The photon energy of a gamma ray with a wavelength of 0.01 A is given by

$$E = \frac{0.0124}{0.01} = 1.24 \text{ MeV}$$

The chart of the electromagnetic spectrum (Table 3-1) shows the general and approximate relationship between photon energy and wavelength for all electromagnetic radiations.

GENERAL SURVEY OF THE ELECTROMAGNETIC SPECTRUM

Only some segments of the electromagnetic spectrum are of direct concern to the radiation biologist. However, a general knowledge of the whole spectrum

---

*The intensity of light (and of any electromagnetic radiation, for that matter) is related to the number of photons emitted by the source per unit of time. For example, a flash light emits about 10^{18} photons/sec.

is extremely useful to every biologist because of the interactions existing between some types of nonionizing radiations and living organisms. Table 3-1 correlates, with different horizontal bands, the wavelengths, frequencies, and photon energies. Note that the scales are logarithmic; for instance, in the energy band, the segment representing the 1 to 100 eV interval is twice as long as the segment representing the 1 to 10 eV interval, rather than 10 times as long. Wavelengths decrease from the left to the right, but frequencies and photon energies increase in that same direction.

There are no sharp border lines between different types of radiations; one type gradually merges into another. The overlapping of two different types of radiation in certain segments of the spectrum finds an explanation in the fact that two radiations may have different origins and therefore different names (for example, X-rays and gamma rays), and yet they may have the same wavelength, frequency, and photon energy.

Radio waves are electromagnetic radiations with wavelengths ranging from a few miles to a few millimeters. They were discovered by H. Hertz and later utilized by

Table 3-2. Emission sources and detectors of electromagnetic radiations

Type of radiation	Emission sources	Detectors
Radio waves	Celestial radio sources (sun, galaxies, nebulas, quasars, etc.) Vacuum tubes, oscillators	Vacuum tubes, transistors Radio telescopes
Microwaves	Magnetrons, klystrons	Magnetrons, klystrons
Infrared	Hot bodies	Infrared photographic films Thermometers, thermopiles, bolometers, radiometers Cutaneous thermoceptors
Visible	Incandescent bodies (including lamps) Fluorescent lamps Lasers Sun, stars, etc.	Photosynthetic systems Retinal cones and rods Photographic films, photoelectric cells, photomultipliers
Ultraviolet	Mercury, hydrogen, and xenon lamps, carbon arc Sun	Photographic films, photoelectric cells, photomultipliers
X-rays	Coolidge tubes, some particle accelerators Electronic transitions resulting from some radioactive decays Sun, and other celestial sources	Photographic films, fluoroscopes, ionization chambers
Gamma rays	Radioactive decay, nuclear fission, other nuclear reactions	Photographic films, ionization chambers, scintillation detectors
Cosmic rays	From outer space	(Same as for gamma rays)

Guglielmo Marconi and others as a means of radio communications. They are produced by radiotransmitters, which, by means of special vacuum tubes, can set free electrons in vibratory motion. The electronic circuits that can accomplish this phenomenon are called "oscillators." The oscillating electrons generate an electromagnetic wave that can be propagated in space by means of an antenna. The frequency of the radiation is identical to the frequency of the electronic oscillation.

Electronic oscillating circuits are not satisfactory for the production of electromagnetic radiation of wavelength shorter than microwaves. See Table 3-2 for the different emitters and detectors of radiation. Infrared radiation is generated by hot bodies. If a metallic body is heated gradually from room temperature to about 200 to 300° C, it will remain invisible in a dark room (it does not generate any light) but will emit infrared radiation, which can be felt by the human body at a distance as a sensation of warmth. Although part of the hot body's heat is transmitted by conduction through the air, even in the absence of air the sensation of warmth can still be felt at a distance. This sensation is due to energy radiated as infrared, which can travel in a vacuum like any other type of electromagnetic radiation.

The hot body is emitting a continuous infrared spectrum of several wavelengths, but the wavelength of maximum radiation intensity is inversely proportional to the temperature of the body.* In other words, as the temperature of the body rises, the infrared energy is radiated at increasingly shorter wavelengths. Several devices are sensitive to infrared radiation and are therefore used to detect and measure this type of radiation. Thermopiles are used for many practical applications; with these instruments, even small variations of temperature can be converted into variable electric currents, which can be accurately measured by galvanometers. They are used, for instance, as detectors in infrared spectrophotometers. Recently, photographic films coated with emulsions sensitive to infrared have been developed, and their practical applications are already numerous. With these films, one can take photographic pictures in complete darkness, as long as the objects to be photographed have at least slightly different temperatures.

Some microscopic sensory organs present in the dermis of the skin (cutaneous thermoceptors) are also sensitive to infrared radiation.

As the temperature of the hot body is further increased, the wavelengths of the radiation become progressively shorter until a new type of radiation, one that can be detected by the human eye, is emitted. The hot body then becomes visible and looks red ("red hot") in a dark room, because it is radiating wavelengths as short as 8000 A. We have approached that narrow segment of the electromagnetic spectrum commonly called "visible light" because it is the only segment that can be detected by the eye. We know that if the red hot body is heated to higher temperature, its red color becomes gradually orange, yellow, and finally white. This change is in perfect accordance with Wien's law: when the body is "white hot," it is emitting not only infrared, which is invisible, but also all the wavelengths between 8000 and 4000 A, that is, all the colors of the visible light segment from red, through orange, yellow, green, and blue, to violet. The "white color" is the sensation produced by the mixture of all those colors. When the hot body has a temperature high enough to emit visible light, it is said to be incandescent. Incandescence is in fact the main cause of emission of visible light. Incandescent lamps produce light because an electric current heats a tungsten filament to incandescence; neon tubes are sources of light because a gas is made incandescent by an electric current. Besides incandescence, the phenomenon of fluorescence can also be the cause of emission of visible light. In fluorescence, electromagnetic radiation of wavelength shorter than 4000 A

---

*Wien's law: $\lambda_m T$ equals a constant. (λ_m is the wavelength at which maximum energy is radiated, and T is the absolute temperature.)

interacts with the atoms of certain substances (fluors) and is converted into visible light. Fluorescent lamps work on this principle. They are said to emit "cold light" because very little heat is evolved in the process.

Besides the cones and rods of the retina, another natural receptor is sensitive to visible light—the photosynthetic system, exemplified by a chloroplast. The chlorophyll molecule present in a chloroplast can absorb very efficiently the energy carried by visible light photons, especially in the red and blue regions, and this energy is what triggers the photosynthetic process.

Examples of artificial detectors of visible light are photoelectric cells, used as exposure meters in photography, and photomultipliers (see Chapter 7).

Ultraviolet radiation encompasses a wavelength range from 4000 A down to about 100 A. A carbon arc and lamps containing mercury vapor, hydrogen, or xenon brought to incandescence by an electric current generate ultraviolet radiation, along with visible light. The bulbs of these lamps are made of quartz or silica because glass is opaque to ultraviolet. It is possible to screen off visible light emitted by these sources with special filters opaque to visible wavelengths but transparent to ultraviolet.

The ultraviolet band of the spectrum is divided by decreasing wavelength, somewhat arbitrarily into near, far, and vacuum ultraviolet. Vacuum ultraviolet is completely absorbed by air and therefore can be propagated only in a vacuum.

Ultraviolet radiation is of special interest in biology because its photon energies are high enough to cause certain biologic effects in living systems. For instance, ultraviolet can convert a provitamin D into the active vitamin; the 2600 A wavelength is absorbed to a great extent by the nucleic acid molecules; some amino acids (phenylalanine, tyrosine, tryptophan) very strongly absorb definite wavelengths between 2500 A and 3000 A. The energy so absorbed may cause chemical changes in the molecules. The 2600 A wavelength is "germicidal" because it is very effective in killing several bacterial species.

Photobiology is the branch of the biologic sciences that is concerned with the effects of visible and ultraviolet radiation on living systems.

The electromagnetic radiations beyond the ultraviolet band (X-rays and gamma rays) are distinctive at least for the following two characteristics: They can easily penetrate through materials that are opaque to most of the other types of radiations, and their photon energies are high enough to dislodge orbital electrons from practically any shell of an atom, thus producing ions. Therefore, they are *ionizing radiations.*

The essential difference between X-rays and gamma rays is in their origin. X-rays originate outside of the atomic nucleus when certain electronic transitions occur or by the bremsstrahlung mechanism, as is explained in Chapter 4. Their classification into "soft" and "hard" refers to their penetrating power, which increases with the photon energy. Gamma radiation is of nuclear origin and sometimes accompanies certain radioactive decay processes and other nuclear reactions. High-energy gamma radiation is also generated by certain interactions of cosmic rays with the earth's atmosphere. These problems are discussed with further details in Chapter 6.

EXCITATION AND ELECTRONIC TRANSITIONS

In Chapter 2 we saw that when an atom absorbs energy it is possible for an orbital electron to be completely removed from the atom (ionization). The energy of photons of certain wavelengths, the kinetic energy of moving particles, such as alpha particles, electrons, etc., may be high enough to ionize an atom. The choice of whether an electron of an inner or of an outer orbit is removed depends mainly on the amount of energy absorbed by the atom; the higher the energy, the more likely will the inner electrons be removed.

In certain conditions, the absorption of energy results simply in a displacement (or "jump") of an electron from an inner orbit to an outer orbit. When this happens, the atom is said to be *excited* or *raised to a higher energy level* because when the displaced electrons slip back to inner orbits, energy is emitted in the form of electromagnetic radiation and the atom reaches a state in which its energy level is lower. If the displaced electrons slip back to their original orbits, the atom returns to its *ground energy state*

The displacements of electrons from inner to outer orbits, or vice versa, are also called *electronic transitions*.

Fig. 3-4 illustrates electronic transitions in a generalized atom. For the sake of simplicity, the existence of subshells within the L- and M-shells has been ignored. Remember that the electron binding energy (E_b) decreases from the K-shell to the M-shell (see p. 29). A certain amount of energy (E_3) absorbed by the atom might be only sufficient to displace an L-electron to the M-shell. To displace a K-electron to an outer allowable orbit, a greater amount of energy is needed because the E_b of the K-electron is greater than that of an L-electron. The K-electron may be displaced to the L-level or to the M-level; the transition $K \rightarrow M$ requires a larger amount of energy (E_1) than the transition $K \rightarrow L$, because the difference $E_{b\,\text{K-electron}} - E_{b\,\text{M-electron}}$ is greater than the difference $E_{b\,\text{K-electron}} - E_{b\,\text{L-electron}}$.

In summary: $$E_1 > E_2 > E_3$$

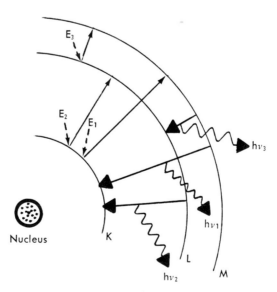

Fig. 3-4. Electronic transitions. Arrows pointed outward represent orbital electrons jumping to outer orbits. Arrows pointing inward represent electrons jumping to inner orbits. E_1, E_2, and E_3 indicate the energies that must be absorbed to displace electrons to outer orbits. Wavy arrows represent photons emitted by atoms in the electronic transitions shown. K, L, and M indicate electronic shells.

When electrons jump back to inner orbits, radiant energy is emitted in the form of photons of definite wavelengths and frequencies. When an electron slips from the M-shell to the L-shell, a photon of relatively long wavelength (and therefore of low frequency and energy) is emitted because the difference between the electron E_b of those two shells is small. Applying the same consideration, one can easily see that the electronic transitions L → K and M → K are accompanied by the emission of photons of shorter wavelengths and therefore of higher frequencies and energies. If we express the photon energy as $h\nu$ (Planck's equation), then

$$h\nu_1 > h\nu_2 > h\nu_3$$

Let us consider some electronic transitions as they occur in the tungsten atom ($Z = 74$). In this atom, there are 74 orbital electrons revolving around the nucleus and distributed in six shells (K through P). Fig. 3-5 represents only the first four shells with their respective electron binding energies in keV. Again, subshells have been ignored; therefore E_L, E_M, and E_N are average E_b values. If these values are given a negative sign, they can represent the different level-energy values of the atom. The reason is that if a tungsten atom is excited, for instance, by the displacement of a K-electron to the L-shell, it must have absorbed an amount of energy equal to the difference between the K-electron E_b and the L-electron E_b (= 69.5 keV − 11.5 keV). If the energy level of the atom with the K-electron in the K-shell is expressed with −69.5 keV, then the energy level of the same atom with the same electron displaced to the L shell can be expressed as: −69.5 + 58 = −11.5 keV. (Remember that −11.5 > −69.5; consequently the level-energy values increase from E_K to E_N). Let us suppose that an electron slips from the L-shell to the K-shell. A photon is emitted, whose wavelength can be predicted by using the formula

$$\lambda \text{ (A)} = \frac{0.0124}{E \text{ (MeV)}} \tag{3-8}$$

which is easily derived from equation 3-7.

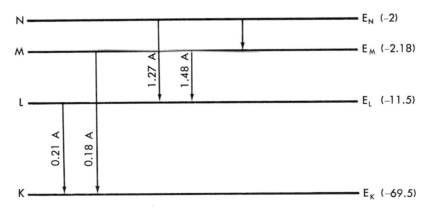

Fig. 3-5. Some electronic transitions in the tungsten atom. Since subshells have been ignored in this diagram, F_K, E_L, E_M, and E_N are average values of sublevel energies. Wavelengths indicated are also average values. Distances between horizontal lines are only roughly proportional to the actual differences between successive level energies.

The *E* of equation 3-8 is the difference between E_L and E_K ($= 58\,keV = 0.058$ MeV). Therefore

$$\lambda = \frac{0.0124}{0.058} = 0.21\ \text{A}$$

One may verify, with the same method, the correctness of the other wavelengths shown in Fig. 3-5.

The wavelengths emitted as a consequence of the electronic transitions shown in Fig. 3-5 correspond to the X-ray band of the electromagnetic spectrum, as one would have expected, considering that the energy differences involved in these transitions are of the order of magnitude of several keV. Therefore, an atom of tungsten will emit X-rays, if a way can be found to remove its innermost electrons from their orbits; when their vacancies are filled by outer electrons, X-rays of definite wavelengths are emitted. In Chapter 4, we will see that the displacement of K- and L-electrons can be accomplished by bombarding tungsten with high-speed electrons.

A filament of tungsten heated to incandescence by an electric current (as in a light bulb) for all practical purposes emits only photons of visible light and infrared radiation because the electronic transitions involved are between the outermost orbits of the tungsten atom, where the energy differences are only a few electron volts or less. This means that the excitation of the tungsten atoms caused by the electric energy supplied to the filament was brought about by displacement of outer electrons only.

The types of excitation discussed so far are referred to as *atomic excitations* because they affect the energy level of the atom itself.

If a nonmonoatomic molecule absorbs energy, possibly the amount of energy will not be sufficient to cause any electronic transition in the atoms of the molecule and therefore no atomic excitation is possible. However, the same amount of energy may be sufficient to raise the energy level of the molecule as a whole. This type of excitation may be called molecular excitation. Molecular excitation may occur in one of the following two ways (Fig. 3-6):

1. Certain infrared frequencies, if absorbed by a molecule, may provide sufficient energy to stretch and bend the individual chemical bonds within the molecule or to increase the magnitude of such stretching and bending. This type of molecular excitation is referred to as *vibrational excitation.* Different types of bonds, or groups of bonds, are

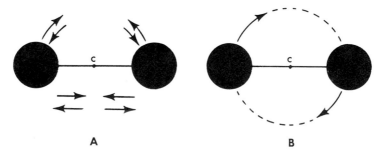

Fig. 3-6. Molecular vibration and rotation. **A,** The two atoms of a diatomic molecule are vibrating back and forth along a straight line passing through the mass center of the molecule, *c*; or the molecule is being bent and stretched around *c*. **B,** The molecule is rotating around the mass center, *c*.

affected by different and well-defined infrared frequencies. The photon energies associated with these frequencies are sometimes called *vibrational energies*.

2. Certain microwave frequencies, if absorbed by a molecule, may provide sufficient energy to cause rotation of the molecule around its center, or to increase the speed of such rotation. This type of molecular excitation is called *rotational excitation*. The microwave frequencies needed to cause rotational excitation are not characteristic of particular bonds in the molecule but are characteristic of the molecule as a whole. The photon energies associated with these frequencies are sometimes called *rotational energies*.

Note in Table 3-1 that vibrational and rotational photon energies are of the order of magnitude of fractions of 1 eV.

PARTICLE PROPERTIES OF PHOTONS

As mentioned earlier in this chapter, the optical effects of interference and diffraction convinced all physicists to abandon definitively the corpuscular theory of light in favor of a wave theory. Indeed, those and other phenomena discovered later like polarization of light, can only be explained by assuming that light behaves like a wave system. Also other types of electromagnetic radiations show the properties of interference, diffraction, and polarization and thus confirm that they, too, like light, behave like waves.

But other phenomena, like photoelectric phenomena and the Compton effect (see Chapter 6), which can be observed when X-rays and gamma rays interact with matter, seem to indicate that in certain conditions photons behave like material particles. In fact, a correct interpretation of those phenomena requires the assumption that a photon has a momentum and therefore a mass (momentum = mass × speed). The very basic principle of the quantum theory, whereby electromagnetic radiation is *quantized*, that is, consists of discrete packages of energy, seems to indicate that perhaps radiation may have a particulate structure.

The best way out of the paradox is to look at radiation as a physical entity with a dual wave-particle nature. Some of the phenomena displayed by radiation (interference, diffraction, polarization) are caused and explained by its wave properties; others (like photoelectric and Compton effects) are caused and explained by its particle properties.

Moreover, as we go through the electromagnetic spectrum from lower to higher photon energies, the particle properties become more evident or at least they can be more readily detected, probably because single photons can be more easily detected with the instrumentation available to us.

REFERENCES

Atkins, K. R. 1965. Physics. John Wiley & Sons, Inc., New York.

Barr, D. L. 1954. A key for use with the chart of electromagnetic radiations. The Welch Scientific Co., Chicago.

Deering, R. A. Dec. 1962. Ultraviolet radiation and nucleic acid. Sci. Amer. **207**:135.

Ewen, H. I. 1955. Radiowaves from interstellar space. In: The new astronomy. Scientific American Book, Simon & Schuster, Inc., New York. P. 235.

Giacconi, R. Dec. 1967. X-ray stars. Sci. Amer. **217**:36.

Jagger, J. 1967. Introduction to research in ultraviolet photobiology. Prentice-Hall, Inc., Englewood Cliffs, N. J.

Koller, L. R. 1965. Ultraviolet radiation. John Wiley & Sons, Inc.

Light. Sept. 1968. Sci. Amer. Vol. 219, a special issue.

Lovell, A. C. B. 1955. Radio stars. In: The new astronomy. Scientific American Book. Simon & Schuster, Inc. P. 229.

Pierce, J. R. 1964. Electrons and waves. Doubleday & Co., Inc.

Semat, H., and White, H. E. 1959. Atomic age physics. Holt, Rinehart & Winston, Inc., New York.

X-rays

HISTORICAL BACKGROUND

Like so many other discoveries recorded in the history of science, the discovery of X-rays was the result of a happy accident and yet it has revolutionized many fields of science and technology. Wilhelm C. Roentgen at the University of Wurtzburg in Germamy accidentally discovered X-rays while experimenting with a cathode-ray tube on the properties of the cathode rays. A cathode-ray tube (or Crookes tube, as it was called at that time) is a glass tube in which a partial vacuum with very small amounts of air or other gas left in it (Fig. 4-1) has been created. Electrodes are mounted in the tube such that a disk-shaped cathode is at the apex of the cone and an anode is on its side, in connection with the negative and positive poles, respectively, of a high-voltage electric supply. When the high voltage is impressed on the electrodes, a beam of invisible rays originates from the cathode (cathode rays); the rays are propagated in a straight line and can make certain substances (such as zinc sulfide) fluoresce. Cathode rays can penetrate a 0.001 mm aluminum film and can be deflected by electric and magnetic fields.

While Roentgen was studying the fluorescence produced by these rays, he noticed that fluorescence could be produced on a screen coated with barium platinocyanide. The screen was placed at a considerable distance from a Crookes tube completely covered with black cardboard. He reasoned that some new kind of rays were being generated within the tube and after passing through the walls of the cardboard box they caused fluorescence in some chemical compounds; he called them "X-rays" (rays of unknown nature). Roentgen and others immediately began experimenting with the new rays and soon discovered their following fundamental properties:

1. They are propagated in a straight line.

2. They are not deflected by electric or magnetic fields and therefore do not transport electric charges.

3. They can penetrate through many materials that are opaque to ordinary light, such as a 1000-page book, a 3.5 mm thick sheet of aluminum, a thin lead foil, and the flesh of the human body.

4. They produce considerable ionization in gases and make them electrically conductive.

5. Screens coated with certain chemical compounds fluoresce in their path.

6. Photographic plates are blackened, even when they are enclosed in light-tight holders.

Throughout the world, practical applications were made of some of the above properties. Most notable was the possibility of observing on a fluorescent screen (fluoroscopy), or of photographing (radiography), the bones of living people and animals and therefore the possibility of detecting fractures and foreign objects in the body, such as bullets and swallowed pins. These medical applications were, perhaps, the first to be exploited. However, most of the early radiologists were completely unaware that X-rays may be harmful when they are handled carelessly. As a result they suffered severe skin burns

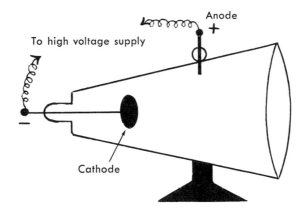

Fig. 4-1. Crookes tube.

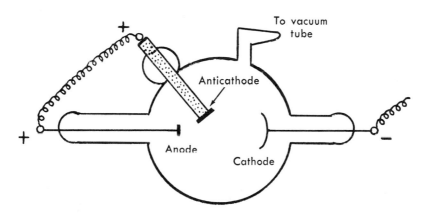

Fig. 4-2. Gas-filled tube.

and cancer, with frequent loss of fingers, hands, and arms. Later a monument was erected at the Roentgen Institute in Hamburg to the memory of those early pioneers who suffered and died from X-ray overexposure incurred in research and medical practice.

At the very beginning of this century, when cathode rays were understood to be nothing but streams of electrons traveling at high speeds, the tube for the production of X-rays was improved. It was observed that X-rays can be produced in a much more efficient manner if electrons are directed against a block of metal. To obtain greater efficiency, a tube as shown in Fig. 4-2 was constructed. A block of aluminum with a concave surface is the cathode. Along with the anode is a block of metal (preferably tungsten) electrically connected with the anode; since its free end is opposite to the cathode, it is called the "anticathode." The tube is partially evacuated through an outlet. When high voltage is applied to the electrodes, the few stray electrons originating in the gas from cosmic ray collisions and other causes are repulsed by the cathode and hurled against the anticathode (posivitely charged). As they travel with accelerated motion, they ionize the atoms of the gas present in the tube, thus producing more electrons which also strike the anticathode at high speeds. X-rays are generated from the free surface of the anticathode.

In this type of tube, electrons are generated in the small amounts of gas that is left in the tube by partial evacuation. For this reason the tube is called a *gas-filled X-ray tube.* Soon, however, this type began to show many undesirable characteristics, such as erratic operation. In 1913, it was generally replaced by the more efficient *Coolidge tube,* invented by W. D. Coolidge. Coolidge tubes of different shapes and sizes are still in use today in our modern X-ray units.

THE MODERN X-RAY UNIT
The Coolidge tube

As mentioned above, since 1913 the "gas-filled" tube has been gradually replaced with the Coolidge tube for most applications. In the gas-filled tube, the intensity of the stream of electrons and the high voltage were difficult to control; this resulted in an unsteady operation. The Coolidge tube produces electrons by thermionic emission, which is the emission of electrons from an incandescent metallic filament. If the incandescence is caused by the passage of an electric current through the filament, then by increasing or reducing the current, it is possible to regulate the intensity of the electron stream and consequently the intensity of the X-rays produced.

Although a variety of Coolidge tubes are built today for different applications, a simple model is shown in Fig. 4-3.

The tube made of Pyrex glass is evacuated to the best vacuum attainable, because even small amounts of air left inside would interfere with stable operation. The cathode is no longer an aluminum rod as in the old gas-filled tube, but rather a tungsten filament that can be heated to incandesence by an electric current, exactly like the filament of an ordinary incandescent lamp. The cathode is the source of electrons produced by thermionic emission. The filament is surrounded by a metal cup that focuses the stream of electrons onto a small area of the target. The anode is a heavy cylindrical block of copper or of some other metal that rapidly disperses heat. The target is located on the free surface of the anode (facing the cathode) and is usually made of a plate of tungsten, or molybdenum, or some other metal of high atomic number. When high voltage is applied to the electrodes, electrons emitted from the cathode are accelerated and directed against the target, from which X-rays are produced. If beaming most of the radiation to one side of the tube is desired, as is usually the case, the target is cut off at an angle. In this way X-rays are directed toward the window, which can be made of a thin layer of beryllium if low-energy radiation

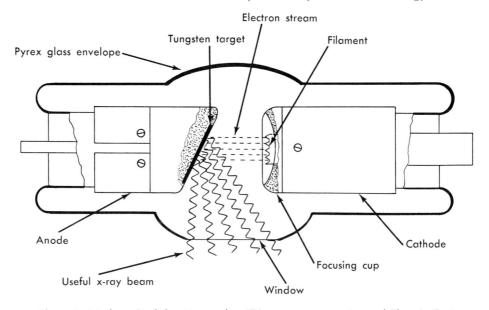

Fig. 4-3. Modern Coolidge X-ray tube. (Diagram courtesy General Electric Co.)

must be allowed to pass, or of a sheet of aluminum if only more energetic radiation is desired.

Associated electric circuits

X-ray tubes must be powered at least by two basic circuits — one to make the cathode filament incandescent, the other to apply a high voltage of thousands of volts between the cathode and the anode.

A typical circuit is illustrated in Fig. 4-4. The current coming from an ordinary AC line is split into two circuits. One of them powers the filament, which requires a voltage of only 10 to 15 volts. Therefore a step-down transformer is required to reduce the 115 or 230 volts of the AC line to the small filament voltage. The voltage applied to the primary coil of the transformer can be controlled so that a variable voltage can be applied to the filament. This indirectly controls the filament current and consequently the intensity of the electron stream emitted.

To obtain the high voltage required across the electrodes, the AC power

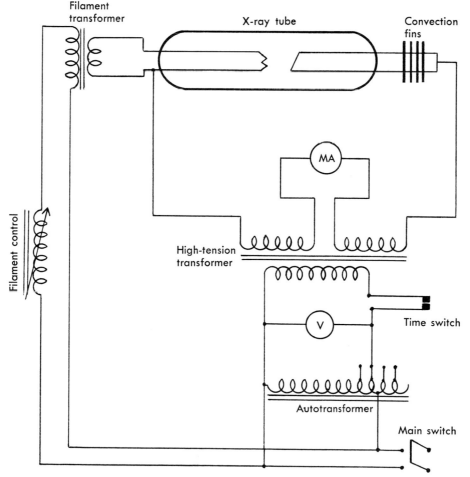

Fig. 4-4. Self-rectified X-ray circuit. **MA,** Milliammeter. **V,** Voltmeter. (From Blatz, H. 1964. Introduction to radiological health. McGraw-Hill Book Co., New York.)

goes through an autotransformer (one coil) and a step-up transformer. With this arrangement, not only 200 to 300 and more kilovolts can be obtained, but it is also possible to vary the voltage applied to the primary coil of the step-up transformer. In this way, a variable voltage can be attained across the tube by connecting one end of the primary to different taps of the autotransformer. Also, with a voltmeter it is possible to set the desired voltage before the time switch is turned on. The voltage is actually applied to the tube when the time switch is closed. A milliammeter measures the electric current across the electrodes when the switch is closed (tube current, to be distinguished from filament current).

Although a DC voltage should be applied to the tube so that the anode is always positive and the cathode negative, in the circuit just described there is no provision (such as rectifying vacuum tubes, etc,) to obtain DC. The circuit is said to be self-rectified because the tube itself rectifies the current; that is, with a 60-cycle AC line, the stream of electrons is allowed to strike the target only during one half of a cycle; during the other half there is no electronic current going across and the tube is idle. For this reason a self-rectified circuit is less efficient, although less expensive, in powering an X-ray tube than a circuit utilizing both halves of the AC cycle.

Control panel

For many commercial X-ray units, controls are housed in a special box or panel that can be located at a considerable distance from the tube, especially when the latter is not sufficiently shielded and when long exposures are used. A remote control panel makes it unnecessary for the operator to be very close to the tube and thus to be exposed to stray radiation, which is always present around an unshielded or partially shielded tube.

The main switch, which sometimes can be turned on only with a key, sends the current from the AC line to the circuits of the unit. It is advisable to turn this switch on some time before an exposure is started to allow for proper warm-up of the various components of the unit. The voltage control sets the voltage to be applied to the tube. This knob is connected to the autotransformer, and the voltage can be read on the voltmeter before the exposure is started. The exposure switch consists of a button for units designed for radiography or fluoroscopy, in which short exposure times are used; the switch remains closed as long as the button is pressed. In units designed for long exposure times (as in X-ray therapy), the exposure switch is connected with a timer that can be set in advance (in minutes or hours). When the preset time has elapsed, the timer shuts the switch off automatically.

The filament control is a knob that regulates the filament current and consequently the tube current. This knob is connected with a milliammeter that shows the tube current while the unit is in operation but is inactive when the exposure switch is off. Therefore, with this meter only, one cannot control the tube current before the exposure is started. For avoidance of this inconvenience, on the control panel of some units there is another meter connected with the filament control; on this meter the filament current can be read before the exposure switch is turned on.

Shielding

In modern X-ray units, the tube is almost completely surrounded by thick steel except for a round window facing the target. This enclosure not only protects the fragile glass tube and other accessories (electric circuits, cooling devices, etc.) but also provides shielding against stray radiation that is always emitted by a tube in all directions. Unshielded tubes are sometimes used for educational purposes to show students their structure and characteristics. If they are to be operated for demonstration, very rigorous precautions should be taken to avoid hazardous exposure to operator and students.

Despite the steel enclosures mentioned above, more shielding is required in some cases when the unit is operated at high voltages and tube currents. It is not sufficient in these cases for the operator to stay out of the beam, because, due to scattering, significant doses can be received also outside of the beam. Some manufacturers sell, with the unit and the control panel, a steel cabinet. The tube is inside of the cabinet where the objects to be irradiated are also placed. The door of the cabinet is equipped with a safety device that interrupts the exposure if the door is inadvertently opened. The exposure dose rate on the external surface of such cabinets is about 1 milliroentgen/hour,* well below the maximum limit recommended. For some high-voltage therapeutic equipment, it is sometimes advisable to keep the unit and the control panel in two separate rooms, with a layer of steel or lead between them. The operator can observe the patient in the treatment room through a window protected by a thick lead glass. More detailed recommendations about the shielding of X-ray machines can be found in the National Bureau of Standards handbooks no. 76 and no. 93.

Efficiency

Not all the electric energy delivered to the anode is converted to X-ray energy. Surprisingly enough, for tubes operated at medium voltages (100 to 250 kV) the X-ray energy obtained is only a small fraction of the total electric energy delivered to the tube. The efficiency of X-ray production is defined as follows:

$$Efficiency = \frac{Total\ radiated\ X\text{-}ray\ energy}{Total\ energy\ delivered\ to\ anode}$$

A useful equation to calculate efficiency is the following†:

$$Efficiency = 1.4 \times 10^{-9}\ VZ \tag{4-1}$$

V is the electric potential in volts applied to the tube, and Z is the atomic number of the target material. Therefore the efficiency is directly proportional to voltage and atomic number of the target material. For example, for a tube with a tungsten target operated at 100 kV,

$$Efficiency = 1.4 \times 10^{-9} \times 74 \times 10^{5} = 1.04 \times 10^{-2} \text{ (a little more than 1\%)}$$

The amount of electric energy that is not converted into X-rays is dissipated as heat in the target. In the example given above, this amount of heat is enormous and unless something is done to remove it, the temperature of the target could rise to the melting point, which process would result in permanent damage to the tube. This would be true also for a tungsten target, although this metal is preferred to others mainly because of its high melting point.

---

*For dose units, see Chapter 9.

† This equation is correct for voltages not exceeding a few million volts.

Several methods are used to cool the anode, depending on the characteristics of the tube and on its operating conditions. If a tube is used only for a few seconds or minutes at a time and is operated with a current of only a few milliamperes (ma), then the anode can take care of dispersing the heat by simple radiation. In other tubes designed to be operated for longer periods of time and at several ma, the cooling is achieved by external convection fins, very similar to those used for the cylinders of air-cooled internal combustion engines (as in motorcycles). In other types of tubes, the target, or the cathode and the target, are cooled by water or oil circulated through them; or the target is cooled by water that boils and condenses in an external metal bulb. An ingenious cooling system is adopted in tubes wherein the source of X-rays is a very small (much less than 1 mm^2) area of the target. In these conditions a large amount of heat would be generated in this small area, and dispersing this heat rapidly and efficiently enough would be extremely difficult. The problem has been solved with a "rotating anode." The target has the shape of a cone or truncated cone with its axis parallel to the longitudinal axis of the tube and is rotated by an electric induction motor, which sometimes is placed inside of the tube. The cathode rays are focused on a small area close to the edge of the cone, but because of the rotation, this small metal surface is rapidly moved out of the cathode rays and replaced by other cooler metal before it has time to melt.

These considerations on heat dispersion should be kept in mind when operating an X-ray unit, so that permanent damage to the tube is avoided. In practice, a good rule is to follow the manufacturer's instructions regarding maximum voltage, maximum current, and any other precaution intended to prevent tube breakdown because of overheating.

ANALYSIS OF THE X-RAY SPECTRUM
Continuous X-rays

What is the wavelength of the X-rays produced by a tube operated at a certain voltage and current?

There is no tube or other type of X-ray–generating equipment that will produce monochromatic X-rays. The X-ray output of a tube is polychromatic; that is, it is made of photons of different wavelengths. An instrument called the *X-ray spectrometer* is capable of analyzing the X-ray beam, just as an optical prism analyzes a beam of white light. The intensity at each wavelength is measured with an ionization chamber; if, on graph paper, wavelengths are plotted on the x-axis against intensities on the y-axis, an *X-ray spectrum* is obtained.

Fig. 4-5 shows spectra of X-rays obtained from a Coolidge tube with a tungsten target by C. T. Ulrey (1918), one of the pioneers in X-ray spectrometric studies. The three curves represent the spectra obtained at three different voltages as indicated.

An inspection of the curves shows that as the voltage is increased (1) the intensity* of the radiation increases at all wavelengths (in fact the curves become higher, and the area they delimit becomes larger), (2) the wavelength with maximum intensity decreases (shifts to the left), and (3) the minimum wavelength present in the beam becomes shorter. Thus, the inspection of these curves tells us immediately how variation of voltage affects the spectrum of the X-rays generated.

These spectra are called *continuous X-ray spectra* because they show a continuum of wavelengths from a maximum to a minimum, without any gaps or sharp peaks.

---

*Intensity is the number of photons emitted by the target per unit of time.

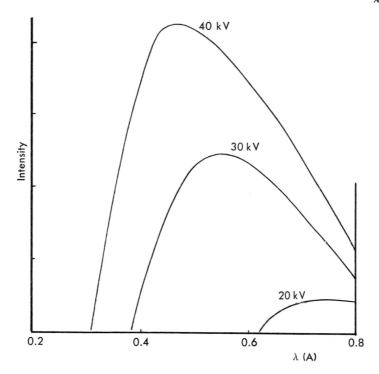

Fig. 4-5. X-ray spectra of tungsten at three different tube voltages. **kV,** Kilovolt.

With regard to the first observation made above, Ulrey found that the area under the curves is proportional to the square of the tube voltage. This relationship means that the total intensity of the continuous X-radiation, considering all wavelengths, increases with the square of the voltage, if other factors remain constant, so that

$$I_{cont.} = kV^2 \qquad (4\text{-}2)$$

where k is a constant.

However, studies made by Ulrey and others also indicated that if the voltage is kept constant, total intensity is proportional to the atomic number of the target, so that equation 4-2 can be generalized as:

$$I_{cont.} = kZV^2 \qquad (4\text{-}3)$$

If spectra of an X-ray beam are taken at constant voltage but at different tube currents, it can be shown that the total intensity is directly proportional to the current, but that there is no shift of the wavelength with maximum intensity or of the minimum wavelength, as there is when the voltage is increased.

A theory developed to explain the origin of the continuous X-rays is summarized here in its simplest form. When the electrons coming from the cathode at high speed strike the target, they collide with its atoms. Some of the electrons may collide head on with atomic nuclei (a highly improbable event), and such a collision means that they will be stopped at once. Others will be decelerated and their path deflected by the strong electrostatic attraction of the positively charged nuclei. The magnitude of the deceleration and deflection depends on the distance at which the electron happens to be from the nucleus. Some electrons will happen to pass at a distance from the nucleus greater than others; they will be decelerated and their path deflected less. The theory predicts that whenever an electron is decelerated, the kinetic energy that is lost appears in the form of an electromagnetic wave. The energy (and thus the wavelength) of this photon depends on the magnitude of the deceleration. This energy is roughly equal to the amount of kinetic energy lost by the electron. Thus the continuous X-rays consist of the electromagnetic waves so generated; radiation produced in this way is known as *bremsstrahlung* (German for "braking radiation"). In Chapter 6 one can find that beta particles are also responsible for the origin of bremsstrahlung when they are absorbed by material of high atomic number.

One can clearly see that X-rays produced in this way will show a continuum of wavelengths. In fact, not all the electrons will lose the same amount of kinetic energy; before coming to rest, some will even be subject to more than one deceleration by passing through the electrostatic fields of many nuclei. The minimum wavelength found in a spectrum will correspond to the maximum amount of kinetic energy lost by an electron, which in turn will vary directly with the kinetic energy possessed by the electron before collision. This kinetic energy is proportional to the tube voltage, and this fact explains why, with a voltage increase, the minimum X-ray wavelength is shifted to the left. The theory also explains why an increase of voltage results in greater radiation intensity. With a higher voltage the electrons strike the target with greater kinetic energy, which enables them to penetrate deeper in the target material. As a consequence, before coming to rest, they will encounter more target atoms and will be subjected to more decelerations and deflections along their path. The occurrence of more deceleration events means production of more X-ray photons and therefore higher intensity.

Likewise, an increase of tube current means a greater "electron flux"; that is, more electrons will strike the target per unit area and per unit time, and again this will result in higher X-ray intensity.

Characteristic X-rays

Fig. 4-6 shows two X-ray spectra plotted with the same coordinates. One is the spectrum obtained with a tungsten target (W) and the other with a molybdenum target (Mo). The tube was operated at the same voltage (35 kV) in both cases. In accordance with equation 4-3, the total intensity of the continuous W spectrum is greater than for the Mo spectrum because the atomic number of W (74) is higher than that of Mo (42). But since the minimum wavelength of a spectrum depends exclusively on the tube voltage used (see equation 4-5), both spectra begin at the same minimum wavelength although the target materials used are different. However, two narrow peaks emerge from the continuous Mo spectrum corresponding approximately to 0.71 and 0.63 A. This means that besides the continuous X-rays there is an intense emission of other X-rays with those wavelengths. If the Mo spectrum is taken with a higher tube voltage, the continuous spectrum will shift to the left, as stated in the preceding section, but the two peaks will not shift. This observation shows that the wavelengths of the peaks are independent of voltage. To see similar peaks for tungsten, one should take the spectrum at higher voltage (for example, 100 kV); peaks will appear with wavelengths between 0.15 and 0.22 A and again they will not be shifted by a voltage increase.

These observations lead to the conclusion that the peaks (or lines, as they are more commonly called) are characteristic of the target material, very much

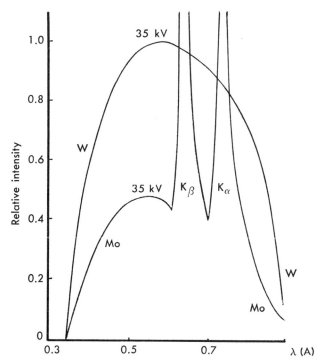

Fig. 4-6. X-ray spectra of molybdenum (Mo) and tungsten (W) with the tube operated at 35 kV.

like the bright lines of the optical emission spectra of hot sodium vapor, mercury vapor, etc. For this reason the system of all the lines appearing in the X-ray spectrum emitted by the target is called the *characteristic spectrum* of the chemical element of which the target is made.

In addition to the strong lines shown in Fig. 4-6, a characteristic spectrum reveals other groups of lines at considerably longer wavelengths (for Mo they appear in the region above 5 A). These groups of lines are called *series* and named with the letters K, L, M, N, etc., in order of increasing wavelengths. So it is customary to speak of K-lines, L-lines, etc. Lines within each series are further identified by using Greek letters as subscripts of the series letters: α is used for the line of the longest wavelength, β for the next to the left, etc. For instance, the lines of the Mo spectrum in Fig. 4-6 would be called K_α (the one on the right) and K_β (the one on the left). Within each series, the intensity of the lines decreases as the wavelengths decrease, so that the α line is the most intense in each series.

With spectrometers of very high resolution, one can see that actually some of the lines are pairs of lines; in this case the two members of an α line would be designated as α_1 and α_2. In our discussion, however, this fact will be ignored to avoid further complications.

The relationships of the characteristic spectra with the atomic number of the target material were firmly established through the work of Moseley during the second decade of this century. The equation correlating the frequency of a line with the atomic number is

$$\sqrt{\nu} = K(Z-\sigma) \tag{4-4}$$

where ν is the frequency of the line, Z the atomic number of the target material, and K and σ constants that depend on the line series. This equation means, in effect, that the frequency of the K-series, L-series, etc., increases systematically (and therefore the wavelength decreases) with increasing atomic number (Moseley's law). As an example, the K-lines emitted by tungsten are found at shorter wavelengths than the same lines emitted by molybdenum. A table of the wavelengths of all the emission lines of the elements with atomic number above 10 can be found in the *Handbook of Chemistry and Physics,* and one should consult it and verify the principles and facts discussed above.

The minimum voltage that must be applied to an X-ray tube to excite the emission of the lines of a series is called the *excitation potential* for that series. From the preceding discussion, the fact should be clear that the excitation potential for each series increases with the Z of the target material. This increase can be seen in Table 4-1 wherein examples of excitation potential are reported in kV.

Once a line has been excited, a further increase in voltage increases its intensity, just as it increases the intensity at all wavelengths of the continuous spectrum. This increase can be verified in Fig. 4-7, wherein X-ray spectra from a tungsten target are shown at different voltages. With the information given in this and the preceding section, note how voltage changes affect both the continuous spectrum and the characteristic spectral lines.

The study of the characteristic spectral lines and especially Moseley's work was instrumental for the formulation of Bohr's theory of atomic structure and in particular of the electronic energy levels. Conversely, the origin of the characteristic X-rays can be explained with this theory.

Unlike continuous X-rays, characteristic X-rays are generated by electronic transitions in the target atoms after ionization. For review see the discussion in the two previous chapters about electron binding energies and electronic transitions and note also Table 2-2 showing the electron binding energies of the molybdenum and tungsten atoms.

Suppose that an X-ray tube is operated with a voltage of 20 kV. Thus the cathodic electrons strike the target with a kinetic energy of 20 keV. If the target is made of molybdenum, some of the impinging electrons may collide with K-electrons of the Mo atoms with an energy sufficient to overcome the binding energy of those electrons. The K-electron will be dislodged from its atom, which is now left as a positive ion. Since the Mo atom has lost one of its innermost electrons, the vacancy left will be filled by an L-electron (less probably

Table 4-1. Excitation potentials, in kV, in relation to the atomic number, Z

Z	Element	K-series	L-series	M-series	N-series
11	Na	1.07	—	—	—
30	Zn	9.65	1.20	—	—
42	Mo	20.0	2.87	0.51	0.06
74	W	69.3	12.1	2.81	0.59
76	Os	73.8	13.0	3.05	0.64
82	Pb	87.6	15.8	3.85	0.89
92	U	115.0	21.7	5.54	1.44

by an M- or N-electron). An electronic transition L → K is the result, with the consequent emission of electromagnetic radiation having a definite wavelength and therefore definite photon energy. Since the transition is from an outer shell to the K-shell, the photon has enough energy to be qualified as an X-ray. Remember that the energy of this X-ray photon is determined exclusively by the type of transition (in this case L → K) and by the atomic number of the target, since the energy levels of the same shells are different for different Z's. It is independent of the voltage applied to the tube; if the tube with a Mo target is operated at 40 instead of 20 kV, the electrons striking the target will have more than enough energy to dislodge a K-electron, but this greater energy will not affect the energy of the X-ray photon emitted as a result of the L → K transition.

Obviously, the photon energy is also independent of tube current; a larger current will only result in more Mo atoms being ionized and consequently in the production of more photons originating by electronic transitions.

The X-rays produced in the manner just described are what we call "characteristic X-rays." X-rays generated by the transition of an outer electron into the K-shell appear on the spectrum as a K-line. In the example presented above, when an L-electron fills the K-vacancy, the photon appears in the K_α-line. The K_β-line is made of characteristic X-rays generated by the transition of M-electrons into the K-shell. Since the difference between energy levels in this case is greater than for the L → K transition, the K_β-line corresponds to a shorter wavelength (and thus to a higher energy) than the K_α-line. However, because

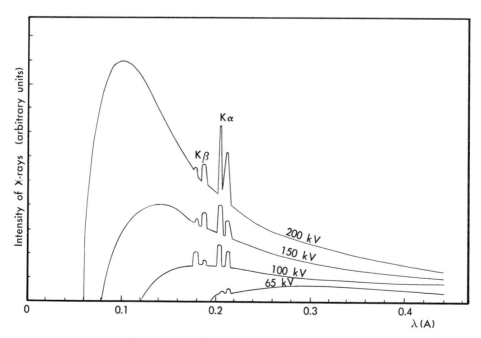

Fig. 4-7. X-ray spectra of tungsten at four different voltages showing continuous and characteristic radiation. Two K_α-lines (α_1 and α_2) and two K_β-lines (β_1 and β_2) are shown in the spectra. (From Blatz, H. 1964. Introduction to radiological health. McGraw-Hill Book Co., New York.)

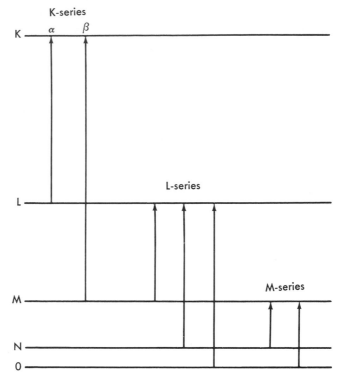

Fig. 4-8. Electronic transitions responsible for the emission of the characteristic X-ray spectral lines.

the M → K transition is less probable than the L → K transition, the intensity of the K_β-line is also less than that of the K_α-line.

The meaning of the excitation potential should be clear now. Twenty kilovolts is the excitation potential for the K-series in Mo because this voltage is the minimum potential necessary to impart enough energy to the cathodic electrons to dislodge a K-electron from the Mo atom. With a lower voltage the K-lines in the spectrum do not appear at all. However, with a lower voltage the cathodic electrons might have enough energy to dislodge an L-electron; the result will be the production of a lower-energy photon, with the appearance of the L-lines only.

With a tungsten target, 69.51 keV are required to dislodge a K-electron from its shell. Consequently, no lines of the K-series will appear in the spectrum, when the tube is operated at 20 kV or even at 35 kV (see Fig. 4-6).

Fig. 4-8 is a simplified diagram showing the correlation between electronic transitions and characteristic spectral lines. Note that the energy difference between K- and L-levels is large compared with that between L- and M-levels, this in turn being larger than that between the M- and N-levels.

Quality

In radiography, radiation therapy, irradiation experiments, and other applications, it is important to have some idea about the "quality" of the X-rays used. The quality of an X-ray beam is determined mainly by the following elements: maximum photon energy and consequently minimum wavelength present in the spectrum, relative intensities at the different wavelengths, and characteristic lines (if any). From what has been said in the preceding two sections, the quality is evidently greatly affected by the tube voltage and by the nature of the metal of which the target is made.

On the basis of their penetrating power, X-rays are often classified for practical purposes as "hard" (with high photon energy and therefore high penetrating power) and "soft" (with low photon energy and therefore limited penetrating power). These are very relative terms and useful in practice. The hardness or softness of a beam is directly related to its quality. For some applications, only soft X-rays are desired, as in the case of X-rays used for the radiography of an oil painting or for the radiation therapy of a skin disease. Hard X-rays would penetrate completely through the canvas or through the skin without any significant absorption. In the first example the photographic film would be uniformly blackened, and in the second no energy would be dissipated in the skin, and thus no therapeutic effect would be obtained. By operating the tube at a voltage between 20 and 50 kV, no hard components will appear in the beam. However, one must remember that the X-ray beam has to go through the glass of the tube before striking the object. Glass does absorb the softest components of an X-ray beam, although it does not show any significant effect on the shorter wavelengths. If very soft X-rays are desired, then one should use tubes with a window made of a thin film of material much less dense than glass. A thin beryllium window is often used for this purpose.

In other instances, only hard X-rays are wanted. If a deeply seated tumor must be treated with radiation therapy, only X-rays of sufficient penetrating power to reach the tumor are desired and needed; soft rays would be absorbed by the more superficial tissues (such as muscles and skin) and cause harmful effects there (such as skin burns). Or, in another example, rats must be irradiated for the purpose of investigating the effect of X-rays on the hemopoietic function of the bone marrow, and no harm should be done to their skin. In these cases, it is not sufficient to increase the voltage applied to the tube because, although a higher voltage does increase the maximum photon energy of the beam and the intensities of the shorter wavelengths (see p. 54 on continuous X-rays), it does not suppress the emission of the least-energetic components.

The problem of the elimination of the soft components is solved "outside of the tube." The hardening of the beam is achieved by *filtration*, that is, by placing in front of the tube window a filter made of a suitable material (Al, Cu, or Pb) of appropriate thickness. Such filter, if carefully designed, will absorb the longer wavelengths of the beam, but not appreciably the harder components. The spectrum of the beam after emerging from the filter is quite different from the spectrum of the beam as it was produced by the target. For this reason we may say that filtration is another factor (besides voltage and nature of the target material) affecting the quality of an X-ray beam.

Filtration of X-rays is a rather complex problem, but also an art that requires great skill. The treatment of the theoretical rationale behind it would be quite long and beyond the scope of this book. The interested reader may consult the more specialized works cited at the end of this chapter. Only the essential principles necessary to understand how filters work in hardening an X-ray beam are discussed briefly.

Filtration is the consequence of selective absorption. The general law governing the absorption of electromagnetic radiation as it passes through matter was formulated for the first time in the eighteenth century by Bouguer and later rediscovered by Lambert. Hence it is known as the Bouguer-Lambert law. It applies to X-rays as well as to gamma rays, ultraviolet, visible light, and infrared. If a beam of radiation of intensity I_0 passes through a certain thickness of absorbing material, the emerging radiation has an intensity I that is a fraction $1/k$ of I_0. The value $1/k$ depends on the nature of the material. If another layer of the same material with the same thickness as the first is added, the emerging radiation shows an intensity equal to $1/k$ of I. If a third layer is added, the emerging radiation is reduced in intensity to $1/k$ of the radiation emerging from the second layer, etc. All this means that electromagnetic radiation is absorbed by matter at an exponential rate. An exponential rate denotes a constant fractional decrease of intensity per unit of absorber thickness. The law therefore can be expressed with an exponential equation, which in the case of X-rays has the following form:

$$I = I_0 e^{-\mu t} \tag{4-5}$$

such that I is the intensity of the radiation emerging from the absorber; I_0 is the intensity of the incident radiation; e is the base of the natural logarithms; μ is a constant called the *linear absorption coefficient;* and t is the thickness of the absorber.

The linear absorption coefficient is constant for a given material and photon energy and is independent of the chemical state of the absorber. Taking the natural logs of both members of equation 4-5, we derive

$$\ln I = \ln I_0 - \mu t$$
$$\ln I - \ln I_0 = -\mu t$$

or

$$\ln \frac{I}{I_0} = -\mu t$$

$$\ln \frac{I_0}{I} = \mu t \tag{4-6}$$

This logarithmic form of equation 4-5 is more convenient for many computations.

The absorption coefficient most commonly measured and used in X-ray work is μ/ρ, whereby ρ is the density of the absorbing material. This quantity is called the *mass absorption coefficient,* and its values for many elements and photon energies are reported in the *Handbook of Chemistry and Physics.* This constant has the advantage of taking into account the absorption per gram, rather than the absorption per cubic centimeter, of absorber. It is therefore independent of the physical state of the material.

If the mass absorption coefficient (mac) is measured for the same absorber at different wavelengths with an X-ray spectrometer and an ionization chamber and its values are plotted against wavelengths, an *X-ray absorption spectrum* of that absorber is obtained. Fig. 4-9 shows the absorption spectra of copper and lead. One can see that, in general, the mac increases with the wavelength, so that the soft rays are absorbed more strongly than the hard rays. However, the variation of the coefficient is not represented by a continuous function. In fact, the spectra show discontinuities, called *absorption edges.* Copper shows an edge corresponding to a wavelength of about 1.3 A, and lead shows a group of three edges between 0.7 and 1 A, plus a small edge at about 0.13 A. For the origin of the

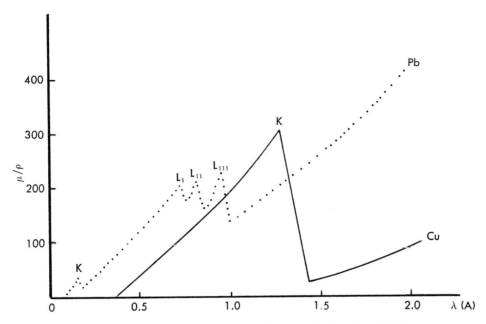

Fig. 4-9. X-ray absorption spectra of copper (Cu) and lead (Pb).

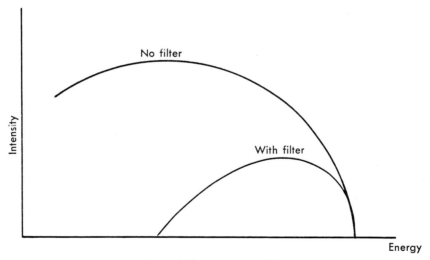

Fig. 4-10. Effect of filtration on an X-ray spectrum.

absorption edges consult the references cited at the end of this chapter (such as Selman and Sproull).

If we disregard the absorption edges, then the mac increases with wavelength. Since the X-ray beam emitted by a tube is polychromatic, a filter will reduce the intensity of low-energy photons more than that of high-energy photons. The actual absorption for each energy will depend, as seen above, on the material used and on its thickness. Fig. 4-10 shows, in rough approximation, the effect of a filter on the energy distribution in an X-ray beam. In practice, when the soft components of a beam are not wanted, an aluminum filter (up to 1 mm thick) is used for equipment operating at 100 kV or less, a copper filter is used for voltages around 250 kV, and a lead filter for very high voltages. A copper filter for low voltages is not advisable because it would absorb significantly also the hard components of the beam.

When a copper filter is used, an aluminum filter is added because of the following reason. The copper filter is efficient in removing the soft components of the beam with a much smaller loss for the high-energy portion. However, the absorption of those wavelengths corresponding to the absorption edges, accompanied by the ejection of peripheral electrons, will cause in the filter material transitions of electrons from outer shells to fill the vacancies left by the ejected electrons. These transitions result in the emission by the filter of X-rays of longer wavelengths and therefore softer than the ones absorbed. This secondary radiation emitted by the filter is a true fluorescent radiation, which makes the filter somehow inefficient in hardening the primary beam. A second filter of aluminum will strongly absorb the copper fluorescence with little attenuation of the remaining primary beam. The aluminum fluorescent radiation is so soft that it is absorbed by a few centimeters of air. If the material to be irradiated is very close to the filter, a Bakelite filter (mostly made of carbon) may be added to the Al filter.

When X-rays are used for irradiation purposes, one often needs to specify in some manner the quality of the beam used. In radiation therapy the radiologist who X-irradiates a brain tumor must inform the neurologist about the quality of the beam and therefore about its penetrating power. The publication of the results of a biologic X-irradiation experiment also requires a description of the quality of the X-rays used.

Obviously, the best way to convey this information would be to take a spectrum of the beam; in a spectrum one finds all the elements (minimum wavelength, characteristic lines, relative intensities, etc.) that determine the quality of a beam. However, the specification of the quality with this method is often cumbersome, time consuming, of difficult interpretation for nonspecialists, and practically impossible. There are simpler, although less accurate, ways to specify X-ray quality. One method commonly used in radiation biology and radiation therapy is the determination of the *half value layer* (HVL).

Let us consider a monochromatic X-ray beam. If a copper filter 5 mm thick reduces its intensity to one half the original value, the beam is said to have a HVL of 5 mm of Cu. The HVL, therefore, is defined as that thickness of a given filter material that reduces the intensity of an X-ray beam to 50% of its initial value.

The HVL of a monochromatic beam can be calculated with the formula

$$HVL = \frac{0.693}{\mu} \tag{4-7}$$

where μ is the linear absorption coefficient of the absorbing material chosen as reference (such as copper) for the wavelength of the beam.

In fact, from equation 4-6, by substituting t with HVL, we derive

$$\ln \frac{I_0}{\frac{1}{2}I_0} = \mu(HVL)$$

and converting to decimal logs

$$2.303 \log 2 = \mu(HVL)$$

and therefore

$$HVL = \frac{0.693}{\mu}$$

The HVL of a polychromatic beam cannot be easily calculated with equation 4-7, because of the many μ values involved, but it can be measured directly. First, the kilovoltage and tube current are selected and kept constant so that the quality of the beam is kept constant throughout the procedure. Then readings are taken of the intensities, with an ionization chamber at a fixed distance from the target, as increasing thicknesses of the same filter material are placed in the beam. An absorption curve is drawn on semilog paper by plotting the intensities on the vertical axis and the filter thicknesses on the horizontal axis. The curve is not a straight line as it would be for a monochromatic beam. The HVL is obtained by drawing a horizontal line from the ordinate that corresponds to an intensity equaling one half the initial value until it intersects the curve. The corresponding thickness (HVL) is read off on the horizontal axis.

Since the HVL is a direct indication of the penetrating power, it is obviously a function of all those variables (target material, voltage, filtration, etc.) affecting the penetrating power of a beam and therefore its quality. For instance, the HVL of Cu for a beam without a filter increases with voltage. With voltage kept constant, the HVL of Cu for the same beam would increase with increasing filtration.

Flux

In the preceding sections, the term "intensity" has been used several times as meaning the number of X-ray photons (of any wavelength) emitted per unit of time by the target, or emerging from the filter, if one is used. One could speak of intensity for any point of an X-ray beam, with the understanding that this intensity varies from point to point within the beam, even though the intensity at the target may remain constant. However, to avoid confusion that inevitably would result from the use of the same term with different meanings, the use of the term *flux* is suggested here in reference to the intensity of an X-ray beam at any point. In this context, the flux of X-radiation may be defined as the quantity of radiation, or number of photons (of any wavelength), flowing through a unit of cross section in the field of a beam, per unit of time.

Knowledge of flux is important whenever X-rays are delivered to an object. From what has been said above, it is evident that the flux at any point in an X-ray beam must be dependent on the tube current, on the voltage applied to the tube, and on the amount and type of filtration used. But unlike the quality and the radiation intensity at the target, the flux varies also with the distance of the object from the target, so that it is not the same at any point within the beam. To understand this type of variation, it is useful to consider briefly the geometry of the beam.

The beam emitted by most X-ray tubes (at least by those with a round window) has the shape of a cone, or better, of a truncated cone (Fig. 4-11), the upper base of which is on the surface of the target and is called the *focal spot*. The focal spot, as explained above, is that portion of the surface area of the target struck by the cathodic electrons and therefore emits the X-ray beam. The area of the focal spot may be very small (less than 0.5 mm^2), such as in the units designed for radiography, or rather large (more than 1 cm^2), such as in radiation therapy units. The lower base of the cone is the plane, perpendicular to the axis of the cone, on which the irradiated object is located. It is called the *irradiation field* and its size varies with its distance from the focal spot.

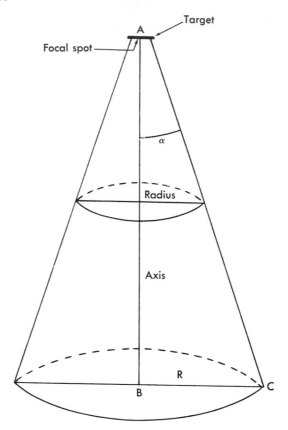

Fig. 4-11. Geometry of an X-ray beam.

The knowledge of the radius of the irradiation field is of importance when large objects or large groups of objects (such as laboratory animals) must be irradiated in a field. Without this knowledge, one may run the risk of leaving part of the object(s) outside of the field. When the angle α is known (usually specified by the manufacturer of the tube), the radius at any distance from the focal spot can be calculated trigonometrically with the formula $R = AB \tan \alpha$. If angle α is not known, the radius can be measured directly by irradiating a fluoroscopic screen at the desired distance from the focal spot (the field will appear as a glowing circle on the screen); or, so that unnecessary exposure to the operator's hands is avoided, the field can be photographed on film, where it will appear as a dark circle after development. Once the radius has been directly measured at a given distance from the focal spot, then it can be calculated for any other distance, after angle α is trigonometrically determined ($\tan \alpha = R/AB$).

The distance of the object from the focal spot is usually called *target-to-object distance* (TOD). The TOD is equal to the distance of the object from the window plus the distance of the window from the focal spot. This latter distance is usually specified by the manufacturer.

The change in radiation flux due to change of distance from the focal spot follows the *inverse square law*. This law, which applies to any type of radiation, simply states that the intensity at any given point in a beam is inversely proportional to the square of the distance between point and source. The reason is that as the photons travel away from their origin, they diverge. If one field p of a beam has a distance from the focal spot twice as long as that of another field q, the area of p is 4 times the area of q. But the same number of photons will pass through both fields per unit of time. This means that one unit of cross section in field p will receive one fourth the number of photons received by one unit of cross section in field q, or the flux has been reduced to one fourth as the distance is increased by 2. The inverse square law can be expressed, therefore, as follows:

$$\frac{\phi_p}{\phi_q} = \frac{D_q^2}{D_p^2} \qquad (4\text{-}8)$$

where ϕ is flux and D is distance.

In the preceding discussion the fact has been implicitly assumed that the two unit cross sections of the two different fields are both on the axis of the beam. But the law is also valid for two unit cross sections of the same field which are at different distances from the axis. In fact, for obvious geometric reasons they are also at different distances from the focal spot. One may easily see that in any field the flux decreases as the distance from the axis increases so that the maximum flux is found on the axis itself.

The inverse square law applies rigorously only when the radiation source is a point source. This is never the case for X-rays. However, significant deviations from the law will be noticed only when the source (focal spot) is fairly large. Other types of deviations from the law are due to physical phenomena such as absorption and scattering, rather than geometric factors.

If point B is farther from the focal spot than point A, radiation must travel through more air to reach B. The added distance results in more absorption of the long wavelengths and means that the flux at B is less than at A, not only because of the greater distance but also because of more absorption.

Scattering is the deflection of a photon from its path. It is caused by Compton recoil (see Chapter 6) in certain conditions when X-rays interact with matter. Scattering, unlike absorption, is responsible for fluxes greater than one would expect on the basis of the inverse square law alone. The flux at a point in the beam is greater than expected if there is an object (especially a metal of high Z) in its immediate vicinity. This point receives not only the photons coming directly from the source, but also those scattered by the object. The flux measured by an ionization chamber located at a certain point varies, depending on whether it is suspended in air or placed on a metallic shelf. Likewise, it is due to scattering if some radiation can be detected outside of the beam. This stray radiation should be taken into account and monitored for safety reasons.

With so many factors affecting the flux, one would not be wise in relying on values obtained through calculations; instead the flux at points of interest should be measured directly.

TYPES OF X-RAY UNITS

A variety of X-ray equipment is built by the manufacturers nowadays. The particular features and specifications are designed in accordance with the purpose for which the equipment is used. In this section the main types of units are classified and briefly surveyed.

For medical applications

Fluoroscopes

In fluoroscopy, the X-ray image of internal organs of the body is observed directly on a fluorescent screen. The patient stands or lies between the X-ray tube and the screen. This technique is used often for diagnostic purposes, whenever radiography cannot provide the desired information, as is the case in many cardiac malfunctions.

The voltage of these units must be such as to produce enough hard radiation capable to pass through the body and give good image contrast. The exposure time can be as long as several minutes so that the heat dispersion problem must be solved accordingly. A fluoroscope is perhaps the type of unit that presents the greatest radiation hazard, because the examining physician is required to be very close to the beam. For minimizing unnecessary exposure various precautions are recommended. Lead shielding should be installed around the beam for protection against scattered radiation. A sheet of lead glass is placed in contact with the fluorescent screen on the examiner's side to stop as much as possible the radiation emerging from the screen. The dose rate can be reduced by means of an electronic device known as the image intensifier. The X-ray shadow of the patient, instead of illuminating a conventional screen, produces an image on a sensing element that can be fed into an electronic amplifying circuit so that the brightness of the resulting image on the screen can be amplified to any degree. With this device, lower tube currents, and consequently lower dose rates, may be used. Usually the brighter image thus obtained makes it unnecessary to work in darkrooms with dark-adapted eyes. Nevertheless, it is recommended that fluoroscopy be avoided whenever the desired information can be obtained by radiography or other techniques.

Radiographic units

In radiography the X-ray image of internal organs of the body is projected on a special X-ray film enclosed in a cardboard cassette to protect it from visible light. The film is then developed and fixed and a radiogram is thus obtained.

Radiographic X-ray units are supplied with an accurate exposure time switch, which allows the operator to preset the exposure time at the desired value for any particular type of radiography. The range of these timers goes from a small fraction of a second to a few seconds. Since the unit is not designed to be on for more than a few seconds at a time, no special heat-dispersing device is necessary. The tube current should be set in advance; hence a filament current meter is usually necessary. These units must have a very small focal spot (1 mm or less). Since the radiation source is practically like a point source, very sharp images on the radiograms can be obtained. The voltage used with these units depends on the region of the body to be radiographed (100 to 400 kV).

Therapeutic units

X-ray therapy, which is but one of many forms of radiation therapy (see Chapter 10), is the application of X-rays in predetermined doses to internal organs or to the skin for the treatment of certain diseases. It is used most frequently for the treatment of cancer and of some nonmalignant skin diseases.

In general, tubes for X-ray therapy equipment do not have a small focal spot, as those for radiographic equipment do. Usually there is little need for a high dose rate; actually extending or fractionating the treatment over a long period of time is often desirable. Treatment times vary from less than a minute in superficial therapy to 20 to 30 minutes for some types of deep-therapy techniques. For these reasons the tubes must be equipped with a good and efficient heat-dispersing system. Since the dose to be administered must be determined quite accurately, special timers are necessary. Because of the frequent use of filters and of their effect on dose rate, not only special devices must be employed to mount and change them, but also monitoring systems are desirable to check whether the correct filter is in position. When very high voltages are used, special shielding is needed for the treatment room and the unit is operated by remote control.

On the basis of the voltage used, and therefore of the penetrating power of the beam, X-ray therapy equipment may be designed for one of the following three different purposes:

1. *Superficial therapy*, with an operating voltage of up to 120 kV. Grenz-ray therapy is a special type of superficial therapy at voltages between 8 and 15 kV and is used to treat superficial disorders of the skin. The X-ray tube is provided with a thin beryllium window to allow the passage of the softest components of the beam.

2. *Deep therapy*, with voltages from 120 to 400 kV, for the treatment of deeper organs of the body. Machines designed for this type of therapy are the most suitable for biologic X-irradiation experiments, for which a voltage of about 250 kV is often required.

3. *Supervoltage therapy*, with operating voltage about 400 kV (sometimes higher than 1000 kV). Special tubes, or a Van de Graaff generator (see p. 70), or a betatron, is needed for this type of therapy, and of course special shielding problems accompany it.

Dental units

These units are designed for dental radiography. They are very much like other types of radiographic equipment, in regard to their physical characteristics. In addition, they must be fairly maneuverable so that the tube can be angled accurately in many positions around the head of the patient. The X-ray beam produced by such tubes is still conical as usual, and thus the radiation is not collimated on the tooth to be radiographed, even though the TOD may be only a few inches. The plastic cone located on the window of the tube serves the purpose of centering the tooth in the beam. Dental units are commonly operated at about 130 kV, in order to limit the exposure to only a small fraction of a second.

For industrial applications

In industry, X-ray radiography or fluoroscopy is used for the detection of internal flaws in metallic or plastic parts, castings or weldings, for the inspection of canned foods, and sometimes for the inspection of internal parts of machines or instruments when direct visual inspection is not practical. Low-voltage units have been used extensively for the radiography of oil paintings and documents. Alteration of the original work has often been revealed by this technique. Paintings may sometimes be proved authentic or fraudulent. In industry, X-rays are also used for the coloring of glass or crystal.

Equipment designed for all these applications is operated with voltage ranging from a few kV (as in the case of radiography of paintings) to over 1000 kV (as in the case of radiography of heavy steel castings).

For X-ray diffraction analysis

Instruments used for this work are essentially X-ray spectrometers. A narrow X-ray beam, properly collimated, impinges on a crystal (usually calcite) at an angle smaller than

90 degrees. The lattice formed by the orderly arrangement of the atoms in the crystal behaves like the artificial gratings used to diffract visible light. Not only is the beam reflected by the crystal, but its individual wavelengths are separated and an X-ray spectrum is thus obtained. By means of suitable filters or a system of slits and by rotating the diffracting crystal, one can isolate one or another wavelength and thus obtain an essentially monochromatic beam. With this beam, the microstructure of crystals and the molecular configuration of important organic compounds can be analyzed. Notable among such applications are the discovery of the steric configuration of deoxyribonucleic acid and of certain proteins (such as hemoglobin).

Tubes for X-ray diffraction equipment employ a special target material whose characteristic spectrum provides a line of the required wavelength. The focal spot is usually in the form of an elongated narrow band. Two, three, or four windows are provided, and they may consist of beryllium, mica, or low-absorption glass, to minimize loss of intensity. Voltages are rather low (30 to 50 kV), but tube currents are high (15 to 20 ma). Target materials most commonly employed are molybdenum, copper, iron, cobalt, and chromium. Since the intensity is high, if an operator inadvertently places a part of his body in the beam path, even for a short time, he would be subject to serious local exposure. For this reason, though accidents involving X-ray diffraction equipment have been rare, those which have occurred have been quite serious.

The fact that the tubes of these instruments are operated for hours without interruption calls for very efficient cooling systems.

Other types of X-ray generators

For the production of high-energy X-rays (above 1 MeV), the Coolidge tube is not adequate, because of the serious problems arising in connection with the high voltages needed and with the larger amounts of heat to be dissipated. Other machines are needed to obtain high-energy X-rays. They are some of the so-called *particle accelerators.*

A particle accelerator may be described as any electric device that increases the velocity (and thus the kinetic energy) of an electrically charged particle (electron, proton, deuteron, ion, etc.) to a value sufficiently high either to cause a conversion of some of the kinetic energy to electromagnetic radiation (such as X-rays) or to cause a nuclear transformation in the target.

Particle accelerators are generally of the following two types: linear, if the particle is accelerated in a straight path, and circular, if the path is nearly circular or spiral. In both cases the particles may be accelerated by either electric or magnetic fields, or a combination of both. In the case of circular accelerators, magnetic fields are used to keep the particle in the proper orbit, while the actual acceleration is achieved by electric fields.

Examples of linear accelerators are the Van de Graaff generator and the Cockroft-Walton accelerator, first developed in the 1930s; examples of circular accelerators are the betatron, the cyclotron, the synchrotron, and the synchrocyclotron. The Van de Graaff generator and the betatron are especially used for the production of high-energy X-rays.

The Van de Graaff generator (Fig. 4-12) consists of a highly insulated electrode, shaped like a hemisphere, and a fast-moving nonconducting belt that conveys a continuous supply of electrostatic charges to the electrode. The belt is made of rubber-impregnated fabric and runs on two pulleys, one grounded and the other located in the high-voltage insulated electrode. A comb of needle points close to the grounded motor-driven pulley is charged to some 20 to 40 kV. These needles spray electrostatic charges on the belt, which moves past them rapidly. The charges are carried to the insulated pulley, which thus becomes more highly charged than the electrode itself. When this pulley acquires a potential of several thousand volts with respect to the electrode, additional

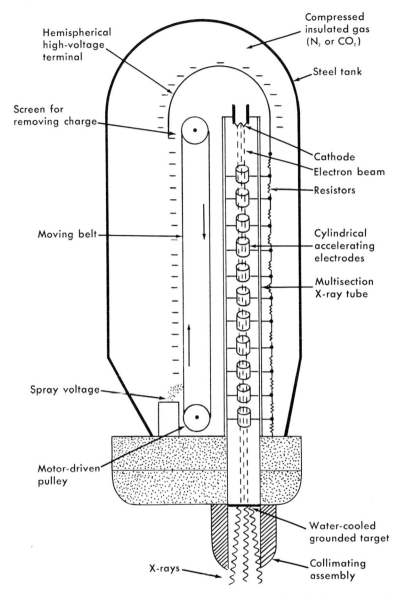

Fig. 4-12. Van de Graaff generator. (From Selman, J. 1960. The basic physics of radiation therapy. Charles C Thomas, Publisher, Springfield, Ill.)

charges brought up by the belt are automatically transferred to the electrode by a comb of needle points placed near the pulley where the belt is arriving. Another comb placed at the opposite side of the pulley sprays the departing belt with charges of the opposite sign, which the belt carries back to the grounded drive pulley. Thus the insulated electrode soon builds up a potential that is limited by the insulation or by corona loss.

The electrons needed for the production of X-rays are generated by a hot filament (cathode) at the top of a cylindrical tube that is located between the hemispherical electrode and the base of the generator. A large difference of potential exists between the two ends of the tube. The accelerated electrons are hurled against a grounded target, and X-rays are thus generated.

In the betatron (Fig. 4-13), magnetic induction is used for the acceleration of the

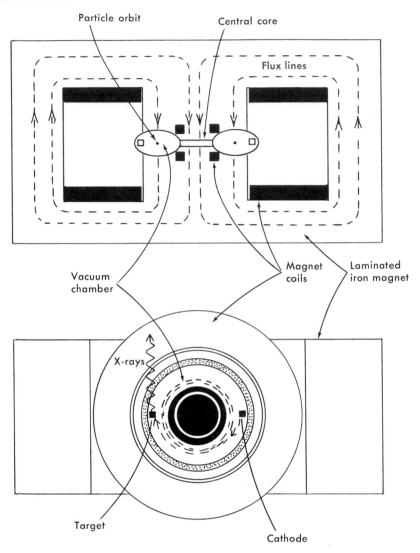

Fig. 4-13. Longitudinal and cross sections of a betatron. (From Blatz, H., editor. 1959. Radiation hygiene handbook. McGraw-Hill Book Co., New York.)

electrons in a circular path. The entire movement of the electrons takes place within a circular vacuum chamber, usually called a "doughnut" because of its shape. Some models of betatrons can easily generate X-rays with peak energies of up to 30 MeV.

Both the Van de Graaff generator and the betatron are widely used in radiation therapy and for a variety of biologic X-irradiation experiments. At the present time, small betatrons, not larger than a desk, are available on the market.

INCIDENTAL SOURCES OF X-RAYS

Certain devices other than X-ray tubes may also become sources of X-rays. Theoretically, any electronic device in which accelerated electrons strike a metal plate is a potential source of X-rays. However, in practice the intensity of radiation generated is negligible and may be ignored if the current is very small and the operating voltage is less than 30 kV. For higher voltages, emission of significant amounts of X-radiation should be suspected and definite steps should be taken to assure protection to personnel. Devices that should be suspected are the following:

1. Cathode-ray tubes and any device that works on the same principle, such as cathode-ray oscilloscopes, radarscopes, and television video tubes. In these tubes a stream of electrons is emitted by an "electron gun," accelerated, and directed onto a fluorescent screen. In a black-and-white television set, the voltage is rather low and the very soft X-rays emitted by the fluorescent screen do not travel very far because they are absorbed by air or by the glass sheet that usually protects the screen. In color sets, the voltage is higher and the possible X-radiation emitted by the screen should be checked.

2. Electron microscopes are operated sometimes with voltages up to 50 to 100 kV. However, the current is small so that the fairly penetrating X-rays generated have a rather low intensity. Nevertheless, shielding for the operator should be considered in the design of these instruments.

3. In any vacuum tube there is a stream of electrons going from the cathode to the anodic plate. Usually differences of potential and currents are very small. However, in rectifier tubes (tubes designed to rectify an alternate current), voltages may be high. X-rays emitted by their anodes may have a significant intensity to warrant protective shielding.

4. Magnetrons and klystrons used in high-power radar equipment definitely should be provided with adequate shielding. A few accidents resulting from the operation of unshielded klystrons, in the form of mild radiation sickness, have been recorded in the scientific literature (see the Lockport accident, p. 412).

X-RAYS IN BIOLOGIC IRRADIATION EXPERIMENTS

In radiation biology, X-rays are used perhaps more than any other type of ionizing radiation because an X-ray unit is within the reach of many research laboratories and is relatively easy to operate and to control; most of all one can obtain radiation of variable quality and intensity. The experiment should be planned carefully according to the material to be irradiated and the nature of the effect to be observed.

The first problem to solve is the type of machine to use. If fairly large biologic specimens are to be irradiated (rats, guinea pigs, whole plants, etc.) and one is interested in the biologic response of internal organs or of the whole organism, a beam of relatively high energy is required. A therapeutic unit with a maximum voltage of about 250 kV is satisfactory in this case, and the maximum or near-maximum voltage is applied to the tube. Lower voltages (about 100 kV) are sufficient for specimens such as small organisms (protozoa, fruit flies, yeasts) or small amounts of enzymes, or when superficial effects are sought (such as effects of X-rays on the skin or fur of rats). The quality of the beam, and therefore its penetrating power, should be carefully planned in advance, and the appropriate voltage and filtration should be chosen accordingly.

Dose rate and total accumulated dose should also be decided upon. The concepts of dose and dose rate are presented and discussed in greater detail in Chapter 9. In the present context, we may provisionally define dose rate as the amount of X-ray energy delivered per unit of time to an object located in the X-ray beam. As such, dose rate is strictly dependent on the radiation flux, which in turn depends, among other factors, on the tube current and TOD.

The TOD is chosen in accordance with the size of the specimen to be irradiated. The minimum TOD that can be used is that corresponding to a radiation field not smaller than the specimen. The dose rate is measured accurately with a dosimeter (such as the Victoreen Condenser R-Meter); the ionization chamber should be placed exactly where the specimen will be placed in the beam. Once the dose rate is known, then the irradiation time can be easily estimated to obtain the desired total dose. For greater accuracy, the total dose should also be measured directly during the actual irradiation of the specimen. Measurement is done by placing a suitable dosimeter (see Chapter 9) in the X-ray beam at a distance from the target identical to that of the specimen. To be sure that both specimen and dosimeter receive exactly the same dose, one often needs to surround both with the same type of scattering material.

In summary, the design of an irradiation experiment involves the consideration of four parameters of merit. Depending on the object to be irradiated and on the nature and purpose of the experiment, the investigator must decide on the quality and intensity of the X-ray beam to use, as well as the dose and dose rate. We have seen through the preceding sections of this chapter that these parameters are functions of a variety of factors. It is useful to summarize at this point how quality, intensity, dose rate, and dose are affected by those factors:

Factors affecting *quality*
  Voltage
  Nature of target
  Filtration (inherent and added)
Factors affecting *intensity*
  Voltage
  Tube current
  Filtration

Factors affecting *dose rate*
  Quality
  Intensity
  TOD
  Scattering
Factors affecting *total accumulated dose*
  Dose rate
  Exposure time

When large specimens or large groups of specimens are irradiated, the investigator is also faced with the problem of the uniformity of absorbed dose. If 10 rats are irradiated at the same time in a tray or cage, the animal that happens to be on the axis of the beam will receive a larger dose than those at a certain distance from it, since the flux, as explained previously, decreases with the distance from the axis in the same field. The problem is solved in different ways for different situations. For instance, it is customary to irradiate a group of rats or other similar specimens in the radial compartments of a round Lucite box, one rat in each compartment. One compartment is reserved for the dosimeter. The box is placed on a rotating turntable. In other situations the problem can be solved equally well by using some ingenuity.

In the written report of an irradiation experiment, when materials and methods are described, the irradiation parameters used should be stated in detail. Data such as type of X-ray unit, voltage applied to the tube, tube current, filtration, HVL, TOD, dose rate, and how they were measured should be reported. Below is an example of such a report:

Rats were placed in a round Lucite box with radial compartments (one animal in each compartment). The box was placed on a rotating turntable. A 250 kV X-ray unit operated at 250 kV and 15 ma tube current and with 0.25 mm of Cu + 1 mm of Al filtration was used for the irradiation of the rats. The HVL was 1.1 mm Cu. The TOD was 65 cm, with a dose rate of 83 R/min, as measured with a Victoreen Condenser R-Meter. The ion chamber was placed in an empty compartment of the Lucite box. For the in vitro irradiation of the ascites lymphoma cells, the same radiation factors were used as for live rats, but total doses delivered were measured with a Fricke ferrous sulfate dosimeter. The concentration of the Fe^{+++} ions produced was measured with a Beckman DU spectrophotometer at 305 nm.

REFERENCES

Blatz, H. 1964. Introduction to radiological health. McGraw-Hill Book Co. New York. Pp. 25-61.

Blatz, H. 1959. Radiation hygiene handbook. McGraw-Hill Book Co., New York. Pp. 6-61 to 6-73 and 13-2 to 13-18.

Bleich, A. R. 1960. The story of X-rays from Roentgen to isotopes. Dover Publications, Inc., New York.

Clark, G. L., editor. 1963. The encyclopedia of X-rays and gamma rays. Reinhold Publishing Corp., New York.

Delario, A. J. 1953. Roentgen, radium and radioisotope therapy. Lea & Febiger, Philadelphia.

Morgan, R. H., and K. E. Corrigan. 1955. Handbook of radiology. Year Book Publishers, Inc., Chicago.

Selman, J. 1960. The basic physics of radiation therapy. Charles C Thomas, Publisher, Springfield, Ill.

Sproull, W. T. 1946. X-rays in practice. McGraw-Hill Book Co., New York.

Radioactivity

THE DISCOVERY

Radioactivity is the property of those atomic nuclei that spontaneously change their nucleonic configuration and/or energy content; the event that brings about the change is known as *radioactive disintegration*, or *radioactive decay*, and is usually associated with the emission of particulate or electromagnetic radiation.

Natural radioactivity (the radioactivity of naturally occurring nuclides) was discovered in 1896 by the French physicist H. Becquerel in uranium (an element known since 1789), when he casually observed that the fluorescent crystals of potassium uranyl sulfate can blacken a photographic plate even in the dark and can discharge a charged electroscope. Becquerel concluded that uranium was the source of some invisible radiation, although he did not immediately consider this radiation as an indication of some sort of atomic transformation.

While studying the radioactivity of pitchblende (a uranium ore extracted in Bohemia), Marie Curie observed that the radioactivity of this mineral was more intense than that of pure uranium metal. With long and tedious analytical procedures, in 1898 she was able to isolate from the mineral two previously unknown elements that turned out to be more radioactive than uranium — radium and polonium. Several other radioelements of high atomic number were discovered within the next few years.

The later scientific contributions to the knowledge of natural radioactivity must be mainly credited to Lord Rutherford and his co-workers. Within about 20 years after the turn of the century the basic following facts were established:

1. The emission of the "rays" by radioactive elements is caused by some sort of "transmutation" of one element to another.

2. The elements resulting from the transmutation of others may be themselves radioactive, so that "family trees" may be constructed. It was found that all known natural radioactive elements can be grouped in three such families, called *radioactive series*, which are the uranium series, the actinium series, and the thorium series.*

3. The duration of radioactive properties varies widely among radioactive elements.

4. The intensity of the rays emitted by a radioactive sample decreases in

---

*Today a fourth series is known — the neptunium series, which begins with the transuranic artificial element neptunium.

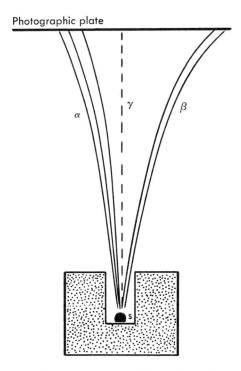

Photographic plate

Fig. 5-1. Magnetic deflection of radiation emitted by radium. The magnetic field is directed toward the paper. The source, **s,** is at the bottom of a lead well so that its radiation is collimated.

such a manner that its reduction to 50% of the initial value takes place within a time interval constant and characteristic of each radioactive element.

5. With the use of strong magnetic fields, Rutherford showed that radium and other radioactive elements emit three distinct types of rays (Fig. 5-1), which he called:

    a. Alpha rays—slightly deflected by the magnetic field and consisting of positively charged particles

    b. Beta rays—strongly deflected by the magnetic field in a direction opposite to that of the alpha particle deflection and consisting of negatively charged particles

    c. Gamma rays—unaffected by the magnetic field and therefore electrically neutral

Further developments in the study of radioactivity led to the understanding of atomic and nuclear structure. These developments are discussed elsewhere in this book.

METHODS OF RADIOACTIVE DECAY

As previously mentioned, when a nucleus disintegrates (or decays), some type of radiation is usually emitted. After the decay event has taken place, the nucleus (daughter nucleus) may be either stable or radioactive.

A radioactive nucleus may decay according to one of the following methods:

1. *Alpha emission.* The nucleus emits an alpha particle, which consists of two protons and two neutrons. Therefore, the mass number *(A)* of the nucleus de-

creases by four units, and the atomic number *(Z)* decreases by two units. If the emission of the alpha particle leaves the nucleus in an excited energy state, the excess energy is liberated in the form of a gamma photon. Alpha emission, therefore, may be accompanied by emission of gamma radiation. This decay method can be schematically represented with the following nuclear equation (*X* is the original nucleus, and *Y* is the daughter nucleus):

$$\ce{^A_Z}X \longrightarrow \ce{^{A-4}_{Z-2}}Y + \ce{^4_2}\alpha \ (+\gamma)$$

Because of the change in *Z*, *Y* is a chemical element different from *X*.

EXAMPLE:

$$\ce{^{226}_{88}}Ra \longrightarrow \ce{^{222}_{86}}Rn + \alpha + \gamma$$

2. *Beta-minus emission.* The nucleus emits a β^- particle, that is, a negative electron (negatron), which originates in the nucleus when a neutron changes to a proton. Therefore, the mass number of the nucleus remains unchanged, and its atomic number increases by one unit. A gamma photon may or may not accompany the emission of the beta particle.

$$\ce{^A_Z}X \longrightarrow \ce{^A_{Z+1}}Y + \ce{_{-1}}e \ (+\gamma)$$

Because of the change in *Z*, again *Y* is a chemical element different from *X*.

EXAMPLE:

$$\ce{^{14}_6}C \longrightarrow \ce{^{14}_7}N + \beta^-$$

3. *Beta-plus emission.* The nucleus emits a β^+ particle, that is, a positive electron (positron), which originates in the nucleus when a proton changes to a neutron. Therefore, the mass number of the nucleus remains unchanged and its atomic number decreases by one unit. A gamma photon may or may not accompany the emission of the beta particle.

$$\ce{^A_Z}X \longrightarrow \ce{^A_{Z-1}}Y + \ce{_{+1}}e \ (+\gamma)$$

Because of the change in *Z*, *Y* is a chemical element different from *X*.

EXAMPLE:

$$\ce{^{22}_{11}}Na \longrightarrow \ce{^{22}_{10}}Ne + \beta^+ + \gamma$$

4. *Electron capture.* An alternative decay method for β^+ emission is electron capture. The nucleus captures one of the orbital electrons (usually a K-electron), which converts a proton into a neutron. Therefore, the mass number of the nucleus remains unchanged and its atomic number decreases by one unit. If the nucleus is left in an excited state, a gamma photon is emitted. An X-ray photon is always emitted as a result of an electronic transition whereby the vacancy left by the captured electron is filled by an outer electron as shown below:

$$\ce{^A_Z}X + \ce{_{-1}}e \longrightarrow \ce{^A_{Z-1}}Y(+\gamma) + \text{X-ray}$$

Because of the change in *Z*, *Y* is a chemical element different from *X*.

EXAMPLE:

$$\ce{^{55}_{26}}Fe + \ce{_{-1}}e \longrightarrow \ce{^{55}_{25}}Mn + \text{X-ray*}$$

---

*Note that in this example no radiation of nuclear origin is emitted as a consequence of the decay process.

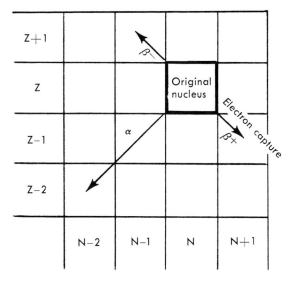

Fig. 5-2. Changes of Z and N in relation to the decay methods.

5. *Isomeric transition.* Isomeric transition is a decay process involving neither the emission nor the capture of a particle. The nucleus simply changes from a higher to a lower energy level by emitting a gamma photon. Therefore, both mass number and atomic number remain unchanged. The daughter nucleus is the same chemical element as the original nucleus. The original nucleus before the transition is said to be in a "metastable" state, which is indicated by adding an m to the mass number in the symbol.

$$_{Z}^{Am}X \longrightarrow {}_{Z}^{A}X + \gamma$$

EXAMPLE:

$$_{38}^{87m}Sr \longrightarrow {}_{38}^{87}Sr + \gamma$$

Other decay methods (such as neutron emission) are rare and of little practical importance. They occur in few artificial radionuclides.

The relationships between original and daughter nucleus, concerning nucleonic composition, can be summarized as follows:

1. In alpha emission, A decreases by four units, as both Z and N decrease by two units.

2. In beta-minus emission, A remains unchanged, as Z increases by one unit and N decreases by one unit.

3. In beta-plus emission and in electron capture, A remains unchanged, as Z decreases by one unit and N increases by one unit.

4. In isomeric transition, A, Z, and N remain unchanged.

These changes can be represented graphically with Fig. 5-2, wherein the neutron numbers have been plotted against the atomic numbers.

The radioactive decay process may also be viewed as an exoergic process, whereby the original nucleus, X, releases a certain determinate amount of nuclear energy as it becomes a daughter nucleus. The energy released is called the *disintegration energy* (E_d), and in most cases it is equal to the sum of the kinetic energy of the particle and the energy of any gamma photon(s) emitted during the disintegration event.* This fact is verified with some examples later in this chapter.

For some radionuclides, all atoms decay according to the same method. For instance, all the atoms of a ^{14}C sample decay by pure beta emission. Other times, different atoms of the same radionuclide may decay according to two or more alternative methods (for example, ^{137}Cs and ^{36}Cl). One can almost always predict the probability of decay by a particular route. This probability is a constant for every radionuclide that decays through more than one route and is expressed as a percentage. For instance, one can predict that, on the average, 98.3% of the atoms of a ^{36}Cl sample will decay by beta-minus emission, and 1.7% by electron capture.

DECAY METHOD IN RELATION TO NUCLEAR STABILITY

Why some radionuclides should decay with one method and others should decay with another method is a complex problem of nuclear physics, and its discussion goes beyond the scope of this textbook. However, one can grasp easily certain general principles that determine the method followed when a radionuclide decays, through the study of the so called "neutron-proton plot" (Fig. 5-3). On linear graph paper let us represent, with dots, all the known nuclides (natural and artificial) according to their N and Z, which are plotted as coordinates. The "$N=Z$ line" joins all the points for which N is equal to Z. If we mark stable and radioactive nuclides with two different colors, we discover that the stable nuclides of low A fall on, or very close to, the $N=Z$ line, but, as the Z and N numbers increase, they fall farther and farther away, and below the $N=Z$ line. Also, the points corresponding to stable nuclides form a band rather than a line, whose width increases with A. All this means that for nuclides of low mass number the condition for nuclear stability is a neutron-to-proton ratio of approximately 1; as the mass number increases, in order for a nucleus to be stable, it must have an excess of neutrons over protons; for some heavy nuclides the neutron-to-proton ratio can be as high as 1.5 (such as stable ^{206}Pb, with 82 protons and 124 neutrons). The imaginary line that runs through the middle of the stable nuclide band may be called the *line of nuclear stability*; it coincides with the $N=Z$ line for low A, but deviates from it as A increases. The graphic vertical distance between the two lines represents the neutron excess.

Any nuclide represented on the graph at a significant distance from the line of nuclear stability must be unstable and therefore radioactive. A nuclide represented below the line has more neutrons than required for nuclear stability; if it decays by beta-minus emission, one of its neutrons changes into a proton and thus it gets closer to the line of nuclear stability. In fact, practically all radionuclides represented on the graph below the line of nuclear stability are beta-minus emitters. On the other hand, a nuclide represented above the line has more protons than required for nuclear stability; if it decays by beta-plus emission or by electron capture, one of its protons changes into a neutron and thus it gets closer to the line of nuclear stability. In fact, practically all radionuclides represented above the line of nuclear stability decay by beta-plus emission or by electron capture. It seems reasonable to believe, therefore, that the

---

*In certain instances, part of E_d will also be used in the recoil of the daughter nucleus, in the creation of a positron-negatron pair, etc., as discussed elsewhere in this book.

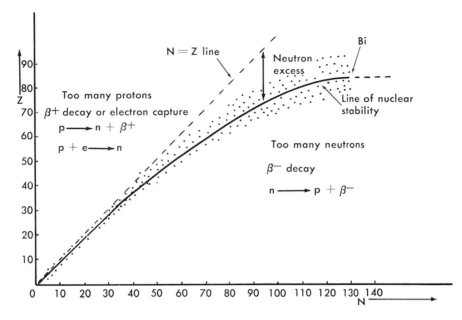

Fig. 5-3. Neutron-proton plot. Only stable nuclides have been represented (*dots*); radio-nuclides are not shown.

decay method is determined at least by the position a radionuclide occupies in relation to the line of nuclear stability.

The nuclides with $Z > 83$ are all radioactive regardless of whether they are close to, or far from, the line of nuclear stability. When the number of nucleons is very large, the nuclear forces responsible for holding the nucleons tightly together are apparently weaker, and this weakness is in itself a cause of instability for very heavy nuclides. To approach nuclear stability, these nuclei should get rid of their excess nucleons, and most of them do so by ejecting alpha particles. Alpha emission, unlike beta emission, is an especially effective method for a nucleus to rid itself of some nucleons. Most alpha emitters are found among the heaviest nuclides. Through the radioactive decay process they become stable in one or more steps, depending on how many nucleons they have in excess. The existence of the natural radioactive decay series, mentioned earlier, proves the validity of this concept. Beta emission and electron capture also occur among the very heavy radionuclides (see a chart of nuclides). They are determined by the same general principles explained above for lighter radionuclides.

DECAY SCHEMES

Most of the facts and data pertinent to the disintegration of a radioactive nuclide can be usefully and briefly summarized graphically with a decay scheme. It is a diagram that basically consists of two or more horizontal lines representing different energy levels of a nucleus. The top line represents the energy level of the nucleus before disintegration occurs; the bottom line represents the energy level of the daughter nucleus. If the decay process occurs in more than one step (such as the emission of a beta particle, followed by the emission of gamma photons), intermediate lines are also needed. The distances between lines are kept approximately proportional to the differences between consecutive energy levels.

Transitions that occur by emission of alpha particles or positrons or by electron capture are shown by straight, oblique arrows directed downward and to the left. Beta-minus emission is shown with an arrow directed downward and

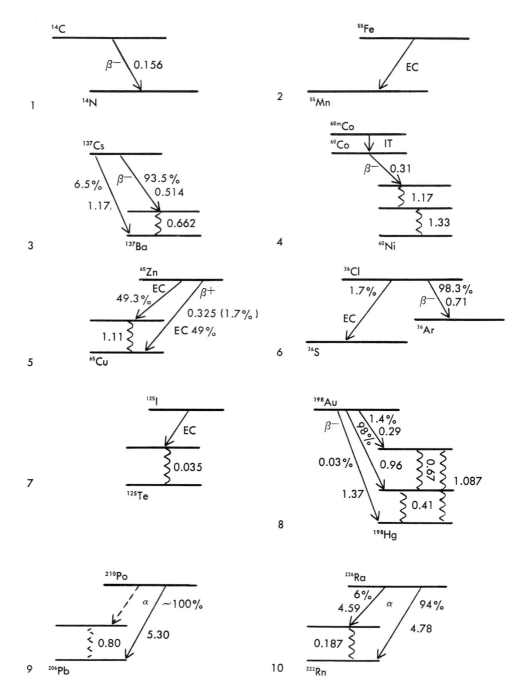

Fig. 5-4. Ten examples of decay schemes.

to the right. Gamma photons are shown with wavy arrows pointed vertically downward. The notation of the daughter nucleus is written below the bottom horizontal line. Alpha-particle energy, gamma energy, and maximum energy (E_{max}) of beta particles* are indicated on the side of the corresponding arrows in MeV. In case of alternate decay routes, their probabilities are also shown as percentages of atoms decaying through each route.

In summary, a decay scheme, even in its simplest form, contains the following information: name and mass numbers of the original and daughter nuclei, decay method, and alternate routes (if any), with their respective probabilities, radiation type, and energy. Sometimes, other constants of importance, such as half-life (see below) and disintegration energy, are also indicated.

Whenever a radionuclide is used in any type of experiment or investigation (experiments on properties of ionizing radiations, tracer techniques, diagnostic tests, etc.), one must get acquainted with its decay scheme beforehand if its physical behavior must be understood and if results must be interpreted correctly. The importance of the study of the decay scheme cannot be overemphasized. In Appendix III the simplified decay schemes of several radionuclides that are likely to be used in investigations and experiments of a biologic nature are reported. Some decay schemes are also reported in Fig. 5-4 and explained below as typical examples.

1. All atoms of ^{14}C decay by emission of a β^- particle, with $E_{max} = 0.156$ McV. The daughter nucleus (^{14}N) is not left in an excited state; no gamma radiation is emitted. ^{14}C is a pure beta emitter.

2. All ^{55}Fe atoms decay by electron capture (EC). The daughter nucleus is a stable isotope of Mn. No gamma radiation is emitted. The only radiation emitted by the atom is an X-ray from an electronic transition.

3. All ^{137}Cs atoms decay by beta emission. However, two routes are possible: (1) 6.5% of the atoms release their disintegration energy (1.17 MeV) in one step, with the emission of a beta particle ($E_{max} = 1.17$ MeV). The daughter nucleus is left in the ground energy state. (2) 93.5% of the atoms release their E_d in two steps: first a β^- particle is emitted ($E_{max} = 0.514$ MeV) and then a γ photon ($E = 0.662$ MeV). Note that $0.514 + 0.662 \approx 1.17 = E_d$. Also note that only 93.5 out of every 100 ^{137}Cs atoms, on the average, emit gamma radiation.

4. ^{60}Co is the daughter nucleus of another nuclide, ^{60m}Co, which decays simply by isomeric transition (IT). Every ^{60}Co atom releases its E_d in three steps. First a β^- particle is emitted and leaves the nucleus in an excited state. The nucleus reaches the ground state with the emission of two consecutive γ photons in cascade. For every 100 atoms decaying, 200 γ photons are emitted. Therefore a sample of ^{60}Co is a more intense source of gamma radiation than a ^{137}Cs sample of equal activity (that is, with the same number of disintegrations per unit of time).

5. Of the ^{65}Zn atoms, 49.3% decay by electron capture, followed by the emission of a γ photon. The other 50.7% reach the ground state, either with the emission of a β^+ particle or by electron capture, not followed by gamma radiation.† Of course, X-rays are also emitted.

6. ^{36}Cl may decay through two routes, which lead to somewhat opposite results. When it decays by EC, the nucleus releases more energy and becomes an isotope of S, with $Z = 16$. The decay by β^- emission leads to ^{36}Ar ($Z = 18$), with a release of less E_d. The alternate decay methods of this radionuclide lead to the formation of two different daughter nuclei.

---

*For the meaning of E_{max} for beta particles, see p. 92.

†The E_d for ^{65}Zn is 1.34 MeV. Yet the β^+ E_{max} is only 0.325 MeV, because part of the E_d is used for the creation of a positron-negatron pair (1.02 MeV). See details in Chapter 6.

7. The decay of ^{125}I occurs always by EC, followed by the emission of a γ photon of very low energy (35 keV).

8. ^{198}Au nuclei are all beta emitters, but most of them (98.6%) will emit beta particles of $E_{max} = 0.96$ MeV, so that the rest of the E_d is released in the form of a 0.41 MeV γ photon. For about 1.4% of the nuclei, the emission of a weak β^- particle is either followed by a high-energy γ photon (1.087 MeV), or by a cascade of two lower energy photons. In both cases, the nucleus will reach the ground energy level. A very low percentage of ^{198}Au nuclei will decay with the emission of a high-energy beta particle ($E_{max} = 1.37$ MeV). Note that whatever route is followed, the sum total of all the particle and photon energies released through the same route is equal to E_d. About 98% of the atoms of a ^{198}Au source will emit a 0.41 MeV γ photon; the intensity of the other two gamma rays is therefore negligible.

9. ^{210}Po is practically a pure alpha emitter because almost 100% of the atoms decay by alpha emission only.

10. ^{226}Ra atoms may decay with the emission of one of two alpha particles of different energy. The emission of the lower energy alpha particle leaves the nucleus in an excited state, from which it reaches the ground state by emitting a γ photon.

THE LAWS OF RADIOACTIVE DECAY AND THE HALF-LIFE

One cannot predict when a single atom of any radionuclide will decay, because the phenomenon of radioactive decay is a random event. Radioactive decay must be studied statistically. In fact, if a very large number of atoms is observed, it is possible to answer such questions as how many atoms decay per unit of time, what the average life-span of all the atoms is, etc.; in other words, it is possible to find an equation expressing the relationship between the number of atoms that have not yet decayed and the time. Experimental evidence shows that the number of atoms decaying per unit of time (decay rate = dN/dt) is proportional to the number of unstable atoms present at any time (N), so that

$$dN/dt \propto N$$

The preceding proportionality relationship can be changed into an equation if a constant (λ) is introduced:

$$-dN/dt = \lambda N \tag{5-1}$$

or

$$\lambda = \frac{-dN/dt}{N} = \frac{-dN/N}{dt} \tag{5-2}$$

The minus sign is used because N decreases with time.

The constant λ is called the *decay constant* (or disintegration constant), which is, in effect, the fraction of atoms present at any time that decay per unit of time. Thus, if $\lambda = 0.0075$ per sec., then 0.75% of all the atoms present in a sample of radionuclide at a certain instant will decay during the next second.

The decay constant varies for different radionuclides but is constant for the same radionuclide and is not affected by any environmental condition (such as temperature or chemical composition). Consequently, it is one of the important and distinctive characteristics of a radionuclide. One consequence of equation 5-1 and of the previous considerations is that the *decay rate* (not to be confused with λ!) of a radioactive sample decreases with time. If we plot time against percentage of N (percentage of radioactive atoms present at any time) on linear graph paper, we will obtain curves as in Fig. 5-5, *A*. For radionuclide *a*, N decreases by a factor of 50% every 4 hours. For radionuclide *b*, N decreases by a

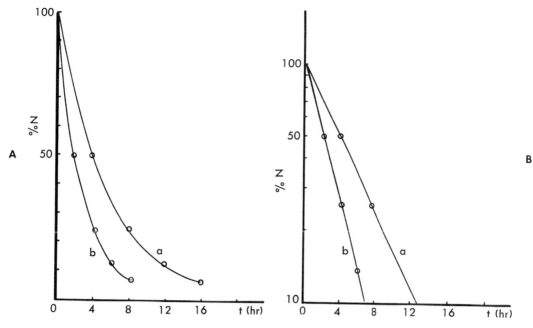

Fig. 5-5. Examples of decay curves. **A,** Plotted on linear graph paper. **B,** Plotted on semilog graph paper. a and b, Arbitrary radionuclides.

factor of 50% every 2 hours. The decay constant is then evidently smaller for a than for b. The same plots on semilog paper would result in straight lines (Fig. 5-5, B), the slopes of which are proportional to the decay constants.

Equation 5-1 tells us that the decay rate is proportional to the number of radioactive atoms still present. In other words, we know the rate at which a change is occurring with respect to time. With the help of integral calculus,* one can find an equation expressing the relationship between the two variables (time and number of radioactive atoms still present). Integrating equation 5-1, we have

$$N_t = N_0 e^{-\lambda t} \tag{5-3}$$

where N_t is the number of radioactive atoms left at time t, N_0 is the number of radioactive atoms present at time 0, and e is the base of the natural logarithms. For practical applications, equation 5-3 is often used in different forms, such as

$$\frac{N_t}{N_0} = e^{-\lambda t} \tag{5-4}$$

or

$$\ln \frac{N_t}{N_0} = -\lambda t \tag{5-5}$$

or (with the more convenient decimal logarithms)

$$2.302 \log \frac{N_t}{N_0} = -\lambda t \tag{5-6}$$

*Integral calculus is concerned with the problem of finding a function when the rate of its change is known.

One can see from the above considerations that the decay constant gives us an idea of how rapidly or how slowly a radionuclide decays, and this information is necessary (and sometimes essential) for many practical applications, such as when a radionuclide is used as a tracer in a biologic experiment. However, the same information can be obtained with another more convenient constant, called *half-life*, which is mathematically related to the decay constant. The half-life (T or $t_{1/2}$) is the time taken for 50% of the atoms of a radionuclide to decay; it is expressed in some suitable time units (seconds, minutes, days, or years). Consult a chart of nuclides to see how different the half-life is for different radionuclides; it ranges from small fractions of a second (for very "short-lived" radionuclides) to thousands of years (for "long-lived" radionuclides). It should be now obvious why the knowledge of the half-life is sometimes essential; a long-range experiment with a very short-lived nuclide used as a tracer would be naturally impossible. One should also keep in mind that the half-life of a radionuclide remains a constant only when applied to a large number of atoms, for the reasons explained at the beginning of this section. Therefore, the statement is not true that of two atoms of a radionuclide with $T = 8$ days, one will decay after 8 days!

The relationship between half-life and decay constant is easily found with some algebraic transformations of equation 5-6. Since, by definition, after one half-life the number of radioactive atoms is reduced by 50%, we can write equation 5-6 as follows:

$$2.302 \log \frac{1}{2} = -\lambda T \tag{5-7}$$

where t has the definite value $T (= \text{half-life})$

$$2.302 \log 2 = \lambda T$$
$$2.302 \times 0.30103 = \lambda T$$
$$0.693 = \lambda T$$

$$T = \frac{0.693}{\lambda} \tag{5-8}$$

and

$$\lambda = \frac{0.693}{T} \tag{5-9}$$

Both equations 5-8 and 5-9 show that the half-life and the decay constant are inversely proportional.

The half-life of a short-lived radionuclide (with a T of up to several days) can be measured quite accurately and directly. A sample is prepared and its activity counted at regular time intervals. A decay curve is then constructed (with time versus activity) on semilog paper, and the time taken for the activity to be reduced to one half is read off on the horizontal axis. This method depends on the fact that the activity (= decay rate) at any time is proportional to the number of radioactive atoms present at the same time (see equation 5-1). Obviously, the same experimental and direct method cannot be used for long-lived radionuclides. In this case, the decay constant is found first (with a long and laborious procedure) and then T is found by using equation 5-8.

THE CURIE

The amount of a radionuclide is not expressed in terms of weight or volume, but in terms of decay rate, because radionuclides are used as emitters of ionizing radiations and it is quite possible that two samples of ^{60}Co each weighing 1 g have different decay rates at the same time, simply because they were produced in a nuclear reactor at different times.

Much more meaningful and useful for practical purposes is the expression of the amount of a radionuclide preparation in terms of disintegrations per unit of time. The basic unit of decay rate is the *curie*, which is defined as the quantity of any radionuclide having a decay rate of 3.7×10^{10} disintegrations per second (dps). The abbreviation for curie is *Ci*. So we have the following:

$$1 \text{ Ci} = 3.7 \times 10^{10} \text{ dps} = 2.22 \times 10^{12} \text{ dpm (disintegrations per minute)}$$

The submultiples most frequently used are the following:

$$1 \text{ millicurie (mCi)} = 3.7 \times 10^{7} \text{ dps} = 2.22 \times 10^{9} \text{ dpm}$$
$$1 \text{ microcurie } (\mu\text{Ci}) = 3.7 \times 10^{4} \text{ dps} = 2.22 \times 10^{6} \text{ dpm}$$

Some multiples are the following:

$$1 \text{ kilocurie (kCi)} = 10^{3} \text{ Ci}$$
$$1 \text{ megacurie (MCi)} = 10^{6} \text{ Ci}$$

The curie is approximately the decay rate of 1 g of radium. This decay rate does not decrease significantly within a few years because of the long half-life of radium (1620 years). For shorter lived radionuclides (with half-lives of a few years or less), the activity expressed in curies decreases significantly with time; so one should obviously specify the activity of samples of such radionuclides for a certain date. For very short-lived radionuclides (with $T =$ a few days), year, month, and day should be stated.

In view of the above considerations, a problem arises when radionuclides are used in several applications (including tracer experiments). What is the activity (A_t) left in a certain solution of a radionuclide, when we know that the activity at a certain date was A_0? From equation 5-5 we derive

$$\ln \frac{N_0}{N_t} = \lambda t$$

and

$$2.302 \log \frac{N_0}{N_t} = \lambda t$$

$$\log \frac{N_0}{N_t} = \frac{\lambda t}{2.302} = 0.434 \, \lambda t$$

and expressing λ as a function of T,

$$\log \frac{N_0}{N_t} = 0.434 \times \frac{0.693}{T} \times t = \frac{0.3t}{T}$$

Since the activity (decay rate) is proportional to N,

$$\frac{A_0}{A_t} = \frac{N_0}{N_t}$$

and therefore:

$$\log \frac{A_0}{A_t} = \frac{0.3t}{T} \qquad (5\text{-}10)$$

which is the equation to use for the solution of the problem stated above.

EXAMPLE:

How many microcuries are left of 50 μCi of ^{35}S after 60 days? The T for ^{35}S is 87 days. Therefore:

$$\log \frac{50}{A_t} = \frac{0.3 \times 60}{87} = 0.206$$

Since the antilog of $0.206 = 1.607$,

$$\frac{50}{A_t} = 1.607$$

and

$$A_t = \frac{50\ \mu Ci}{1.607} = 31\ \mu Ci$$

REFERENCES

Lederer, C. M., J. M. Hollander, and I. Perlman. 1967. Table of isotopes. John Wiley & Sons, Inc., New York.

Romer, A. 1960. The restless atom. Doubleday & Co., Inc., Garden City, New York.

U. S. Department of Health, 1960. Table of isotopes. In Radiological health handbook. Washington, D. C.

CHAPTER 6

The properties of some ionizing radiations and their interactions with matter

ALPHA PARTICLES
Nature and origin

The nature of alpha particles was discovered by W. Ramsay and F. Soddy in 1903, when they observed that alpha-emitting radionuclides invariably produce helium gas. Previous experiments had shown that these particles are electrically charged because they are deflected by magnetic fields and that their charge is positive and twice the magnitude of the charge of the electron. Therefore an alpha particle can be considered as a helium nucleus stripped of its orbital electrons. Its mass is about 4 amu and its charge is $+2$. It is made of two protons and two neutrons kept together by very strong nuclear forces.

In Chapter 5 we saw that alpha emission is a decay method whereby an unstable heavy nucleus reaches a temporary or permanent degree of stability by ejecting a certain number of nucleons (two protons and two neutrons); in fact, for heavy atomic nuclei, the main cause of their instability is an excessive number of nucleons, with the consequent weakening of the nuclear forces. This fact explains why alpha emitters, with few exceptions, are found among radionuclides with $Z > 82$. Since the chemical elements of very high Z are of little biologic importance, alpha emitters are of less interest to the radiation biologist than many beta and gamma emitters of lower Z, at least in terms of their use as radiotracers.

Energy, range, and velocity

The statement that alpha particles, unlike beta particles, are monoenergetic is correct only if it is properly explained. When radionuclide X (such as ^{238}U) decays to radionuclide Y (^{234}Th) by alpha emission, the alpha particles emitted by all the atoms of X have the same energy (are monoenergetic) as long as the daughter nuclei are all of nuclear species Y (see definition of radionuclide in

Chapter 2). This condition is not true in beta emission, wherein, for the same transition from X to Y, different atoms of X emit beta particles of different energies (see details below). The energy range of alpha particles for most emitters is 3 to 8 MeV. These relatively high energy values are accounted for by the large mass of the alpha particle, rather than by the magnitude of its velocity $\left(\text{recall that } KE = \dfrac{ms^2}{2}\right)$.

The large mass also explains why the particle energy is significantly less than the disintegration energy, even though the whole transition may be accomplished by alpha emission only. The disintegration energy is the total energy liberated by a nucleus during a transition. In alpha decay, the emitted particle has a mass large enough to cause a measurable recoil of the nucleus (cf. the third law of dynamics). The recoil energy of the daughter nucleus is then the difference between the disintegration energy and the particle energy; for example, ^{210}Po decays to ^{206}Pb by emission of an alpha particle, whose energy is 5.3 MeV. No gamma radiation is emitted in this transition, yet the disintegration energy is 5.4 MeV. The difference is dissipated in the recoil of the daughter nucleus.

The speed of an alpha particle is, of course, a function of its kinetic energy, as it is for any type of particulate radiation. For most alpha particles it is in the range of 1.4 to 2×10^9 cm/sec, or about 14,000 to 20,000 km/sec, which is less than one tenth the speed of light. These figures are velocities at the exit of the particle from the nucleus.

The range of a particle is the distance traveled by the particle in a given type of matter. The range is dependent on the particle energy and the number of interactions between the particle and the absorbing material, and therefore on its nature and density. Shortly after the discovery of radioactivity by alpha decay, it was noticed that alpha particles are easily stopped by an ordinary sheet of paper. Since the epidermis of the skin has approximately the same thickness and density of ordinary paper, it also stops alpha particles completely. Consequently, external sources of alpha particles do not present any biologic hazard. The story is completely different for internal alpha emitters. The range of a 5 MeV alpha particle in aluminum can be shown to be about 24 μm.

Several equations (Geiger equations) have been suggested to calculate the range of alpha particles in air, and they are applicable for different energy ranges. However, if a high degree of precision is not required, the following general and approximate equation can be used:

$$R = E \qquad\qquad (6\text{-}1)$$

where R is the range, in air, in cm and E is the particle energy in MeV. Thus an 8 MeV alpha particle would have a range, in air, of about 8 cm.

Ionization

As an alpha particle travels through an absorbing material, it gradually dissipates its kinetic energy by interacting with the material, either through the process of excitation (displacement of orbital electrons into higher energy levels) or through the process of ionization (complete removal of orbital electrons from atoms), or, less frequently, by head-on collisions with nuclei. When all its energy has been dissipated, the particle comes to rest and becomes a helium atom after capturing two stray electrons from the environment.

Because of their high electric charge and relatively low velocity, alpha par-

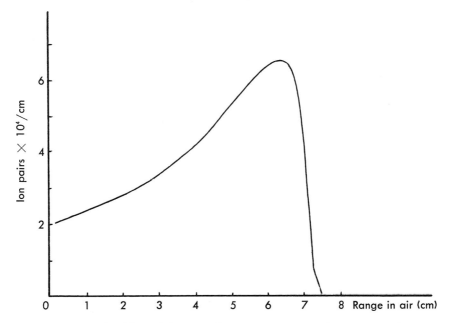

Fig. 6-1. Bragg curve for ^{214}Po alpha particles (E = 7.69 MeV) showing specific ionization as function of distance traveled from the source in air.

Fig. 6-2. An alpha track.

ticles are powerful ionizers. This fact can be better appreciated by studying the particle's *specific ionization,* which is the number of ion pairs produced by the particle per unit length of path. The specific ionization of an alpha particle, although always very high, increases as the particle loses energy along its path; as it slows down, it spends more time in the vicinity of atoms and consequently has a higher probability of ionizing them. Fig. 6-1 shows the variation of specific ionization with the distance traveled. The number of ion pairs formed per centimeter of path increases as the particle travels farther from the source, slowly at the beginning and very sharply toward the end of the range, until a high peak is reached, after which it drops to zero within a distance of a few millimeters.*

Some of these facts can be verified experimentally with a cloud chamber.† The alpha tracks that can be seen in the chamber are very thick and dense (Fig. 6-2) because of the high specific ionization (the number of droplets of condensed vapor is equal or at least proportional to the number of ion pairs produced). The thickness of the track increases as the end of the range is approached—an indication of an increase of specific ionization. The thin branches (delta tracks) sometimes visible along the main track are the result of secondary ionizations; they are produced by secondary ion pairs formed by the primary ions, which may have enough energy to ionize more atoms. Alpha tracks (unlike

---

*The energy required to form one ion pair in air is approximately 32.5 eV.
†See details on pp. 142 to 144.

beta tracks) are almost straight lines, because of the large mass of the particle. The crooked appearance at the end of the track is a consequence of straggling; when the energy of the particle is reduced to a very low value, quite probably the particle is deflected from its path by the electrostatic repulsion of a nucleus of the absorbing material.

BETA PARTICLES
Nature and origin

Beta particles are electrons ejected by unstable atomic nuclei during that radioactive disintegration process commonly called *beta decay* (see Chapter 5). They may be either negatrons (beta-minus) or positrons (beta-plus), depending on whether their electric charge is -1 or $+1$. According to the basic principles of Fermi's theory of beta decay, beta particles do not exist in the unstable nucleus before they are emitted; they are produced as a result of subnuclear transformation whereby a neutron changes to a proton (a β^- particle is emitted) or a proton changes to a neutron (a β^+ particle is emitted).

Although their interactions with matter are basically the same, an important difference between the two types of beta particles is in their ultimate fate when they come to rest after losing all their kinetic energy. A negatron will continue to exist as such and very likely will be captured by an atom into one of its electronic orbits, whereas a positron will combine with a negatron (its antiparticle) and both will be annihilated so that their mass will be converted into two γ photons (for more details see p. 103). The rest mass of a beta particle is the same as the mass of the electron ($= 0.00055$ amu) and is therefore much smaller than the mass of an alpha particle. Its speed is a function of its energy. Because of the very small mass involved, speeds of high-energy beta particles can be very close to the speed of light. With such high speeds, the mass of a beta particle in motion is a function of its speed, in accordance with the relativistic equation discussed in Chapter 1:

$$m = \frac{m_0}{\sqrt{1 - \dfrac{s^2}{c^2}}}$$

(6-2)

As an example, the mass of a beta particle with a speed 0.99 times the speed of light is equal to 7.07 times its rest mass.

Certain types of particle accelerators can produce high-intensity beams of electrons with energies considerably higher than those normally found associated with beta particles. In radiation biology these beams are often used to investigate the effects of electron radiation on living systems whenever high intensities and/or high energies are desirable.

Energy

Unlike alpha particles, the beta particles emitted by a radionuclide are not monoenergetic. This statement means that if a very large number of atoms of the same radionuclide are observed (for example, with a magnetic or scintillation spectrometer) while they are decaying, the beta particles they emit do not have the same energy even though the disintegration energy is the same for all the atoms. The beta energies range from a minimum of practically zero to a maximum (E_{max}), which is typical of the radionuclide in question. When, for in-

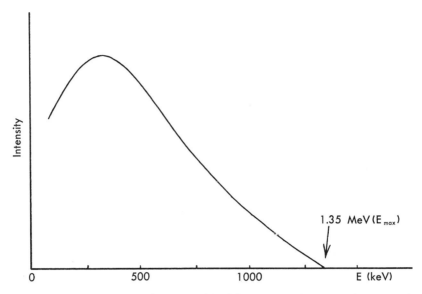

Fig. 6-3. Beta-particle spectrum of ^{40}K. (Modified from B. Dželepow et al. 1946. Phys. Rev. **69**:538. Reprinted in Lapp, R. E. and H. L. Andrews. 1954. Nuclear radiation physics. Ed. 2. Prentice-Hall, Inc., Englewood Cliffs, N. J.)

stance, we say that the beta-particle E_{max} for ^{32}P is 1.71 MeV, we mean that the beta particles emitted by a very large number of ^{32}P nuclei have a continuous spectrum of energies varying from 0 to 1.71 MeV. This continuous energy distribution can be seen graphically by plotting the number of beta particles emitted with energy E as ordinate, against E as abscissa. A *beta spectrum* is thus obtained. The beta spectrum of ^{40}K is shown in Fig. 6-3. For this radionuclide $E_{max} = 1.32$ MeV. Very few beta particles are emitted at E_{max}. For energies lower than E_{max}, the number of particles increases up to a peak that is more or less close to 0 MeV for different radionuclides. The average energy (E_{av}) of the beta particles emitted by the nuclei of the same radionuclide is roughly equal to 1/3 E_{max}. Values of E_{max} vary from 0.0186 MeV for tritium to 4.81 MeV for ^{38}Cl. One should familiarize himself with the energies of the most common beta emitters, frequently used in radiotracer techniques.

For a long time the continuous energy distribution shown by the beta particles emitted by different atoms of the same nuclear species has been a baffling problem in nuclear physics. How can the law of energy conservation still hold true if two atoms of the same radionuclide are converted to the same daughter nuclear species by emitting beta particles of different energies? The solution to the problem was suggested by W. Pauli, who postulated that, in beta decay, the nucleus simultaneously emits not one, but two particles—a beta particle and a *neutrino*. For theoretical reasons, the neutrino was assumed to have no electric charge, an infinitesimal mass, but a measurable amount of energy*—a very elusive and strange particle indeed! According to this hypothesis, the disintegration energy would be shared between the beta particle and the neutrino in equal or unequal parts, but always in such a way that beta energy + neutrino energy = disintegration energy. Consequently, the higher the energy of the beta particle, the lower the energy of the neutrino and vice versa. Investigators also expected that because of its properties the neutrino would have a great penetrating power, would not

---

*In Italian, *neutrino* literally means "small, electrically neutral body."

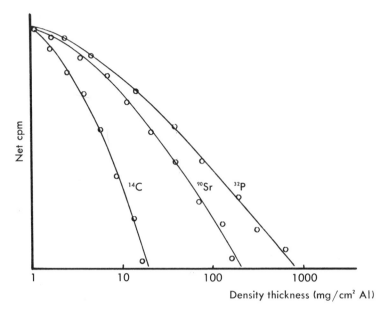

Fig. 6-4. Absorption curves for ^{14}C, ^{90}Sr, and ^{32}P beta particles.

interact with matter appreciably, and thus would be almost undetectable. Recently, however, this particle has been detected, and we have direct evidence of its existence.

Absorption

When beta particles travel through an absorbing material, they gradually dissipate their energy by interacting with the atoms of the absorber. If their energy is completely dissipated within the absorber, they will be stopped by it (will be absorbed). Evidently, absorption must be a function of beta-particle energy, as well as of absorber thickness and density.

In absorption studies, it is convenient to unify the two properties of the absorber, density and thickness, and to consider rather its density-thickness, which is measured in mg (or g)/cm^2. Absorption curves of beta emitters are obtained by plotting count rate as function of density-thickness.* Fig. 6-4 shows absorption curves of different beta emitters. One can see that the absorption curves intersect the axis of the abscissas at different points; as the beta E_{max} increases, from ^{14}C to ^{32}P, more density-thickness of absorbing material is needed to reduce the count rate to zero.

Interactions with matter

The absorption of beta particles in matter, as noted above, is the consequence of the dissipation of their energy, which is caused by various types of interactions between the particles and the atoms of the absorbing material.

It is possible to see beta tracks in a cloud chamber. They are much longer than alpha tracks because beta particles can travel farther from the source. The tracks also show that their path is not a straight line, but rather a tortuous one,

---

*See experiment no. 4 in *Ionizing Radiation and Life: Laboratory Experiences.*

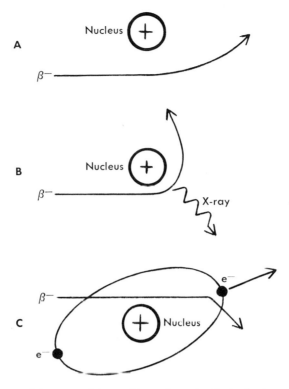

Fig. 6-5. Interactions of beta particles with matter. **A,** Rutherford scattering. **B,** Bremsstrahlung. **C,** Ionization.

apparently because of frequent deflections resulting from multiple scattering. The interactions of beta particles with atoms may be called, in general, *collisions*, although very seldom are these collisions of the head-on type. Usually they result in deflection of the beta particle from its straight-line path, as a consequence of coulombic forces (of attraction or repulsion) between the particle and an atomic nucleus or its orbital electrons. *Collisions* may be of the following two types:

1. *Elastic*, if there is no change in the internal energy or total kinetic energy of the colliding particles

2. *Inelastic*, if it results in a change in the internal energy of one or more of the colliding systems . . . as well as in the total kinetic energy of the systems*

The following three types of *interactions* are of some importance and occur rather frequently (see Fig. 6-5):

1. *Rutherford scattering* is an elastic collision of the beta particle with an atomic nucleus. The particle is simply deflected from its straight-line path by the electrostatic force of attraction (for negatrons) or repulsion (for positrons) but does not lose any of its kinetic energy. This type of interaction is mainly responsible for the phenomenon of backscattering.

2. If the Z of the absorbing material is high and the beta energy is also

---

*From Chase, G. D., and Rabinowitz, J. L. 1962. Principles of radioisotope methodology. Ed. 2. Burgess Publishing Co., Minneapolis. P. 114.

high, another type of collision with an atomic nucleus is quite probable. This is an inelastic collision, because, as a result of the coulombic force, the beta particle is decelerated; the energy lost by the particle is converted to a photon of electromagnetic radiation, similar to an X-ray in wavelength. This radiation is more commonly called *bremsstrahlung* (German, for "braking radiation"). The intensity of bremsstrahlung produced is roughly a direct function of the beta E_{max} and of the Z of the absorbing material. Significant bremsstrahlung can be observed, for instance, when the beta source is ^{32}P and the absorbing material is lead. This type of radiation, like X-rays, is more penetrating than the beta particles themselves. Consequently, an efficient shielding of beta emitters is obtained with the use of low Z materials (such as Lucite), which will minimize the probability of bremsstrahlung production.

3. *Ionization* is the interaction between a beta particle and an orbital electron. The electrostatic force of attraction or repulsion causes deflection of the beta particle, with transfer of some of its energy to the orbital electron, which is completely ejected from its atom. The beta particle is decelerated, and the lost energy appears as kinetic energy of the ejected electron. This is an inelastic collision. The atom that has lost one of its orbital electrons is now an ion.

Range

The range of beta particles is defined as the density-thickness of absorber sufficient to stop all the beta particles emitted by the radionuclide in question. It is a function of E_{max} and is nearly independent of the Z of the absorber. The range-energy relationship is linear at least for energies above 1 MeV. If E_{max} is not known, the range can be determined quite accurately with a number of methods, such as the Feather analysis. It can be roughly estimated with an absorption curve; the range is somewhere near the point of intersection of the absorption curve with the axis of the abscissas. If E_{max} is known, the range can be calculated with Glendenin's formula:

$$R\left(\frac{mg}{cm^2}\right) = 542\, E_{max} - 133 \qquad (6\text{-}3)$$

which, however, is valid only for $E_{max} \geq 0.8$ MeV. In many reference books* a variety of graphs are available, with range-energy curves for a wide spectrum of beta energies. Actually, for ranges of less than 500 mg/cm², the use of these empirical curves is preferable to the use of equations.

Specific ionization

The beta tracks that can be seen in a cloud chamber are much thinner than the alpha tracks, because beta particles have a much lower ionizing power as a result of their higher velocities, smaller mass, and electric charge. Specific ionization values in air (ion pairs/cm) are of the order of magnitude of 50 to 400, compared with 20,000 to 60,000 for alpha particles. For beta particles, it is customary to show graphically the specific ionization as a function of E_{max} (Bragg curve, Fig. 6-6). The specific ionization decreases with the increase of

---

*Cf., for example, Radiological health handbook. 1960. U. S. Department of Commerce, Washington, D. C.

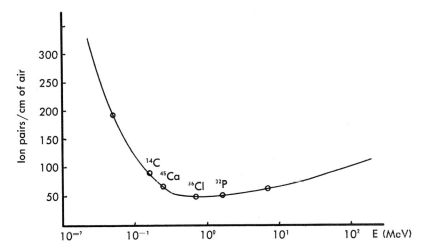

Fig. 6-6. Beta Bragg ionization curve, showing specific ionization as a function of beta particle E_{max}.

E_{max} up to about 1 MeV. Beyond this energy value there is a slow increase of specific ionization.*

Conversion electrons

As noted above, beta particles are of nuclear origin. However, some unstable nuclides emit negatrons of extranuclear origin in the process of radioactive decay. These negatrons will behave exactly like β^- particles, as they travel through matter. As suggested by Rutherford in 1914, there is the possibility that a gamma photon emerging from the nucleus of a beta-gamma emitter may produce a kind of photoelectric effect (see p. 100) with one of the orbital electrons of the same atom. The whole energy of the photon is transferred to the electron, which is ejected from the atom with a $KE = E_\gamma - E_b$ (E_b is the electron binding energy). The gamma ray ceases to exist and is said to be *internally converted*. The ejected electron is called a *conversion electron*, or *Auger electron*. The electrons of any shell can be ejected by the gamma photons, but internal conversion occurs with a higher probability in the K-shell than in the L- and M-shells. In any event, the KE of an L-conversion electron will be higher than that of a K-electron, because its E_b is lower (the difference $E_\gamma - E_b$ is greater), and the KE of an M-conversion electron is even higher. Since, for a given gamma emitter, the KE of its conversion electrons is dependent exclusively on the shell from which they are ejected, conversion electrons, unlike beta particles, are monoenergetic. The fraction of gamma photons producing conversion electrons is called the *internal conversion coefficient*.

Conversion electrons may be "seen," if a beta spectrum of the radionuclide that emits them is taken. They appear as prominent peaks (lines) superimposed on the continuous spectrum produced by the nuclear beta particles. A magnetic spectrometer is preferable for this purpose, because of its high resolution. But a scintillation spec-

---

*The explanation of this interesting phenomenon is outside the scope of this book. A discussion of it can be found in Lapp, R., and H. Andrews. 1954. Nuclear radiation physics. Prentice-Hall, Inc., Englewood Cliffs, N. J. Pp. 170–171.

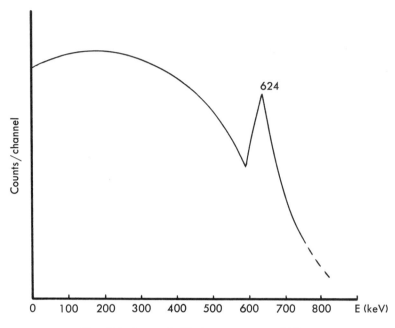

Fig. 6-7. Beta scintillation spectrum of ^{137}Cs.

trometer is equally useful, if the internal conversion coefficient is high. ^{137}Cs is a typical example of a beta-gamma emitter with conversion electrons (Fig. 6-7). On the slope of its beta spectrum there is at least one sharply rising peak corresponding to the emission of K-conversion electrons. The energy corresponding to this peak is 624 keV ($= E_\gamma - E_b = 663$ keV $- 39$ keV).

GAMMA RAYS
Nature and origin

Unlike alpha and beta radiation, gamma rays are electromagnetic radiations. As such, they constitute a segment of the electromagnetic spectrum, as discussed in Chapter 3. Their speed is independent of their wavelength, frequency, and energy, and it is the same as that of all other types of electromagnetic radiations (about 3×10^{10} cm/sec). Their wavelength, frequency, and energy are correlated by the following equations*:

$$c = \lambda\nu \qquad \text{and} \qquad E = h\nu$$

Gamma-energy values for the most commonly used gamma emitters vary roughly between 0.2 and 1.5 MeV. If we consider the high-energy X-ray equipment (Van de Graaff accelerator, cyclotron, etc.) available at the present time, there is no essential difference between X-rays and gamma rays in terms of their photon energies. The essential difference is in their origin. X-rays are of extranuclear origin, that is, either they are generated by electronic transitions or by the mechanism of bremsstrahlung—in either case outside of the nucleus. A gamma photon originates from an unstable nucleus through a radioactive decay process. An unstable nucleus may reach a stable energy configuration

---

*For the meaning of the symbols and further details see Chapter 3.

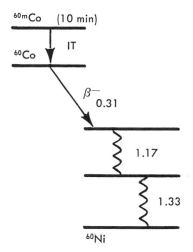

Fig. 6-8. Decay scheme of ^{60}Co. Beta and gamma energies are in MeV.

by simply emitting a particle (alpha or beta), or by capturing an orbital electron. Such is the case for ^{214}Po, ^{14}C, ^{32}P, ^{55}Fe, and others. Other times, the emission of a particle or the capture of an orbital electron is not sufficient to bring the nucleus to the ground energy level; the nucleus relieves itself of the excess energy by radiating it in the form of one or more gamma photons. The case of ^{60}Co is rather typical. The nucleus emits first a 0.31 MeV beta particle, thus becoming ^{60}Ni. This daughter nucleus, however, is left in an excited state. Within a very small fraction of a second it reaches a stable energy level by emitting two gamma photons, one after the other, with energies of 1.17 and 1.33 MeV respectively (Fig. 6-8). For some radionuclides (pure gamma emitters), the only radiation emitted through the decay process is gamma radiation.

Another important difference between gamma radiation and X-rays (at least those artificially generated) of practical value consists of the fact that an X-ray beam is always polychromatic and therefore polyenergetic (Chapter 4), whereas a gamma ray beam is either monoenergetic (^{137}Cs) or consists of few discrete energies (^{60}Co). This advantage makes a gamma irradiator preferable over an X-ray machine in certain types of biologic irradiation experiments.

Absorption

As gamma radiation travels through matter, it dissipates its energy by interacting with the atoms of the absorbing material, according to the mechanisms discussed below. Since gamma rays and X-rays have identical nature and properties, the absorption of gamma rays by matter follows the same laws and patterns described for X-rays in Chapter 4. One should review the pertinent sections. It suffices here to repeat briefly that the absorption obeys Lambert's law, which is expressed by the equation

$$I = I_0 e^{-\mu t} \tag{6-4}$$

wherein I_0 is the intensity of incident radiation, I is the intensity of radiation emerging from absorber, μ is the linear absorption coefficient of absorber, t is the thickness of absorber, and e is the base of the natural logarithms. The concepts of linear absorption coefficient, mass absorption coefficient, and HVL (half value layer) apply also to the absorption of gamma rays.

Interactions with matter

Several types of interactions between gamma photons and the atoms of the absorbing material are known. The degree of probability for each type depends on a number of variables, such as gamma energy or Z of the absorbing material. The interactions may occur between the photon and the orbital electrons or the nucleus, or the electrostatic field of the nucleus, and may be elastic or inelastic collisions.

Some of these interactions are not of sufficient importance in radiation biology to warrant a detailed description and discussion here.* Only the photoelectric effect, Compton scattering, and pair production deserve our attention, because of their large contribution to gamma absorption and because they provide an explanation to many facts of relevance to the radiation biologist and to the user of radionuclide techniques.

Photoelectric effect

Before gamma rays were discovered, the fact was already known that certain types of electromagnetic radiations may eject electrons upon striking suitable materials. For instance, photons of visible light will eject electrons from a film of selenium, and the number of electrons ejected is dependent on the intensity of the incident light. Photographic exposure meters and photomultipliers are based on this phenomenon, which is called the "photoelectric effect." Also ultraviolet and infrared radiations may show a photoelectric effect with appropriate target. Since the ejected electrons are the orbital electrons of the target atoms, the electromagnetic radiation must evidently possess a photon energy at least equal to the electron binding energy to produce a photoelectric effect. Any energy in excess will appear as kinetic energy of the ejected electron.

Fig. 6-9 shows a generalized relationship between the frequency (and therefore the energy) of the incident photon and the maximum KE of the electron, for a given target. The threshold is the minimum frequency required to produce a photoelectric effect in the target considered. For different targets, the threshold can be found in the infrared, visible, ultraviolet, X-ray, or gamma-ray ranges of the electromagnetic spectrum. One can also see that as the frequency increases above the threshold the kinetic energy of the ejected electron also increases.

The photoelectric effect caused by gamma photons is then a particular case of a more general phenomenon. The probability of its occurrence is high for low-energy gamma photons and high Z of absorber. Upon striking the electron, the gamma photon disappears completely (Fig. 6-10, *A*), and its energy is transferred to the electron (usually a K-electron), which is ejected from the atom, with the consequent formation of an ion pair. The ejected electron is usually

---

*Some interactions can be simply mentioned as follows: A very high energy gamma photon (7 to 15 MeV) may be completely absorbed by an atomic nucleus, with the ejection of a neutron (photodisintegration). Or the nucleus absorbs the energy of the photon. An isomer is thus formed. The excited nucleus then decays by gamma emission (Mössbauer effect). Or gamma rays may be reflected by the surface of crystals, like X-rays, according to the Bragg equation (Bragg scattering).

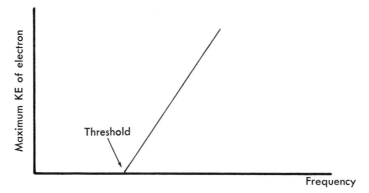

FIG. 6-9. Photoelectric effect. Relationship between the frequency of the impinging photon and the maximum KE of the photoelectron.

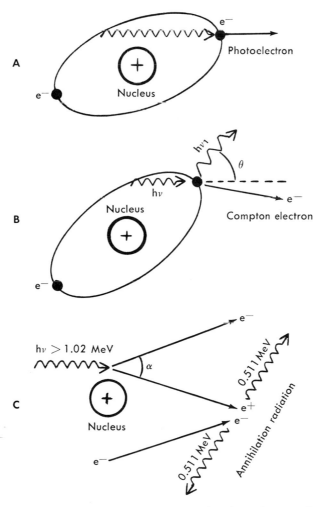

Fig. 6-10. Interaction of gamma rays with matter. **A,** Photoelectric effect. **B,** Compton scattering. **C,** Pair production.

called a photoelectron; its kinetic energy is given by Einstein's photoelectric equation:

$$KE_e = 1/2ms^2 = h\nu - E_b \qquad (6\text{-}5)$$

wherein $h\nu$ is the gamma photon energy and E_b is the electron binding energy. Photoelectrons have enough kinetic energy to become sources of ionizations, such that along their paths they will strip orbital electrons from other atoms of the absorber, thus producing secondary ion pairs.

Compton scattering

Unlike the photoelectric effect, Compton scattering is a type of interaction that does not occur with electromagnetic radiation of lower energy than X-rays and gamma rays. It was discovered by A. H. Compton in 1922. The incident gamma photon collides with an orbital electron, without disappearing (Fig. 6-10, *B*); it is simply deflected (scattered) from its path at an angle θ, and after the collision it has a longer wavelength and therefore lower frequency and energy than the incident photon. The energy lost by the photon is imparted to the electron, which is usually one of the outermost orbits; the electron (*Compton electron*) is ejected from the atom with a certain amount of kinetic energy and thus with the ability of producing secondary ion pairs. Taking into account the laws of conservation of energy and momentum, we can derive the following equation to find the relationship between the energies of the incident and scattered photons*:

$$E_{\gamma'} = \cfrac{E_\gamma}{1 + \cfrac{E_\gamma}{m_0 c^2}\,(1 - \cos\theta)} \qquad (6\text{-}6)$$

wherein $E_{\gamma'}$ is the energy of the scattered photon, E_γ is the energy of the incident photon, m_0 is the rest mass of the electron in grams, c is the speed of light, and θ is the angle of deflection (Fig. 6-10, *B*). The kinetic energy of the Compton electron is given thus:

$$KE_e = E_\gamma - E_{\gamma'} \qquad (6\text{-}7)$$

For all practical purposes no fraction of E_γ is used to overcome the E_b of the electron because the electron ejected is one from the outermost orbits, and therefore its E_b is negligible (a few electron volts) when compared with the energy of the incident gamma photon, unlike what happens in the photoelectric effect.

The angle θ is independent of E_γ; it can have any value from 0 to 180 degrees for the same incident gamma photon. An inspection of equation 6-6 shows, however, that the larger the θ, the more energy is lost by the incident photon, and therefore the greater the energy of the electron. For this reason, a beam of monoenergetic gamma rays (such as those emitted by ^{137}Cs) traveling through an absorber will produce scattered photons with a continuous spectrum of energies. Although this effect has a high probability of occurrence in materials of low Z (as water and Lucite), the Z of the absorbing material does not affect the difference $E_\gamma - E_{\gamma'}$.

---

*The derivation of this equation assumes only some knowledge of algebra and trigonometry. One can find it, for instance, in Sproull, W. T. 1946. X-rays in practice. McGraw-Hill Book Co. P. 86.

In Compton scattering, the gamma photon seems to behave like a particle. The interaction between the incident photon and the electron results in the same type of effects as the collision between two particles. The discovery and explanation of the Compton effect led for the first time to the conclusion that electromagnetic waves present some properties of particles.

Pair production

Pair production is a type of interaction that cannot take place unless the photon energy is at least 1.02 MeV, and its probability increases with the Z of the absorber. Pair production is a rather startling phenomenon because it involves the conversion of energy into matter, as predicted by the general theory of relativity (see Chapter 1). When a high-energy gamma photon passes very close to a nucleus of high Z (Fig. 6-10, *C*), it interacts with the strong nuclear electrostatic field and disappears. Its energy is entirely converted into two material particles, a negatron and a positron (hence pair production). This creation of particles from energy is sometimes called "materialization of energy." For the formation of one e$^-$ and one e$^+$, at least 1.02 MeV of energy is required. In fact, for the creation of one electron, we apply the equation

$$E = mc^2 \qquad \text{or} \qquad E \text{ (MeV)} = 931 \times mass \text{ (amu)*}$$

The mass of the electron is 5.5×10^{-4} amu. Therefore the energy needed for the creation of 1 electron $= 931 \times 5.5 \times 10^{-4}$ MeV $= 0.511$ MeV. For the creation of an e$^-$ and an e$^+$, $0.511 \times 2 = 1.022$ MeV are needed. Anything in excess of this value is converted into kinetic energy of the two particles. Thus the conversion of the total energy of the gamma photon is distributed as follows:

$$E_\gamma = h\nu = 2\ m_0 c^2 + KE_{e-} + KE_{e+} \tag{6-8}$$

Quite frequently, the two electrons possess enough kinetic energy to produce secondary ions along their paths. When the negatron has dissipated all of its energy, it will come to rest and continue to exist either as a free electron or as an orbital electron of an atom. When the positron comes to rest, it combines with a negatron and both together disappear (annihilation of matter). Their matter is converted to energy, and two gamma photons (annihilation radiation) originate from the masses of the two annihilated particles, each having an energy of 0.511 MeV and traveling in opposite directions. Pair production can be easily observed with gamma spectrometry, using high-energy gamma emitters (such as ^{22}Na or ^{65}Zn).†

Two or all three of the interactions described above contribute to the absorption of gamma radiation by matter. The percent contribution of each interaction depends on the gamma photon energy and on the Z of the absorber.

NEUTRON RADIATION

A beam of free neutrons traveling at high speed is what we may call *neutron radiation*. Although this type of radiation has an indirect ionizing power, it attracts the interest of the radiation biologist for several reasons. First, neutron radiation causes, in living systems, biologic effects that are in part similar to, and in

---

*See Chapter 1.

†See the experiments on gamma spectrometry in *Ionizing Radiation and Life: Laboratory Experiences.*

part different from, the effects caused by other types of ionizing radiations. Only with the knowledge of the physical properties of neutrons is it possible to explain the differences. Secondly, some radiation accidents are caused by neutron radiation from nuclear reactors or atomic explosions. When living systems (as well as other materials) are irradiated with a neutron beam, they may be radioactivated (made radioactive) and thus become sources of ionizing radiation. Finally, neutron activation analysis, which has been used extensively by chemists for a number of years, is now being used more and more in the biologic sciences as a powerful and sensitive analytical method for the detection, identification, and quantitative measurement of trace elements present in organisms.

Some properties of the neutron

The neutron is one of the two types of elementary particles present in every atomic nucleus, except the lightest isotope of hydrogen. It is electrically neutral and its mass is slightly greater than the mass of the proton (1.0086 amu). Although the existence of a particle with these properties had been postulated for a long time, the neutron was discovered only later by J. Chadwick in 1932. A neutron within an atomic nucleus is a stable particle. But when it is free (as in neutron radiation), it becomes unstable and therefore undergoes radioactive decay, unless it is captured by another nucleus before it decays. When a neutron decays, it emits a beta-minus particle, thus becoming a proton:

$$n \longrightarrow p + \beta^-$$

Its half-life is about 13 minutes.

Sources

Only a few very short-lived, artificial radionuclides emit neutrons in the process of radioactive decay (such as ^{17}N and ^5He). For obtaining a fairly intense beam of neutron radiation, more convenient sources are commonly used. Such sources are nuclear reactors, which use fission, and particle accelerators and nuclear howitzers, which utilize nuclear reactions other than fission.*

In nuclear reactors, neutron radiation is generated by the nuclear fission of uranium or plutonium (see Chapter 8). The shielding material of the reactor has one or more portholes to allow the passage of the radiation, with appropriate facilities for the irradiation of biologic objects or other materials. Research reactors are operated for the utilization of their neutron radiation.

The devices utilizing nuclear reactions other than fission are particle accelerators and neutron howitzers. In some particle accelerators it is possible to produce deuterons and accelerate them at high speed. If the deuterons are allowed to strike a target containing tritium, a nuclear reaction occurs whereby the deuterons eject neutrons out of the tritium atoms and take their place. Tritium thus becomes helium. The reaction can be represented by the equation

$$^3_1H + ^2_1H \longrightarrow ^4_2He + ^1_0n$$

or in short notation

$$^3H \, (d, n) \, ^4He$$

---

*Those who are not familiar with nuclear reactions are urged to read the appendix on p. 113.

Fig. 6-11. Neutron generator of the accelerator type. (Courtesy Accelerators, Inc., Austin, Texas.)

Several neutron generators of the accelerator type are available on the market for educational or research purposes (Fig. 6-11).

In neutron howitzers the bullets used to dislodge neutrons from the target atoms may be either alpha particles or gamma photons. The target is usually beryllium. The alpha source should have a relatively long half-life, so that it need not be changed frequently, and should not emit high-energy gamma radiation, which would pose a serious shielding problem and would result in a high background count rate in the detectors used to measure the neutron radiation. Two transuranic radionuclides are commonly used as alpha sources: plutonium 239, with a half-life of 24,360 years, and americium 241, with a half-life of 458 years. In practice, the target material (Be) and the alpha source are mixed together as powders; the mixture is the neutron source (Pu-Be source or Am-Be source). The nuclear reaction involved in this method of neutron production is

$$\ce{^{9}_{4}Be} + \ce{^{4}_{2}He} \longrightarrow \ce{^{1}_{0}n} + \ce{^{12}_{6}C}$$

or in short notation

$$^{9}\text{Be}(\alpha, \text{n})^{12}\text{C}$$

A neutron-emitting reaction may be induced in certain nuclei by high-energy gammas. This process can take place only if the energy of the gamma photons is greater than the binding energy of the neutrons in the target nuclei. This threshold energy is 1.664 MeV for beryllium. The interaction of the gamma photon with the target nucleus is an example of photodisintegration (see the footnote on p. 100); the neutrons emitted are referred to as photoneutrons.

The most commonly used gamma emitter for these neutron sources is antimony 124. Its half-life is 60 days and its gamma energy is sufficiently high to overcome the neutron binding energy of beryllium.

The neutrons emitted by photoneutron sources are almost monoenergetic, because most of the gamma sources emit single-energy gamma photons above the threshold and because all the gamma energy is used in the (γ,n) reaction. For an Sb-Be source, the energy of 95% of the neutrons emitted is 25 keV.

Whereas, with (α, n) sources, there is an essential need that the radioactive material be mixed intimately with the target because of the short range of the alpha particles, such is not the case for the photoneutron sources, since the target material simply surrounds the gamma source; this configuration facilitates the manipulation of the whole system and greatly reduces the fabrication costs. In the case of the Sb-Be source, the gamma source must be changed rather frequently due to the relatively short half-life of ^{124}Sb.

A neutron howitzer consists basically of a tank of glass, Lucite, steel, or aluminum, filled with water or paraffin (for the shielding of the neutrons). The neutron source can be accomodated in the center of the tank and is removable. A number of portholes provide access to the source for sample irradiation (Fig. 6-12).

The yield of a neutron source (of any type) is defined as the number of neutrons emitted per second. The flux (ϕ) is the number of neutrons per square

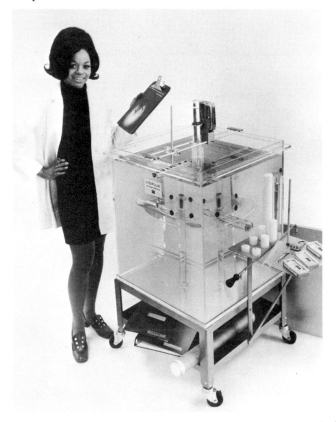

Fig. 6-12. Neutron howitzer. (Courtesy Reactor Experiments, Inc., San Carlos, Calif.)

centimeter per second. Nuclear reactors have high neutron fluxes (about 10^{12} neutrons/cm²/sec) and therefore provide the most intense neutron radiation, although access to their use is not always possible. For neutron generators of the accelerator type fluxes can be as high as 10^8 n/cm²/sec. For an Am-Be neutron howitzer with 1 Ci of Am, the neutron flux is about 10^4 n/cm²/sec, which is a satisfactory flux for several types of experiments done for educational purposes.

Classification

Neutrons are classified according to their energy and velocity, as follows:
1. Slow neutrons – energy below 1 keV
    a. Thermal neutrons (average E = 0.025 eV at 20°C), which have the same average kinetic energy as the atoms of the surrounding matter
    b. Resonance neutrons (E = 1 to 100 eV)
2. Fast neutrons – energy above 1 keV
    a. Intermediate neutrons (E = 1 to 500 keV)
    b. Relativistic neutrons (E above 20 MeV)

Neutron howitzers and neutron sources of the accelerator type emit neutrons covering energy ranges up to 10 MeV. Fission neutrons are fast neutrons with energies up to 15 MeV. Fast neutrons can be slowed down (moderated) to the thermal range with the use of suitable moderators such as paraffin or water. The behavior of neutrons and their interactions with matter are very much dependent on their energy.

Absorption and interaction with matter

Since neutrons have no electric charges, they can travel long distances through matter without interacting with orbital electrons or the coulombic field of nuclei. As a consequence, their ranges are much longer than those of other particulate radiations, like alpha and beta. The only way a neutron can dissipate its energy is by contact collisions with nuclei (very much like collisions between billiard balls). These collisions are more probable in materials of low Z, because of their higher nuclear density (more nuclei are present per unit of volume than in materials of high Z). For this reason materials made of elements of low Z (water, paraffin, graphite, etc.) are more effective in slowing down neutrons; in fact they are ideal neutron moderators and are widely used in nuclear reactors and neutron howitzers. The following three types of neutron-nucleus interactions are briefly described:

1. *Elastic scattering* occurs when the neutron collides with a nucleus, without leaving it with excess energy or in an excited state. The neutron loses some of its energy and is slowed down. The lost energy is transferred to the nucleus as recoil energy. On the average, a neutron loses 60% of its energy when it collides elastically with a hydrogen nucleus. After repeated collisions of this type, the neutron can be completely stopped in the absorber. Moderation of neutrons or their shielding with materials of low Z is achieved mainly by this type of interaction.

2. *Inelastic scattering* occurs when the neutron collides with a nucleus, leaving it in an excited state, from which the nucleus recovers usually by emission of a gamma photon. Again the neutron loses part of its energy as a result of the collision.

3. *Capture* occurs when the incident neutron is captured by the target nucleus and a compound nucleus is thus formed. The compound nucleus is usually in an excited state and/or in an unstable configuration (in terms of nucleon composition). Immediately after the capture, the compound nucleus emits a gamma photon, or two neutrons, or a proton, or an alpha particle, depending on the energy of the incident neutron. What is left after this "prompt" transition is usually a radioactive nucleus of a species different from the original target nucleus. It will decay with its characteristic half-life.

Neutron capture results in radioactivation of the target material, which, as mentioned above, is an important difference between irradiation of materials with neutrons and irradiation with other types of ionizing radiation.

Examples of neutron capture are the following:

$$^{127}\text{I (n, }\gamma\text{) }^{128}\text{I} \quad \xrightarrow[\searrow\beta^-]{T = 25 \text{ m}} \quad ^{128}\text{Xe} \qquad (T = \text{half-life})$$

$$^{59}\text{Co (n, }\gamma\text{) }^{60}\text{Co} \quad \xrightarrow[\searrow\beta^-,\,\gamma]{T = 5 \text{ y}} \quad ^{60}\text{Ni}$$

$$^{14}\text{N (n, p) }^{14}\text{C} \quad \xrightarrow[\searrow\beta^-]{T = 5730 \text{ y}} \quad ^{14}\text{N}$$

$$^{6}\text{Li (n, }\alpha\text{) }^{3}\text{H} \quad \xrightarrow[\searrow\beta^-]{T = 12 \text{ y}} \quad ^{3}\text{He}$$

The probability that a given neutron will be captured is a function of its energy, as well as of the nature of the target material, and this function is called the *absorption cross section* (σ). Tables have been compiled with values of absorption cross sections for different neutron energies and different target materials. Graphs are also available for each chemical element, for which the neutron-absorption cross section is plotted against neutron energy. Many artificial radionuclides are prepared by the neutron-capture reaction. A stable element is irradiated with a neutron beam (usually in a nuclear reactor) with the formation of the desired radionuclide, whose yield will be high if the cross section of the target element for the neutron energy used is high (see the examples above).

As already mentioned above, a neutron traveling through matter does not interact with atomic orbital electrons and therefore is unable to produce ion pairs directly. However, in elastic and inelastic collisions of a neutron with a nucleus, the nucleus receives enough of the neutron's kinetic energy to recoil (like collisions of billiard balls). The kinetic energy transferred by the neutron to the nucleus is considerable, if the nucleus is of small mass like that of hydrogen. As the nucleus recoils, it may ionize other atoms by interacting with their orbital electrons with its positive charge. The ionizing power of neutron radiation is therefore indirect. Since in organisms light nuclei (like hydrogen or carbon) are quite abundant, it follows that neutron radiation traveling through living matter will exhibit a high specific ionization.

Neutron-activation analysis

Neutron-activation analysis is a method of elemental analysis, both qualitative and quantitative, based on neutron capture and radioactivation. It became a powerful analytical tool when very intense sources of neutron radiation, like nuclear reactors and the others listed above, became easily available. It finds its most useful application in the identification and quantitative measurement of trace amounts of chemical elements. Its sensitivity is by far superior to that of other analytical methods, such as the gravimetric, colorimetric, and spectrophotometric methods. If the neutron flux of a reactor is used, neutron-activation analysis can detect and measure an impurity present with a concentration as low as $10^{-8}\%$.

If the purpose is simply the identification of the trace element, the technique is briefly as follows: First, the best and most appropriate nuclear reaction is selected. Thermal neutrons usually will produce reactions of the (n, γ) type, whereas fast neutrons are more likely to produce (n,p), (n,d), or (n,α) reactions. Most activation analysis today is done with thermal neutrons. Also the absorption cross sections of the main elements present in the sample should be considered. For example, if the trace to identify is mixed with an organic material, and thermal neutrons are used, carbon and hydrogen (abundant in organic molecules) have a small cross section for thermal neutrons. If the trace element has a much larger cross section for the same type of neutrons, the trace will be radioactivated more readily than the carbon and hydrogen of the surrounding organic material. The sample is prepared in a suitable manner and then enclosed usually in a Lucite capsule. Depending on the sensitivity required, the sample is irradiated either in a nuclear reactor, or with an accelerator-produced neutron beam, or in a neutron howitzer. The irradiation time is a function mainly of the neutron flux, cross sections involved, and the half-lives of the radioisotopes produced. Concerning the last factor, it is obviously useless to irradiate the sample for a long period of time, if the radionuclide resulting from the activation of the element we wish to identify has a very short half-life; in this case, after a short irradiation period, the number of radioactive atoms produced is equal to the number of atoms that decay during the same time (thus saturation has been reached).

After irradiation the trace element is identified by finding the half-life, type, and energy of the radiation produced by its radioisotope. Among all known radionuclides, no two of them have the same half-life and radiation energy.

An example will serve as an illustration of the general technique. Suppose we have an aluminum foil containing traces of an unknown element. Aluminum is made of only one stable isotope, ^{27}Al. If thermal neutrons are used for irradiation, the (n, γ) reaction will convert aluminum to ^{28}Al (half-life = 2.3 minutes; gamma energy = 1.78 MeV) and the unknown impurity to a radioisotope that must be identified. The absorption cross section of aluminum is very small; let us assume that the cross section of the impurity is significantly larger. As a consequence, a much larger percentage of the trace element than of aluminum will be radioactivated. After irradiation the activity of the sample is assayed at frequent intervals (of a few minutes) and a decay curve is plotted on semilog paper. The curve will not be a straight line, because it is a composite of two decay rates. By careful extrapolation, one can isolate the two decay curves and read off the half-lives of the two components. One will be the half-life of radioaluminum, and the other, the half-life of the radioactivated impurity. Let us assume that this latter is about 64 hours. While the activity of the sample is being assayed for the construction of the decay curve, the gamma spectrum of the sample is taken with a scintillation detector and a multi-channel pulse-height analyzer (see pp. 153 to 163). Since every gamma emitter has its own characteristic gamma spectrum, the spectrum of the sample will actually be a composite of the ^{28}Al spectrum and of the spectrum of the trace element. Suppose that the spectrum shows a prominent peak (photopeak) in correspondence with a photon energy of about 0.411 MeV. With the help of suitable tables that list radionuclides in order of increasing half-life and gamma photon energy, it is possible to identify the gamma emitter of our example (half-life = 64 hours; gamma energy = 0.411 MeV) as ^{198}Au. Therefore the trace element present in the aluminum foil was gold. If the gamma energy of the unknown cannot be identified because its spectrum is obscured by the gamma spectrum of ^{28}Al, one is advised to let the latter decay and then take the spectrum of the sample again. After some time, the spectrum will be practically a pure gamma spectrum of ^{198}Au, and the identification of its gamma energy will be much easier.

Once the impurity has been identified, its quantitative determination can be done directly (by means of the so-called activation equation), if neutron flux, absorption cross section and other more easily measurable variables are exactly known (absolute method). But because the first two variables are frequently uncertain, the comparative method must be used sometimes. An aluminum foil control containing a known amount of gold is irradiated with another sample, and both are processed exactly in the same way. The activity of radiogold in the sample is compared with the activity of radiogold in the control to obtain the weight ratio, from which the weight of gold in the sample is calculated.

Perhaps one of the most spectacular applications of neutron-activation analysis was the discovery of arsenic in Napoleon's hair in 1961. Neutron irradiation of his hair revealed amounts of arsenic that could be accounted for only if he had been poisoned. Other cases of poisoning have been detected through activation analysis of blood, urine, fingernails, etc.

The same analytical method is being used more and more for the measurement of minor and trace elements in biologic specimens and also for the clarification of the role played by the trace elements in normal and abnormal metabolism. More recently, a few groups of investigators are using neutron activation analysis with living human subjects for the determination of the total amount of iodine in the thyroid, of the total-body calcium content, etc. Regarding the radiation hazard involved in this kind of studies, the results seem to indicate that the total absorbed neutron dose is no greater than that from clinically acceptable diagnostic procedures using radioisotopes.

COSMIC RADIATION
Nature and composition

Cosmic radiation is a composite radiation, made of several types of particles and photons that strike the surface of the earth from outer space. It was discovered at the turn of this century, although its true nature and possible origin

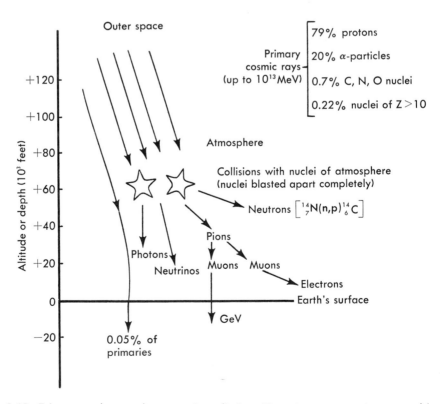

Fig. 6-13. Primary and secondary cosmic radiation. Pions (or π-mesons) are unstable particles decaying with an extremely short half-life. They may be either electrically neutral or charged. The muons (particles lighter than pions) are the decay products of pions. They are also unstable; when they decay, they give rise to electrons and neutrinos.

were understood only in the late 1920s. What surprised the early investigators and continues to intrigue and fascinate nuclear physicists today is the enormous energy linked with some components of cosmic rays (up to 10^{18} eV!). Even with our modern particle accelerators, we are still unable to attain energies of such magnitude; consequently, research in high energy physics still depends on the knowledge of cosmic radiation.

Today it has been definitely established that cosmic radiation is made of two components — primary and secondary rays (Fig. 6-13). The primary cosmic rays are high-energy particles coming from some source outside the earth's atmosphere. They consist mainly of protons and atomic nuclei stripped of their orbital electrons. It is estimated that 79% of the primary particles are protons (H nuclei), 20% alpha particles, 0.7% nuclei of C, N, O, and 0.22% nuclei of $Z > 10$.

As the primaries enter the earth's atmosphere, they interact with molecules of air. At very high altitudes (100,000 feet and higher) these interactions are improbable because of the very low densities of the upper atmospheric layers; the primaries do not lose any significant amounts of energy. But at lower altitudes (around 60,000 feet), with the increase in the density of air, a large percentage of primary particles collide with the atomic nuclei. Because of the very high energy possessed by the particles, the target nuclei are literally blasted apart, with the liberation of particles, such as neutrons, electrons, neutrinos, mesons, and gamma photons. The secondary component of the cosmic radiation is made of all these elements. The secondaries may have enough energy to collide with other atoms of the atmosphere; cascades of secondaries are thus formed.

One may conveniently distinguish in the secondary cosmic radiation a soft component (photons, electrons) from a penetrating component (mesons) with energies up to several GeV (giga electron volts). These secondaries and a few primaries that escape collisons with atmospheric atoms can, in fact, be detected at the depth of several hundred feet within the earth's crust.

The intensity of the cosmic radiation near or on the surface of the earth is a function of several factors, which have been intensively investigated for the past 40 to 50 years. The most significant factors are as follows:

1. *Altitude effect.* The intensity increases with altitude up to 12 miles, beyond which it decreases up to 25 miles. Above 25 miles only primaries exist and their intensity remains constant up to the highest altitudes that have been reached by the detecting instruments (carried by rockets and satellites).

2. *Latitude effect.* The intensity of cosmic radiation at sea level increases with latitude, with a minimum at the magnetic equator and a maximum around each of the magnetic poles. This effect is attributed to the influence of the earth's magnetic field on the charged particles present in the cosmic radiation. At latitudes above 60°, the radiation level is about 15% higher than at the equator.

3. *Barometric effect.* For the same location on the earth, it has been observed that the intensity of radiation decreases with an increase in the barometric pressure. When this pressure is high, the air is also more dense and therefore absorbs more radiation.

Physicists have calculated that the total amount of cosmic radiation energy arrives at the surface of the earth at the rate of about 3 to 4 million kw. Although this amount may seem considerable, it is approximately equal to the amount of energy arriving to the earth in the form of starlight.

Origin

The problem of the origin of the cosmic radiation is still far from a complete solution. Investigators formerly thought that the sun was the sole, or at least the main source. But they observed that for the same location there was no significant variation in radiation intensity between daytime and nighttime; obviously this finding is not in agreement with the hypothesis of a solar origin. Attempts to find variations of intensity with solar flares and with the cyclic activity of the sunspots also failed. Nevertheless, we believe at the present time that possibly no more than 0.1% of the radiation comes from the sun. Certainly, the bulk of the cosmic radiation must come from somewhere else in space. The hypothesis of the supernovae (unstable stars that undergo a gigantic explosion, with production of heavy nuclei) within our galaxy as sources has enjoyed the favor of many for several years. More recently, some investigators have suggested that sources might be found in certain strong radio sources (quasars) located in extragalactic space. It is also possible that cosmic radiation originates from many different sources in the universe.

Whatever the sources, what is still completely obscure is the mechanism of production. Possibly, somewhere in space, there must be some natural systems resembling our man-made cyclotrons and synchrotrons, capable of accelerating protons and other atomic nuclei to the fantastic energy of the primary cosmic rays. How these systems work and how the particles themselves are generated remains to be seen through future research.

How is the biosphere affected by the cosmic radiation? What effect does cosmic radiation have on the lives of organisms and of man? Undoubtedly, this problem is of enormous interest to the radiation biologist and is dealt with in later chapters. In the meantime, it is useful at this point to mention that we owe the existence of ^{14}C in the atmosphere, and therefore in the organisms, to cosmic radiation. It is believed that this isotope of carbon is formed in the upper atmosphere by the neutrons present in the secondary cosmic rays, through an (n, p) reaction. In laboratory experiments it has been found that nitrogen has a large cross section for thermal neutrons. When the neutrons of the secondary radiation have been slowed down to thermal energies through several interactions, the following reaction with nitrogen nuclei of the atmosphere becomes highly probable:

$$^{14}_{7}N + ^{1}_{0}n \longrightarrow ^{14}_{6}C + ^{1}_{1}p \quad \text{or} \quad ^{14}N(n, p)^{14}C$$

This reaction is the foundation of the whole theory behind the radiocarbon dating method.* Furthermore, the same neutrons of the secondary radiation are held responsible for initiating the spontaneous fission of uranium and the chain reaction, as soon as the critical mass is exceeded (see Chapter 8).

SUMMARY OF PROPERTIES AND INTERACTIONS

The purpose of this chapter is the review of the fundamental properties of some types of ionizing radiations and their interactions with matter, which are of particular interest in radiation biology. The student should familiarize himself with this material if he wants to grasp quickly the information that is presented in later chapters. Tables 6-1 and 6-2 are summaries intended to facilitate this task. Certain interactions, which have not been discussed in this chapter,

---

*For further details see Libby, W. 1955. Radiocarbon dating. University of Chicago Press, Chicago.

Table 6-1. Summary of some properties of ionizing radiations*

Radiation	Charge	Approximate energy range	Approximate range		Primary source
			In air	In water	
Particulate					
Alpha	+2	3 to 9 MeV	2 to 8 cm	20 to 40 μm	Some nuclei of high Z
Beta	±1	0 to 3 MeV		Up to a few mm	Nuclei with n/p ratio higher or lower than a certain average value
Neutrons	0	0 to 10 MeV	0 to 100 m	0 to 1 m	Nuclear reactions
Electro-magnetic					
X-rays	None	A few eV to several MeV	A few mm to 10 m	Up to a few cm	Interactions of accelerated electrons with nuclear field Orbital electron transitions
Gamma rays	None	10 keV to 10 MeV	A few cm to 100 m	From a few mm to several cm	Nuclear transitions

*Modified from Chase, G. D., and J. L. Rabinowitz. 1962. Principles of radioisotope methodology, Ed. 2. Burgess Publishing Co., Minneapolis.

are included in Table 6-2 only for the sake of completeness; they are of no importance in radiation biology and may be disregarded.

APPENDIX: Nuclear reactions

In 1919, Rutherford discovered that when nitrogen is bombarded with alpha particles, the alpha particles disappear as such and other particles of longer range originate from nitrogen. Further investigation indicated that the new particles were simply free protons and that nitrogen had been transformed in an isotopic form of oxygen (^{17}O). For the first time, the old and long-awaited dream of the medieval alchemists had been realised—the artificial transmutation of one chemical element into another. Although nitrogen had not been transformed into gold and although only a few atoms were involved in that first transmutation, Rutherford's discovery showed that transmutation of elements was possible.

The explanation of the phenomenon is that the nitrogen nucleus absorbs the alpha particle so that a new compound nucleus is formed. Since the compound nucleus is very unstable, its life is ephemeral (only 10^{-5} sec), and immediately it emits a proton, thus becoming the nucleus of an oxygen atom.

This series of events can be classified as a *nuclear reaction*, which in many respects is analogous to a chemical reaction. In a chemical reaction, molecules are the reagents; atoms are rearranged and certain chemical bonds are broken, with the formation of new bonds and new molecules, while the atoms involved remain unchanged in their nature and number. A chemical reaction is represented with a chemical equation, which must be balanced; that is, it must show the same kinds and numbers of atoms in both members, in compliance with the law of conservation of matter. In a nuclear reaction, atomic nuclei and nuclear particles are the "reagents"; nucleons are rearranged, certain nuclear bonds are broken, with the formation of new bonds and new atomic species, while the

Table 6-2. Summary of interactions of ionizing radiations with matter*

Incident radiation	In collision with	Type of collision		
		Elastic	Inelastic	Complete absorption
Alpha	Nucleus	**Rutherford scattering**	Bremsstrahlung (negligible)	Transmutation
	Orbital electron	(Negligible)	**Ionization and excitation**	None
Electrons (including beta-plus and beta-minus particles)	Nucleus	**Rutherford scattering**	**Bremsstrahlung production**	Electron capture
	Orbital electron	Causes some scattering	**Ionization and excitation** (production of characteristic X-rays)	**Annihilation** (for positrons)
Neutrons	Nucleus	**Moderation of neutrons**	**Resonance scattering**	**Radioactivation and other nuclear reactions**
	Orbital electron	(Negligible)	(Negligible)	None
Photons (X or gamma)	Nucleus	Thomson scattering (negligible)	Mössbauer effect	Photodisintegration
	Orbital electron	Rayleigh scattering	**Compton effect**	**Photoelectric effect and internal conversion**
	Field	Delbruck scattering	(Negligible)	**Pair production**

*Modified from Chase, G. D.; and J. L. Rabinowitz. 1962. Principles of radioisotope methodology. Ed. 2. Burgess Publishing Co., Minneapolis.

nucleons involved remain unchanged in their nature and number. Like chemical reactions, nuclear reactions may be exoergic or endoergic; one significant difference is that in chemical reactions energies of only a few eV are involved, whereas nuclear reactions take place with the absorption or release of a few or several MeV of energy. A nuclear reaction is represented with a *nuclear equation*, which shows the "reagents" in one member and the "products" in the other member. All terms carry a notation of their A and Z numbers. The equation must be balanced; the sum total of the A numbers in the first member must be equal to the sum total of the same numbers in the second member. The same rule applies to the Z numbers.

The equation for Rutherford's transmutation can be written as follows:

$$^{14}_{7}N + ^{4}_{2}He \longrightarrow ^{18}_{9} \text{ (Compound nucleus)} \longrightarrow ^{17}_{8}O + ^{1}_{1}H$$

In nuclear reactions of this type, the nucleus that is bombarded is often referred to as the "target," and the particle used to bombard the target as the "bullet."

In Rutherford's transmutation, the new element produced (^{17}O) was a stable (that is, nonradioactive) atomic species. In 1934, another transmutation was obtained by

F. Joliot and I. Joliot-Curie (the daughter of Pierre and Marie Curie), but in this case the new element produced was a radioactive atomic species. In this reaction the target was aluminum and the bullets were high-energy alpha particles; the product was a radioisotope of phosphorus and the particle emitted was a neutron instead of a proton. The equation for this reaction can be written as follows (the formation of a compound nucleus is not shown):

$$^{27}_{13}Al + ^{4}_{2}He \longrightarrow ^{30}_{15}P + ^{1}_{0}n$$

^{30}P is a β^+ emitter, with a half-life of about 2 minutes.

During the following years, and especially after the advent of the atomic age (1945), several other types of nuclear reactions were obtained that led to the creation of a large number of artificial nuclides, stable and radioactive. These reactions became practically feasible, when a variety of bullets (neutrons, protons, deuterons, etc.) of a wide range of energies could be easily obtained with nuclear reactors and particle accelerators. Probably the greatest scientific importance of these reactions lies in the fact that they have made it possible to produce most of the artificial nuclides available today.

A short-hand system for the representation of nuclear reactions of the type described above is widely used. With this system, the symbol of the target is written first, with its A number only, since the Z number is really redundant. Parentheses follow, in which the bullet and the particle emitted are indicated and separated by a comma. The parentheses are followed by the symbol of the nuclear species produced, with the indication of its A number. Rutherford's transmutation would be written in short-hand notation as

$$^{14}N \ (\alpha, p) \ ^{17}O$$

which would be interpreted as follows: a nitrogen nucleus absorbs an alpha particle; a proton is ejected and a nucleus of the isotope oxygen 17 is thus formed.

The transmutation obtained by Joliot and Joliot-Curie is represented by the short-hand notation

$$^{27}Al \ (\alpha, n) \ ^{30}P$$

A selection of other significant nuclear reactions is listed below, with nuclear equations and shorthand notations. The examples chosen are reactions used for the production of radionuclides widely employed in tracer methodology.

Target bombarded with neutrons

$^{6}_{3}Li + ^{1}_{0}n \longrightarrow ^{3}_{1}H + ^{4}_{2}He$	$^{6}Li \ (n, \alpha) \ ^{3}H$
$^{41}_{19}K + ^{1}_{0}n \longrightarrow ^{42}_{19}K + \gamma$	$^{41}K \ (n, \gamma) \ ^{42}K$
$^{23}_{11}Na + ^{1}_{0}n \longrightarrow ^{24}_{11}Na + \gamma$	$^{23}Na \ (n, \gamma) \ ^{24}Na$
$^{197}_{79}Au + ^{1}_{0}n \longrightarrow ^{198}_{79}Au + \gamma$	$^{197}Au \ (n, \gamma) \ ^{198}Au$
$^{32}_{16}S + ^{1}_{0}n \longrightarrow ^{32}_{15}P + ^{1}_{1}H$	$^{32}S \ (n, p) \ ^{32}P$
$^{58}_{26}Fe + ^{1}_{0}n \longrightarrow ^{59}_{26}Fe + \gamma$	$^{58}Fe \ (n, \gamma) \ ^{59}Fe$
$^{35}_{17}Cl + ^{1}_{0}n \longrightarrow ^{35}_{16}S + ^{1}_{1}H$	$^{35}Cl \ (n, p) \ ^{35}S$
$^{44}_{20}Ca + ^{1}_{0}n \longrightarrow ^{45}_{20}Ca + \gamma$	$^{44}Ca \ (n, \gamma) \ ^{45}Ca$
$^{64}_{30}Zn + ^{1}_{0}n \longrightarrow ^{65}_{30}Zn + \gamma$	$^{64}Zn \ (n, \gamma) \ ^{65}Zn$
$^{59}_{27}Co + ^{1}_{0}n \longrightarrow ^{60}_{27}Co + \gamma$	$^{59}Co \ (n, \gamma) \ ^{60}Co$
$^{14}_{7}N + ^{1}_{0}n \longrightarrow ^{14}_{6}C + ^{1}_{1}H$	$^{14}N \ (n, p) \ ^{14}C$
$^{40}_{20}Ca + ^{1}_{0}n \longrightarrow ^{37}_{18}Ar + ^{4}_{2}He$	$^{40}Ca \ (n, \alpha) \ ^{37}Ar$
$^{12}_{6}C + ^{1}_{0}n \longrightarrow ^{11}_{6}C + 2 \ ^{1}_{0}n$	$^{12}C \ (n, 2n) \ ^{11}C$

Target bombarded with protons

$^{7}_{3}Li + ^{1}_{1}H \longrightarrow ^{8}_{4}Be + \gamma$	$^{7}Li \ (p, \gamma) \ ^{8}Be$

Target bombarded with deuterons

$$^{24}_{12}\text{Mg} + {}^{2}_{1}\text{H} \longrightarrow {}^{22}_{11}\text{Na} + {}^{4}_{2}\text{He} \qquad {}^{24}\text{Mg} \,(d, \alpha)\, {}^{22}\text{Na}$$

$$^{10}_{5}\text{B} + {}^{2}_{1}\text{H} \longrightarrow {}^{11}_{6}\text{C} + {}^{1}_{0}\text{n} \qquad {}^{10}\text{B} \,(d, n)\, {}^{11}\text{C}$$

REFERENCES

Burbridge, G. Aug. 1966. The origin of cosmic rays. Sci. Amer. **215:**32.

Cachon, A., A. Daudin, and L. Jauneau. 1965. Cosmic rays. Walker & Co., New York.

Corliss, W. 1964. Neutron activation analysis. U. S. Atomic Energy Commission, Division of Technical Information, Washington, D. C.

Ginzburg, V. L. Feb. 1969. The astrophysics of cosmic rays. Sci. Amer. **220:**51.

Hughes, D. 1959. The neutron story. Doubleday & Co., Inc., Garden City, N. Y.

International Atomic Energy Agency. 1967. Nuclear activation techniques in the life sciences. Vienna.

Lapp, R., and H. Andrews. 1954. Nuclear radiation physics. Prentice-Hall, Inc., Englewood Cliffs, N. J. Chapters 5, 6, 7, and 12.

Wahl, W. H. and H. H. Kramer. April 1967. Neutron-activation analysis. Sci. Amer. **216:**68.

CHAPTER 7

Radiation detection and measurement

GENERAL CONSIDERATIONS

Detection and measurement techniques

Any study or investigation involving ionizing radiations would obviously be impossible without some device capable of indicating their presence. Any instrument that can detect the presence of ionizing radiation is called a *radiation detector*. Within the detector, an interaction occurs between the radiation and the matter filling the detector. Sometimes this interaction produces a visible effect (as in cloud chambers); other times it can be detected by means of some auxiliary devices.

In most cases, the mere detection of ionizing radiation is not sufficient. What we also need is its measurement, that is, some sort of quantitative appraisal of its intensity, or of some other parameter somehow dependent on its intensity. For this reason, most detectors are coupled with one or the other of a variety of measuring apparatuses; a detector associated with a measuring apparatus constitutes a *detection system*.

Radiation detection systems can measure ionizing radiation in one of two ways, as follows:

1. Some systems are designed to measure the intensity of radiation by giving an indication of the activity (such as number of disintegrations per unit of time) of a radioactive source, or an indication of the number of particles or photons emitted per unit of time by an artificial source (such as an X-ray unit). The operation of these systems may be of pulse type or nonpulse type.

    a. In the pulse type of operation (also called "nonintegrating" operation) the interactions occurring within the detector are detected separately as single and distinct events because the system is so designed that the output of the detector is a series of discrete pulses. If these pulses are counted by means of mechanical and/or electronic devices, the system is referred to as a *radiation counter*.

    b. In the nonpulse type of operation (also called "integrating" operation) the system does not resolve the distinct interaction events occurring in the detector, but rather measures directly the average effect due to many of those interactions. For instance, if the effect of the interactions is the discharge of an electroscope, the system measures the rate of this discharge, which of course must be proportional to the number of interactions per unit of time and therefore to the intensity of the radiation.

2. Other systems are designed and calibrated to measure, not the activity of a radiation source, but the energy delivered to or absorbed by an object exposed to radiation. This energy is called the *radiation dose* and the systems that give a measurement of the dose are called *dosimeters*. The dosimetric systems are described in Chapter 9.

Most detecting systems and radiation counters can not only detect and measure the intensity of radiation, but also give important information about its energy. Detectors are basically characterized by the type of interaction that takes place within them and that is utilized for the detection of radiation. Three main types of interactions are used for radiation detection, and accordingly there are three major groups of radiation-detecting systems, which are the following:

1. Radiation produces ion pairs, either directly or indirectly, in the sensitive volume of the detector. The ions may produce some visible effect, like condensation of supersaturated vapor, or they may be collected with an electric potential gradient, thus producing a detectable electric current. Electroscopes and electrometers, proportional counters, Geiger-Müller (G-M) counters, semiconductor detectors, and cloud and bubble chambers belong to this group of detectors.

2. Radiation causes excitation of atoms or molecules of certain materials (fluors). When the excited atoms or molecules return to the ground state, they fluoresce, that is, they emit flashes of visible light (scintillations) that can be viewed and counted directly, either with the help of a microscope or by means of special electronic devices, after being transformed into measurable electric pulses. The spinthariscope and solid and liquid scintillation systems belong to this group.

3. Radiation induces specific chemical reactions in photographic emulsions similar to those induced by visible light. When the emulsion is developed and fixed, the areas affected by ionizing radiations will appear black. Radiography and radioautography are detecting techniques based on this type of interaction.

Sample activity

In the preceding section, I mentioned that an indication of the intensity of radiation emitted by a radioactive source could be given by the measurement of its activity. If an instrument could measure the number of decay events occurring within a sample per unit of time, it would measure what is commonly called *absolute activity*. Unfortunately, no detecting system that can directly measure this type of activity exists. At most, a few sophisticated instruments in certain conditions can count all the particles or photons emitted by a radioactive sample. This number may or may not be identical to the number of radioactive decay events, since it depends on the decay scheme. For instance, if the radionuclide involved is one that decays by beta emission or electron capture and the counter being used is only sensitive to beta particles, the number of particles counted is less than the number of disintegrations. However, absolute counting is often a tedious and time-consuming process, essential only in nuclear physics, in the preparation of radioactive standards, and in radiochemistry. In most of the biologic work done with radionuclides, measurement of the *relative activity* of a sample, which is only a fraction of the true disintegration rate, is sufficient. When a series of samples is to be measured, the relative activities can be kept proportional to the absolute activities, if all the samples are measured in the same identical conditions with the same instrument.

For several reasons, analyzed below, most instruments can measure only a fraction of all the particles and/or photons emitted by a sample. The relative activity measured by a detecting system with the pulse type of operation is given usually in counts per minute (cpm). The cpm is therefore only a fraction (Y) of the number of disintegrations per minute (dpm), so that

$$cpm = dpm \times Y \tag{7-1}$$

Therefore

$$dpm = \frac{cpm}{Y} \tag{7-2}$$

and

$$Y = \frac{cpm}{dpm} \tag{7-3}$$

The fraction Y is called the *yield* of a counter. It can be easily calculated if dpm and cpm are known. If the dpm value is unknown, the measurement of Y is

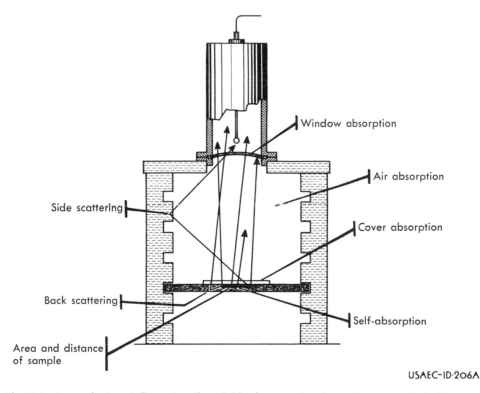

USAEC-ID-206A

Fig. 7-1. Some factors influencing the yield of a counter (counting geometry). (Courtesy AEC.)

more difficult, though possible. The yield of a counter depends on several factors (Fig. 7-1) of which the most significant are listed below:

1. *Distance of the sample from the detector.* Since ionizing radiations are emitted by a sample in all directions, a detector external to the sample will receive only some of the radiation. Furthermore, the longer the distance of the detector from the sample, the smaller is the fraction of the radiation emitted that will be received by the detector.

2. *Counter efficiency.* Not all ionizing radiations entering a detector will actually interact in it. Noninteracting particles or photons are lost to the count rate. The counted fraction of the radiation entering the detector is the *efficiency* of the counter. Several factors are responsible for the efficiency, for instance:

    a. Detector size. The larger the useful volume of the detector, the larger the number of particles or photons that are likely to interact within it.

    b. Radiation type. Because different types of ionizing radiations have different properties, different types of detectors may be more or less sensitive to different radiations. For instance, a Geiger-Müller detector is more sensitive to, and thus more efficient for, beta particles than to gamma radiation.

The efficiency of a counter for a given type of radiation is expressed in terms of percentage. A 60% efficiency for gamma radiations means therefore that a detector will respond to 60% of the gamma photons that enter its sensitive volume. Consequently the efficiency of a counter should not be confused with its yield. Efficiency is only one of the several factors that affect the yield of a counter.

3. *Backscattering and sidescattering.* If a beta-emitting sample is backed by some type of material of high atomic number (such as lead), a significant fraction of the beta particles emitted in a direction opposite to the detector may be deflected 180° and enter the detector, thus increasing its yield. Similar deflections may occur if the particles strike an object located on the side of the sample.*

4. *Self-absorption.* Weak beta particles may be absorbed by the sample itself and never reach the detector if the sample is not very thin. Whenever low-energy beta emitters (such as ^{14}C) must be counted, the samples must be prepared in such a way as to minimize self-absorption and thus increase the yield of the counter.

5. *Decay mode.* The number of particles or photons emitted per minute by the sample may not be identical to the number of disintegrations per minute. For instance, a gamma counter has a better yield for ^{125}I than for ^{137}Cs, everything else remaining equal, because every disintegration of ^{125}I results in the emission of a gamma photon, whereas only about 93 out of every 100 decaying atoms of ^{137}Cs emit a gamma photon (see the decay schemes in Fig. 5-4).

Factors 1, 3, 4, and a few others not considered here are collectively referred to as *counting geometry.*

In summary, the following factors affect the yield of a counter:

1. Sample-detector distance
2. Counter efficiency
3. Backscattering and sidescattering
4. Self-absorption
5. Decay mode

---

*See the experiments on backscattering in *Ionizing Radiation and Life: Laboratory Experiences.*

Even if the random nature of the radioactive disintegration process is ignored, the count rate registered by a radiation counter is seldom a true representation of the actual average rate due to the activity of the sample. The difference between the registered count rate (gross cpm) and the actual count rate (net cpm) is caused by several factors, of which the most important in actual practice are coincidence loss and background radiation.

In the case of *coincidence loss,* if two particles or photons enter a detector and produce interactions exactly at the same time, they will be counted as one; but even when the two interaction events follow each other within a time interval shorter than a certain minimum limit, they will be counted as one event. The minimum time interval necessary for the resolution of two events occurring in the detector is known as the *resolving time* of the counter. During this time, the counter is insensitive to another interaction event. For different counters resolving times vary from less than 1 μsec to 200 or 300 μsec. The loss of counts due to coincidence (or quasi-coincidence) results in a registered activity lower than the actual activity. The higher the activity of the sample, the larger the number of events lost.

Except when the activity is very low (a few hundred cpm), it is advisable to correct the registered activity for coincidence loss. One can show that true activity and registered activity are correlated by the equation

$$R = \frac{r}{1 - rt} \tag{7-4}$$

wherein R is the true activity of the sample in counts per second (cps), r is the registered activity in cps, and t is the resolving time of the counter in seconds.

For example, with a resolving time of 200 μsec:

1. If the observed count rate is 900 cpm, the actual count rate is

$$R = \frac{15}{1 - 15 \times 0.0002} = 15.05 \text{ cps} = 903 \text{ cpm}$$

2. If the observed count rate is 15,000 cpm, the actual count rate is

$$R = \frac{250}{1 - 250 \times 0.0002} = 263 \text{ cps} = 15,780 \text{ cpm}$$

In the first case the correction for coincidence loss could have been omitted for most practical purposes, whereas in the second case it was necessary.

Coincidence loss has several sources in a detecting and counting system. Detector, electronics of the pulse amplifier, and mechanical and electronic counting devices have their own characteristic resolving times. In certain systems the resolving time of the detector is much longer than that of the associated amplifying and counting apparatus; in others, the opposite is true. When the observed count rate is corrected for coincidence loss, the longest resolving time must be used for the calculation.

If a detecting system is turned on, with no radioactive source in or close to the detector, a few cpm will be registered. These counts are due to what is generally called *background radiation* (BG). Consequently, BG is recorded also when the counter is measuring the activity of a radioactive sample. The result is that the observed activity is higher than the actual activity of the sample. The following are sources of background radiation:

1. Cosmic rays. Because of the enormous penetrating power of some of the components of cosmic rays (see Chapter 6), keeping them from interacting in the detector is extremely difficult. Heavy lead shielding reduces

their effect but does not suppress it. More sophisticated methods used to reduce the BG count rate are not discussed here.

2. Natural radioactivity of the materials surrounding the sensitive volume of the detector. This activity is due to natural radionuclides present as impurities in many building materials (radium, radiolead), in glassware used in connection with the counting procedure (^{40}K), or in the body of the operator (^{40}K, ^{137}Cs). Very little can be done about the reduction of this component of BG, except for the use of glassware of low potassium content.

3. Chemicals on stockroom shelves, or radioactive samples kept in the vicinity of the detector. This source of BG can be easily eliminated.

4. Wristwatches, with luminescent dials (usually radioactive), worn by the operator.

5. Radioactive contamination of counting equipment, which may result from improper and careless handling of the radioactive samples to be counted. Rigorous precautions should be taken to prevent such contamination. Decontamination of a detector may be a tedious procedure. In certain instances, detectors cannot be decontaminated and consequently must be discarded.

6. Radioactive fallout caused by nuclear explosions in the atmosphere (see Chapter 8).

7. Spurious pulses originating in some detectors or, more often, in the electronic devices associated with them.

Background radiation should be kept to a minimum according to the simple precautions just suggested. If one assumes that these precautions are taken, the BG count rate registered by a detecting system is dependent on several variables, such as the nature and size of the detector, its sensitive material, its efficiency, and the type of radiation to which it is sensitive. BG count rates vary from less than 20 cpm for small G-M and proportional detectors to better than 100 cpm for solid scintillation gamma detectors.

When low-activity samples are measured, a correction of the observed count rates for BG is always necessary. With no radioactive sample in or near the detector, the counter is operated for a suitable period of time (at least 5 minutes), and the BG count rate is recorded in cpm and subtracted from the observed activity of the sample. If several samples are measured for prolonged periods, the BG count rate is taken at regular intervals.

IONIZATION INSTRUMENTS

The first group of radiation-detecting systems, as indicated in the preceding section of this chapter, consists of those instruments based on the detection of primary or secondary ionizations produced by radiations. If the ion pairs are produced in the absence of an electric potential gradient, they will quickly recombine by electrostatic attraction. However, before recombination occurs, in certain conditions, they may produce some sort of visible effect in the medium where they are formed. Cloud chambers and bubble chambers are operated in these conditions (see below).

In other ionization instruments the ion pairs are not allowed to recombine because they are produced between two electrodes kept at different electric potentials, so that the negative ions (electrons) are attracted to the positive

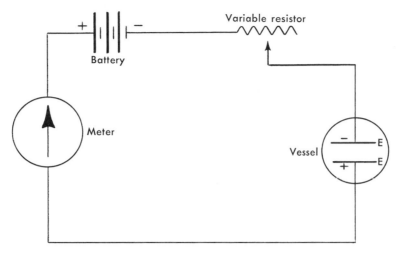

Fig. 7-2. Simplified circuit of a gas-filled ionization chamber. **E,** Parallel-plate electrodes.

electrode (anode) and the positive ions are attracted to the negative electrode (cathode). A short pulse of electric current is thus produced between the two electrodes for every ionization event. An electronic apparatus can detect, amplify, and record the pulses in an integrating or nonintegrating manner. In this type of ionization detector, the medium in which the ionizations take place may be a gas that fills a detector chamber or a solid with electric semiconducting properties. Accordingly, the ionization instruments that are operated with an electric potential gradient are classified into gas-filled ionization chambers and semiconductor detectors.*

Gas-filled ionization chambers
General principles

The operation of the gas-filled ionization chambers can be better understood if we consider the somewhat idealized and simplified circuit of Fig. 7-2. A vessel contains two parallel plates that function as electrodes. An electric potential gradient can be impressed to the electrodes by means of a battery, and a variable resistor is used to increase the voltage between the electrodes from zero to the highest possible value. A suitable meter is inserted in the circuit to measure any electric current that might flow through the circuit. If a voltage is applied to the electrodes, the meter will not detect any current flow, because the air or any other gas present between the two plates has insulating properties and will not allow the passage of any electric charges, unless the voltage used is very high. If, however, ionizing radiation is allowed in the vessel, ion pairs are formed between the electrodes, and thus the gas present in the vessel will lose its insulating properties. In these conditions, it is extremely useful to study the behavior of the ion pairs in relation to the voltage applied to the electrodes. This study can be done by examining the curves obtained when we plot the

---

*Strictly speaking, in semiconductor detectors, ionizing radiation does not produce ion pairs, but electron-hole pairs. Nonetheless, since these charge carriers are equivalent to ion pairs and are collected by electrodes, the inclusion of these detectors among the ionization instruments would seem appropriate, if the term "ionization" is accepted here in a rather broad sense.

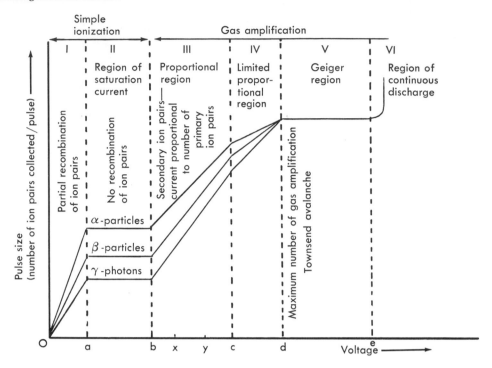

Fig. 7-3. Variation of pulse size with applied voltage in an idealized ionization chamber (voltage–pulse size plot).

applied voltage against the size of the current pulse, as measured by the meter, for three types of ionizing radiations (alpha and beta particles and gamma photons) (Fig. 7-3). For these curves, only arbitrary values can be given for the voltage and pulse size, because the actual values depend on a variety of factors, such as size of the ionization chamber, distance between the electrodes and their shape, and nature of the gas filling the chamber. One should also remember that the value of the pulse size measured by the meter is proportional to the number of ion pairs collected by the electrodes.

Now let us assume that the voltage applied to the electrodes is zero and that a particle or photon is passing between the electrodes. In these conditions, ion pairs are formed, but all of them recombine because of the electrostatic attraction existing between the two members of each pair. No ions are attracted to the electrodes, and therefore the meter detects no pulse. At zero voltage, we have, therefore, a total recombination of ion pairs. Now let us apply a low voltage to the electrodes. Ion pairs are still being formed in the chamber, but, because of the electric potential gradient existing between the electrodes, some of the negative ions are attracted to the anode and an equal number of positive ions are attracted to the cathode. The meter will detect and measure a current flow. Because of the low voltage values used, many ion pairs will be able to recombine. For a certain voltage range, this is the region of *partial recombination* of the ion pairs (region I). In this region, the number of ion pairs collected relative to the total number of ion pairs produced depends on the actual voltage value, and consequently the pulse increases with the voltage. An inspection of the curves,

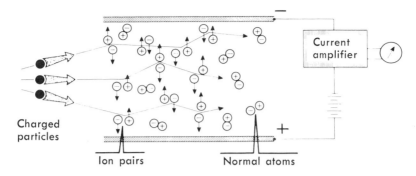

Charged particles

Ion pairs   Normal atoms

Incoming particles ionize atoms

Electrodes attract ions

Arrival of ions constitutes current

Current is measure of particles

Fig. 7-4. Saturation current. (Courtesy AEC.)

however, shows that for the same voltage the pulse size is largest for an alpha particle, smaller for a beta particle, and even smaller for gamma radiation. The explanation is to be sought in the different specific ionizations of these three types of radiation (see Chapter 6). The specific ionization of alpha particles in air can be as high as 10^5 ion pairs per cm, whereas it can be as low as 10^2 ion pairs per cm for beta particles, and for gamma photons even lower. As a result, at the same voltage, the absolute number of ion pairs collected by the electrodes is highest for alpha particles, lower for beta particles, and even lower for gamma radiation.

Beyond a certain voltage value a, the pulse size reaches a peak and remains constant while the voltage is increased up to a value b. Apparently, this peak results because voltage a is high enough to prevent any recombination of ion pairs and all ions produced are attracted to the electrodes, so that in region II a *saturation current* is obtained (Fig. 7-4). The saturation current remains constant for the same type of radiation with a moderate increase of voltage, because obviously the number of ions collected per pulse remains constant. An increased voltage within this region has only the effect of increasing the kinetic energy of the ions as they are accelerated toward the electrodes. In regions I and II, the voltage is high enough to cause the collection of some or all the ion pairs produced by the ionizing particle or photon (primary ion pairs). For this reason, these two regions are called regions of *simple ionization;* that is, the only ions responsible for the production of the pulse of ionization current measured by the meter are the primary ions.

If the voltage is increased beyond a certain value b, a new phenomenon takes place in the chamber. The primary negative ions (electrons) may acquire enough kinetic energy to ionize more atoms or molecules of the filling gas thus producing secondary ion pairs; the secondary electrons in turn are accelerated by the electric potential gradient and produce still more ion pairs, and so on. The formation of a cascade or avalanche (Townsend avalanche) of secondary ion pairs is thus triggered by each primary negative ion. This process is known as

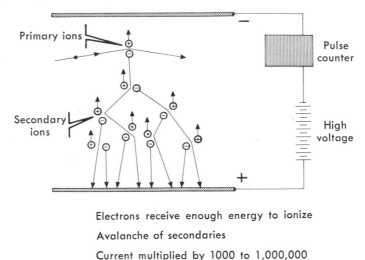

Primary ions

Secondary ions

Pulse counter

High voltage

Electrons receive enough energy to ionize

Avalanche of secondaries

Current multiplied by 1000 to 1,000,000

Fig. 7-5. Gas amplification. An ionizing particle travels through the sensitive volume between two electrodes. If the voltage applied to the electrodes is sufficiently high, the electron of the primary ion pair shown in the diagram is accelerated toward the anode and gains enough kinetic energy to produce another (secondary) ion pair in the gas. About 30 to 35 eV are required to produce an ion pair in gases. Secondary electrons may cause further ionizations. As shown in the diagram, eight negative ions are collected by the anode and eight positive ions are collected by the cathode. The amplification factor is eight. (Courtesy AEC).

gas amplification (Fig. 7-5) because the electric pulse originated by one single primary ion pair is amplified (that is, made larger) with the formation and collection of all the secondary ion pairs. The *gas amplification factor* (the *A* value) may be defined as the total number of ions (of either sign) collected at each electrode for each primary ion pair produced by the ionizing particle or photon. The amplification factor depends on a variety of factors, but for the same ionization chamber and for the same filling gas it depends exclusively on the voltage applied to the electrodes. As an example, an *A* value of 10^3 means that 1000 secondary ion pairs are produced for each primary electron. *A* values vary with voltage from 10^0 in region II to a maximum of 10^8 or 10^9 in the vicinity of the voltage value *d*.

Let us consider the pulse size, as the voltage is increased from *b* to *c* (region III). The number of primary ion pairs produced by an alpha particle remains the same throughout this region and is identical to the number produced in region II. However, because of the gas amplification, which increases above voltage *b*, the pulse size caused by the same alpha particle increases with the increasing voltage. For instance, if an alpha particle produces 10^5 primary ion pairs at a voltage *x* between *c* and *d* and the *A* value is 10^3, a total of 10^8 ion pairs will be collected on the electrodes. With a higher voltage (*y*), the number of primary ion pairs is still 10^5, but if now the *A* value is 10^5, a total of 10^{10} ion pairs will be collected on the electrodes and therefore the pulse size is larger. Within region III, for a given voltage, the pulse caused by an alpha particle is larger than that produced by a beta particle, because the number of primary ion pairs produced by the two particles is different although the *A* value is the same;

consequently, the total number of ion pairs collected is different (and thus the pulse size is different). The same consideration applies to the pulses caused by gamma photons. It follows that in this region the pulse size is proportional, at any voltage, to the number of primary ion pairs; for this reason, region III is known as the *proportional region*. Within it, as in region II, the discrimination of alpha from beta particles and gamma photons is still possible because the voltage–pulse size curves are still distinct and parallel. The added advantage of region III over region II is the larger size of the pulses produced (see the practical implication of this advantage below).

The curves show that, beyond a certain voltage, the pulse size increases with voltage more slowly than in the proportional region and that there is no longer a strict proportionality between the pulse size and the number of primary ion pairs produced. For this reason, region IV is known as the *limited proportional region*. At these voltages, the gas amplification factor is approaching the maximum value allowed by the constructional characteristics of the chamber. These characteristics set a limit to the total number of ion pairs that the chamber can accommodate. If we assume, for example, that the highest number of ion pairs that the chamber can accommodate is 10^{11} and that the A value obtained at a certain voltage in this region is 10^8, then an alpha particle that forms 10^6 primary ion pairs will produce only a total of 10^{11} (with an actual amplification of only 10^5), whereas a beta particle producing only 10^2 primary ion pairs could be amplified by a factor of 10^8, with a total of 10^{10} ion pairs collected. One should understand, therefore, that in these conditions there is no strict proportionality between the number of primary ion pairs and pulse size. In this region the possibility of distinguishing between different types of ionizing radiations gradually vanishes; this change is shown by the gradual convergence of the three curves. When the voltage limit d is reached, all types of ionizing radiations, regardless of their specific ionization, will produce the same total number of ion pairs, which is the maximum number the chamber can accommodate. Consequently, the size of the pulse will be the same, whether the pulse is triggered by an alpha or a beta particle or by a gamma photon. The pulse size will remain constant within a wide voltage range. Region V is known as the *Geiger region;* in this region discrimination between different types of ionizing radiation is no longer possible.

Beyond the Geiger region, a slight increase in voltage results in a sharp increase of pulse size, for reasons not analyzed here. Region VI is the *region of continuous discharge.*

The preceding discussion should not be construed as meaning that the same ionization chamber could be operated in any region by simply changing the voltage. Note that the chamber in Fig. 7-2 has been referred to as an idealized chamber. Whether a chamber behaves according to the characteristics of one region or another depends not only on the voltage applied between its electrodes, but also on its size, shape of the electrodes, shape of the electric potential gradient, gas filling the chamber, and several other factors. Ionization chambers are designed and constructed to be operated in a particular region under specific conditions. Electroscopes and electrometers are designed to be operated in the region of saturation current, proportional counters are operated in the proportional region, and Geiger-Müller counters are operated in the Geiger region. The fact that the operating voltage of a particular G-M detector might be lower than the operating voltage of a certain proportional detector should not be surprising in view of what has just been stated; voltage is not the only factor characterizing the region in which a certain detector functions.

In the preceding discussion, attention has been focused on the pulse triggered by one single ionizing particular or photon, and the variation of the size of this pulse with the increasing voltage has been analyzed. When an ionization chamber is exposed to a radioactive source, a shower of particles or photons will interact, within the chamber, with the filling gas, and a series of pulses will be triggered. If the pulse size is rather small, as in the region of saturation current, either the pulses are amplified electronically outside of the detector in the associated circuits and counted separately (nonintegrating mode of operation), or, more easily, what is measured is the continuous current flow between the electrodes, which obviously is proportional to the number of ionization events occuring in the detector per unit of time (integrating mode of operation). In proportional and Geiger detectors, the pulses are amplified within the detector itself (gas amplification), so that very little external electronic amplification is needed. The result is that, for these counters, the nonintegrating mode of operation is generally easy and preferred.*

Electroscopes and electrometers

Electroscopes and electrometers are ionization chambers operated in the region of saturation current. The voltage applied to the electrodes is relatively low (100 to 200 volts), and the current flow is exclusively caused by the collection of the primary ion pairs produced by the ionizing particles or photons. No gas amplification takes place, and as indicated above, discrimination of alpha particles from beta particles and gamma photons is possible.

These instruments are constructed and operated basically in one of the following two ways:

1. In the condenser or capacitor type of instruments, the two electrodes of the chamber behave like the plates of a capacitor. An electric charge is momentarily applied to the electrodes. If ionizing radiation is present between the electrodes, the capacitor (that is, the chamber) is discharged, slowly or rapidly, depending on the intensity and type of radiation. The rate of discharge is measured with an appropriate device. These instruments are all operated in the integrating mode.

2. In other instruments, the two electrodes are permanently kept connected to a source of electric potential during their operation, and what is measured is the current flow between the electrodes. These instruments may be operated in the integrating or nonintegrating mode. In this latter case, sophisticated electronic circuitry is needed to amplify sufficiently the weak output pulses of the chamber.

There is considerable confusion among authors as to the specific connotation of the names "electroscopes" and "electrometers," when they are used to indicate ionization chambers operated in the region of saturation current. For present purposes, I shall call *electroscopes* the instruments constructed and operated as explained in 1 just above and *electrometers* the instruments constructed and operated as explained in 2.

One of the oldest radiation detectors, used by Becquerel, Mme. Curie, and

---

*For a thorough understanding of the operation of all gas-filled ionization chambers, described in the following sections, the reader is earnestly advised to review the analysis of the voltage–pulse size curves described above.

others, was the simple gold-leaf electroscope (see any elementary textbook of physics for details). When an electric charge is applied to the electroscope, the gold leaves, because of electrostatic repulsion, diverge at an angle that is roughly proportional to the charge and remain divergent, unless the charge itself is removed, or leaks off. If a source of ionizing radiation is in the vicinity of the electroscope, the charge will leak off, because it is carried away by the ions produced in air by the radiation, and the electroscope will be discharged. The rate of discharge is proportional to the number of ions produced per unit of time and therefore is an indication of the radiation intensity. The electroscopes used at the present time as radiation detectors are nothing but refined versions of the primitive gold-leaf electroscope.

Consider the simple circuit shown in Fig. 7-6. The two metallic plates are separated by air, which is an electric insulator. If an electric charge is applied to the plates with a power supply by closing the switch momentarily, the electric charges of opposite sign will remain on the plates, which will behave like the plates of an electric capacitor. If a light quartz fiber is attached with one end to the positively charged plate, it will also carry a charge of the same sign; therefore, it will be deflected away from the plate by the electrostatic force of repulsion. When ionizing radiation enters the space between the plates of the capacitor, the ion pairs formed in the air discharge the capacitor slowly or rapidly, depending on the number of ion pairs formed per unit of time. The positive ions are attracted to the negative plate and the electrons to the positive plate, thus reducing their electric potential. With a reduction of charge on the positive plate, the quartz fiber drifts gradually back toward the plate; the drift rate should be proportional to the rate of discharge of the capacitor, and therefore to the number of ion pairs formed per unit of time.

One of the earliest instruments constructed and operated according to the simple circuit illustrated in Fig. 7-6 is the Lauritsen electroscope, invented by C. C. Lauritsen and T. Lauritsen in 1937. However, another more widely used type is described below. It is the Landsverk electroscope manufactured by the Landsverk Electrometer Company, with the name "Model L-75D Isotope Analysis Unit"; this electroscope differs from the Lauritsen electroscope only in some minor details, which are mentioned later.

The Landsverk electroscope (Fig. 7-7) consists of a cylindrical ionization

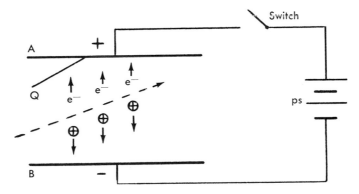

Fig. 7-6. Simplified diagram of an electroscope of the condenser type. **Q,** Quartz fiber. **ps,** Power supply. **A** and **B,** Metallic plates.

chamber, whose wall is equivalent to the negatively charged plate of the one in Fig. 7-6. The anode (or "collecting electrode") is a thin metallic rod located in the center of the chamber and electrically insulated from the wall by means of a polystyrene disk. A sensitive quartz fiber is attached to the anode; its drift can be viewed through a microscope against a 100-division arbitrary scale. The

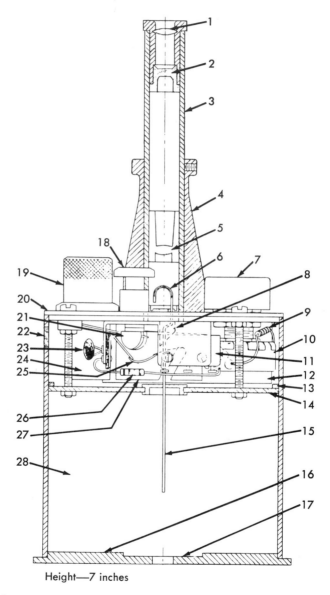

Height—7 inches

Fig. 7-7. Model L-75D Isotope Analysis Unit. **1,** Eyepiece. **2,** Reticle. **3,** Microscope barrel. **4,** Microscope mounting turret. **5,** Objective lens. **6,** Quartz fiber electrometer. **7,** Zero-control potentiometer. **11,** Microswitch. **12,** Transistor power supply. **13,** Lockring. **14,** Separator plate. **15,** Collecting electrode. **16,** Base plate. **17,** Recess for planchet holder. **18,** Charging button. **19,** Battery holder cap. **20,** Top plate. **21,** Charging button plunger. **22,** Window. **23,** Capacitor. **24,** Battery holder bottom. **25,** Electrode contact. **26,** Resistor. **27,** Shield electrode. **28,** Ion chamber. (Courtesy Landsverk Electrometer Co., Glendale, Calif.)

base of the ionization chamber is in the form of a separate aluminum disk, which also functions as a sample holder. The sample is therefore within the sensitive volume of the chamber, and this position contributes to the relatively high efficiency of the instrument. The volume of the chamber is about 200 cm³. A transistorized voltage supply, which is used to charge the electroscope, is built in an enclosed space directly above the chamber. The charger is powered by a single mercury cell. When the instrument is zeroed, it is charged to about 120 volts. The full-scale drift of the quartz fiber, from 0 to 100, corresponds to a reduction of voltage of about 37 volts. With this instrument, measurement of radiation intensity is obtained by the fiber-drift rate in terms of scale divisions per minute. Different samples of the same radionuclide can thus be compared. The instrument is particularly useful with alpha emitters, for which its efficiency can be very close to 100%.

The Lauritsen electroscope is different from the Landsverk electroscope essentially in two respects. First, the delicate quartz fiber is located within the sensitive volume of the chamber. In order to leave the fiber undisturbed, the radioactive sample is located outside of the chamber. The base of the chamber consists of a thin window that allows the passage of the ionizing radiation emitted by the external sample. The instrument is usually equipped with a set of interchangeable windows that are of different material and thickness and can be used for different types of radiations.

As discussed in Chapter 9, many types of dosimetric instruments are constructed according to the same principles of the condenser type of electroscope. They are calibrated in dose units, and the charging device is usually a separate unit. Some of them are rugged and miniaturized, like the pocket dosimeter used for personnel monitoring; others, like the Victoreen condenser R-Meter, are more sophisticated precision instruments, used for the accurate measurement of the dose delivered by an X-ray machine.

As mentioned above, electrometers are ionization instruments that are also operated in the region of saturation current; but what is measured, directly or indirectly, is the current flowing between the electrodes of the chamber, which are kept constantly connected to a source of electric potential. Especially if the instrument is operated in the nonintegrating mode, a considerable electronic amplification of the very weak output pulses is required. The actual measurement of the amplified current is done with a variety of sensitive devices, such as the vibrating-reed electrometer. Ionization chambers connected with these detecting systems are often designed in such a way as to allow the possibility of introducing a radioactive gas into their sensitive volume. $^{14}CO_2$ and ^3H-labeled gaseous compounds can thus be radioassayed with practically 100% efficiency. These instruments are often used in kinetic studies of respiratory metabolism, in which, for instance, a ^{14}C-labeled substrate is administered to an animal, and the $^{14}CO_2$ evolved is measured.* Another noteworthy feature of these instruments is their wide range of sensitivity. They can measure activities as low as 5×10^{-5} μCi, or as high as 1 mCi. No other type of radiation detector has this degree of versatility.

Proportional counters

A proportional detector is an ionization chamber that is operated in the proportional region, that is, in the region of gas amplification in which the size of the current pulse produced by an ionizing event is proportional to the number of primary ion pairs formed by the ionizing particle or photon. As a result, it

---

*See an example of this type of experiment in Wang, C. H., and D. Willis. 1965. Radiotracer methodology in biological science. Prentice-Hall, Inc., Englewood Cliffs, N.J.

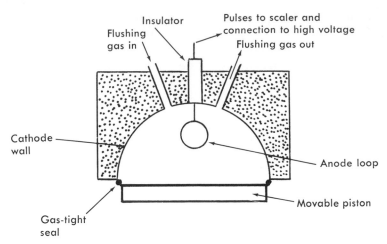

Fig. 7-8. Windowless proportional detector. (Redrawn from Wang, C. H., and D. L. Willis. 1965. Radiotracer methodology in biological science. Prentice-Hall, Inc., Englewood Cliffs, N. J.)

is possible to distinguish alpha particles, with a high specific ionization, from beta particles with a much lower specific ionization. Operation of these detectors in the proportional region is made possible by some constructional details. The anode is much thinner than in electroscopes and electrometers and often in the shape of a loop. The gas filling the chamber is not air, but rather a mixture designed for the control of the gas amplification through the regulation of the voltage applied to the electrodes; a commonly used proportional gas is a mixture of 90% argon and 10% methane, or 10% of some other gaseous hydrocarbon of low molecular weight. The counting gas is not sealed in the chamber, but flows through it; hence these detectors are known as *gas-flow detectors*. Special simple devices are associated with the detector, to allow a smooth flow of the gas and to control the rate of flow (see Fig. 7-8). If the detector is operated with an external radioactive sample, its base consists of a very thin window made of Mylar film, to allow the passage of alpha particles and of very weak beta particles. To increase the counting yield, with some of these detectors one can introduce the sample within the sensitive volume of the chamber. In this case, no window is necessary (windowless detector); the base of the chamber is a plate that functions also as sample holder and is connected to the rest of the chamber with a gastight seal. In other varieties of proportional detectors, the chamber is spherical and the sample can be accomodated in the center of the sphere supported by a very thin plastic film; with this arrangement, counting yields very close to 100% can be achieved for some radionuclides.

With most proportional counters, gas amplification factors of the order of 10^3 to 10^4 are easily obtained. Their efficiency for alpha particles and weak beta particles is very high. The efficiency for gamma radiation is rather poor; hence proportional counters are not normally used to count the activity of gamma emitters.

One distinct advantage of proportional detectors is the very short resolving time, which increases with voltage to values of only about 10 μsec (compare

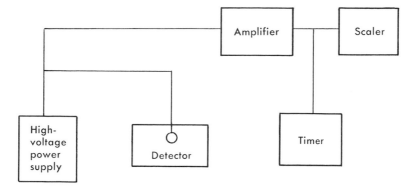

Fig. 7-9. Block diagram of a proportional counting assembly.

with 200 to 300 μsec of some G-M tubes). Also the background count rate is very low, in part because of the poor efficiency of these instruments for gamma radiation, which is a significant component of the background radiation.

The proportional systems are operated in the nonintegrating mode. Due to the gas amplification, which occurs inside of the ionization chamber itself (internal amplification), the output current pulses of the detector are fairly large. Therefore only a moderate electronic amplification of the pulses is needed.

The circuitry associated with a proportional detector consists at least of the following parts (Fig. 7-9):

1. An amplifier, for the electronic amplification of the pulses coming from the detector. The amplifier incorporates an electronic device known as a *pulse discriminator*, whose function is to stop pulses of sizes smaller than a certain threshold.

2. A scaler, whose function is to tally electronically and/or mechanically the pulses within a given counting time and to display the accumulated counts visibly.

3. A timer, for the presetting of the counting time. When the preset time has elapsed, the timer automatically stops the scaler.

4. High-voltage power supply, for the application of the appropriate electric potential gradient to the electrodes (anodic wire and wall) of the detector. In proportional counters, the high-voltage output must be very stable, because the ionization chamber is operated on a steep slope (see voltage–pulse size curves), on which a small variation of voltage results in a significant variation of pulse size.

Usually, amplifier and scaler are associated together in one chassis. Some manufacturers incorporate all parts (including the detector) in one compact unit.

A correct understanding of the operation of a proportional counter and of its use can be best attained by examining its *characteristic curve* (Fig. 7-10) This curve is obtained by plotting count rate as a function of the voltage applied to the chamber. This is done by recording the count rate of an alpha-beta emitter (such as ^{210}Pb) at regular voltage intervals, starting with the minimum voltage available. (CAUTION: This curve should not be confused with the voltage-pulse size curves analyzed above. What is plotted as ordinate in a characteristic curve is not the size of a pulse, but the number of counts or pulses recorded by the scaler per unit of time.)

As the voltage is increased from zero, a certain value will be reached at which the scaler will just start showing a few counts *(alpha threshold)*. A small increase of voltage will be accompanied by a rapid increase in count rate, which then will remain approximately constant over a wide voltage interval *(alpha*

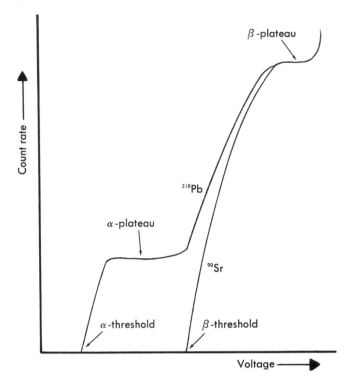

Fig. 7-10. Characteristic curves of a proportional detector.

plateau). The terms "alpha threshold" and "alpha plateau" refer to the fact that, at these voltages, the scaler is counting only alpha particles. Because of the high specific ionization of alpha particles, the alpha pulses are large enough to pass through the discriminator built in the scaler and thus are counted, whereas the beta particles, producing fewer primary ion pairs, do not produce, with the same gas amplification factor, pulses large enough to pass through the discriminator and thus are not counted. The existence of an alpha plateau is explained by the monoenergetic nature of the alpha particles; accordingly, they would be expected to generate pulses of almost equal size.

A slight increase of voltage beyond the alpha plateau will cause a sudden increase in the count rate. This sudden increase means that a gas amplification factor has been attained, whereby the most energetic beta particles are able to generate pulses large enough to be counted *(beta threshold)*. As the voltage is further increased, the amplification factor is likewise increased, and consequently beta particles of lower and lower energy are able to produce pulses large enough to be counted. This increase in count rate continues until there is a beta plateau which occurs when the least energetic beta particles that were able to pass through the window of the detector are counted. The beta plateau is not as well defined as the alpha plateau because of the nonmonoenergetic nature of the beta particles (recall that the beta particles emitted by the same radionuclide cover a wide spectrum of energies from zero to E_{max}). Beyond the beta plateau, a further increase of voltage simply results in a "continuous discharge," that is, in the production of additional pulses caused by the breakdown of the insulating properties of the counting gas. These spurious pulses, obviously, are not generated by ionizing events caused by the incident radiation.

If a pure beta emitter is counted with the same system, the scaler starts counting at a voltage corresponding to the beta threshold and the characteristic curve will show only the beta plateau. With an alpha-beta emitter, it is evident then that at a voltage correspondent to the alpha plateau, the scaler is counting only alpha particles, and at the voltage of the beta plateau, it is counting both alpha and beta particles. Consequently, if only alpha particles are to be counted, the operating voltage should be adjusted to the center of the alpha plateau. To obtain the beta count rate, one must set the voltage in correspondence with the beta plateau and subtract the alpha count rate from the count rate recorded at the beta plateau.

In summary, proportional detectors have a high efficiency for alpha and weak beta particles, provided that they are windowless or with a very thin window. Whenever pulse operation is required, the proportional counter is the system of choice for counting alpha emitters. This same counter is also preferred to any other system (except liquid scintillation systems, as mentioned further) for weak beta emitters, such as ^{14}C. The possibility of discriminating between alpha and beta particles is a desirable feature in certain types of work. The very short resolving times of proportional detectors allow the recording of very high count rates without significant coincidence loss.

Geiger-Müller counters

The G-M counters are probably the most widely used and versatile radiation-detecting instruments. The G-M detector is essentially a gas-filled ionization chamber designed for operation in the Geiger region. The "end window" G-M tube with a thin window of mica at one end, is a typical detector, widely used for radiation measurements of a certain degree of precision (Figs. 7-11 and 7-12). It consists of a metal shell that functions as the cathode of the chamber. The anode is located concentrically inside the tube and consists of an electrically insulated copper or tungsten rod. The tube is filled with an inert gas, such as helium or argon, with about 1% of another gas used as a quencher (see below). The window prevents the escape of the gas and yet is thin enough to allow the passage of radiation. However, a G-M tube is most sensitive to beta particles of medium or high energy; alpha particles (except the most energetic ones) are usually absorbed by the thicknesses of the windows used for these tubes; gamma photons, because of their high penetrability and low specific ionization, usually travel through the sensitive gas without interacting with it appreciably. In general the efficiency of G-M tubes for gamma radiation is only about 1%.

As mentioned in a previous section, the Geiger region is characterized by maximum gas amplification; consequently the output pulses of the detector are all of the same size, regardless of the nature and energy of the particle interacting with the counting gas. This means that a G-M tube, unlike proportional detectors, cannot distinguish between alpha and beta particles.

The quenching gas added to the counting gas prevents the spreading of the Townsend avalanche (the avalanche of secondary ion pairs) beyond the formation of the electric pulse generated by the ionizing particle, and thus it prevents the production of spurious pulses. Tubes may be quenched with

Fig. 7-11. X-ray picture of an end-window G-M tube. The metallic lining of the tube is the cathode; the axial rod is the anode.

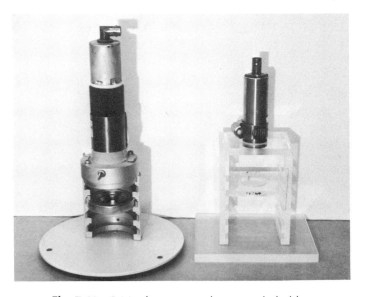

Fig. 7-12. G-M tubes mounted on sample holders.

a halogen (such as chlorine) or with an organic gas (such as butane). Halogen-quenched tubes have a much longer life because the biatomic molecules of the halogen, dissociated by the Townsend avalanche, recombine spontaneously. The resolving time of halogen-quenched tubes can be as long as 200 to 300 μsec. This long duration is perhaps one of the most serious shortcomings of these tubes. The resolving time of the organic-quenched tubes is considerably shorter.

Beside the end-window type described above, G-M tubes are manufactured in a variety of different shapes and sizes suitable for different special applications. For instance, rugged tubes with a thick end window or side window are made for the probes of portable Geiger counters or rate meters often used for prospecting or monitoring purposes. These probes are, at most, somewhat sensitive only to energetic beta particles, or only to gamma radiation if the window is shielded with a metallic cover. Other G-M detectors are of the gas-flow type (like most proportional detectors). Since the gas is continuously renewed, an organic-quenched gas can be used without shortening the life of the detector and with the advantage of a short resolving time. Moreover, because the gas pressure inside the tube is practically equal to the atmospheric pressure (in sealed tubes it is considerably smaller), the use of very thin windows with improvement of counting yield is possible.

The circuitry associated with a G-M tube is practically the same as for proportional detectors, although certain requirements are less stringent. Since operation in the Geiger region allows greater internal amplification, less external electronic amplification of the pulses is required and considerable savings in construction and maintenance result. The high-voltage power supply need not be highly stable, because in the Geiger region voltage fluctuations do not affect pulse size. Portable instruments are supplied with a rate meter, rather than a scaler. With a rate meter, the average count rate is indicated by a pointer on a dial but not registered.

The characteristic curve of a G-M tube (Fig. 7-13), obtained by plotting count rate as a function of voltage, is much simpler than the characteristic curve of proportional detectors. With a radioactive sample placed near the window, as the voltage applied to the tube is slowly increased, a voltage at which the scaler just starts counting (*starting potential*) is reached. At lower voltages, the pulses are too small to pass through the discriminator and therefore are not detected. With rising voltage, the count rate increases rapidly up to a certain voltage known as the *threshold potential*. At this point, maximum gas amplification is attained, with A values of about 10^8. Beyond the threshold potential,

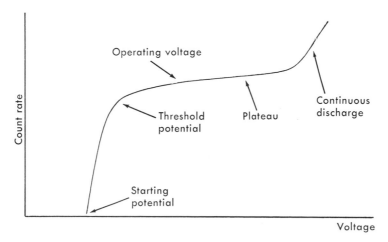

Fig. 7-13. Characteristic G-M curve.

a significant increase of voltage results only in a slight increase of count rate. This part of the curve is known as the *G-M plateau* (although it is not exactly flat), and, in good tubes, it should be reasonably level for a range of at least 200 to 300 volts. Beyond the plateau, a further increase of voltage is accompanied by a rapid increase of the count rate, caused by the breakdown of the insulating properties of the counting gas and consequently by pulses that are not generated by ionizing particles (continuous discharge). Operation at these high voltages causes permanent damage to the tube.

The actual values of starting and threshold potentials depend very much on the size and configuration of the tube and on the composition of the counting gas. In general, one can say that the larger the tube, the higher the starting potential. An organic-quenched tube has a higher starting potential than a halogen-quenched tube of comparable size and shape. The characteristic curve of a given tube should be plotted at regular intervals to check its performance. A tube should be operated at a voltage selected on the plateau far from the threshold potential and from the region of continuous discharge, so that small voltage fluctuations caused by the AC line or by the high-voltage power supply will not affect the count rate significantly.

In summary, with the exception of highly sophisticated models constructed for special purposes, G-M tubes have a satisfactory efficiency for medium- or high-energy beta particles and are used routinely for the measurement of the activity of many beta emitters whenever a high degree of accuracy and reliability is not required. However, they cannot distinguish between ionizing particles of different nature and energy, because, in the Geiger region, pulses are all of the same size, regardless of the particle that generates them. Another disadvantage is the relatively long resolving time; in most cases, the maximum count rates that can be reliably recorded are well below those of proportional detectors. The BG count rate for G-M tubes varies with their size, but, in general, it is always higher than for proportional detectors.

Semiconductor detectors

Logically, the description of semiconductor detectors should be given at this point, because, as indicated above, in these detectors ionizing radiations produce free charge carriers, which are collected by means of a potential gradient applied to the electrodes with the formation of discrete pulses of current. However, the reader who is not yet familiar with the other detecting methods described later in this chapter, is advised to read this section after he has studied all other types of detectors if he wants a better appreciation of the capabilities of the semiconductor detectors. Only the theoretical principles upon which the semiconductor detectors are based are discussed here, in a very elementary form, along with some considerations of a practical nature. A more detailed treatment would present a certain degree of difficulty and therefore would not be compatible with the introductory nature and with the scope of this textbook. Furthermore, these detectors are not likely to be found in most radioisotope laboratories designed for educational purposes. The interested student may consult the more specialized literature cited at the end of this chapter.

From the point of view of electric resistivity (see Chapter 1), all materials may be classified as conductors, semiconductors, and insulators. In conductors, the resistivity is of the order of 10^{-5} ohm cm. Metals are good electric conductors, because their resistivity is of this order of magnitude. If a difference of potential is applied to the two ends of a metallic wire, an electric current flows through it, because of the presence within the wire of numerous "conduction electrons," which are either free or very loosely bound to the atoms in their outermost orbits. In certain other conductors, like molten or dissolved salts, the electric current is carried by ions, rather than by electrons. In insulators (such as glass and porcelain), the electric resistivity is between approximately 10^{14} and 10^{22} ohm cm. They do not conduct electric currents because there are no free or loosely bound electrons or ions.

The properties of the semiconductors have been investigated only recently. The understanding of their behavior has made possible such important applications as transistors and semiconductor radiation detectors. The electric resistivity of semiconductors is between 10^{-2} and 10^9 ohm cm and therefore intermediate between conductors and insulators. One important difference between conductors and semiconductors is that in the latter the electric conductivity increases with temperature, whereas in the former exactly the opposite occurs. A variety of semiconducting materials is known today; but, for the study of their properties and behavior, we will consider only those two that have been suitable for most applications—silicon and germanium. These chemical elements have atomic numbers of 14 and 32, respectively, and therefore the atoms of both have four valence electrons in their outermost shells.

Let us consider a crystal of pure silicon (or germanium). Its atoms are arranged in the form of an orderly crystal lattice (Fig. 7-14) and kept firmly together by covalent bonds, which result from the sharing of pairs of valence electrons. In these conditions, a crystal of pure Si or Ge is practically an insulator because it does not contain any free or loosely bound electrons. However, ordinary heat energy can disrupt some of the covalent bonds, thus freeing some electrons, which are now available as charge carriers.

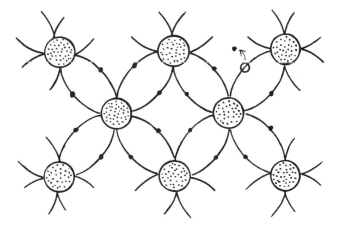

Fig. 7-14. Si or Ge crystal structure, showing the orderly arrangement of the atoms (large circles), the valence electrons (dots), an electron raised to the conduction band (indicated with an arrow), and the hole left behind (small circle).

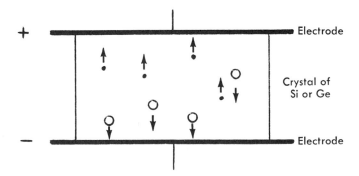

Fig. 7-15. Diagram of a semiconductor detector. *Dots,* Conduction electrons. *Circles,* holes.

More technically, we say that heat has lifted these electrons from the *valence band* to the *conduction band*. The crystal now can conduct an electric current, if a difference of potential is applied across it. An ionizing particle traveling through the crystal transfers its kinetic energy to some of the valence electrons and thus lifts them to the conduction band. The empty space left behind by the electron is referred to as a *hole;* this point of the lattice has a net positive charge, and for many practical purposes (see below), it can be considered as a positively charged particle. Therefore, an ionizing particle passing through a crystal of Si or Ge produces *electron-hole pairs.* This phenomenon is analogous to what happens in a gas-filled ionization chamber, in which an ionizing particle produces ion pairs, that is, pairs of electrons and positively charged ions. However, the average energy required for the production of one electron-hole pair in Si or Ge is about 3 eV, that is, only one tenth the energy required for the production of an ion pair in a gas (about 30 eV).

Fig. 7-15 is a schematic representation of a simple semiconductor detector. If a suitable electric potential is applied to the two electrodes in contact with two parallel faces of the crystal, the electrons raised to the conduction band by the ionizing radiation will move toward the anode and thus create a pulse of electric current. But also the movement of holes toward the cathode contributes to the pulse. A hole moves through the crystal when it is filled with an electron from a neighboring bond. This electron leaves a hole behind it, which can be filled with another electron. If the hole-filling electrons are moving toward the anode (as it happens when the voltage is applied to the electrodes), successive holes are created closer and closer to the cathode. This is tantamount to a "movement" of holes toward the cathode.

The crystal detector described above is analogous, therefore, to a gas-filled ionization chamber with the gas being replaced by a solid. However, nothing comparable to the phenomenon of internal amplification takes place in a semiconductor detector. The only *electron-hole* pairs contributing to the pulses are the primary pairs, just as, in an ionization chamber operating in the region of saturation current, the only *ion* pairs contributing to the pulses are the primary pairs. Consequently, it would seem that the size of the pulses should be very small, like in electroscopes and electrometers; but one should remember that for the same amount of energy dissipated by an ionizing particle 10 times as many electron-hole pairs are produced in a semiconductor detector as primary ion pairs produced in a gas chamber. Therefore, the pulse size is not as small as one would expect, in spite of the absence of internal amplification.

However, in a semiconductor detector some electrons and holes may not reach the electrodes, either because they recombine with other electrons and holes, or because they are trapped at certain points of the crystal lattice. Trapping may be caused by imperfections of the lattice or by impurities; nothing similar happens in an ionization chamber. Electrons and holes, which either recombine or are trapped, are lost to the current.

The type of detector described above, made of a crystal of pure Si or Ge between two electrodes, may be designated as a *homogeneous* or *bulk* semiconductor detector. It has been useful to us for the illustration of the basic principles of operation of these detectors. However, in practice, it has several disadvantages and undesirable features, which have been partly or totally eliminated by more or less sophisticated additions and improvements. Work on these improvements was begun in the late 1950s and is still in a developmental stage. Such names as "surface barrier detectors," "lithium-drifted detectors," "junction-type detectors," and "totally depleted detectors" apply to a variety of semiconductor detectors that are used for different purposes. No attempt is made here to describe them. Some years ago, the construction of these detectors was an art practiced only in research laboratories. Today, practically every type of semiconductor detector is available commercially; they can be made of any size and shape. In a typical detector the sensitive volume consists basically of a wafer of Si or Ge (Fig. 7-16), which can be less than 1 mm thick.

Semiconductor detectors present at least the following *advantages* over other types of radiation detectors:

1. Small size. Because of the much greater density of a solid material, compared with that of a gas, the sensitive volume can be made much smaller than the gas chambers

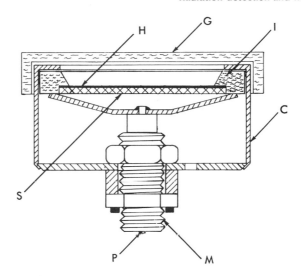

Fig. 7-16. Cross section of a typical surface barrier semiconductor detector. **H,** Sensitive surface coated with a thin layer of gold. **S,** Circular silicon wafer mounted in a ceramic ring (**I**) whose back and front surfaces are metallized. **C,** Grounded metal case. **M,** Connector. **P,** Center electrode of the connector which is used for the signal output and voltage connection. The cap, **G,** serves as a protective cover when the detector is not in use. (Courtesy Oak Ridge Technical Enterprises, Oak Ridge, Tenn.)

of ionization instruments. A 5 MeV alpha particle is completely stopped and therefore dissipates its entire energy in a thickness of only 30 μm of silicon, whereas its range in a gas is about 5 cm. A 1 MeV beta particle is completely absorbed by 1.5 mm of silicon.

2. Very short pulse duration (about 5 nsec) and no measurable dead time. The resolving time is therefore limited by the electronics associated with the detector. Very high count rates can be handled by these instruments, without need of correction for coincidence loss.

3. The energy required to produce one electron-hole pair is much lower than the energy required to produce one ion pair in gases (see above). This advantage makes semiconductor detectors superior to electroscopes and electrometers for alpha counting.

4. Linear proportionality between pulse size and energy dissipated in the detector. This property is shared with proportional and scintillation detectors but not with G-M tubes. It makes pulse height analysis and spectrometry possible.

5. Much better resolution than obtainable with scintillation detectors. Semiconductor detectors are definitely the instruments of choice for alpha spectrometry, which is not practically possible with alpha fluors. But even in the beta and gamma spectra obtained with these detectors, bands appear much narrower and sharper than in scintillation spectra. For instance, it is possible to see, in the beta spectrum of ^{137}Cs, the K- and L-conversion electrons as separate and distinct bands, whereas a scintillation spectrum shows only one rather broad band. It is mainly their remarkable resolution that makes these detectors so appealing to a number of investigators, who prefer them to scintillation detectors for any type of spectrometric work. Their resolution can be matched only by magnetic spectrometers, which, on the other hand, can be used only for electrically charged particles.

6. Better efficiency. If the right type and size of detector is used for the specific ionizing radiation under investigation, the efficiency can be very close to 100%. Weak radioactive samples can thus be counted for relatively short times.

7. Thin windows. Since the sensitive volume is occupied by a solid, not a gas, these detectors may be used with windows of negligible thickness, or with no window at all for certain configurations. This means little or no energy dissipated by radiation before it reaches the sensitive region.

Along with desirable features, however, semiconductors possess at least the following *disadvantages,* which make some investigators hesitant in using them:

1. Prolonged exposure of these detectors to radiation may cause deterioration of their desirable properties. Apparently radiation produces lattice imperfections, which in turn can act as trapping centers, with resultant deterioration of resolution, pulse size duration, and linear proportionality between pulse size and energy dissipated. This damage reduces the useful life of the detector.

2. Semiconductor detectors produce small output signals, which require considerable external amplification with special preamplifiers and amplifiers.

3. The operating conditions vary with temperature and other environmental factors. This inconvenience can be attenuated by operating the detectors at constant temperature.

4. All semiconductor detectors are sensitive to light.

5. The best of these detectors are less sensitive to gamma radiation than fairly large NaI scintillation crystals. This lesser sensitivity is perhaps one reason why the most significant work done in gamma spectrometry during the past few years has been done with scintillation detectors.*

Silicon detectors have the advantage of greater versatility over germanium detectors, because they may be used for practically any type of ionizing radiation. More is known about construction methods and about their properties. They can be operated at room temperature for most applications, whereas certain types of germanium detectors must be operated in liquid nitrogen and stored in dry ice. However, Li-drifted Ge detectors are more efficient for gamma spectrometry because of the higher atomic number of germanium, and, at the present time, they are used mainly for this purpose. We feel that, in the future, germanium detectors will replace silicon detectors for many applications, because of certain fundamental advantages of Ge over Si. Among other reasons, Ge is much more readily available commercially.

Semiconductor detectors can be used for neutron detection by placing an appropriate converter foil made of hydrogen-rich material, boron or lithium, in front of the sensitive surface. The products of the interactions between the incident neutrons and the foil material will be detected and identified by the sensitive volume of the detector.

Cloud chambers

A cloud chamber can be broadly considered as a gas-filled ionization chamber; but since no electric potential gradient is applied, the ion pairs are not collected and therefore no pulses are generated. Yet the paths of ionizing particles are made visible because condensation of a vapor occurs around the ions that are formed along the path of the particle. Therefore the cloud chamber is a true radiation detecting device, although the counting of the particles is usually impractical. Since the cloud chamber was invented by C. T. R. Wilson in 1911, it has been an invaluable tool for the study of important properties of ionizing radiations and of complex events, like cosmic radiation, meson production, uranium fission, etc.

The instrument is based on very simple principles. Imagine an enclosed space in which air is saturated with aqueous vapor. The volume of the space can be increased or decreased by means of a movable piston. If the volume is abruptly increased, the air within the chamber will suddenly expand (adiabatic expansion) with a consequent drop of its temperature. At this lower temperature, that volume of air contains more water vapor than is necessary for saturation; the air is said to be supersaturated. The excess vapor condenses in the

---

*See, for instance, R. L. Heath's outstanding work on gamma spectrometry: Scintillation spectrometry and gamma ray spectrum catalogue. 1964. U. S. Department of Commerce, Washington, D. C.

form of fine droplets, forming a sort of mist or cloud around nuclei of condensation, if they are available; dust particles are good condensation nuclei. Also the ion pairs produced along the path of an ionizing particle can function as condensation nuclei. Therefore, if an ionizing particle happens to travel through the chamber while the air is supersaturated, a white, misty trail will be seen in correspondence with its path; the track is actually a linear cloud, made of a row of numerous water droplets. The thickness and density of a track is an indication of the number of ion pairs produced per unit of path length. Visual observation of the tracks is done best in a dark room by illuminating the chamber with appropriate floodlight sources. Photographs of tracks can be taken automatically by coupling the piston mechanism with the shutter of a camera. The cloud chamber thus described is of the "expansion" type; its main disadvantage is that tracks are visible only during the short expansion phase, whereas a chamber of the "diffusion" type allows continuous observation of the events as they happen. Its construction is much simpler, and the mechanism used to produce supersaturation is completely different.

The diffusion cloud chamber (Fig. 7-17) is a cylindrical box; the side wall and the top are made of glass, and the bottom is an aluminum pan. A small amount of methanol is introduced in the pan, and the whole chamber is placed on dry ice. In these conditions a steep temperature gradient is created between the top and bottom layers of air within the chamber. Consequently, the methanol vapor is saturated in the upper warmer layers, from where it gradually diffuses downward toward colder layers of air, in which it becomes supersaturated. Because of convection currents, this supersaturation persists as long as there is liquid in the pan. The chamber is continuously sensitive in the lower region; tracks will be always visible when ionizing events occur. The side wall of the chamber has a hole for the insertion of radioactive sources or of a plain stopper.

With this type of cloud chamber, the following tracks can be easily seen if the operating conditions are optimal:

1. Alpha tracks are thick and dense, about 5 to 6 cm long, and straight. The

Fig. 7-17. Cloud chamber of the diffusion type. The radioactive source is at tip of the needle.

thickness increases as the alpha particle travels farther from the source, because its specific ionization increases. Occasionally, short and thin branches can be seen coming from the sides of the track (delta tracks, see Fig. 6-2); they are caused by secondary ionizations.

2. Beta tracks are much thinner and show frequent bending—an indication of a much lower specific ionization of beta particles and of their many collisions with air molecules. The deflections from a straight path are explained by their very small mass.

3. Cosmic tracks can be seen without any radioactive source inserted in the chamber, and they appear to originate from every possible direction. Not infrequently, double or multiple tracks of cosmic showers may be seen; they may look even thicker than the alpha tracks and originate from the interaction of cosmic radiation with the wall of the chamber.

Since cloud chambers are not designed to measure the intensity of ionizing radiation but only for their detection and for the study of some of their properties, they are very useful tools for the nuclear physicist, but of little value in radiation biology and for the user of radiotracer techniques. Diffusion cloud chambers are often used for educational purposes, because probably no other device provides the student with the same sense of direct perception of ionizing radiations. In addition, with a cloud chamber one can very easily acquire a first-hand knowledge of cosmic rays and of the essential differences between alpha and beta particles.

SCINTILLATION INSTRUMENTS
General principles

The simplest scintillation detectors were among the earliest instruments used to detect and measure ionizing radiation. As soon as X-rays were discovered, researchers learned that certain materials like barium platinocyanide emit visible light (fluoresce) when they are struck by X-rays. After the discovery of radioactivity, they found that nuclear radiation also can make some materials fluoresce. One of the materials widely used at that time was zinc sulfide. Most of the classical experiments with alpha particles were done by Rutherford and co-workers with the use of a screen coated with ZnS; an alpha particle impinging on the screen produces a faint flash of light, which can be seen with the help of a microscope. The assembly of the fluorescent screen and the microscope is called a *spinthariscope.** With this primitive instrument and with great patience, the flashes of light, and therefore the alpha particles impinging on the screen, can be counted visually, after proper dark-adaptation of the observer's eye. Long and endless hours were spent in a dark room for this kind of work, and probably these difficulties stimulated Geiger (one of Rutherford's co-workers) to invent the Geiger counter.

While ionization instruments were being developed and improved, scintillation devices were left aside, mostly because no method was known, other than the direct visual method, to count the flashes of light. But in the late 1940s the photomultiplier was invented, an instrument that can "see" faint flashes of light and convert them into electric pulses that can be counted by a scaler. This breakthrough stimulated renewed interest in scintillation instruments.

---

*Spinthariscopes are still being constructed today for demonstration purposes.

Various other materials were found to have the same fluorescent properties as zinc sulfide, with even more desirable features. Soon it was recognized that scintillation detectors are often preferable to ionization chambers for many applications.

Any material that is capable of converting the energy of ionizing radiation into visible light is called a *fluor*. When an ionizing particle or photon interacts with a fluor, a flash of light is generated within the fluor itself; its duration is very short (of the order of a few nanoseconds). The flash is commonly referred to as a *scintillation*.

In the long search for fluors suitable for radiation detection, investigators have accepted only those that present at least the following characteristics:

1. A high efficiency in the ionizing radiation–to–visible light conversion process. It is desirable that for a given amount of radiation energy dissipated within the fluor a scintillation of the highest possible intensity be obtained.

2. Transparency of the fluor to its own fluorescent light so that the scintillation can be viewed and received by the photomultiplier with as little loss of intensity as possible.

3. Prompt scintillation of short duration. The scintillation must occur immediately after the interaction of the ionizing radiation with the fluor and must last for a very short fraction of a second. This characteristic is essential in a good scintillation detector in order for the detector to distinguish and detect separately two nonsynchronous ionizing particles or photons. If the emission of visible light is delayed and lasts for seconds or hours, the phenomenon is called "phosphorescence"; materials that behave in this manner are called "phosphors" and are not suitable as scintillation detectors.

4. Density adequate to the type of radiation to be detected. For gamma radiation, with a much greater penetrating power than alphas and betas, a material of high density is required, so that the incoming photons have a chance to dissipate within the fluor as much of their energy as possible. Along with the density, this chance is increased also with the size of the fluor.

Fluors may be classified as solid or liquid; they may be either inorganic or organic materials. The duration of the scintillation is considerably shorter in organic than in inorganic fluors. Among the inorganic solids, crystals of sodium iodide, lithium iodide, and cesium iodide are very efficient for gamma detection, because of their high density and atomic numbers. However, they fluoresce only if "activated" with an impurity, like thallium for NaI and CsI, and europium for LiI. The notation "NaI(Tl)" designates a crystal of sodium iodide activated with traces of thallium; this crystal is the most widely used for gamma radiation, although the other two types exhibit certain more desirable features.

As mentioned previously, ZnS activated with traces of silver is another solid inorganic fluor used mainly for the detection of alpha particles. It cannot be obtained in the form of large crystals; hence it is used as a powder coating a transparent plastic film. Solid organic fluors are mostly hydrocarbons of the aromatic series, like crystals of anthracene, *trans*-stilbene, *p*-terphenyl; all very efficient for beta particles, although anthracene is the most commonly used. Plastic fluors are made of plastic material, like polystyrene, impregnated with one of the organic substances mentioned above. They are inexpensive, but exhibit poor energy resolution, and therefore cannot be used for spectrometry.

Several liquid fluors have been tried with different success. All of them consist of solutions, with a solvent and two solutes. Solvents normally used are toluene, dioxane, and others; solutes are *p*-terphenyl or 2,5-diphenyloxazole (PPO), with 4-bis-2-(5-phenyloxazolyl)-benzene (POPOP) as second solute. A typical solution is made of 4 g of PPO, and 100 mg of POPOP in 1 liter of toluene. Liquid fluors are commonly used for weak beta emitters (^{14}C, ^{3}H, etc.) but also less frequently for X-ray emitters (nuclides that decay by electron capture), because, in most cases, no other types of detectors possess the same high efficiency for these emitters.

In liquid scintillation counting, the radioactive sample is incorporated in the fluor, and the mixture is contained in a vial (internal scintillation counting). whereas in solid scintillation counting the sample is external to the fluor.

In summary, the fluors most commonly used can be briefly classified as follows:

Fluors
    Solid
        ZnS(Ag) coated screen (for alpha particles)
        Anthracene crystal (for beta particles)
        NaI(Tl) crystal (for gamma photons)
    Liquid
        Toluene + PPO + POPOP (for low-energy radiation)

The mechanism of the scintillation process is still imperfectly understood. It seems that the phenomenon of fluorescence is the effect of both excitation (molecular or atomic) and ionization occurring in the fluor when ionizing radiation interacts with it. To simplify the problem, let us consider the interaction of a gamma photon with a NaI(Tl) crystal. We know that a gamma photon may interact with matter mainly by means of photoelectric effect, Compton scattering, or pair production. In all cases, electrons are ejected from atoms of the absorbing material. These electrons may excite neighboring atoms. It is when these excited atoms return to the ground state that fluorescent light is emitted. The wavelength of the fluorescent light covers a rather narrow spectrum in the violet and near ultraviolet, between 3500 A and 4500 A; the wavelength of maximum emission varies for different fluors. For a NaI(Tl) crystal it is about 4200 A.

A single particle or photon traveling through a fluor may interact more than once with it. After one or more interactions, it can be completely absorbed, or it may emerge from the fluor with some residual energy. In the first case, all of its energy has been

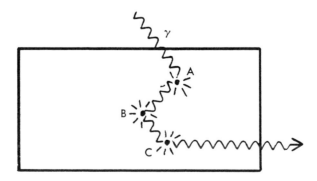

Fig. 7-18. The scintillation process. A gamma photon (γ) enters a NaI(Tl) crystal, interacts at three different points (*A, B,* and *C*) and then emerges, thus dissipating only part of its energy within the fluor. A minute flash of light is generated as a result of each interaction. The scintillation is a composite of all three flashes. Its intensity is proportional to the energy dissipated by the gamma photon in the crystal.

dissipated within the fluor; in the second case only part of its energy is dissipated. A minute flash of light is actually emitted as a result of every single interaction (Fig. 7-18), so that what is known as scintillation is really a shower of partial flashes originating at different points of the fluor. However, because the partial flashes occur within a very short time interval, the eye or the photomultiplier sees them as one single flash (scintillation), the intensity of which is the sum total of the intensities of the partial flashes.

In a good fluor, the light intensity of the scintillation is directly proportional to the energy dissipated by the particle or photon within the fluor. Consequently, if two 0.66 MeV gamma photons enter the same NaI(Tl) crystal but one dissipates all its energy in it, without emerging, and the other emerges after partial dissipation of its energy, the scintillation produced by the first photon is more intense than the scintillation produced by the second one. As we shall see further, this important property of the scintillation detectors makes spectrometry possible.

NaI(Tl) crystals are grown under carefully controlled conditions and cut in the shape of cylinders; they are available in sizes of 1, 2, 3, or more inches in diameter. Since they are hygroscopic and very fragile, they are encased in an aluminum can (Fig. 7-19), which has also the purpose of shielding the crystal from external light and from alpha and most beta particles possibly emitted by the radioactive sample. The surface of the crystal facing the photomultiplier is protected by an optical window, which consists of a transparent disk of Lucite (light pipe). To prevent partial absorption of the fluorescent light by the opaque material of the can, the can itself is coated internally with a layer of MgO, or Al_2O_3, which acts as a reflector.

An anthracene crystal, being nonhygroscopic, presents simpler mounting problems. It is only necessary to protect it from ambient light. However, if the crystal is canned, the surface facing the radioactive source must be protected with a very thin, opaque film (such as thin aluminum foil) to allow the passage of beta particles, for which this fluor is specifically designed and used.

Solid scintillation systems
Simple solid scintillation systems for integral counting

As indicated previously, in a scintillation system, the scintillations that take place in the fluor are viewed by a photomultipler (PM) and converted into electric pulses. In fact, the purpose of the photomultiplier is the transformation of

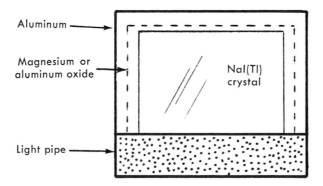

Fig. 7-19. A mounted NaI(Tl) crystal (lateral view).

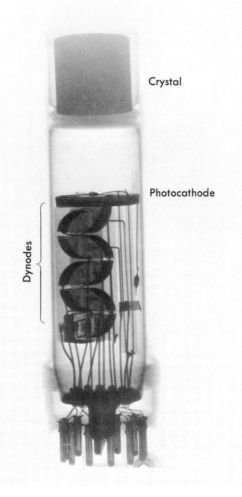

Crystal

Photocathode

Dynodes

Fig. 7-20. Radiogram of a solid scintillation detector.

each scintillation into a voltage pulse of an amplitude* proportional to the intensity of the scintillation. A solid fluor is optically coupled to the sensitive surface of the PM, by means of a suitable silicone cement. The PM is a glass tube with a window at one end (the sensitive surface), and a series of pins for electric connections at the other (Fig. 7-20). The inside surface of the window is coated with a special cesium-antimony alloy; this layer is called the *photocathode* (Fig. 7-21). It behaves like the cathode of a simple photoelectric cell (Fig. 3-3). When the light generated in the fluor by the scintillation process strikes the photocathode, electrons are emitted from its surface (photoelectric effect); their number is proportional to the intensity of the scintillation. The maximum sensitivity of a Cs-Sb photocathode is for light with a wavelength of about 4900 A, which is considerably close to the wavelength of the light generated in most fluors.

The PM tube differs from a simple photoelectric cell, because the minute

---

*The terms "size" and "height" are also used as synonyms for "amplitude."

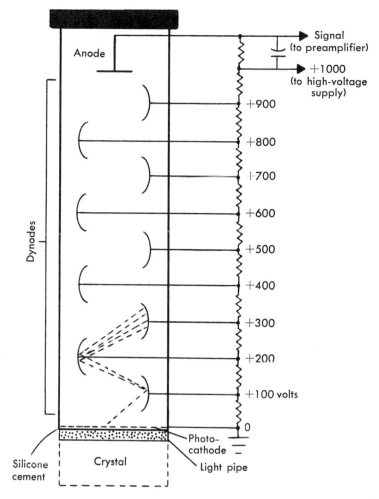

Fig. 7-21. Diagram of photomultiplier.

shower of electrons emitted by the photocathode, when it is struck by the faint scintillation light, is multiplied many (up to a million) times within the tube itself, with the result of an output pulse of considerable amplitude. This electron multiplication is obtained by means of a series of concave metallic plates, called *dynodes*. A high DC voltage is applied to the PM, so that the photocathode is negative (and grounded). Therefore, the electric potential at the photocathode is zero. By means of a series of resistors (voltage dividers), the voltage applied to the PM is divided between the dynodes in equal parts. If one assumes that the voltage applied to a PM tube with 9 dynodes is 1000 volts, the first dynode (facing the photocathode) will be at an electric potential 100 volts higher than the photocathode, the second dynode 200 volts higher, etc. With this arrangement, there is a strong electrostatic field between the photocathode and the first dynode, the first dynode and the second, etc.; consequently, the few electrons emitted by the photocathode are pulled and accelerated toward the first dynode, which they strike with enough kinetic energy to dislodge from it a number of electrons that is three to five times the number of the impinging electrons. The electrons emitted by the first dynode strike the second dynode, which is at a potential 100 volts higher, thus dislodging more electrons, which are pulled by the third dynode, etc. The process of electron multiplication continues, until the electrons emitted by the last dynode (about a million times more than those emitted by the photocathode) are collected by the anode, where they generate a pulse with an amplitude of a few millivolts; this pulse can be further amplified and counted.

The electron multiplication factor or *gain* (1 million, in our example) is dependent on the voltage applied to the photomultiplier. Evidently, if the voltage is lower than 1000 volts, the difference of potential between successive dynodes is less than 100 volts. Consequently, the electrons emitted by one dynode strike the next with less kinetic energy, thus dislodging from it less electrons than they would have with a higher difference of potential. For a given voltage, the PM gain remains constant, which means that the number of the original electrons emitted by the photocathode is multiplied by the same factor. But since the number of the original electrons depends on the intensity of the scintillation light, the pulses collected at the anode are obviously of different amplitude, which is proportional to the intensity of the scintillation light and therefore to the energy dissipated in the fluor by the ionizing photon or particle. In order for this linear proportionality to be maintained it is essential that the high voltage applied to the PM be stable. In fact, a very stable high-voltage supply is required for scintillation systems.

An electronic device, called a "preamplifier," is usually connected to the PM tube, or sometimes incorporated in it. It does not amplify the output signals of the PM but simply matches them with the electronics of the amplifier and prevents them from being lost in the several feet of cable connecting the detector to the other components of the system.

In conclusion, the output of a scintillation detector (fluor + photomultiplier + preamplifier) is a series of pulses of different amplitudes and of the order of magnitude of a few millivolts; the pulse amplitudes are directly proportional to the energy dissipated in the fluor by the ionizing radiation.

The purpose of the amplifier is to magnify these pulses to a magnitude of a few or several volts so that they can drive a scaler and be counted. However, for most practical applications, it is desirable that all the output pulses coming from the detector be magnified by the same factor so that the strict proportionality between pulse amplitude and energy dissipated in the fluor is maintained. An amplifier that fulfills this requirement is called a linear amplifier; its use is essential for scintillation spectrometry (see below). The amplification factor, or *amplifier gain* (not to be confused with PM gain), can be made adjustable. If, for instance, the amplifier gain is 8000, a 3-millivolt input pulse will become a 24-volt output pulse, a 5-millivolt input pulse will become a 40-volt output pulse, etc. Like the amplifiers used for proportional and Geiger counters, the amplifiers used for scintillation systems are also equipped with a pulse discriminator, whose function is to block the very small input pulses not generated in the detector by ionizing radiation. If all the amplified pulses are to be indiscriminately counted, they are directly fed to a scaler or rate meter. All the components of this simple scintillation system are shown in Fig. 7-22.

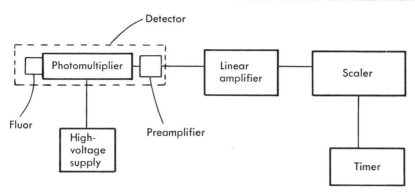

Fig. 7-22. Block diagram of a scintillation integral counting system.

The consecutive events occurring in the scintillation counting system just described above may be summarized as follows:

1. The energy of an ionizing particle or photon is partially or totally absorbed by the fluor, resulting in the excitation and ionization of some of its atoms or molecules.

2. Through the fluorescence process, the energy dissipated in the fluor is converted into light energy.

3. The light photons strike the photocathode of the PM tube.

4. The light photons are absorbed by the photocathode, resulting in emission of photoelectrons (photoelectric effect).

5. The electron multiplication process occurs within the PM tube.

6. The PM output pulses, the amplitude of which is proportional to the energy dissipated in the fluor, are amplified linearly by the amplifier.

7. The amplified pulses are counted by a scaler.

Let us assume that the fluor used in a scintillation system is a NaI(T1) crystal and that a gamma source is placed near the crystal. If the gross count rate is plotted as a function of PM voltage, a curve similar to curve *A* of Fig. 7-23 is obtained. This curve vaguely resembles the characteristic voltage–count rate curve of a G-M tube, but it does not have the same meaning. The inspection of the curve shows that below a certain voltage no pulses are coming into the scaler, because with the PM gain obtained at those low voltages any pulse is too small to pass through the discriminator of the amplifier. Above that minimum voltage, only the pulses generated by complete or almost complete absorption of the gamma photons in the crystal are strong enough to pass through the pulse discriminator and thus to be counted by the scaler. With increasing voltage, the PM gain is likewise increased, and therefore also the pulses generated by partial absorption of the gamma photons in the crystal are counted. There is no true plateau, because, at least theoretically, between total absorption and no absorption at all every degree of partial absorption is equally possible. However, depending on the gamma energy involved and on the size of the crystal, certain types of absorption are statistically more probable than others. This explains the existence of small peaks or bumps on curve *A*.

Furthermore, unlike what happens for a G-M characteristic curve, the shape of the voltage–count rate curve and the starting potential obtained with the same detector vary from one gamma emitter to another. For instance, if the gamma

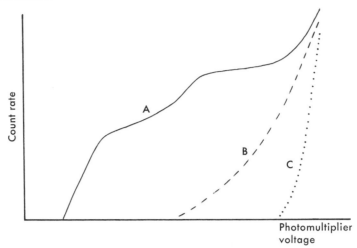

Fig. 7-23. Curves for count rate against photomultiplier voltage.

photon energy is low, the starting potential is high, because it takes a high PM gain to make the pulses generated by total absorption large enough to be counted. In addition, the probability of total absorption of a low-energy photon in a crystal of a given size is higher than for a high-energy photon. This fact is shown by a rather prominent shoulder in correspondence to a voltage just above the starting voltage. Other analogous considerations will thoroughly explain why curve *A* is, to a certain extent, characteristic of the gamma source.

If the system is operated with no radioactive source near the detector, the voltage–count rate curve looks like *B* (Fig. 7-23). Apparently, this curve shows the variation of the background-radiation count rate with voltage.* The steep rise of the curve at high voltages is caused by small pulses originating within the PM tube itself; when the tube is operated at high voltage, the dynodes get hot and, by thermionic effect, they emit electrons that are unrelated to the ionizing radiation that interacts with the fluor. These pulses constitute what is commonly called the PM *thermal noise*. The existence of the noise pulses can be shown, if the photomultiplier tube is operated with no crystal connected with it; in these conditions, any pulse detected must be unrelated to ionizing radiation. Curve *C* shows the relation of the noise-pulse count rate with the voltage applied to the PM tube. It is evident, then, that curve *B* is a composite of BG count rate and noise-pulse count rate, and curve *A* is a composite of the count rates of the sample, BG, and noise pulses.

For some applications, it is sufficient to count a gamma-emitting sample with the simple scintillation system described above. This method is called *integral counting*, because most or all gamma photons interacting with the crystal are counted, regardless of the energy they dissipate in it.

For integral counting, the detector is operated with a voltage at which the ratio of the sample net count rate to the BG count rate (signal-to-noise ratio) is the highest. This voltage may vary for different gamma emitters, because their voltage–count rate curves

---

*BG count rates for NaI(Tl) crystals are considerably higher (up to 100 or 200 cpm) than for G-M tubes. Consequently heavy shielding is required for these scintillation detectors.

are different (see above), but, at any rate, it is always well below the voltage at which the PM noise begins.

Integral beta counting with an anthracene crystal is done practically with the same system and the same techniques discussed above. The voltage–count rate curves for different beta emitters are rather uniform in shape although the starting potentials may be different, because of different E_{max}. The BG count rate for anthracene crystals is considerably lower than for NaI(T1) crystals; hence detector shielding is a less stringent requirement in beta scintillation counting.

The resolving time of scintillation detectors is very short (slightly above or below 1 μsec); it is shorter for anthracene than for NaI(T1) crystals. The resolving time of the whole system is therefore almost always limited by its electronics. Consequently, very high count rates can be handled without the necessity of correction for coincidence loss.

Scintillation spectrometry

As stressed previously, the output pulses of the amplifier in a scintillation system have an amplitude that is proportional to the energy dissipated in the fluor by ionizing radiation. It is possible, and very often useful, to count with a scaler not all the pulses, regardless of their amplitude, but only those that fall within preselected amplitude intervals. Such counting is accomplished with an electronic instrument called a *pulse-height analyzer* (PHA).

The PHA is an ingenious circuit placed between the linear amplifier and the scaler. It incorporates an electronic device, called a *lower discriminator*, which blocks all the pulses of amplitude smaller than a preselected value. When the lower discriminator is set on zero, all the output pulses of the amplifier will be passed on to the scaler by the PHA and counted. As the lower discriminator is raised above zero, only pulses of amplitude above the value at which the discriminator is set pass on to the scaler. Another discriminator, the *upper discriminator*, can be set; its function is to block any pulse of amplitude greater than a certain value. If both discriminators are operative, then only those pulses whose amplitudes are between the limits set by the discriminators are counted. The amplitude interval between the two limits is often called the "window"; its width (magnitude of the interval) can be made variable, but once it is set, it remains constant and follows the lower discriminator, as the latter is raised or lowered.

Let us illustrate these concepts graphically. In Fig. 7-24 we assume that the PHA has been fed by the amplifier with 23 pulses during the time interval of 1 minute. Since they are of different heights (or amplitudes), they are represented with vertical segments of different lengths. If the lower discriminator is set at pulse height unit (phu) 3 and the upper discriminator is disabled (Fig. 7-24, *A*), the scaler will count only 18 of the 23 pulses, because the height of five of them is smaller than the limit set by the discriminator; these pulses are thus rejected.

In Fig. 7-24, *B*, both discriminators are operative; the lower one has been set at pulse height 1 and the upper one at pulse height 5 so that the window width is 4 phu. In these conditions, only those pulses (7) whose amplitude falls within the window (that is, between pulse height 1 and 5) are counted by the scaler; all others are rejected.

If we raise the lower discriminator to phu 9 (Fig. 7-24, *C*) and leave the window width unchanged, the upper discriminator will be automatically raised

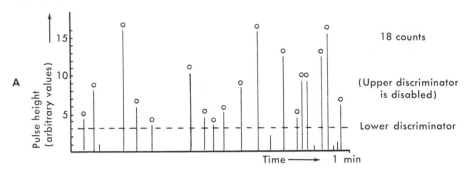

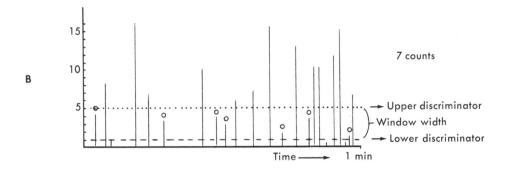

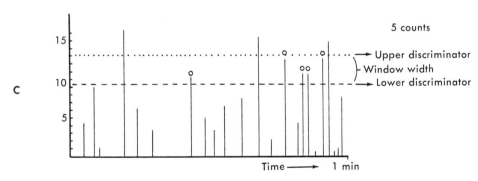

Fig. 7-24. Illustration of the function of a pulse-height analyzer. Small circles indicate the pulses that are counted.

to phu 13. This time pulses of larger amplitude are counted, that is, five with an amplitude between phu 9 and 13.

If we set the window width at 4 phu, then it is possible to sort out (that is, count separately) the pulses whose amplitudes fall within the 0 to 4 phu interval, the 4 to 8 phu interval, the 8 to 12 phu interval, etc., by setting the lower discriminator first on zero and counting for a certain time interval, then on 4 and counting for the same time, etc. Each pulse height interval is often referred to as a *channel*. If we decide to scan the series of pulses coming from the amplifier from 0 to 100 phu, then the number of channels depends on the window width used; with a 4 phu width we need 100/4 or 25 channels, with a 2 phu width we need 100/2 or 50 channels.

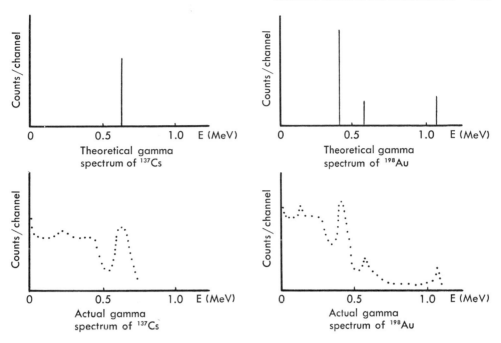

Fig. 7-25. Some theoretical and actual gamma spectra.

The data collected from the analysis done with this method consists of a series of count rates, one for each channel. If we plot on graph paper the channel number, or pulse height units, as abscissa and the count rates as ordinates, the resulting curve is a spectrum of distribution of the pulses within the channels *(pulse-height spectrum)*. With appropriate corrections (such as subtracting the small noise pulses generated by the PM tube itself), one can obtain the *energy spectrum* of the radiation emitted by the source.

Spectrometry is the spectral analysis of the energy dissipated in a detector by the radiation emitted by a source; if the detector is a scintillation fluor, we speak of *scintillation spectrometry*. We will now concern ourselves with gamma spectrometry and beta spectrometry.

Gamma spectrometry. The gamma spectrum of a gamma emitter is obtained by plotting on graph paper the count rate per channel as a function of pulse-height units. It is possible to calibrate a spectrometer with a monoenergetic gamma emitter of known gamma energy (such as ^{137}Cs) and thus convert phu into energy units (keV or MeV).* In this case the *spectrum is a plot of count rate per channel as a function of gamma energy*.

Let us assume that a ^{137}Cs source is placed near the detector and that (1) every gamma photon (662 keV) is *completely* absorbed by the NaI(Tl) crystal through one or multiple interactions, (2) no other extraneous radiation enters the crystal, and (3) no thermal noise pulses are generated by the PM tube. In this case the gamma spectrum of ^{137}Cs would show only a line corresponding to 662 keV (Fig. 7-25), indicating that the only pulses received by the PHA were those generated by the dissipation in the crystal of 662 keV of energy. The

---

*See the calibration procedure in the laboratory manual.

length of the spectral line would depend on the activity of the source used. Similarly, if we use a [198]Au source and make the same assumptions as above, its gamma spectrum would look like that in Fig. 7-25. The decay scheme of this radionuclide shows that it emits three gamma rays of different energy (0.41, 0.67, and 1.08 MeV) and also with different abundance, the most abundant being the 0.41 MeV gamma ray. Consequently, the spectral line corresponding to 0.41 MeV is the tallest of the three.

If these line spectra were the actual response given by a gamma spectrometer, then it would be extremely easy, for instance, to identify an unknown gamma emitter from its spectrum; it would suffice to count the lines, measure their relative heights, and read off the energies they correspond to, and with the help of reference sources, decay schemes, etc., the unknown gamma emitter could be quickly identified. Unfortunately, the response actually given by a gamma scintillation crystal is quite different, mainly (but not exclusively) because many gamma photons do not dissipate all their energy in a crystal. Let us consider the actual spectrum of a ^{137}Cs source (Fig. 7-25). The tall peak at the far right of the spectrum is caused by the total absorption of the 662 keV gamma photons. It is commonly called *photopeak* or full-energy peak and is a distinguishing characteristic of all spectra. A gamma emitter with two or more gamma photons (such as [198]Au, Fig. 7-25) would show two or more photopeaks. The uneven plateau to the left of the photopeak in the cesium spectrum is an indication of the existence of pulses of amplitude lower than that corresponding to 662 keV. These pulses are generated by photons that dissipated in the crystal only part of their energy by means of Compton interactions, followed by the escape (from the crystal) of the scattered photon and/or the Compton electron. The magnitude of the partial energy thus dissipated in the crystal covers a range from zero to a maximum that is well below the photopeak energy. Consequently, these partial-energy dissipations appear in the spectrum as a continuum (Compton smear) encompassing a wide energy range, rather than as a sharp peak. The steep rise of the spectrum to the extreme left is mainly

Fig. 7-26. An integrated spectrometer, incorporating high-voltage power supply, amplifier, pulse-height analyzer, scaler, and timer. The detector is not shown. (Courtesy Baird-Atomic, Inc., Bedford, Mass.)

caused by barium X-rays (32 keV), that is, by the X-rays generated in the daughter atoms of ^{137}Cs (^{137}Ba) when their K-electrons are dislodged by some of the Cs beta particles (see Chapter 4 on the origin of the characteristic X-rays).

The actual spectra obtained with a spectrometer (Fig. 7-26) are therefore much more complex than the theoretical spectra. The complications increase, if the source (such as ^{60}Co or ^{198}Au) emits more than one gamma photon and if the gamma photon energies are above 1.02 MeV. The interpretation of such spectra is often difficult, and one needs considerable training and experience to explain certain particular features shown by gamma spectra. Nevertheless, despite all complications and distortions, each gamma-emitting radionuclide still gives a characteristic spectrum, which can be distinguished from others mainly because of the number and position of the photopeaks on the energy scale. The gamma spectrum is the fingerprint of a gamma emitter. Detailed discussion of the interpretation of gamma spectra is beyond the scope of this introductory textbook, and only some considerations are suggested here to help the student in their correct interpretation.

The general shape of a spectrum is determined, first of all, by the gamma-ray energy itself. Besides the obvious fact that the photopeaks of gamma photons of different energies are found in different positions of the energy scale, the gamma energy also affects the shape of the spectrum to the left of the photopeak. Fig. 7-27* shows the spectra, plotted on the same paper, of four gamma rays

---

*For a more exhaustive discussion of the spectra shown in Fig. 7-27, see the masterly and concise treatment presented by R. L. Heath. 1964. Scintillation spectrometry and gamma ray spectrum catalogue. U.S. Department of Commerce, Washington, D.C. Vol. 1, pp. 6–7.

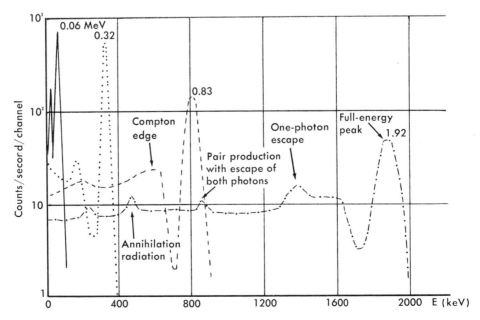

Fig. 7-27. Pulse-height response of a 3″ × 3″ crystal to four monoenergetic gamma rays. (Redrawn from Heath, R. L. 1964. Scintillation spectrometry and gamma ray spectrum catalogue. U. S. Department of Commerce, Washington, D. C. Vol. 1, p. 7.)

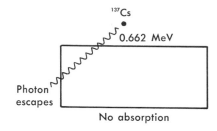

A

No interaction

Energy absorbed by crystal = 0 MeV

No flash of light

No pulse

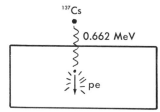

B

One photoelectric interaction

Energy absorbed by crystal = full energy of the incident gamma photon

One flash of light

Pulse counted under photopeak energy (=0.662 MeV)

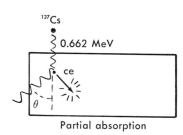

C

One Compton interaction with escape of scattered photon

Energy absorbed by crystal = KE of Compton electron < 0.662 MeV

Flash of light less bright than in **B**

Pulse counted under energy lower than that of the photopeak

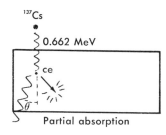

D

One Compton interaction with escape of scattered photon

Energy absorbed by crystal < energy absorbed in **C** (because of smaller θ)

Flash of light less bright than in **C**

Pulse counted under energy lower than in **C**

Fig. 7-28. Examples of partial and total absorption of gamma energy in a NaI(Tl) crystal. **pe,** Photoelectron. **ce,** Compton electron. **KE,** Kinetic energy. **e⁺,** Positron. **e⁻,** Negatron.

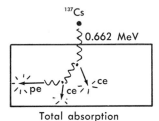

Total absorption

E

Two Compton interactions followed by a photoelectric interaction

Energy absorbed by crystal = 0.662 MeV

Three flashes of light almost simultaneous

Intensity of composite flash as in **B**

Pulse counted under photopeak

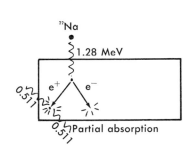

Partial absorption

F

Pair production with escape of both annihilation photons (double escape)

Energy absorbed by crystal = 1.28 MeV − 1.02 MeV = energy dissipated in crystal by negatron and positron

Composite flash of light less bright than in **H**

Pulse counted under energy = 1.28 MeV − 1.02 MeV

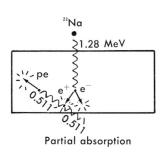

Partial absorption

G

Pair production with escape of one annihilation photon (single escape)

Energy absorbed by crystal = 1.28 MeV − 0.511 MeV = KE of pair + KE of photoelectron

Composite flash of light brighter than in **F** but still less bright than in **H**

Pulse counted under energy = 1.28 MeV − 0.511 MeV

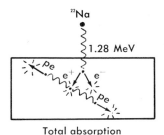

Total absorption

H

Pair production with both annihilation photons absorbed

Energy absorbed by crystal = 1.28 MeV = KE of pair + KE of both photoelectrons

Composite flash is produced of intensity greater than in **F** and **G**

Pulse counted under photopeak

Fig. 7-28, cont'd. For legend see opposite page.

emitted from four different monoenergetic sources, with energies of 0.060, 0.320, 0.830, and 1.92 MeV, *all of equal intensity.*

The spectrum of the 0.060 MeV gamma ray is very simple; at this low energy, a very high percentage of incident photons dissipate their entire energy in the crystal, and consequently the spectrum is made almost entirely of the photopeak. As the energy increases (see the 0.320 and 0.830 MeV gamma rays), the intensity of the photopeak decreases, because the percentage of photons dissipating only part of their energy in the crystal increases and the Compton smear appears in the spectrum. The spectrum of the 1.92 MeV gamma ray is even more complex, because, at this high energy, a photon may interact in the crystal also by means of pair production (see Chapter 6). This type of interaction may result in partial dissipation of energy if one or both of the annihilation photons (0.511 MeV) escape from the crystal. These events are shown in the spectrum by small peaks superimposed on the long Compton smear; the partial absorption with the escape of both photons is noted at an amplitude equal to 1.92 minus 1.02 MeV, and the partial absorption with the escape of one photon is noted at an amplitude equal to 1.92 minus 0.511 MeV. The diagrams of Fig. 7-28 are pictorial representations of some types of interactions between gamma radiation and the NaI(Tl) crystal.

The amplifier gain affects the shape of a spectrum. Suppose that we take the spectrum of a ^{137}Cs source with the same spectrometer twice, once with an amplifier gain a and then with a gain $2a$. By doubling the amplifier gain, we double the amplitude of all the pulses presented to the PHA. Consequently, the pulses generated by the total dissipation of the gamma-photon energy are counted in a channel whose pulse height is twice as large as the pulse height of the same pulses counted with gain a. Doubling the gain results in a sort of stretching of the whole spectrum to the right (see Fig. 7-29). With gain $2a$ the intensity of the photopeak is lower, because the pulses of all sizes are distributed through twice as many channels as with gain a. However, the areas

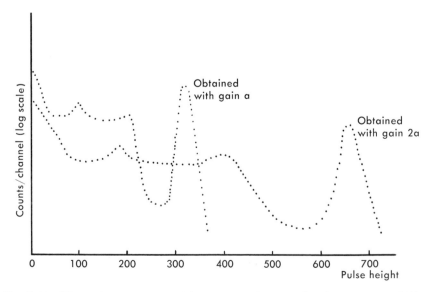

Fig. 7-29. Effect of increased amplifier gain on the pulse-height spectrum of ^{137}Cs.

under the two spectra are equal (the total number of pulses in all channels is graphically represented by the surface area under the spectral curve). The practical lesson to learn from the above considerations is that for low-energy gamma rays the amplifier gain should not be too low, if we do not want a spectrum "squeezed" to the left, which would result in the difficulty of reading off the photopeak energy and of distinguishing other peculiar spectral features that might aid in the identification of an unknown source. On the other hand, one should not think that a very high amplifier gain will always result in the appearance of more details in the spectrum. The quality of a spectrum is limited by other factors, such as crystal size and resolution, intensity of the source, etc.

The high voltage supplied to the photomultiplier tube affects the shape of a spectrum in a manner similar to the amplifier gain. In fact, by increasing the high voltage, we are increasing the electron multiplication factor (the PM gain), and consequently the pulses presented to the amplifier have larger size. The effect on the spectrum is the same as above; the spectrum stretched to the right. However, the application to the PM of a voltage above a certain limit set by the manufacturer is not recommended. In general, it is preferable, in spectrometry, to use a low PM gain and a high amplifier gain.

The window width also affects the general shape of a spectrum. The use of a narrow window results in the possibility of using more channels within the same pulse-height interval and therefore in a better definition of spectral details.

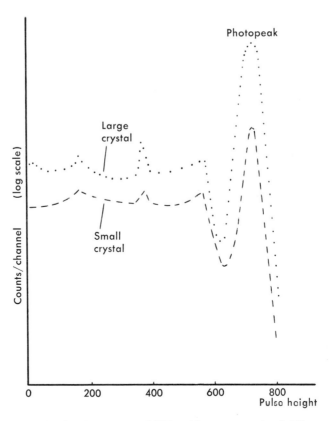

Fig. 7-30. Pulse-height spectra of ^{65}Zn with two crystals of different size.

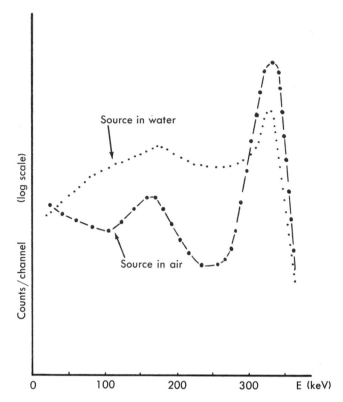

Fig. 7-31. Gamma spectrum of ^{51}Cr in air and in water.

However, again the choice of window width is limited by the quality and size of the crystal and by the intensity of the source (with a weak source and with short counting times a statistically insignificant number of counts would be collected per channel, if the number of channels is large). For most applications, one can say that a window width equal to 2% of the total pulse height scale is perfectly satisfactory with crystals of medium size.

The size of the crystal affects the shape of a gamma spectrum, because the larger the crystal, the higher the probability that a gamma photon dissipates all its energy in the crystal, with the result of a more intense photopeak and of a less intense Compton smear, as illustrated in Fig. 7-30. There is a definite advantage in using crystals as large as possible. Unfortunately, the price of very large crystals of high quality is often prohibitive for many laboratories.

Extra counts are added in the Compton region of the spectrum when the crystal is surrounded by some material (such as shielding). These extra counts may originate from (1) gamma rays that strike the crystal after being reflected by the shielding material, in which they dissipated part of their energy through a Compton scattering event, (2) gamma rays that are completely absorbed in the surrounding material and cause fluorescent X-rays from the material to enter the crystal, and (3) gamma rays that scatter out of the crystal, enter the surrounding material, and backscatter into the crystal at a lower energy.

The type of material surrounding the source also affects the shape of its gamma spectrum. Fig. 7-31 shows the spectra of ^{51}Cr (gamma energy, 320 keV), when the source is surrounded by air and by water. The total counts (area under the curve) for the source surrounded by water have increased considerably over those of the bare source. When the source is surrounded by water, many gamma rays are deflected into the detector through Compton scattering, which is more probable in water than in air. On the other hand, some gamma rays that would have struck the detector at their original energy and would have dissipated their energy completely in the detector are deflected by water and continue on at lower energy. This deflection explains the lower intensity of the photopeak when the source is surrounded by water. These facts should be borne in mind when the source under investigation is localized in an organ of the body; in these conditions, the activity is surrounded by tissues and therefore by a large amount of water.

The general advantages presented by scintillation systems in gamma counting over other types of radiation-detecting instruments have been pointed out in preceding paragraphs. Spectrometry and differential counting are preferable to integral counting in certain situations and for the solution of a number of problems. The gamma spectrum of an unknown gamma emitter may lead to the identification of the source. Even the composite spectrum of a mixture of two or more unknown emitters can be unscrambled and can lead to the identification of the single components, if the gamma-ray energies are sufficiently different to show distinct photopeaks in the spectrum. This is a problem very often encountered in activation analysis (see Chapter 6) and in medical studies involving, for example, the accidental ingestion of unknown radioactive materials. If the source is known, in many cases one may prefer to count its activity differentially under the photopeak, that is, to collect only the counts generated by the photons totally absorbed in the crystal. With this method, the signal-to-noise ratio is greatly improved, because in the channels encompassing the photopeak the thermal noise is nonexistent and the background count rate is very low. Differential gamma counting is the method of choice in many biologic tracer experiments and in medical diagnostic tracer studies involving gamma emitters (such as radioiodine-uptake studies). If the experiment or study involves the simultaneous use of two or more radiotracers, they can be counted separately, by counting the activity successively under the photopeaks of the different components of the mixture.

Good, sophisticated spectrometers are available on the market today. They incorporate all components of the system (except the detector) in one single cabinet and offer a number of desirable features, such as the automatic passage from one channel to the next, automatic recording on paper of the count rates for each channel, automatic subtraction of background from the count rate of each channel, and automatic plotting of spectra on graph paper (Fig. 7-26).

The spectrometer described so far incorporates a *single-channel pulse-height analyzer,* because it counts in one channel at a time. It is perfectly satisfactory if the relatively long time needed to take a spectrum is not objectionable and if the half-life of the source is not too short. If the half-life of the source is very short (a few seconds), then one obviously cannot take a reliable spectrum with a single-channel analyzer, because it is quite possible that by the time the spectrometer is counting in the last channels, the activity has practically disappeared. This is a problem often encountered in activiation analysis. In situations of this type, a *multichannel analyzer* should be used. This instrument (Fig. 7-32) analyzes all the channels simultaneously, so that if the activity of the source is fairly high the whole spectrum can be taken in 1 minute or less. The spectrum is displayed on a fluorescent screen similar to that of an oscilloscope, and the count rates for all the channels can be printed out on paper or punched on tape. The spec-

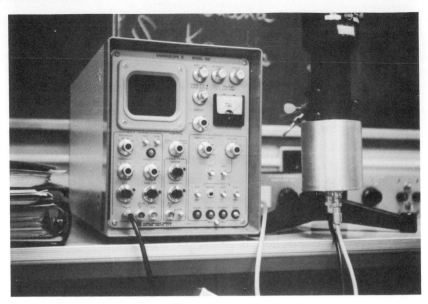

Fig. 7-32. Multichannel pulse-height analyzer. The pulse-height spectrum can be displayed on the fluorescent screen (upper left corner). The detector is to the right of the analyzer.

trum can be stored in the memory of the instrument for a long time, so that the printing-out of the data and the automatic plotting of the spectrum can be done a long time after the source has been analyzed. Some multichannel analyzers can split the whole pulse-height range into several hundred channels and analyze them simultaneously.

Beta spectrometry. Although magnetic spectrometers are generally preferred for beta particle spectrometry because of their better resolution, scintillation spectrometers can be useful at least for educational purposes, and, in addition, they present the advantage over the former of not requiring sources of very high activity. For beta scintillation spectrometry, of course, an anthracene crystal is used instead of the NaI(Tl) crystal, as explained above for beta integral scintillation counting. All other components of the system are identical to those used for gamma spectrometry. Unlike gamma differential counting, beta differential scintillation counting does not find any practical applications and does not present any advantage over integral counting. Likewise, the usefulness of beta scintillation spectrometry is rather limited and practically confined only to the study of beta particle properties and behavior. (Note a beta spectrum in Fig. 6-7.)

Liquid scintillation systems

In 1950 G. T. Reynolds and H. Kallman reported that certain organic solutions fluoresce when such solutions absorb the energy of ionizing radiation, and they suggested, therefore, that these solutions could be used as scintillation detectors for radiation counting.

The radioactive sample can be mixed with a volume of liquid fluor in a vial, which is coupled with a photomultiplier. Whenever an ionizing particle or photon interacts with the surrounding fluor, a scintillation is produced, whose intensity is proportional to the amount of energy dissipated. The photocathode converts the flashes of light into bursts of photoelectrons, which undergo the multiplication process through the dynodes of the PM, thus resulting in electric pulses, whose amplitudes are proportional to the energy dissipated in the flour by radiation. These pulses can then be handled as in other scintillation systems.

There are several advantages offered by liquid fluors over solid scintillators. First, the very fact that the radioactive sample is incorporated in the fluor improves the counting geometry considerably. In many instances, the sample can be dissolved in the fluor, with the result of the most intimate contact possible. In other cases it can only be suspended, or deposited on some type of supporting material (commonly filter paper), which is then immersed in the fluor. In any event, there is always a better contact between sample and sensitive volume of the detector than in most of the other radiation-detecting systems. Consequently, none of the radiation energy is absorbed by any material interposed between the source and the detector, and usually there is no self-absorption either. This advantage is of great importance when very low energy beta emitters, such as ^{3}H and ^{14}C, are counted. Another advantage consists of the very short resolving times possible with the liquid fluors (as short as 0.001 μsec against 0.3 μsec of many solid scintillators). Obviously, to take full advantage of this desirable property, it is necessary that the electronic components associated with a liquid scintillation detector (such as amplifiers and scalers) be as "fast" as possible. With short resolving times, relatively high count rates can be easily handled without need of correction for coincidence loss.

Because of all these desirable characteristics, a liquid scintillation system is the system of choice for accurate work with weak beta emitters, such as ^{45}Ca, ^{35}S, and in particular ^{3}H and ^{14}C, which are so often used in biologic tracer methods. Except for autoradiography (see p. 171), a liquid scintillation system is, in effect, the only device that can detect ^{3}H. Radionuclides that decay exclusively by electron capture (such as ^{55}Fe) and emit only a soft X-ray in the decay process can be counted efficiently only with a liquid scintillation system. Recently, the system has been used for the radioassay of alpha emitters with counting efficiencies better than those obtainable with other instruments.

With good liquid scintillation systems the counting efficiency for ^{14}C can be as high as 90% and for tritium 40%. On the other hand, a liquid fluor is less efficient than a solid scintillator in the conversion of the energy of the ionizing radiation into light energy. The efficiency in question here is the percentage of radioactive energy dissipated in the fluor that is actually converted into light energy. This efficiency is never 100%, because part of the radiation energy is converted into heat or other useless forms. This relatively poor conversion efficiency of the liquid fluors results in weaker output pulses from the PM, as compared with the pulses generated in solid scintillation detectors by equal amounts of energy absorbed. This disadvantage can be corrected with a higher electronic amplification of the pulses. Another disadvantage of liquid fluors is, obviously, the fact that they must be discarded after use. However, at the present time, the costs of the scintillation solutions are well within reasonable limits, so that this disadvantage is not a serious objection against liquid scintillation counting.

A typical scintillation solution is made of a primary solvent (with or without a secondary solvent) with a primary and a secondary solute. The scintillation mechanism in these solutions is still imperfectly understood, or at least is not as clear as in solid fluors. It seems, however, that the energy of an ionizing particle or photon is directly absorbed by the abundant molecules of the solvent, which are either excited, ionized, or dissociated. From the solvent molecules the energy is transferred to the molecules of the primary solute, which are thus raised to a higher energy level. When they return to the ground state, they emit energy in the form of a violet or near-ultraviolet photon (fluorescence). The primary solute is therefore a fluor. The short wavelength (about 3500 A) of the fluorescence photons is not the one to which the photocathode of the PM is most sensitive. The peak sensitivity of most photocathodes is in the 4000 to 4500 A range. The purpose of the secondary solute is to shift the short wavelength of the light photons emitted by the primary solute to a longer wavelength (about 4200 A), that is, to light of a wavelength that can be converted most efficiently into photoelectrons by the photocathode.

In summary:

Radiation energy $\longrightarrow$ solvent $\longrightarrow$ primary solute $\longrightarrow$ fluorescence (short λ) $\longrightarrow$
  secondary solute $\longrightarrow$ photon of longer λ $\longrightarrow$ photocathode $\longrightarrow$ photoelectrons.

There is a wealth of information in the literature on the search made by several investigators for the best possible solvent and solutes suitable for scintillation solutions. The ideal solvent should be one that is transparent to the fluorescence light emitted by the solutes, so that the light can reach the photocathode with as little loss as possible; it should be able to transfer the absorbed energy promptly to the primary solute and to dissolve a wide variety of radioactive samples. Unfortunately, many organic solvents that are satisfactory in every other respect are not good solvents of many radioactive chemicals. After considering all the unfavorable and favorable characteristics, most investigators are still using toluene as primary solvent in spite of its poor solvent power.

Among the hundreds of substances tested as primary fluors, only two have gained wide acceptance for their desirable characteristics — p-terphenyl and the more soluble 2,5-diphenyloxazole (PPO).* The counting efficiency of a primary solute is roughly directly proportional to its concentration in the scintillation solution, up to a certain limit, which for PPO is 2 g/liter of toluene.

The most widely used secondary fluor is 4-bis-2-(5-phenyloxazolyl)-benzene (POPOP).* Its fluorescence has a peak intensity at about 4200 A, which matches the wavelength to which the photocathode is most sensitive. The use of a secondary solute is particularly advisable for counting tritium. A typical composition of a toluene-PPO-POPOP solution has been given on p. 146.

Special glass vials of about 20 ml capacity are used for the fluor solution — sample mixture. Common ordinary glass is always radioactive, because it contains a large amount of potassium and therefore of the natural radioisotope ^{40}K. To reduce the BG count rate, special glass with low potassium content is used for the vials employed in liquid scintillation counting. Available on the market are also vials of polyethylene; although they are inexpensive and present no problem with potassium, they seem to be permeable to the solvents commonly used in scintillation solutions.

---

*Structural formulas of PPO and POPOP (P = phenyl, O = oxazole):

PPO   **POPOP**

Liquid scintillation counting presents special problems that must be faced and solved correctly, if this counting method is to be efficient, useful, and reliable.

One will recall that in the output of a PM tube there are small pulses that are caused by the so-called thermal noise generated in the tube itself. These noise pulses present no problem in gamma scintillation counting, because they do not interfere with the pulses of larger amplitude generated by gamma interaction in the crystal. In a liquid scintillation system used to count beta particles as weak as those of ^{3}H or ^{14}C, the noise pulses may have an amplitude not too different from that of the pulses generated by beta particles, and thus they may interfere with the counting. The thermal noise can be reduced if the photomultiplier is kept at low temperature during the operation. Many commercial liquid scintillation counters are sold with a deep freezer that houses the photomultiplier and the sample. The choice of the temperature is limited by the freezing point of the scintillator used. For toluene solutions the temperature can be set at $-8°$ C.

Another way to solve the problem of the thermal noise is to block the noise pulses with a pulse discriminator incorporated in the electronics of the system. This discriminator (also called low gate) can be set at a variable pulse-height level, depending on the maximum or average energy of the beta particles to be counted. The setting of the low gate can be quite problematic in counting tritium.

A more ingenious method to minimize the same problem is the inclusion of a coincidence circuit in the system. On the opposite sides of the vial con-

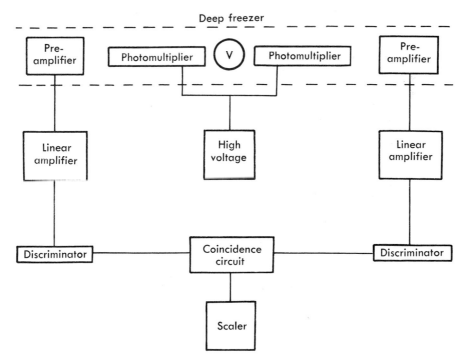

Fig. 7-33. Block diagram of a simple liquid scintillation counter. **V,** Vial containing the scintillator-sample mixture.

taining the scintillator-sample mixture there are two PM tubes, which are connected, via their preamplifiers, to two amplifiers. The output pulses of the two amplifiers are fed to a special electronic circuit, called a "coincidence circuit," that rejects any pulse coming from one amplifier only, whereas it accepts and passes on to the scaler pulses that arrive simultaneously (in coincidence) from both amplifiers (see Fig. 7-33). Since thermal noise pulses are random pulses, the probability that they originate simultaneously in both PM tubes is very small. Consequently, most of them will be blocked by the coincidence circuit. On the other hand, pulses generated by the scintillations occurring in the fluor must be simultaneous in both tubes, because both tubes "see" a scintillation at the same time. These pulses are accepted by the coincidence circuit and counted by the scaler.

Another problem often encountered in liquid scintillation counting is fluorescence quenching (which should not be confused with the quenching of G-M and proportional detectors). Fluorescence quenching occurs whenever a substance present in the scintillator interferes with the scintillation mechanism discussed above. This interference results in decreased efficiency of the energy conversion (radioactive energy → light energy) and consequently in a reduced light output per ionizing particle or photon absorbed. The quencher may absorb the energy transferred from the solvent and make it unavailable for the primary fluor (chemical quenching), or if a quencher is a colored substance, it may absorb part of the violet or ultraviolet light emitted by the fluors and make it unavailable for the photocathode (color quenching). Possible quenchers may be found among components of the scintillation solutions or among the radioactive samples themselves (such as blood and carotenoids). This problem can be corrected only by preventing it. If quenching cannot be prevented for any reason, the next best thing that can be done is to be aware of it, try to estimate its extent, and find ways to stabilize it, so that it remains constant in all the samples to be counted. A discussion of these techniques is outside the scope of this book.

All the components of a simple liquid scintillation system can be briefly summarized with a block diagram (Fig. 7-33). A high-voltage power supply impresses a high voltage to the PM tubes, which are at the opposite sides of the sample vial. The output pulses of the PM tubes are fed to two linear amplifiers that are connected with the coincidence circuit, as explained above, through two discriminators. The pulses not rejected by the discriminators and by the coincidence circuit are counted by the scaler.

More versatile instruments are supplied with other discriminators that allow the simultaneous counting within two windows (channels) of different pulse-height intervals; the width and the location of the channels on the pulse-height unit scale (1000 units) are variable. For instance, the instrument can be set in such a way as to count simultaneously, but separately, the pulses falling in the channel 50 to 400 phu and in the channel 400 to 900 phu. Obviously, two scalers are needed for the two channels. The choice of the channels is an art that can be mastered only after some experience. It is dictated by the nature of the radionuclide to be counted, by the degree of efficiency required, by the difference in amplitude between the noise pulses and the pulses generated by the radiation involved, and by other considerations. With two channels, one can count simultaneously a mixture of two different radionuclides, or a doubly

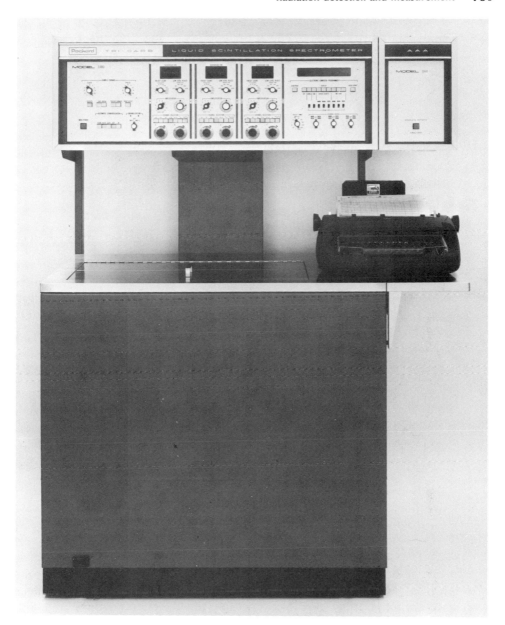

Fig. 7-34. Modern liquid scintillation spectrometer. (Courtesy Packard Instruments Co., Downers Grove, Ill.)

labeled sample, such as a compound that has been labeled with ^{14}C at one point of its molecules and with tritium at another. This simultaneous double counting is possible whenever the radiation energies of the two nuclides in question are quite different in magnitude, such as the case for ^{14}C and tritium; the beta particles from tritium (of lower energy) are counted in the "lower" channel, and the beta particles from ^{14}C are counted in the "higher" channel.

Several types of instruments offer additional automatic devices for speedier operation, such as automatic background subtraction from the sample count rates, automatic sample changer, which can handle up to 200 or 300 vials, automatic standard source to check the counting efficiency, etc. (Fig. 7-34).

In conclusion, because of its high efficiency in counting weak beta emitters and its reliability and versatility, the liquid scintillation counter has become very popular among biologists using radiotracer techniques in their research, and today it is a necessary piece of equipment for any laboratory, where extensive work is done with radiocarbon and tritium.

RADIOGRAPHY AND RADIOAUTOGRAPHY

As soon as X-rays and radioactivity were discovered in 1895 and 1896 respectively, the investigators noticed that a photographic emulsion is sensitive to both X-rays and radiations emitted by radioactive materials and that therefore it could be used to detect these radiations (see Chapters 4 and 5). The first X-ray pictures of hands and other objects were taken about 1895. We may recall that Becquerel was led to the discovery of the radioactivity of his uranium salts by noticing that those salts produced a latent image on a photographic plate, exactly as visible light does.

The fact is well known that a photographic emulsion is basically made of grains of silver halide (such as AgBr) suspended in a gelatin medium. If the emulsion is exposed to visible light, ultraviolet rays, or ionizing radiation, the photon or particle initiates a reduction of the silver halide molecules, through a mechanism that is still imperfectly understood. However, no darkening is visible in the emulsion, until the latter is developed, that is, treated with special reducing agents (developers) that complete the reduction of the exposed silver halide to metallic silver (black). The latent image thus becomes visible. To fix and make the image permanent, one treats the emulsion with a fixer, which is a chemical that removes all the unexposed, unreduced silver halide. Since the degree of darkening of the emulsion is, within limits, a function of the amount of energy dissipated in the emulsion, one can, to a certain extent, use these methods quantitatively in some instances. However, quantitative determinations are not as easy and reliable as with other radiation-detecting techniques.

Photographic emulsions can be used, in connection with ionizing radiations, for one of the following three different purposes:

1. If an object is not uniformly opaque to a certain type of ionizing radiation and yet its internal structure is not visible (such as an organ of the body or a sealed instrument), it is possible to have an image of its internal structure on a photographic film, by interposing the object between the film and a source of X- or gamma rays. With this method, the purpose is not to detect the possible presence of ionizing radiation or to measure its intensity; radiation is simply used as a means to detect the internal structure of the object. This technique, widely used in medical sciences, in industry, and occasionally in biology, is called *radiography,* and the image of the object on the film is a *radiogram* (X-ray picture). In a radiogram the parts of the object that were less transparent to the radiation used (such as the bones of the hand) appear less dark than the more transparent parts.

2. If an object contains radioactive materials and therefore sources of ionizing radiation, it should be possible to detect its radioactivity (and identify its location, if it is not uniformly distributed) by simply placing the object in close contact with a photographic emulsion. Obviously, no external source of

radiation is needed. Since the object, in effect, is "taking a picture of itself" by using its own source of radiation, this technique is called *radioautography*, and the image of the object obtained with this technique is a *radioautogram*. In a radioautogram, the parts of the object that are more radioactive appear darker than the less radioactive ones.

3. Finally, a photographic emulsion can be used to detect the passage of an ionizing particle, interactions between elementary particles, fission, or other similar events. After development the emulsion will show dark tracks in correspondence to the paths of the particles. These techniques are widely used by nuclear and high energy physicists; they need not concern us here.

Radiography

The applications of radiographic techniques have been mentioned above and in Chapter 4. Since they have little use in biologic work, only some additional information is given below.

The choice of radiation source depends very much on the object that is radiographed and on its opacity and thickness. Although for the radiography of oil paintings 20 kV X-rays are perfectly satisfactory, 130 kV X-rays are needed for dental radiography, 250 kV for deeply seated organs of the body, and in industry X-rays of higher energy must be used for the radiography of heavy metallic objects. Recently, very intense gamma sources (thulium 168) have come into use for this purpose.

Special photographic films (such as Kodak No-Screen X-ray film) are commonly used. They differ from ordinary camera films essentially in two ways; they are coated with sensitive emulsion on both sides so that a sharp image can be obtained regardless of which side is in contact with the object; and they are faster in order to reduce exposures to very short times, especially in medical and dental radiography.

Medical radiography presents special problems of interest to the radiation biologist. They are dealt with elsewhere in this book.

Radioautography

Radioautography (or autoradiography, as it is sometimes called) may be defined as "a method for locating radioactive substances by use of modified photographic techniques."*

It seems that the first published radioautogram of a biologic material was obtained by E. S. London, in 1908. It was the radioautogram of a frog that had been kept for some time in a solution containing radium. The whole frog was then placed on a photographic plate. The radioautogram showed that the skin of the animal had uniformly absorbed radium.

Since 1908, radioautography has been extensively used by biologists, and the techniques have been greatly refined, while a variety of photographic emulsions suitable for specific needs has become increasingly available. The radioautographic method has a special appeal to the biologist interested in radiotracer techniques because no sophisticated instrumentation is required, and mostly because with this method it is possible to locate precisely the radioactivity in both macroscopic and microscopic biologic materials. This result cannot be obtained with any other radiation-detecting device.

The most desirable characteristic of a radioautogram is good resolution. Resolution is defined as the minimum distance between two point sources of radioactivity that still allows them to be seen as distinct points on the radio-

*See W. Gude. 1968. Autoradiographic techniques. Prentice-Hall, Inc., Englewood Cliffs, N. J. P. xii.

autogram. For all practical purposes, good resolution means sharp images. The degree of resolution is dependent on several factors, such as thickness of the specimen and of the emulsion, type of contact between the specimen and the emulsion, and specific ionization and range of the radiation emitted by the specimen. Therefore, no matter what special technique is used, one must assure the closest possible contact between emulsion and specimen. In regard to the characteristics of the radiation, the radiations most suitable for radioautography are those with a high specific ionization and short range, since they can dissipate most or all of their energy within a thin layer of emulsion, with the largest number of interactions. Consequently, alpha emitters are particularly suitable for radioautography, although actually they are not extensively used because of their limited applications as biologic radiotracers. The degree of resolution obtained with beta emitters is roughly inversely proportional to the E_{max} of the beta particles; thus ^3H and ^{14}C are most suitable for radioautography, followed by ^{35}S, ^{45}Ca, ^{90}Sr, and others with beta E_{max} in the same range. A strong beta emitter like ^{32}P (beta $E_{max} = 1.71$ MeV) gives ill-defined images. X-ray emitters (^{55}Fe) may give satisfactory results in certain conditions. Gamma emitters are little suitable for radioautography, but if they emit also betas of relatively low energy (like ^{131}I), they can be used with some success.

Several types of emulsions are available at the present time for radioautographic work, and they are sold with or without a supporting base (glass plate or celluloid film). The most commonly used emulsions manufactured by Eastman Kodak are those of the NT (nuclear track) Series. Type NTA has fine grain and low sensitivity; it is designed for particles of very strong ionizing power, such as protons and alpha particles. Other types are specifically suitable for beta emitters; such types are NTB, with low sensitivity and excellent resolution; NTB_2, with intermediate sensitivity and very good resolution; and NTB_3, with high sensitivity and satisfactory resolution. The exposure time necessary to obtain a good radioautogram decreases from NTB to NTB_3. Type NTE (Nuclear Track Electron) is an extremely fine-grain emulsion with very low sensitivity, recommended for electron-microscopic specimens.

One can seldom determine or calculate exactly in advance the time needed to expose the specimen to the emulsion in order to obtain a good radioautogram. The determination of the exposure time is empirical in most situations; nevertheless, it is advisable, for this determination, to take into account the specifications of the emulsion used, the manufacturer's recommendations, and the estimated activity of the sample. It is good practice to prepare several specimens of the same material and expose them to the emulsions for different time intervals. In many instances, exposure times must be as long as several days or weeks.

Distinction must be made between gross and microscopic radioautography. If the technique is used to locate the radioactivity in parts or organs of the specimen that are visible and distinguishable with the naked eye, then the whole specimen is radioautographed by placing it in contact with the emulsion supported by a glass plate or celluloid film (gross radioautography). The location of the activity can be seen on the developed radioautogram with the naked eye. The gross radioautograms of Fig. 7-35 were made to find out whether a radionuclide is absorbed uniformly by a leaf and by the body of a mouse, or rather selectively in the veins of the leaf and in the skeleton of the mouse. On the other hand, the purpose of the technique may be to locate the activity at a microscopic level, that is, in parts of the specimen, such as tissues or cellular organelles

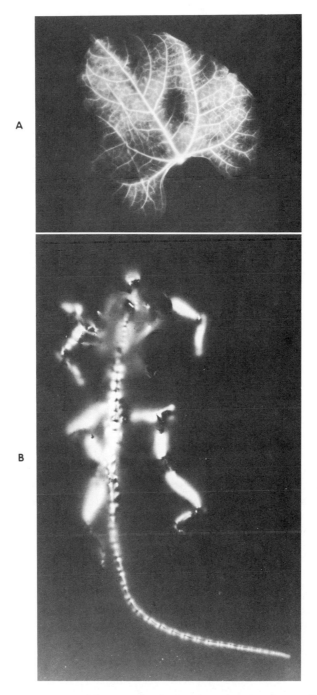

Fig. 7-35. A, Radioautogram of leaves of a bean plant that has been allowed to absorb [45]Ca through the root system. **B,** Radioautogram of the skeleton of a mouse that had been injected with a solution of [90]Sr. Light areas correspond to high activity because the illustrations are contact prints.

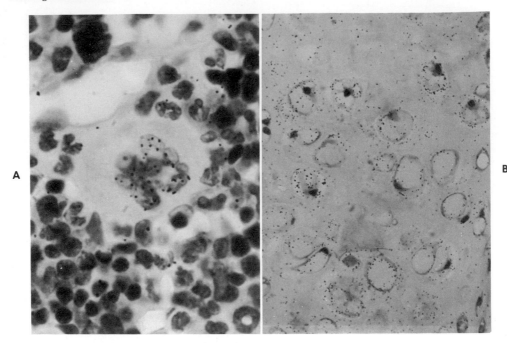

Fig. 7-36. A, Radioautogram of rat bone marrow labeled with tritiated thymidine. The large cell in the center of the field is a megakaryocyte in mitosis. The label was injected intraperitoneally about 1 hour before the animal was killed. Exposure to NTB$_2$ emulsion was for 15 days. After development and fixing, the tissue was stained in Harris' hematoxylin and eosin and mounted. **B,** Radioautogram of a section of rat tracheal cartilage labeled with ^{35}S sulfate. Mucopolysaccharides in the cartilage cell cytoplasm have incorporated the radioisotope. The section was prestained with Feulgen-fast green before the NTB$_2$ liquid emulsion was applied for a 15-day exposure. (From Gude, W. D. 1968. Autoradiographic techniques: localization of radioisotopes in biological material. Prentice-Hall, Inc., Englewood Cliffs, N. J.)

that can be seen only with a microscope (Fig. 7-36). The location of the activity on the radioautogram (which sometimes is superimposed on the specimen) can be seen only with a microscope. More sophisticated methods are needed to achieve this purpose (microscopic radioautography) and to obtain the best possible resolution.

The technique for gross radioautography is relatively simple. For best results, the specimen should be thin (or made thin, if possible), and dried. Leaves are easily dried by keeping them under pressure between sheets of absorbent paper. In a darkroom, the specimen is placed in contact with a sheet of No-Screen X-ray film, or other film suitable for radiography; both are placed in a lightproof film holder (Fig. 7-37). A weight is laid on top of the film holder to ensure the best possible contact between specimen and emulsion. If one suspects that the specimen is not completely dry, he would be wise to place a sheet of thin plastic film between the specimen and the emulsion to prevent a possible chemical reaction of any juices squeezed out of the specimen with the emulsion. Even when the radionuclide involved is a beta emitter as weak as ^{14}C, the plastic film should not stop most of its beta particles. After the desired exposure time has elapsed, the film is removed from the holder in the darkroom, developed, fixed (with X-ray developers and fixers), washed, and dried. The radioautogram can be examined most conveniently with an X-ray viewer (a sheet of white glass illuminated from behind). The matching of the radioautogram with the specimen is sometimes necessary to locate

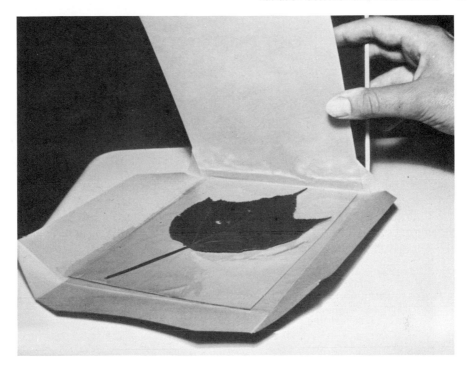

Fig. 7-37. Technique for gross radioautography. The specimen is laid on a No-Screen X-ray film in a film holder.

exactly the radioactivity. The same technique is also used to obtain radioautograms of chromatograms containing spots of radioactive compounds (radiochromatograms).

In microscopic radioautography, the problem of obtaining a good resolution and definition of details is even more stringent. The technique, therefore, must involve the use of sections of material as thin as possible, the use of very thin layers of emulsion, and, most of all, the best contact between specimen and emulsion, especially if the labeling isotope is tritium. These results can be achieved with variable success with one of the following four basic methods:

Method of simple apposition. The method of simple apposition does not differ substantially from the method used for gross radioautography. After the specimen has been fixed and the paraffin sections have been obtained, a microscopic slide with the sections mounted on it is placed against the emulsion supported by a glass plate or celluloid film, and clamped with it to ensure good contact (Fig. 7-38, *A*). After the desired exposure time in a lightproof box, the emulsion is developed and fixed, and the slide with the sections is processed as usual (removal of paraffin, hydration, staining, dehydration, and mounting). This method is satisfactory only in a limited number of situations, because the contact achieved between sections and emulsion is not the most intimate possible and because specimen and radioautogram are separate, so that the matching of the dark areas on the emulsion with the tissues, cells, or intracellular structures of the specimen becomes somewhat problematic.

Mounting method. The paraffin sections are mounted directly on a microscopic slide that has been previously coated with the emulsion. The mounting is best done by floating the sections on warm water (Fig. 7-38, *B*) and picking them

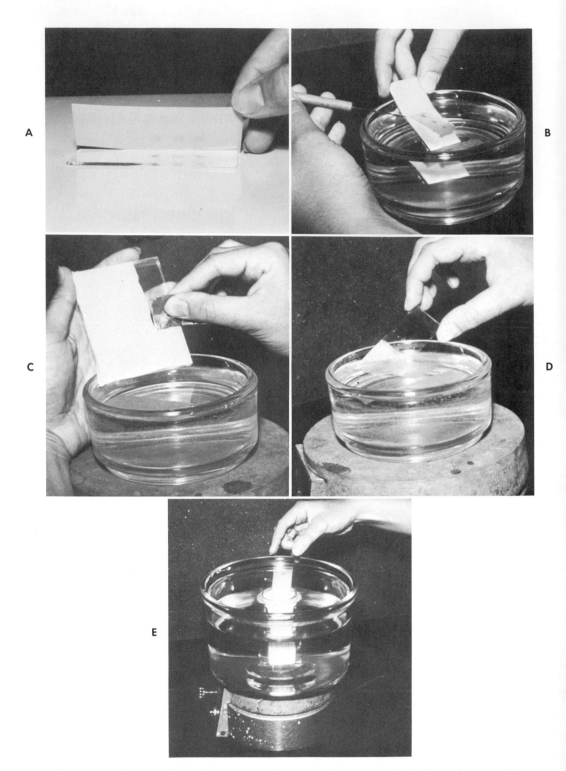

Fig. 7-38. Techniques for microscopic radioautography. **A,** Method of simple apposition. **B,** Mounting method. **C** and **D,** Stripping film method. **E,** Dipping method.

from the water with the slide and a teasing needle. After drying, the slide is placed in a lighttight box for the exposure. When the desired exposure time has elapsed, the paraffin is removed as usual. The slide is first developed and fixed, then stained if desired, and mounted with a cover slip.

With this method, specimen and radioautogram remain permanently together on the same slide and are viewed simultaneously through the microscope, with the advantage of seeing the dark areas of the radioautogram in correspondence with the radioactive structures of the specimen that produced them. Notice that with this method the emulsion (and thus the radioautogram) is between the supporting glass slide and the specimen.

Stripping film method. For the stripping film method, a so-called stripping film plate is needed, which can be purchased from the manufacturer. The stripping film plate is a glass plate loosely supporting a layer of emulsion that can be stripped easily from the plate with a razor blade. The sections of the specimen are mounted on an ordinary microscopic slide and deparaffinized. A layer of emulsion of suitable size is stripped from the supporting plate, floated on warm water, and wrapped around the slide, which supports the sections (Fig. 7-38, C and D). After the slide is dry, it is kept in a lightproof box for exposure and then developed, fixed, stained, and mounted. Staining can be done before the stripped film of emulsion is placed in contact with the sections.

Again, with this method, specimen and radioautogram remain permanently together on the same slide, with the advantage already pointed out for the second method. Notice, however, that with the stripping film method the specimen is between the supporting glass slide and the radioautogram.

Dipping method. The dipping method is similar to the stripping film method, except that the slide supporting the deparaffinized sections is coated with the emulsion, by dipping it into a jar containing the emulsion, which has been previously melted in a warm water bath (Fig. 7-38, E). A much closer contact between sections and emulsion is achieved with this technique. The emulsion is sold by the manufacturer as a gel, in a bottle. After the coated slide is dried, it is handled as with the stripping film method. Again, there is a permanent contact between specimen and radioautogram, and again the specimen is between the slide and the radioautogram.

Of course, all the steps just described, involving the handling of the emulsion before it is developed and fixed, must be carried out in a darkroom with a safelight (a 15-watt bulb with a special filter).

With microscopic radioautography, one can in many instances determine quantitatively the activity of a microscopic structure (chromosome, nucleolus, etc.) and therefore the relative amount of radioisotope present in it, by counting the black silver grains that are seen on the radioautogram in correspondence with the structure. For this purpose a high-power microscope with a good resolution should be used.

While examining a radioautogram, one should be aware of certain artifacts that might lead to erroneous conclusions if they are not detected. Like any other radiation-detecting device, any emulsion used in radioautography is sensitive to background radiation. The BG causes the same effect in the emulsion as the radiation under study. Some black silver grains in a radioautogram may be produced by BG. To minimize this BG artifact, one would be wise to use fresh emulsions, which have been stored only for limited periods of time in a refrig-

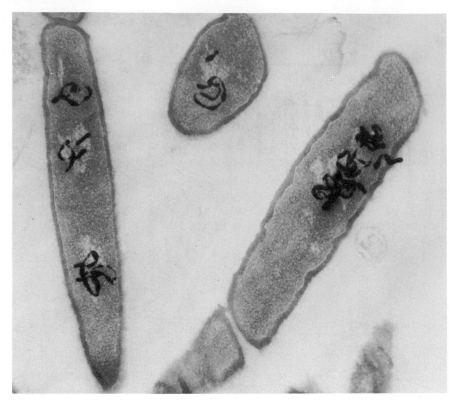

Fig. 7-39. Thin section of *Bacillus subtilis* labeled with tritiated thymidine. Stained in uranyl acetate. The grains seem closely associated with the nuclear region. (38,000×.) (From Caro, L., and R. P. Van Tubergen. 1962. High resolution autoradiography. J. Cell Biol. **15:**173-188.)

erator and away from radioactive sources. The radioautographic emulsion is perhaps the only item that should never be stored in a radioisotope laboratory. Also, very long exposures of the emulsion to the specimen should be avoided. Other artifacts may be caused by overdevelopment of the emulsion, or by chemical reactions between some substances present in the specimen and the emulsion (chemical fogging).

Recently, radioautographic techniques have been applied in electron microscopy. Special emulsions with a very fine grain (such as the Eastman Kodak NTE emulsion) are available for this purpose. The grid supporting the specimen is coated with the emulsion, exposed, and then developed, fixed, and stained (with uranyl or lead stains). The electron micrograph of Fig. 7-39 is a remarkable example of the results obtainable with this technique. With the very high magnifications of the electron microscope, the silver grains reveal their true (spaghetti-like) shape.

With special methods, it is possible to distinguish radioautographically two compounds labeled with different radionuclides, such as ^{3}H and ^{14}C, present in the same specimen. It is also possible to study the metabolic fate of a doubly labeled compound.*

---

*See, for instance, Trelstad, R. L. 1965. Double isotope autoradiography. Exp. Cell Res. **39:**318–328.

Two layers of emulsion, separated or not by an inert film of celloidin, can be deposited on the specimen (with the stripping film or dipping method). The tritium beta particles, with a shorter range, produce silver grains only in the first layer, whereas some of the ^{14}C betas (the most energetic) are recorded in the second layer. With proper focusing, one can possibly distinguish the two layers and thus the two nuclides. However, for a correct analysis and interpretation of the results, one should bear in mind that many ^{14}C betas (the least energetic) are recorded in the first layer, along with the tritium betas. Nevertheless, the technique can be reliable, if used cautiously.

Radioautographic techniques have been used fruitfully in a wide variety of areas of biologic research, and specifically in histology, cytology, cytogenetics, cell physiology, and molecular biology. The solution of many basic problems in these fields has been possible through the application of these techniques. The synthesis of nucleic acids in the cellular nucleus and cytoplasm has been clarified by exposing the cells to tritium-labeled precursors of nucleic acids (for example, tritiated thymidine) and radioautographically following their fate within the cells. We have been able to understand the behavior of chromosomes and other cellular organelles during the mitotic cycle and to solve many problems of bacterial and viral genetics, DNA duplication, and intracellular protein synthesis. The development of new, more refined and sophisticated radioautographic techniques is a fascinating area of endeavor for biologists interested in radioisotope tracer methods.

SAMPLE PREPARATION TECHNIQUES

Earlier in this chapter I stated that in the biologic work done with radiotracers measurement of the relative activity of a sample is sufficient and that when a series of samples must be radioassayed the relative activities can be kept proportional to the absolute activities if all the samples are assayed with the same instrument in identical conditions. In other words, the relative activity data for a series of samples are meaningful if the counting yield is kept constant. For satisfaction of this condition, in most cases the samples must be manipulated and transformed in some suitable physical form.

Suppose that we have designed a ^{131}I-uptake experiment to show that different organs of a rat take up different amounts of an injected dose of radioiodine, that these different amounts are independent of the size of the organ, and that the minuscule thyroid takes up much more radioiodine than any other single organ of larger size. Suppose that we have decided to assay several organs by counting the radioiodine beta particles with a G-M counter, rather than the gamma radiation with a scintillation counter. It would be foolish to place organs as such, or chunks of organs, in front of the G-M tube, record the count rates, and assume that they are proportional to the absolute activities present in the samples! The different sizes, shapes, water contents, textures, etc., of the samples cause different degrees of self-absorption and backscattering, and affect differently other factors that determine the counting yield. The counting yield is not kept constant, and consequently if the thyroid shows a count rate 10 times larger than that of the liver, this ratio does not mean at all that the thyroid has taken up 10 times more radioiodine than the liver. It should be clear then that in an experiment of this type the samples should have been prepared in such a way as to ensure at least a constant counting geometry, which is one condition to obtain a constant counting yield throughout the radioassay of a series of samples.

A variety of sample-preparation techniques is available; although several of them are standard and commonly used in many situations, many others can be developed to suit special counting problems, and their number is only limited by the ingenuity of the investigator. However, before a sample-preparation technique is decided upon, the investigator should answer the following questions:

1. *Does this technique convert the sample into a form suitable to the type and energy of the radiation to be counted?*

If the sample contains ^{14}C, is the sample thin enough so that self-absorption is negligible? On the other hand, if gamma radiation is counted, the problem of self-absorption obviously does not exist.

2. *Is the sample form obtained with this technique suitable for the type of detector to be used? To its size? To its efficiency?*

Concerning this point, however, one must say that, on occasions, the sample form available or obtainable is what determines the type of detector to be used.

3. *Does the sample form obtained with this technique ensure the highest possible counting yield?*

The highest possible counting yield is of vital importance when samples of very low activities are involved. The highest possible count rates will mean more statistically significant data. If the counter used is a proportional detector with a 2-inch window, a circular sample 2 inches in diameter might give a better counting yield than a sample 1 inch in diameter. With a windowless proportional detector, a sample of a weak beta emitter in gaseous form might give a better counting yield than if it is in solid form. In liquid scintillation counting, better counting yields may be obtained if the sample is dissolved, rather than suspended, in the fluor.

4. *Does the sample form obtained with this technique ensure a constant counting yield, and therefore reproducibility, for a series of samples, so that their count rates are really proportional to their disintegration rates and thus to the amounts of radioactive material they contain?*

If an affirmative answer can be given to all of the preceding questions, then the sample preparation technique chosen is appropriate. Details concerning these techniques can be found in specialized laboratory manuals. Here only some information of a general nature is given.

In many instances, it is convenient to place the samples in planchets, which are metallic dishes (of aluminum or stainless steel) of various diameters and depths (Fig. 7-40). A wide variety of planchets suitable for different counting problems is available on the market.

To deposit on a planchet very small, measured quantities of liquid, micropipettes, which can measure volumes as small as 5 to 10 μl, are used. Suction of radioactive material is never done by mouth, but with special suction devices (propipettes, micropipettors). If materials of high activity (several millicuries) must be handled, lead shielding should be placed between the material and the operator for protection against radiation hazards. Pipetting is done with special remote-control devices, while the operator looks at the objects he is handling through a mirror.

In the beta counting of samples deposited on planchets, there is a considerable degree of backscattering, which depends on the thickness of the planchet and on the nature of the metal. This backscattering is at times desirable because it improves the counting yield, as long as it is the same in magnitude

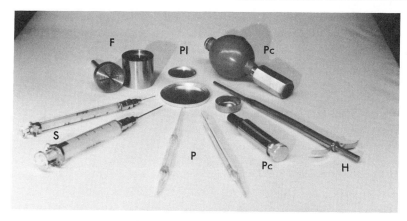

Fig. 7-40. Some basic equipment for the preparation of radioactive samples. **Pl,** Three planchets of different size and depth. **P,** Two micropipettes. **Pc,** Two suction devices for micropipettes. **S,** Syringes. **F,** Disassembled filtration apparatus used for the uniform deposition of precipitates on disks of filter paper. **H,** Device to hold planchets and prevent contamination of the operator's fingers.

for all the samples of the same series. If for some reason the backscattering effect is objectionable, the sample can be prepared on a thin plastic film supported in a hole cut in the center of a square piece of cardboard (card mounts).

Active material in solution is usually assayed in solid form, after evaporation of the solvent. Unless very low activities are expected, only small volumes (a few microliters) of the solution are pipetted on planchets and dried. In this way very thin samples are obtained and self-absorption is reduced to a minimum. Drying of liquid or wet samples deposited on planchets or card mounts is best done with the heat of an infrared lamp. To ensure uniform drying of a series of samples, a rotating turntable may be used.

For gamma counting, a solution could be assayed as such, even in relatively large volumes, because the problem of self-absorption does not exist. However, unless there are serious reasons to do otherwise, the radioassay of liquid samples should be discouraged as a rule, because of the danger of spilling, with possible contamination of the detector and other equipment. If counting of liquid samples is done, precautions should be taken to prevent spillage and to ensure reproducibility. Scintillation crystals are available for gamma counting, with a "well" drilled in for accommodation of special plastic tubes containing liquid samples. With these crystals, the counting yield is improved because the fluor surrounds the sample, and therefore more gamma photons are detected. To ensure reproducibility, however, one must use the same volume for all the samples of the same series, and this volume should not exceed the volume of the well. With a crystal of the well type, gamma counting of small organs can be done.

Active materials in suspension can be precipitated uniformly on a disk of filter paper with a filtration apparatus (see Fig. 7-40, *F*). The sample is dried on the disk, which is then placed in a suitable planchet and counted.

Beta counting of materials that are already in solid, dry form is done after reducing them to a fine powder. The powder is placed into a planchet and distributed uniformly by tapping the planchet gently against the laboratory bench until a thin and uniform layer in the center of the planchet is obtained. This technique can be used for the assay of active spots on a thin-layer chromatogram; spots are carefully scraped with their supporting silica base from the glass plate and placed in planchets. Active spots on paper chromatograms can easily be assayed after cutting the sections of paper supporting the active material. Instruments are available that assay the active spots of a radiochromatogram directly (radiochromatogram scanners) and record the activity data in graphic form. The efficiency and reliability of some of these instruments are, however, questionable.

As mentioned previously, beta counting of organs and tissues cannot be done with reliable results, without transforming them into some more suitable form. The problem can be solved in different ways, although only two methods are indicated in most cases. One, usually called *wet ashing,* consists of disintegrating the organ or tissue in a small volume of hot concentrated nitric acid or of some other acid mixture. Even bones can be completely dissolved with this method. The samples are then poured into porcelain dishes of suitable size or into watch glasses (not into metallic planchets!), dried, and counted. Otherwise, tissues or organs may be homogenized (for example, with a Waring blendor), and the homogenate is dried and counted. With either method, it is absolutely necessary to obtain finished samples of the same size and thickness to ensure good geometry and reproducibility.

Radioactive gaseous samples (such as $^{14}CO_2$ exhaled by an animal injected with a ^{14}C-labeled metabolite) can be assayed as such, if they can be mixed with the counting gas of an electrometer, proportional counter, or G-M detector. With this method, the efficiency is very high because the sample is introduced in the sensitive volume of the detector. Or the sample can be transformed into a solid by means of some chemical reaction. For instance, $^{14}CO_2$ can be converted to insoluble $Ba^{14}CO_3$, in the presence of barium hydroxide; the barium carbonate precipitate is then treated as described above. In certain cases, the gas is simply dissolved in an appropriate solvent and radioassayed in solution, if the sample is a gamma emitter.

Special techniques for the preparation of samples for liquid scintillation counting have been described previously in another section of this chapter.

SUMMARY OF RADIATION DETECTORS AND OF THEIR CHARACTERISTICS

For review purposes, all the radiation-detecting instruments and techniques described in this chapter are summarized and classified in the following outline:

Detection instruments and techniques
    Ionization instruments
        Electric potential gradient
            Gas-filled ionization chambers
                Electroscopes and electrometers
                Proportional counters
                Geiger-Müller counters
            Semiconductor detectors
        No electric potential gradient (cloud chambers and bubble chambers)
    Scintillation instruments
        With solid fluors (solid scintillation systems)
            Alpha fluor
            Beta fluor
            Gamma fluor
        With liquid fluors (liquid scintillation systems)

Photographic emulsions
    Radiography
    Radioautography
        Gross
        Microscopic
            Simple apposition
            Mounting method
            Stripping film method
            Dipping method

Certain characteristics of several types of radiation-detecting instruments and techniques are summarized in Table 7-1. This information should help the student in the choice of the appropriate detector or technique when he is confronted with a particular counting problem. Most of the figures given for the amplification factors and resolving times are only approximate; they may vary

Table 7-1. Summary of radiation detector characteristics*

Detector	Energy discrimination	Detection medium	Detector amplification factor	Resolving time (μsec)	Background	Relative efficiency		
						α	β	γ
Electroscopes and electrometers	Yes	Gas	1	—	Low	Very good	Good	Poor
Proportional detectors	Yes	Gas	10^2 to 10^4	5 to 50	Low	Good	Good	Poor
Geiger-Müller detectors	No	Gas	10^7	10 to 1000	Medium	Fair	Good	Poor
Semiconductor detectors	Yes	Solid	1	Negligible	Medium	Very good	Very good	Fair
Cloud chambers	No	Gas	1	—	Medium	Very good	Good	Poor
Bubble chambers	No	Liquid	1	—	Medium	Very good	Good	Poor
Solid scintillation detectors								
NaI crystal	Yes	Solid	10^6†	About 0.3	High	Fair	Fair	Good
Anthracene	Yes	Solid	10^6†	About 0.03	Low	Fair	Good	Poor
Liquid scintillation detectors	Yes	Liquid	10^6†	About 0.001	Low	Very good	Very good	Fair
Radioautography	No	Solid	1	—	Medium	Very good	Variable‡	Poor

*Modified from Wang, C. H., and D. L. Willis: 1965. Radiotracer methodology in biological science, Prentice-Hall, Inc. Englewood Cliffs, N.J.
†Refers to the electron multiplication factor in the photomultiplier.
‡Depends on beta E_{max}; very good for low E_{max}.

somewhat for each different instrument. Likewise, the terms used to rate backgrounds and efficiencies are general and relative, since they may vary for different instruments of the same type.

APPENDIX: Whole-body counters

A whole-body counter can detect and measure the radioactivity present in the whole body of a live animal or human individual. These instruments are essentially designed for the detection of gamma radiation emitted by the body, and therefore they are scintillation counters, which use NaI(Tl) crystals or liquid fluors as detectors. They can be sensitive enough to detect activities of small fractions of a microcurie in a relatively short counting time.

Several types of such counters are available commercially for the radio-assay of small animals (mice, rats); they all consist of a small chamber, into

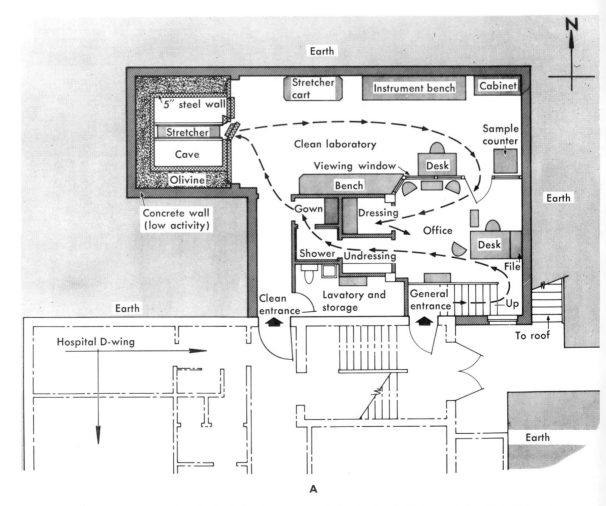

A

Fig. 7-41. Low-level whole-body counter of the Medical Division of the Oak Ridge Associated Universities, supported by the U.S. Atomic Energy Commission. (Courtesy G. A. Andrews, Oak Ridge Associated Universities, Oak Ridge, Tenn.)

which the animal is introduced. The chamber is surrounded by a number of solid scintillation detectors, or by a jacket filled with a liquid fluor. To reduce the background radiation to a minimum, heavy shielding is also required.

Whole body counters for humans are obviously much larger and subject to more stringent requirements. They are useful in many radiobiologic and medical studies; they reveal the kind and amounts of radioactive material accumulated in the body of human subjects as a consequence of radioactive fallout, of accidental ingestion, or of radioisotopes given for diagnostic or therapeutic purposes; they can supply information on the potassium content of the body

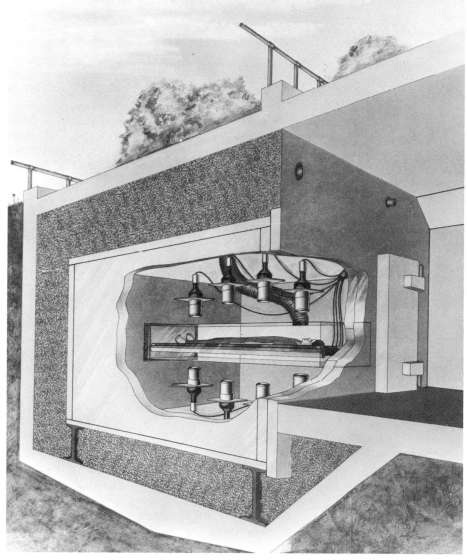

B

continued.

Fig. 7-41. B, Cutaway of the cave.

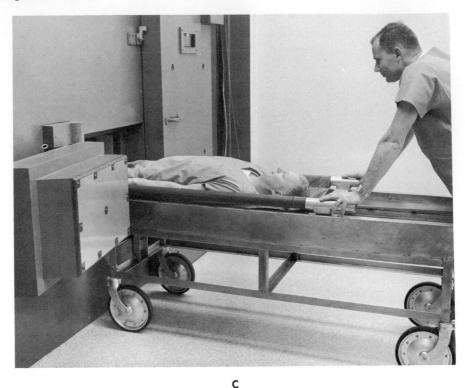

C

Fig. 7-41. C, Introduction of a subject into the cave.

and on several clinical disturbances related to potassium metabolism, such as muscular dystrophy and others.

There are basically the following two types of human counters — the Argonne counter and the Los Alamos counter. The first was developed and built for the first time at the Argonne National Laboratory, and it uses one or more large NaI(Tl) crystal detectors; the subject sits or lies down in a well-shielded room, while the detectors view his body from a short distance. The second was originally built at the Los Alamos Scientific Laboratory. The subject is introduced into a cylindrical chamber surrounded by a jacket filled with 140 gallons of liquid scintillation solution; such large volume of liquid fluor can absorb and detect a significant percentage of gamma radiation emitted by the subject. The fluor is surrounded by 108 large photomultipliers; the whole counter is shielded by a large lead cylinder.

With both types of counters, counting is done differentially, so that gamma spectra of the subject are obtained. The Los Alamos type gives a better counting yield, because although the "sample" is external, that is, not in direct contact with the fluor as in ordinary liquid scintillation counting, it is completely surrounded by the fluor. Therefore relatively short counting times (2 to 3 minutes) are usually sufficient to obtain count rates well above background. On the other hand, the same counter does not have a resolution as good as the NaI(Tl) crystals used in the Argonne type, with the result of poor separation of photopeaks. The Argonne type offers better resolution, but lower counting yield; consequently longer counting times (several minutes) are needed.

The low-level whole-body counter of the Medical Division of the Oak Ridge Associated Universities (ORAU) (Fig. 7-41) is of the Argonne type and consists mainly of a counting chamber, or "cave," and an instrument room. The cave is about 2.5 meters long and 2.5 meters high. Its walls are made of thick layers of steel, cement, and lead; all these materials were subjected to a number of tests and found to be practically free of radioactive impurities. In the center of the cave there is a Plexiglass envelope, into which the subject to be counted is introduced on a stretcher, through a small opening. The detecting system consists of eight 5 × 4 inch NaI (TI) crystals with their photomultipliers, arranged in two lines of four units, one above and one below the horizontal subject. Signals from the eight detectors are combined into one input, which is fed to a multichannel pulse-height analyzer located in the instrument room. An air-filtration system prevents the entrance of radioactive contaminants into the cave. The normal counting time is 20 minutes. Data handling consists of storing the pulse-height analyses in 200 channels. Normally a 2 MeV full scale is used; calibration is obtained by simultaneous measurement of ^{137}Cs and ^{60}Co standard sources. The sorted counts are recorded on tape, typed digitally, and plotted in the form of a gamma spectrum with an automatic X-Y plotter. Background is subtracted automatically from the gross counts. All the equipment needed for the handling of data is located in the instrument room, along with the multichannel analyzer.

The usual procedure for a subject to be counted is to enter an undressing room, place his clothes in a basket, and then enter a showering room to remove any superficial radioactive contaminants from his body. After the shower he goes into a gowning room where he dries and dons foot scuffs and a scrub suit. He then enters the instrument room and reclines on the transport stretcher, which is rolled into the counting chamber. Then the heavy-steel shielding door is closed. For the subject's comfort, the equipment inside of the chamber includes a ventilation system, intercommunication system, lighting, high-fidelity music, and a "panic button," which can be pressed in case of discomfort of any kind. After the counting is completed, the subject is rolled out of the cave and then he goes back to the dressing room.

Typical gamma spectra of noncontaminated subjects obtained with this counter are shown in Chapter 9. Likewise, the information obtained with whole-body counters is also discussed in several other chapters.

REFERENCES

GENERAL
Hine, G., editor. 1967. Instrumentation in nuclear medicine. Academic Press, Inc., New York.
Low, F. H. 1960. Radioisotope measurement in nuclear medicine. Picker X-ray Corp.
Nokes, M. C. 1958. Radioactivity measuring instruments. Philosophical Library, Inc., N.Y.
Price, W. J. 1964. Nuclear radiation detection. McGraw-Hill Book Co., New York.
Snell, A. H. 1962. Nuclear instruments and their uses. John Wiley & Sons, Inc., New York.

SEMICONDUCTOR DETECTORS
Dearnaley, G., and D. C. Northrop. 1963. Semiconductor counters for nuclear radiation. John Wiley & Sons, Inc., New York.
Taylor, J. M. 1963. Semiconductor particle detectors. Butterworth & Co. (Publishers) Ltd., London.

SCINTILLATION COUNTERS AND SPECTROMETRY
Baird Atomic Co. 1960. A handbook of scintillation spectrometry. Cambridge, Mass.
Bell, C. G., and F. N. Hayes, editors. 1958. Liquid scintillation counting. Pergamon Press, Inc., New York.
Birks, J. B. 1964. Scintillation counters. Pergamon Press, Inc., New York.
Crouthamel, C. E. 1960. Applied gamma ray spectrometry. Pergamon Press, Inc., New York.
Curran, S. C. 1953. Luminescence and the scintillation counter. Academic Press, Inc., New York.
Harshaw Chemical Co. 1960. Harshaw scintillation phosphors. Cleveland.
Hayes, F. N. 1953. Liquid solution scintillators. U.S. Atomic Energy Commission, LA-1639, Washington, D.C.
Heath, R. L. 1964. Scintillation spectrometry and gamma ray spectrum catalogue. U.S. Department of Commerce, Washington, D.C.
Rapkin, E. 1964. Liquid scintillation counting A review; 1957–1963. Int. J. Appl. Radiat. **15:**69.
Schram, E. 1963. Organic scintillation detectors. Elsevier Publishing Co., Amsterdam.
Siegbahn, K. 1966. Alpha-, beta- and gamma-ray spectroscopy. North-Holland Publishing Co., Amsterdam.

RADIOAUTOGRAPHY

Boyd, G. A. 1955. Autoradiography in biology and medicine. Academic Press, Inc., New York.

Caro, L. G., and R. van Tubergen. 1962. High resolution autoradiography. J. Cell Biol. **15:**173.

Crafts, A. S., and S. Yamaguchi. 1964. The autoradiography of plant materials. Manual 35, University of California Press, Berkeley, Calif.

Gude, W. 1968. Autoradiographic techniques. Prentice-Hall, Inc., Englewood Cliffs, N.J.

Rogers, A. W. 1967. Techniques of autoradiography. Elsevier Publishing Co., Amsterdam.

WHOLE-BODY COUNTERS

International Atomic Energy Agency. 1962. Whole body counting. Vienna.

International Atomic Energy Agency. 1966. Clinical uses of whole-body counters. Vienna.

Morris, A. C., C. C. Lushbaugh, R. L. Hayes, H. Kakehi, and D. W. Gibbs. 1964. Whole-body counters. Research Report of the Medical Division of the Oak Ridge Institute of Nuclear Studies. P. 111.

Woodburn, J., and F. Lengemann. 1964 Whole-body counters. U.S. Atomic Energy Commission Division of Technical Information, Washington, D.C.

CHAPTER 8

Fission

THE PHENOMENON OF NUCLEAR FISSION

The discovery

In the 1930s, shortly after the discovery of the neutron, several nuclear physicists were interested in bombarding stable elements with neutrons, because they hoped to transform them into new elements. Chapter 6 mentions that some of these experiments were successful (nuclear transmutations). In fact, it was even possible to transform stable atoms into radioactive ones.

At the same time, E. Fermi and his associates were wondering what would happen if the heaviest element (uranium) were bombarded with neutrons; their hope was to obtain elements of atomic number higher than 92 (transuranic elements). Their experiments showed that uranium, which is an alpha emitter, after being bombarded with neutrons, manifested some beta activity; there were three or four groups of beta-particle activities with different half-lives. Fermi's interpretation was that a $^{238}_{92}$U nucleus, after capturing a neutron, becomes ^{239}U, which decays by beta emission; the daughter nucleus should have an atomic number of 93 ($^{239}_{93}$X); this element could also be a beta emitter, and its daughter nucleus would have an atomic number of 94. It seemed therefore that the beta activities detected were coming from these transuranic unstable elements.

The problem of identifying these suspected elements became the concern of O. Hahn and F. Strassmann in Germany. Using techniques of chemical qualitative analysis, they reached the conclusion that the new elements produced by the bombardment of uranium with neutrons were not elements with atomic number greater than 92, but rather isotopes of radium ($Z = 88$). According to their interpretation, this type of transformation could only occur if the capture of a neutron by a uranium nucleus caused the simultaneous emission of two alpha particles ($92 - 2 - 2 = 88$). However, there was no evidence of this unusual alpha-emission activity.

At the same time, Mme. Joliot-Curie was carefully analyzing a product obtained by the action of neutrons on uranium; she identified it as an element of atomic number much lower than 88 and probably not too different from lanthanum ($Z = 57$). This finding, soon confirmed by other investigators, generated the suspicion that a neutron striking a uranium nucleus may break it up into two fragments of similar masses, which would be obviously much smaller than the mass of uranium. No similar nuclear reaction had ever been encountered until that time.

In 1939, L. Meitner and O. R. Frisch could indeed confirm that the uranium nucleus is split by a neutron in two (or sometimes more) fragments. This phenomenon was called *nuclear fission*. In a matter of a few months, in many European and American laboratories the formation of fission fragments was demonstrated directly with several methods, and their tracks were even seen in cloud chambers. As expected, they showed much larger mass, kinetic energy, and ionizing power than any of the particles previously known. At the same time, the investigators found that the fission fragments resulting from the fission of uranium are not always the same elements. A number of elements were identified as being produced in the fission process, all of them, however, of medium mass and atomic number.

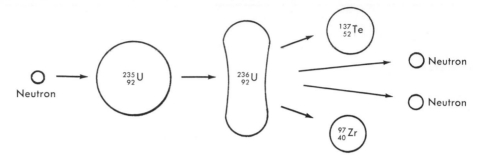

Fig. 8-1. Fission of a ^{235}U nucleus. Two neutrons are liberated in this fission. Zirconium and tellurium nuclei are the fission fragments. Note that the sum of their atomic numbers is 92 and the sum of their mass numbers is $236 - 2 = 234$.

Fissionable materials

The element uranium, as it exists in nature, is chiefly made of two isotopes, both radioactive, ^{235}U and ^{238}U, with natural abundances 0.7% and 99.3%, respectively. A third isotope (^{234}U) is also found in natural uranium but in very insignificant amounts.

After the discovery of fission, which had been caused by bombarding uranium with slow neutrons, investigators saw that what is really fissioned in these conditions is ^{235}U. They postulated that when the ^{235}U nucleus captures a neutron a composite nucleus (^{236}U) is formed (as it happens in any neutron-capture reaction) and the composite nucleus has a very short life before splitting into the fission fragments. They also showed that two or three free neutrons are emitted by the fissioned nucleus. Therefore, a general equation for the fission reaction can be written as follows:

$$^{1}\text{n} + {}^{235}\text{U} \longrightarrow {}^{236}\text{U} \longrightarrow \text{f}_1 + \text{f}_2 + 2 \text{ or } 3 \ {}^{1}\text{n}$$

where f_1 and f_2 represent the *fission fragments*. The sum of their atomic numbers must be equal to 92, and the sum of their mass numbers must be equal to 236 minus 2 or 3, depending on the number of free neutrons emitted during the reaction (see Fig. 8-1).

^{235}U is particularly fissionable with slow (thermal) neutrons, although even neutrons of higher speed can cause it to fission. Also $^{239}_{94}$Pu (plutonium, a transuranic element) and the artificial ^{233}U are fissionable with neutrons of any speed. Other fissionable materials are ^{238}U, thorium, and protactinium ($Z = 92$, 90, and 91, respectively), but their fission can be obtained only with neutrons of energy exeeeding 1 MeV (fast neutrons).* The same nuclei can be fissioned with other high-energy particles, such as deuterons, alpha particles, protons, and even with very high energy gamma photons (photofission).

In 1947 at the Lawrence Radiation Laboratory it was possible for the first time to fission elements of lower atomic number, such as bismuth, thallium, mercury, and gold, with particles accelerated to energies above 100 MeV. When elements of even lower

---

*When ^{238}U is bombarded with slow neutrons, it becomes ^{239}U, which decays by beta emission, thus yielding $^{239}_{93}$Np (neptunium, a transuranic element), which is also a beta emitter with a half-life of a few days. The daughter nucleus of neptunium is plutonium 239, an alpha emitter with a half-life of 24,360 years.

atomic number are bombarded with particles in the 400 to 500 MeV energy range, their nuclei are literally disintegrated in 20 to 30 fragments. This phenomenon is known as *spallation*.

It is interesting to note here that neutrons present in cosmic radiation are responsible for the fission of small amounts of natural uranium. Moreover, we know that uranium undergoes also spontaneous fission. However, the half-life for this spontaneous fission is about 10^{16} years (much longer than the radio-active-decay half-life). Consequently, in one gram of natural uranium, about 25 nuclei undergo spontaneous fission every hour.

Although many elements are fissionable with slow or fast neutrons and other particles, the only fissionable materials that are still used today for practical applications (atomic weapons, nuclear reactors) are uranium 235, uranium 238, and plutonium.

Energy liberated

What is remarkable about the phenomenon of nuclear fission is the large amount of energy liberated by the process. When one uranium nucleus is fissioned, the amount of energy released is about 200 MeV, the exact value depending on the nature of the fission fragments formed and on the number of neutrons liberated. Energy of this magnitude is not comparable with the energies liberated by other nuclear processes, like radioactive disintegration, whereby only a few McV of energy are released. The energy liberated in fission derives from a loss of mass. In fact, the total mass of fission fragments and free neutrons is less than the total mass of the original uranium nucleus and of the neutron that caused fission. This "missing" mass appears as fission energy.

Let us assume that the fragments resulting from the fission of a ^{235}U nucleus are $^{95}_{38}$Sr and $^{139}_{54}$Xe and that two neutrons are liberated by the fissioned nucleus. The fission fragments are unstable; through a chain of other unstable nuclei they finally become stable ^{95}Mo and ^{139}La. The sum total of the atomic mass of ^{235}U and of the neutron that reacted with it is

$$235.043 + 1.009 = 236.052 \text{ amu}$$

The sum total of the masses of ^{95}Mo and ^{139}La and of the two neutrons is

$$94.905 + 138.906 + 2.018 = 235.829 \text{ amu}$$

The mass decrement is then

$$236.052 - 235.829 = 0.223 \text{ amu}$$

This mass is equivalent to an energy of approximately 207 MeV, which is liberated in the type of fission considered.

The energy liberated by the fission process appears in several forms; their approximate distribution is summarized in the following table:

Kinetic energy of fission fragments	168 MeV
Energy of fission neutrons (fast neutrons)	5
Instantaneous gamma rays emitted by the fission process	5
Gamma rays from fission products	6
Beta particles from fission products	7
Neutrinos from fission products	10
Total	About 200

All these various forms of energy are eventually dissipated as heat. This is the origin of the heat generated in nuclear reactors and in the explosion of atomic weapons (see details below).

Fission fragments and fission products

As stated above, the two fission fragments produced in the fission of ^{235}U are not always the same nuclides, but the sum total of their mass numbers is always equal to 236 minus the number of neutrons (2 or 3) liberated in the process. If a large amount (for example, 1 g) of ^{235}U is completely fissioned, it is possible to measure, or at least to derive indirectly, the concentrations of the various fission fragments produced, and therefore the percentage of fissions yielding each particular fragment. This value is called *fission yield*. When the fission yield for the different fragments is plotted as a function of their mass number, a curve similar to the one represented in Fig. 8-2 is obtained. This curve shows that the mass numbers of the fragments cover a range from about 70 to 160, or that the fragments are all of medium mass. However, they are not produced with the same frequency. There are two groups (corresponding to the two peaks of the curve) that are produced with the highest frequency; their mass numbers are around 90 and 140. On the other hand, the yield of fragments with mass number around 118 (= 236/2) is significantly low. Thus the curve shows, in contrast with the first expectations of the physicists who discovered fission, that most fissions are *asymmetrical*, that is, the uranium nucleus is usually split in two fragments of unequal size. *Symmetrical* fissions, with fragments of approximately equal size (that is, with mass numbers $\cong 236/2$) are relatively rare. Why fission is usually asymmetrical and not symmetrical is still a problem under investigation.

Since there are approximately 90 mass numbers covered by the curve, this range means that the uranium nucleus can be fissioned in about 45 different

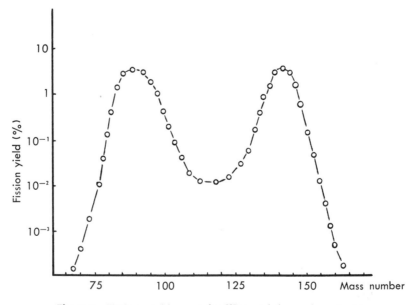

Fig. 8-2. Fission-yield curve for ^{235}U and thermal neutrons.

ways (types or modes of fission). The amount of energy liberated depends on the type of fission, although its value is always very close to 200 MeV. As previously mentioned, most of the fission energy appears as kinetic energy of the fission fragments. A speed of about 10^9 cm/sec is associated with their energy. Although their mass is considerably large, they show an appreciable penetrating power because of their high energy. The range in air is about 1.9 cm for the heaviest, and about 2.5 cm for the lightest fragments.

When fission occurs, the uranium nucleus is probably stripped of almost one half of its orbital electrons. Consequently, even the fission fragments have more protons in their nuclei than orbital electrons; therefore they behave as strongly positive ions. Their large electric charge explains their enormous ionizing power (much greater than that of alpha particles), as evidenced by the very thick tracks they generate in cloud chambers.

Since the neutron-proton ratio in fission fragments is considerably higher than the value compatible with nuclear stability, every fragment initiates a short or long chain of beta-minus emitters that ends with a stable nucleus (*fission decay chain*). All the members of a chain are isobars, with the same mass number as the original fission fragments. Hence every fission decay chain can be identified with a mass number. Within the same chain, of course, the atomic number increases by one unit from each member to the next, unless two consecutive members are isomers. Because of the change of atomic number, the members of each chain are also different chemical elements.

All the nuclides produced directly (fission fragments) or indirectly (other members of the decay chains) in the process of fission are designated with the cumulative name of *fission products*. The total number of fission products is well above two hundred.

Examples of fission decay chains, identified by their mass numbers (A), are given below, with the type of radiation and half-life associated with each transition. Note that some chains are shorter than others; also note the wide range of half-life values among the fission products:

HEAVY GROUP

$A = 143$

$$^{143}_{54}\text{Xe} \xrightarrow[\beta^-]{1 \text{ s}} {}^{143}_{55}\text{Cs} \xrightarrow[\beta^-]{2 \text{ s}} {}^{143}_{56}\text{Ba} \xrightarrow[\beta^-]{12 \text{ s}} {}^{143}_{57}\text{La} \xrightarrow[\beta^-,\gamma]{14 \text{ m}} {}^{143}_{58}\text{Ce} \xrightarrow[\beta^-,\gamma]{33 \text{ h}}$$
$$^{143}_{59}\text{Pr} \xrightarrow[\beta^-]{13\text{d}} {}^{143}_{60}\text{Nd} \text{ (stable)}$$

$A = 140$

$$^{140}_{54}\text{Xe} \xrightarrow[\beta^-]{16 \text{ s}} {}^{140}_{55}\text{Cs} \xrightarrow[\beta^-,\gamma]{66 \text{ s}} {}^{140}_{56}\text{Ba} \xrightarrow[\beta^-,\gamma]{13 \text{ d}} {}^{140}_{57}\text{La} \xrightarrow[\beta^-,\gamma]{40 \text{ h}} {}^{140}_{58}\text{Ce} \text{ (stable)}$$

$A = 139$

$$^{139}_{54}\text{Xe} \xrightarrow[\beta^-,\gamma]{41 \text{ s}} {}^{139}_{55}\text{Cs} \xrightarrow[\beta^-,\gamma]{9.5 \text{ m}} {}^{139}_{56}\text{Ba} \xrightarrow[\beta^-,\gamma]{83 \text{ m}} {}^{139}_{57}\text{La} \text{ (stable)}$$

$A = 137$

$$^{137}_{52}\text{Te} \xrightarrow[\beta^-]{\text{short}} {}^{137}_{53}\text{I} \xrightarrow[\beta^-]{24 \text{ s}} {}^{137}_{54}\text{Xe} \xrightarrow[\beta^-,\gamma]{4 \text{ m}} {}^{137}_{55}\text{Cs} \xrightarrow[\beta^-]{30 \text{ y}} {}^{137\text{m}}_{56}\text{Ba}$$
$$\xrightarrow[\text{I.T.}]{2.6 \text{ m}} {}^{137}_{56}\text{Ba} \text{ (stable)}$$

$A = 133$

$$^{133}_{51}\text{Sb} \xrightarrow[\beta^-]{4 \text{ m}} {}^{133}_{52}\text{Te} \xrightarrow[\beta^-,\gamma]{50 \text{ m}} {}^{133}_{53}\text{I} \xrightarrow[\beta^-,\gamma]{21 \text{ h}} {}^{133}_{54}\text{Xe} \xrightarrow[\beta^-,\gamma]{5.3 \text{ d}} {}^{133}_{55}\text{Cs} \text{ (stable)}$$

$A = 131$

$$^{131}_{50}\text{Sn} \xrightarrow[\beta^-]{<2 \text{ m}} {}^{131}_{51}\text{Sb} \xrightarrow[\beta^-,\gamma]{25 \text{ m}} {}^{131}_{52}\text{Te} \xrightarrow[\beta^-,\gamma]{1.2 \text{ d}} {}^{131}_{53}\text{I} \xrightarrow[\beta^-,\gamma]{8 \text{ d}} {}^{131}_{54}\text{Xe} \text{ (stable)}$$

LIGHT GROUP

$A = 99$

$$^{99}_{42}\text{Mo} \xrightarrow[\beta^-, \gamma]{67 \text{ h}} {}^{99m}_{43}\text{Tc} \xrightarrow[\text{I.T.}]{6 \text{ h}} {}^{99}_{43}\text{Tc} \xrightarrow[\beta^-]{2 \times 10^5 \text{ y}} {}^{99}_{44}\text{Ru (stable)}$$

$A = 97$

$$^{97}_{36}\text{Kr} \xrightarrow[\beta^-]{\text{short}} {}^{97}_{37}\text{Rb} \xrightarrow[\beta^-]{\text{short}} {}^{97}_{38}\text{Sr} \xrightarrow[\beta^-]{\text{short}} {}^{97}_{39}\text{Y} \xrightarrow[\beta^-]{6 \text{ s}} {}^{97}_{40}\text{Zr} \xrightarrow[\beta^-, \gamma]{17 \text{ h}}$$

$$^{97m}_{41}\text{Nb} \xrightarrow[\text{I.T.}]{1 \text{ m}} {}^{97}_{41}\text{Nb} \xrightarrow[\beta^-, \gamma]{72 \text{ m}} {}^{97}_{42}\text{Mo (stable)}$$

$A = 95$

$$^{95}_{38}\text{Sr} \xrightarrow[\beta^-]{0.8 \text{ m}} {}^{95}_{39}\text{Y} \xrightarrow[\beta^-, \gamma]{11 \text{ m}} {}^{95}_{40}\text{Zr} \xrightarrow[\beta^-, \gamma]{65 \text{ d}} {}^{95}_{41}\text{Nb} \xrightarrow[\beta^-, \gamma]{35 \text{ d}} {}^{95}_{42}\text{Mo (stable)}$$

$A = 90$

$$^{90}_{36}\text{Kr} \xrightarrow[\beta^-, \gamma]{33 \text{ s}} {}^{90}_{37}\text{Rb} \xrightarrow[\beta^-, \gamma]{2.9 \text{ m}} {}^{90}_{38}\text{Sr} \xrightarrow[\beta^-]{28 \text{ y}} {}^{90}_{39}\text{Y} \xrightarrow[\beta^-, \gamma]{64 \text{ h}} {}^{90}_{40}\text{Zr (stable)}$$

Some of the fission products (for example, ^{137}Cs, ^{90}Sr, and ^{131}I) are radionuclides of considerable importance in radiation biology and radiotracer methodology.

THE CHAIN REACTION
The mechanism

The possibility of a chain reaction was recognized as soon as investigators knew (1939) that a fissioned uranium nucleus releases free neutrons. They thought that in a mass of uranium the free neutrons released by one fissioned nucleus could trigger the fission of neighboring nuclei; from these nuclei other neutrons would be liberated and made available for the fission of still other nuclei, and so forth. This self-propagating chain reaction would be very similar to the combustion of a fuel, in which the high temperature generated by the combustion of a few molecules triggers the combustion of other molecules.

The possibility of the release of enormous amounts of energy was seen in a chain reaction. If the fission of only one uranium nucleus releases as much as 200 MeV, it is not difficult to calculate the total amount of energy released by the complete fission of one gram of uranium. The amount is approximately 2.3×10^4 kilowatt-hours. To obtain the same amount of energy by combustion would require more than 3 tons of coal! A self-sustaining chain reaction would certainly occur if *all* the neutrons released in the fission process would actually fission other nuclei. If we assume that the first fissioned nucleus releases two neutrons, and these neutrons fission two other nuclei with the release of a total of four neutrons, which in turn fission four more nuclei, etc., then the chain reaction would be self-sustaining. Actually, for a chain reaction to occur, it is not necessary that *all* the released neutrons fission other nuclei; the minimum requirement is that on the average at least one of the neutrons released by every fission be used to fission another nucleus. In this instance, there would be at any time a constant number of neutrons available for fission.

The condition required for a self-sustaining chain reaction is conveniently expressed with a figure of merit called the *multiplication factor* (k), which is defined as the ratio of the number of fissioning neutrons of any generation to the number of fissioning neutrons of the preceding generation. If the k number is greater than 1, it means that the number of fissioning neutrons is increasing as the fissions take place; the reaction is thus self-sustaining. If $k = 1$, it means that the number of fissioning neutrons is neither increasing nor decreasing as fissions occur; the reaction is still self-sustaining, although it will propagate itself more slowly than in the first case. If k is less than 1, it means that the number of fissioning neutrons is decreasing as the fission proceeds; the reaction will not sustain itself for long.

In many systems containing uranium a chain reaction is not possible because the multiplication factor is less than 1. In a system made of natural uranium and other elements (such as a system made of a uranium compound) the free neutrons released in fission may be involved in one of the following processes (Fig. 8-3):

1. Some neutrons escape from the system through its surface. They are lost for further fission. However, if the system is surrounded by an appropriate reflector, these neutrons can be brought back into the system and made available for other fissions.

2. Some neutrons are captured by the abundant ^{238}U atoms present in the system (nonfission capture), as a result of a (n, γ) reaction. They are lost for further fission.

3. Some neutrons are captured by other elements present in the system, impurities, and fission fragments (parasitic capture). They are also lost for further fission.

4. Some fast neutrons fission ^{238}U nuclei, and some slow or fast neutrons fission ^{235}U nuclei (fission capture). (Recall that the neutrons released in fission are fast neutrons, but they may be slowed down through elastic or inelastic collisions with other atoms, before being eventually captured.)

The first three processes obviously remove neutrons from the system. The fourth process also removes neutrons from the system, but at the same time produces two or three times as many neutrons. If the first three processes occur to a great extent, the k factor is less than 1 and any fission initiated in the system does not trigger a self-sustaining chain reaction. The system is said to be

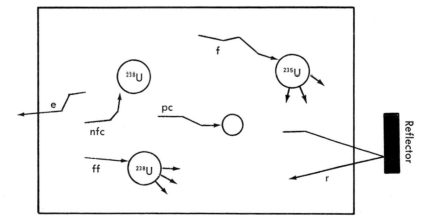

Fig. 8-3. Diagram showing the different fates of neutrons in a thermal chain reaction. **e,** Neutron escape (first process). **r,** Neutron reflection. **nfc,** Nonfission capture by ^{238}U (second process). **pc,** Parasitic capture by impurity (third process). **f,** Fission of ^{235}U (fourth process–note the free neutrons produced by this process). **ff,** Fast fission of ^{238}U (fourth process–note the free neutrons produced by this process).

subcritical. However, it is possible to minimize the first three processes, so that the number of free neutrons produced by the fourth process is equal to, or exceeds, the number of neutrons lost by escape, fission, and nonfission capture. If the number of neutrons gained is exactly equal to the number of neutrons lost, the *k* factor is equal to 1 and the system is *critical.* If the number of neutrons gained exceeds the number of neutrons lost, the *k* factor is greater than 1 and the system is *supercritical.*

To obtain a critical or supercritical system, one must meet several conditions. First, let us consider the fact that the last three processes are dependent exclusively on the composition of the system, whereas the first process is mainly influenced by its geometry, and in particular by its size, shape, and surface-to-volume ratio. If the system is large and spherical, then its surface-to-volume ratio is relatively small; the number of neutrons lost by escape are only a small fraction of the number of neutrons available in the system. Evidently, a self-sustaining reaction is not possible if the size of the system is less than a certain minimum (critical size). The critical size is not a constant, but depends on several factors, such as the isotopic composition of the fissionable material, the amounts of impurities and other neutron-moderating or -absorbing materials present in the system, and the physical arrangement of all the components. As I mentioned above, it is possible to reduce the escape of neutrons from the system by surrounding the latter with a reflector.

To minimize parasitic capture, one should make the system as free as possible of impurities and other chemical elements that would capture neutrons. Metallic uranium would be an ideal fissionable material for a critical system, although some compounds like uranium oxides are also compatible with criticality.

To minimize nonfission capture by ^{238}U, one should reduce its percentage by increasing the percentage of ^{235}U (enriched uranium).

If the amount of energy must be or is to be released slowly and for a long time by the chain reaction in a critical or supercritical system, one must control the chain reaction itself in such a way as to keep the *k* factor equal to 1 or slightly above 1. For obtaining a variable rate of energy release, some device must be incorporated in the system so that the *k* factor may be changed. This is done by introducing in the system materials that absorb neutrons (neutron absorbers) and therefore increase the extent of parasitic capture (as in the third process described previously). Rods of neutron-absorbing material (control rods) are used for this purpose; they can be introduced partially or totally into the system. As they are slowly withdrawn from the system, less absorbing material is available and the *k* factor increases. If the chain reaction is to be interrupted, one need only push the control rods entirely into the system; the *k* factor will immediately drop below 1, the system will become subcritical, and the chain reaction will no longer be self-sustaining. A system like the one generally described above, in which the chain reaction can be controlled as explained, is called a *nuclear reactor.*

If, on the other hand, the total amount of energy from the complete fission of a system must be released in a very short time interval (a small fraction of a second), then it is necessary to bring the fissionable system to criticality as quickly as possible and to reach the highest possible value of the *k* factor in the shortest time possible. To obtain this condition, one should prevent the occurrence of all processes that remove neutrons from the system, so that virtually all the neutrons produced in fission will cause further fissions. Moreover, fission must be caused by fast neutrons, if the whole chain reaction is to be completed in a very short time interval. All of these requirements may be fulfilled if the system consists of a supercritical mass of pure ^{235}U or ^{239}Pu and contains no material capable of absorbing or slowing down neutrons. If the system is enclosed in a small space, the release of a very large amount of energy in a small fraction of a second results in a formidable explosion, just as the rapid combustion of a large amount of gasoline enclosed in a container is accompanied by an explosion. A system whereby the total fission energy from a supercritical mass is released explosively is usually called a *nuclear fission bomb*, or atomic bomb.

Controlled chain reaction – the nuclear reactor
Construction

The first nuclear reactor was assembled by E. Fermi, assisted by the nuclear physicists A. Compton, L. Szilard, and others, in a squash court beneath the Stagg football field of the University of Chicago. It was operated supercritically in the afternoon of Dec. 2, 1942. It was the first time that man had initiated a self-sustaining nuclear chain reaction. That date marks the beginning of the

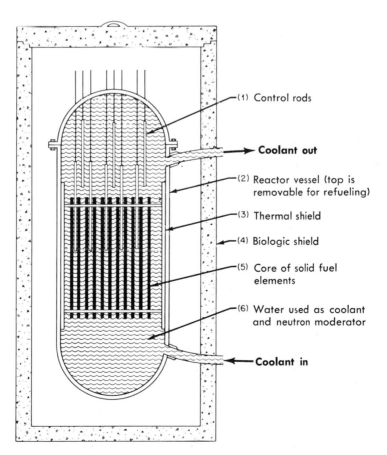

Fig. 8-4. Diagram of nuclear reactor, pressurized water type. (Courtesy AEC.)

atomic age. Since then, several nuclear reactors with a variety of purposes and sizes have been built in several countries; almost all of them utilize nuclear energy for peaceful purposes.

Six basic components of every nuclear reactor (Fig. 8-4) are the following:

1. *"Fuel"* (that is, fissionable material). Fuel is uranium (natural or enriched with ^{235}U) or plutonium. A typical enriched uranium fuel may contain anywhere between 10% and 90% of ^{235}U. The richer the fuel is in fissionable material, the more compact the reactor can be.

Some reactors use a liquid fuel, such as an aqueous solution of enriched uranium; but usually the fuel is solid and in the shape of plates, rods, or "slugs" (fuel elements). Several fuel elements, such as plates, may be sandwiched together and constitute a subassembly. A reactor may be operated with a variable number of subassemblies, depending on the power desired.

Solid fuel elements incorporate what is known as cladding, which takes the form of a protective coating of aluminum, stainless steel, or zirconium, which prevents direct contact and thus chemical reactions between the fuel material and the reactor coolant.

2. *Moderator.* In most nuclear reactors it is desirable to thermalize (that is, slow down) the fast neutrons produced by fission, because ^{235}U can be fissioned more easily with thermal neutrons. The thermalization of neutrons is accomplished with materials that do not absorb neutrons significantly but slow them down through a series of elastic collisions (neutron moderators). The best moderators are heavy water, ordinary water, graphite (a variety of pure carbon), and beryllium. They are usually interposed between the fuel elements or subassemblies in a variety of geometric configurations, or intimately mixed together with the fuel.

3. *Reflector.* The reflector surrounds the fuel-moderator system (core of the reactor) and is made of one of the materials that can also be used as neutron moderators. One may recall that the function of the reflector is to send back into the core the neutrons that escape from it.

4. *Control system.* The control system is made of rods of cadmium or boron steel, that is, of materials with high neutron absorption cross sections. The control rods may be inserted more or less deeply into the core of the reactor to regulate the k factor, as explained above. A reactor is equipped with at least two sets of control rods, which are regulating rods for routine control purposes, and safety rods, which are inserted rapidly and automatically into the core to shut down the reactor in an emergency.

The operator starts a reactor by removing from the core the safety rods first and brings it to criticality by gradually withdrawing the regulating rods, while watching the gradual increase of the k factor. This procedure is done by monitoring the neutron density with neutron counters placed at different points of the core surface. The neutron density is an indication of the fission rate.

5. *Cooling system.* The purpose of the cooling system is to remove from the core the heat generated by the chain reaction. Its construction and coolant used depend on the temperature at which the reactor is operated. The temperature, in turn, is dependent on the rate of the chain reaction. Operating temperatures may range from 80° to 250° C. Coolants commonly used are air, helium, carbon dioxide, water, or liquid sodium. In some reactors, water serves as coolant and moderator.

Fig. 8-5. Argonaut training reactor (Argonne National Laboratory). Lateral view showing the portholes for the introduction of the samples to be irradiated. The bar protruding from the center of the panel contains a small neutron source; it is pushed into the core of the reactor for the start-up operation.

6. *Radiation shield.* The radiation shield protects the personnel from the high-level neutron and gamma radiation emitted by the core. It takes the form of several feet of high-density concrete surrounding the reactor.

In a reactor with a critical configuration, the chain reaction could be initiated either by the neutrons released in a few spontaneous fissions, or by stray neutrons present in cosmic radiation. However, for easier start-up operation, many reactors have a built-in neutron source (such as an Am-Be source), which can be pushed automatically into proximity with the core when the reactor is started (Fig. 8-5).

The power of a reactor is the amount of energy released by the chain reaction per unit of time. It depends on a variety of factors and can be calculated with special formulas. Reactors may be operated with powers ranging from a few watts to several megawatts.

It is very unlikely that a nuclear reactor could explode if it went out of control, even when all the automatic and manual safety devices fail to operate. If the rate of the chain reaction becomes too high, the dissipation of energy in the form of heat would be so great as to break the whole system apart. If this happened, the system would automatically be reduced to fragments of subcritical size and the propagation of the chain reaction would automatically come to an end. Of course, the danger to personnel from falling debris and from neutron and gamma radiation emitted by the exposed fissionable material would be considerable.

Types and uses

The products put out by a nuclear reactor are numerous; they are high-level neutron and gamma radiation, energy, fission products (many of which are artificial radionuclides), etc. The design and construction of a reactor are adapted to the particular product desired. The following is a brief survey of the main types and uses of reactors.

Research reactors. Research reactors are of particular interest to radiation biologists and investigators using radiotracer techniques. A research reactor is defined by the American Standards Association as "one designed to provide a source of neutron and/or gamma radiation for research into basic or applied physics, biology, or chemistry, or to aid in the investigation of the effects of radiation on any type of material . . . ." It is evident that what is utilized of the output of a research reactor is its neutron and gamma radiation.

These reactors are widely used by nuclear physicists for the study of nuclear reactions, by solid-state physicists for the determination of crystal structures with neutron diffraction techniques, by radiation chemists for the study of the effects of radiation on certain chemical reactions and on the properties of some materials, by analytical chemists for the identification of trace impurities with activation analysis techniques, and by radiation biologists for the study of the effects of neutron and gamma radiation on biologic systems.

A research reactor is built in such a way as to allow the most efficient use of neutron and gamma radiation. A *beam tube* is a hollow tube that passes through the reactor shielding from a position in or near the core and provides a path for neutrons to irradiate materials outside the shield. When the beam tube is not in use, a shutter prevents the exit of radiation from the core (Fig. 8-5). If a narrow beam of neutrons is desired, a collimator is adapted to the opening of the tube. Special *filters* (of lead or bismuth) are also used to stop the gamma component and allow the passage of neutrons only. A so-called *rabbit* provides a means for rapidly inserting specimens into the reactor core itself when a high neutron flux is needed. It consists of a carrier (such as a plastic tube) containing the specimen, which is driven in a pipe by compressed air from a laboratory area to the reactor core and vice versa. A *thermal column* is similar to a beam tube but is filled with a neutron moderator (such as graphite) to obtain a beam of thermal neutrons. Rotary specimen racks ("lazy Susans") surround the core of many research reactors. Research reactors are often surrounded by large work areas to accommodate equipment needed for several simultaneous experiments.

One of the most important specifications of a research reactor is its neutron flux (number of neutrons emitted per square centimeter of core surface per second). Neutron flux is related to the power level at which the reactor is operated and also to several design factors. The neutron fluxes obtained are in the order of 10^{12} neutrons/cm^2/sec. These reactors are operated at relatively low temperatures; consequently, they do not require elaborate cooling systems.

Many major universities and scientific laboratories in the United States own and operate research reactors (Fig. 8-6).

Power reactors. Power reactors are intended for production of power. The heat generated by the fission process in the core is used, for instance, to produce steam at high temperature and pressure. The steam may drive a turbogenerator, thus producing electric energy; or it may drive a steam engine for the propulsion of submarines or surface ships. Small power reactors are being designed for the propulsion of spacecraft. Power reactors are operated at relatively high temperatures.

Fig. 8-6. A "swimming-pool" research reactor. The core of the reactor is at the bottom of the pool. Note the control rods going down into the core and the glow that is caused by Cerenkov radiation. The pool is filled with water. (Courtesy University of Michigan.)

Production reactors. Production reactors are designed and intended mainly for the production of several artificial radionuclides, which are abundantly found among the fission products. The fuel is enclosed in special slugs. After the fuel has been "burnt," the slugs are ejected or withdrawn from the reactor core and opened by remote control behind thick lead-glass plates, and the fission products are manipulated chemically for the extraction and purification of the radionuclides they contain. Common radioisotopes widely used in radiobiologic and radiotracer work, like cesium 137, strontium 90, and iodine 131, are produced in these reactors.

Some production reactors are specifically designed for the production of plutonium, used for the fabrication of nuclear weapons and for research. The fuel is natural uranium. The geometric configuration of the moderator in these reactors is such as to make available as many neutrons as possible in the energy range in which they are most likely captured by ^{238}U present in the fuel (nonfission capture). As mentioned above, this capture results in the formation of ^{239}U, which decays to ^{239}Np and then to the long-lived ^{239}Pu.

Uncontrolled chain reaction — the nuclear fission bomb
Construction

As previously mentioned, in a fission bomb the chain reaction propagates itself through a critical mass of fissionable material within a very short time. The liberation of enormous amounts of energy in a very short time and within a small, enclosed space results in an explosion of tremendous proportions and destructive effects.

If 1 kg of ^{235}U is completely fissioned, the energy liberated is equivalent to the energy released by the explosion of almost 20,000 tons of TNT! It seems that the smallest nuclear bomb cannot contain much less than 1 kg of fissionable material; with a smaller amount, the attainment of a critical configuration would be difficult or impractical.

The first nuclear bomb was the end product of a long and secret wartime project known as the "Manhattan Engineering District Project," and was exploded in the desert of Alamagordo, New Mexico, on July 16, 1945. The following month, two nuclear bombs were detonated over two Japanese cities, Hiroshima and Nagasaki, for hostile purposes and with enormous casualties and loss of life among the civilian population. Since then, many other nuclear devices have been exploded in several parts of the world for testing purposes, in air as well as underwater and underground. At the present time (1971), some of the nuclear powers have agreed to explode nuclear devices exclusively underground.

Distribution of energy in a nuclear explosion

The total fission energy released by the explosion of a nuclear device in air appears in different forms, as shown in Fig. 8-7. The explosion itself is accompanied by a considerable increase of the temperature of the bomb materials (up to 10 million degrees centigrade!), so that the bomb materials become extremely hot gases. The sudden expansion of these gases is responsible for the shock wave, or blast wave, which can be propagated for several miles from the point of explosion. About 50% of the fission energy is released in this form.

The high temperature generated at the point of explosion accounts for a

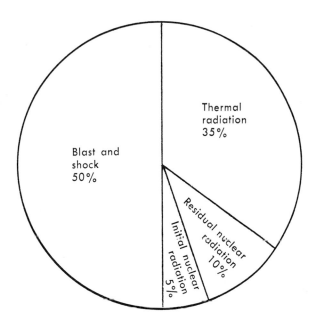

Fig. 8-7. Energy distribution in a nuclear explosion.

"heat wave" (thermal radiation). This radiation is about 35% of the total energy. The warmth of this wave can be felt as far as 70 miles from the explosion.

The initial nuclear radiation consists of penetrating neutron and gamma radiation emitted instantaneously by the fission process itself. For an idea of the distance at which this type of radiation can be propagated, it suffices to say that at the distance of 1 mile from the explosion of a 1-megaton fission bomb (equivalent to 1 million tons of TNT), a person would probably need the protection of about 1 foot of steel or 4 feet of concrete to be relatively safe from this type of radiation. The initial nuclear radiation is about 5% of the total fission energy. The residual nuclear radiation (about 10% of the total) consists of delayed neutrons, gamma rays, and beta particles emitted by the fission products within minutes after the explosion. Gamma radiation is also generated by nuclear reactions (n, γ) between neutrons and atomic nuclei (of air, bomb materials, etc.), or from materials radioactivated by neutrons. Alpha particles are also present in the residual nuclear radiation; they are emitted by the fissionable material that escaped fission.

Description of a nuclear explosion

For descriptive purposes, it is convenient to distinguish the following four types of explosions, depending on where the bomb is detonated:
1. Airburst, occurring when the bomb is detonated in the air at an altitude of at least several hundred feet
2. Surface burst, occurring when there is a detonation on the surface of land or water, or a few feet above it (such as on top of a tower)
3. Underground burst, occurring when there is a detonation at least several feet below the surface of the ground
4. Underwater burst, occurring when the bomb is detonated several feet below the surface of a natural body of water.

Fig. 8-8. Fireball at Nevada test site, 1957. The device was exploded at an altitude of 500 feet and was suspended from a plastic balloon. (Courtesy AEC.)

The phenomenology of nuclear explosions has been described in considerable detail by S. Glasstone.* Here only a brief description of an airburst (the one that causes most damage to human populations) is presented, with occasional references to other types of bursts.

At the instant of the explosion, a spherical mass of very hot and incandescent gases appears at the point of the explosion; it is called the "fireball" (Fig. 8-8). Its size increases tremendously within a few milliseconds, and for bombs of high yield its diameter can become as large as a few miles. The fireball from a 1-megaton bomb can appear about 30 times brighter than the sun to an observer 60 miles away! The fireball is an intense source of visible light, as well as infrared and ultraviolet radiation.

As the fireball cools, it loses its brightness and gradually takes the appearance of a cloud ("atomic cloud"), which is chiefly made of vaporized bomb materials (Fig. 8-9). The atomic cloud is red at first, then reddish brown, and then white. These colors are attributed to gaseous compounds (nitrogen oxides) produced by chemical reactions between the constituents of the air at those very high temperatures.

*Glasstone, S. 1962. The effects of nuclear weapons. U. S. Atomic Energy Commission, Washington, D. C.

Fig. 8-9. The atomic cloud begins its ascent to a height of about 40,000 feet. The device was exploded on a 500-foot tower. Nevada test site, 1957. (Courtesy AEC.)

The high temperature of the fireball and of the cloud causes a vigorous updraft in the surrounding air, which starts lifting the cloud upward. The cloud drags with it a considerable amount of debris from the ground surface. The amount of ground material that is incorporated in the cloud depends mainly on the yield of the bomb, the altitude of the explosion, and the nature of the terrain. The atomic cloud generated by a bomb exploding at high altitude (a few miles or more) presumably does not contain any earth debris.

At first, the ascending cloud has the shape of a column. Its speed of ascent decreases with time from about 300 mph during the first minute to about 35 mph when it has reached an altitude of about 15 miles. The actual height reached by the cloud depends chiefly on the energy yield of the bomb, on the meteorologic conditions existing at the time of the explosion, and on other contingent physical properties of the air around the cloud. In any event, when the upward lift ceases, the materials that make up the cloud begin to spread horizontally and radially, so that the cloud takes the well-known shape of a mushroom. The height reached by the cloud formed by a bomb of moderate yield exploded at moderate altitude (several hundred feet) is about 5 to 10 miles, and the complete ascent lasts 8 to 10 minutes. It is important to keep in mind that the atomic cloud is made of pulverized or vaporized bomb material (fission products, unfissioned uranium or plutonium, and bomb jacket) and soil fragments of variable size (dust as well as gross debris) partly fused with the fission fragments and partly radioactivated by the intense neutron radiation. Even as the cloud rises, the fragments of larger size start falling back to the ground in the immediate vicinity of *ground zero* (the point on the ground surface closest to the point of explosion). Evidently in a surface burst much more ground material becomes incorporated in the atomic cloud. Another characteristic of a surface burst is the formation of a crater around the point of explosion. Its size and depth depend on the yield of the bomb.

The cloud generated by an underwater explosion contains large amounts of aqueous vapor and of liquid water dispersed in very fine droplets. The same type of explosion generates a system of concentric, gigantic water waves (up to 150 feet high). Their height gradually diminishes as they travel away from the point of explosion.

In an underground burst several thousand feet deep, no fireball or atomic cloud is seen above the surface. The only effect felt by the human population is an earthquake whose intensity is obviously a function of the bomb yield and depth of detonation.

In airbursts, if there are combustible materials (forests, wood frame houses, etc.) within a few miles from ground zero, the thermal radiation generated by the explosion may have a temperature high enough to ignite those materials and start extensive fires. Fires were generated in this manner in Hiroshima and Nagasaki. The spreading of these fires depends on the winds blowing at the time of the explosion. However, even when the atmosphere is calm, the intense heat generated by the fires may cause radial movements of large masses of air converging from the surroundings toward the burning area (fire storm). These winds may blow with speeds up to 35 mph and in certain conditions may prevent the spreading of fires. A fire storm occurred at Hiroshima.

Immediate damage to human population

Several types of injuries are caused to human populations by nuclear explosions; some are immediate (or short-term) injuries, others manifest themselves several days or years after the explosion (long-term injuries). For the present purpose, "immediate injury" is that which manifests itself immediately or within 1 to 2 days after the explosion.

The only source of information available to us concerning immediate injuries is the Japanese experience. It is estimated that the number of casualties in Japan was 70,000 killed or missing at Hiroshima and 36,000 at Nagasaki, with 70,000 injured at Hiroshima and 40,000 at Nagasaki. Although the yield of the two bombs exploded over the two cities was the same (20 kilotons), there were less casualties at Nagasaki, because of several factors, one of them being the hilly configuration of the terrain, which provided some shelter against blast wave and thermal radiation.

The immediate injuries can be conveniently classified into the following three groups: (1) blast injuries, caused by the shock wave, (2) flash and fire burns, and (3) nuclear radiation injuries, caused by the initial nuclear radiation emitted by the exploding bomb. All long-term injuries are exclusively caused by the initial and residual nuclear radiation. All injuries caused by nuclear radiation are dealt with extensively in other chapters of this book. Here we shall concern ourselves only with blast injuries, and flash and fire burns.

Blast injuries caused by nuclear explosions are not qualitatively different from those caused by the explosion of conventional bombs; however, they affect a larger number of individuals because the shock wave travels a longer distance.

Direct blast injuries are caused by the impact of the shock wave with the body. Very close to ground zero the body can be literally crushed, or at least some internal organs may be seriously damaged. Ruptures of eardrums have been reported in individuals located at greater distance from ground zero. Temporary loss of consciousness is another possibility, and it seems to be caused by the shock wave. In the most serious cases of blast casualties, there may be damage to the nervous system, heart failure, suffocation caused by lung hemorrhage, and gastrointestinal hemorrhage.

Indirect blast injuries are caused by falling buildings, flying missiles, etc.

Burns may be caused by the infrared and ultraviolet components of the radiation emitted by the fireball (flash burns), by certain components of the nuclear radiation (beta and gamma burns), or by fire (fire burns).

Flash burns were responsible for a large number of casualties in Japan. Serious flash burns occurred as far as $2\frac{1}{2}$ miles from ground zero. They were frequent among people who were in the open, or in buildings with windows open. Otherwise, walls and other structures provided sufficient protection against this type of burn. Interestingly enough, some protection was also provided for people who were in the open by light-colored clothes, by hats, or by parts of clothes of light color (as in plaid shirts); in this last case, often the pattern of the burns on the skin matched the color pattern of the clothes (Fig. 8-10). Even some natural protuberances of the body (chin, earlobes, and nose) provided some protection in some cases. These strange effects, of course, can be easily explained by recalling that electromagnetic radiation (and therefore infrared and ultraviolet) travels in a straight path and that it is differently absorbed by different materials.

The severity of flash burns ranged from simple erythema to second- and third-degree burns. However, they were rather more superficial than ordinary fire burns, because of the very short duration of the flash. Nonetheless, many flash burns became sources of infection, because of lack of proper care and adequate sanitation facilities. The degree of infection was probably also responsible for the appearance of keloids (thick overgrowths of scar tissue) on healed burns. The influence of nuclear radiation on the formation of keloids, first suspected some time ago, is now denied.

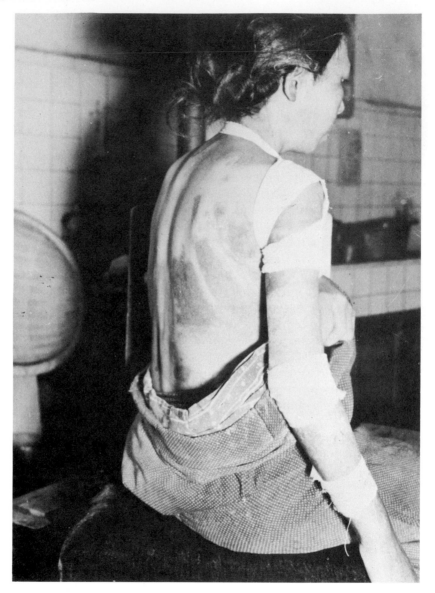

Fig. 8-10. Flash burns produced through the clothing of a Japanese woman by atomic bomb explosion, Nagasaki, 1945. (Courtesy AEC.)

The flash was also responsible for many eye injuries. Retinal burns were caused by the focusing of large amounts of thermal energy on the retina by the eye lens. Flash blindness was also reported, along with the formation of blind spots on the retina in people who happened to be looking at the fireball. Keratitis (of the cornea) also resulted in some cases.

Fallout

The composition of the atomic cloud formed by air and surface bursts has already been described. After the gross debris have fallen back onto the ground, a finely dispersed material, consisting of particles ranging in size from that of grains of sand or small flakes to that of invisible dust particles, remains suspended in the atmosphere for periods of a few hours to several years. When the finely dispersed material falls back on the ground, it is called "fallout."

Fallout is radioactive because it consists of fission products fused more or less with earth materials, of unfissioned uranium or plutonium, and to a much lesser extent of radioactivated ground particles. This last component of fallout may be quite considerable if the explosion was a surface burst; it is quite rich in ^{24}Na, resulting from the radioactivation of ^{23}Na, which is abundant in rocks.*

Because of its radioactivity, fallout is the main source of the residual nuclear radiation that is emitted by a nuclear explosion. It is responsible, therefore, for many delayed radiation injuries to the human population; for this reason problems concerning fallout are of considerable interest to the radiation biologist.

The amount of fallout generated by an explosion is very much dependent on the energy yield of the bomb, the height of the explosion, the nature of the terrain around ground zero, and the meteorologic conditions existing at the time of the explosion. The largest particles of the atomic cloud start falling back on the ground as soon as the cloud has reached its highest altitude. This is the *local fallout*, and it occurs within a few hours after the explosion. Local fallout is particularly heavy after surface or near-surface bursts. Its pattern of distribution around ground zero is determined by the direction of the prevailing winds after the explosion. If no winds were blowing for several hours after the explosion, the local fallout would be distributed on a roughly circular area around ground zero. The radius of this area would be a few miles or more, depending on the energy yield of the bomb and the altitude of the explosion. With winds blowing, the area of distribution of the local fallout takes the shape of an ellipse overlapping in part a circular area, whose center is approximately ground zero (Fig. 8-11). The length of the elliptic area increases with time and reaches values of 200 miles or more.

The heaviest local fallout ever recorded was that which followed a test explosion of a multimegaton thermonuclear device† on the Bikini atoll (Marshall Islands) on March 1, 1954. The explosion was a near-surface burst, because the device had been set up on a barge near a coral reef. Before the explosion, meteorologists had predicted that any fallout would have been pushed by the winds westward or northwestward, toward an area of the ocean with no inhabited islands. However, it seems that shortly after the explosion the winds shifted eastward, and heavy fallout began, like a shower of white particles, over two atolls inhabited by some American servicemen and Marshallese natives and on a Japanese fishing boat that was, at the time, slightly north of one of the islands. The fallout began about 7 hours after the explosion and extended eastward for about 220 miles. Many of the Americans, Marshallese natives, and Japanese fishermen were

---

*Radioactivation of atmospheric nitrogen, leading to the formation of ^{14}C, is another important phenomenon accompanying nuclear explosions. With so many bombs detonated in air since 1945, it is believed that the amount of radiocarbon in the atmosphere (and therefore in organisms) has increased considerably. Before the advent of the atomic age, radiocarbon was produced only by cosmic radiation. The new equilibrium will last for a long time, even if no nuclear devices will ever be exploded again, because of the very long half-life of ^{14}C (5730 years).

†In a thermonuclear bomb (popularly known as the H-bomb) nuclear energy is released by the "fusion" of light atomic nuclei, such as hydrogen. Since fusion can occur only at very high temperatures, a thermonuclear bomb is combined with a fission bomb; the high temperature generated by the explosion of the latter triggers the fusion process. The radioactive fallout produced by a thermonuclear explosion originates exclusively from fission.

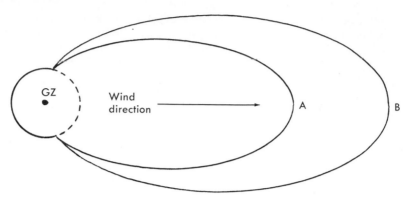

Fig. 8-11. Generalized pattern of distribution of local fallout. **A,** A few hours after the explosion. **B,** Several hours later. **GZ,** ground zero.

victims of injuries like beta burns, caused by the beta emitters present in the fallout dust that was deposited on their skin for several hours or days, hematologic effects caused by gamma radiation of internal emitters inhaled with air or ingested with food and water, and other injuries that are discussed in more detail elsewhere in this book.

The atomic cloud generated by a nuclear explosion in the megaton range can reach an altitude well above the upper limit of the troposphere (about 30, 000 to 50,000 feet). The heaviest radioactive dust is left suspended in the troposphere, whereas the lightest dust particles are injected into the stratosphere. This dust will fall back on the earth's surface after several months or years, and at distances of hundreds and thousands of miles from ground zero. This is the so-called *worldwide fallout;* this name, however, should not be taken as meaning that this fallout will occur everywhere on the surface of the globe.

The distribution pattern of the tropospheric component of the worldwide fallout is again determined by the meteorologic conditions after the explosion, and especially by rains and winds. Because of the continuous turbulence existing in the troposphere and the frequent formation of rain, the tropospheric radioactive dust does not remain suspended in the air for long, and the fallout begins at most a few days after the explosion. On the contrary, the dust material injected into the stratosphere remains "stored" there for long periods (up to a few years) and may travel horizontally several thousand miles before being trapped in the troposphere and before descending to the earth's surface (stratospheric fallout). High-yield bombs exploded at considerable altitudes produce little or no local fallout, but only worldwide fallout.

The behavior and movements of the stratospheric radioactive dust are not yet completely understood. It seems, however, that the minute debris injected into the stratosphere by explosions occurring near the equator drift slowly toward both poles, although prevalently toward the pole closer to the explosion. Months or years later they descend into the troposphere through certain gaps of the tropopause; this "sinking" seems to be accelerated in late winter or early spring and by some weather conditions.

Theoretically, all the several radionuclides that make up the fission-decay chains (see above) could be present in fallout. However, many of them are so short lived that they have practically decayed before any local fallout begins.

Those fission products whose half-lives are of a few days are found only in local fallout and in tropospheric fallout that occurs within a few weeks from the explosion. Longer-lived radionuclides are found in any type of fallout.

Three fission products present in fallout are of particular interest because they represent the most significant health hazard. They enter the food chains quite rapidly and eventually end up in the human body. Iodine 131 (half-life, 8 days) is present in significant amounts in local and early tropospheric fallout. It is deposited on the surface of plants that are eaten by dairy animals, who secrete it in milk, which is then consumed by the human population. Iodine 131 is accumulated in the thyroid gland.

Cesium 137 (half-life, 30 years) is one of the longest-lived fission products; consequently it is found in any type of fallout in significant amounts. The pathway of this nuclide from fallout to man is the same as that for iodine 131, but it distributes itself in the body more uniformly, because the chemical properties of cesium are similar to those of potassium, whose distribution in the body is rather uniform.

Strontium 90 (half-life, 28 years) is also found in every type of fallout. It is ingested by man with milk, and since it is chemically similar to calcium, it is accumulated in bones and teeth with calcium.

Other radiobiologic aspects of these radionuclides are discussed in Chapter 9.

In air and surface bursts, injuries to human populations caused by shock wave, thermal radiation, and initial nuclear radiation can be completely avoided if the bomb is detonated in an area uninhabited for thousands of square miles. But the possible injuries from worldwide fallout cannot be prevented with any practical means. For this reason, air and surface explosions for testing purposes should be discouraged and even forbidden by international treaties, unless it is decided that the benefit to society from such tests is well worth the biologic cost that society has to pay.

REFERENCES

Eisenbud, M. 1963. Environmental radioactivity. McGraw-Hill Book Co., New York.

Fowler, J. M. 1960. Fallout. Basic Books, Inc., Publishers, New York.

Glasstone, S. 1958. Sourcebook on atomic energy, ed. 2, D. Van Nostrand, Co., Inc., Princeton, N.J. Chapters XIII to XV.

Glasstone, S. 1962. The effects of nuclear weapons. U.S. Atomic Energy Commission, Washington, D.C.

Hahn, O. Feb. 1958. The discovery of fission. Sci. Amer. **198**:76.

Leachman, R. B. Aug. 1965. Nuclear fission. Sci. Amer. **213**:49.

Smyth, H. D. 1945. Atomic energy for military purposes. Princeton University Press, Princeton, N.J.

The following booklets of the "Understanding the Atom" series, published by the Atomic Energy Commission, Division of Technical Information, are also extremely useful:

The first reactor
Nuclear reactors
Research reactors
Fallout from nuclear tests

CHAPTER 9

Radiation dose and radiation exposure of the human population

RADIATION DOSE
General considerations

In radiation biology we are interested in the study of several types of effects caused by ionizing radiations in biologic systems. This study, of course, cannot be done properly without a quantitative measurement of the amount of ionizing radiation responsible for the production of a given effect.

The quantity of ionizing radiation energy delivered to, or absorbed by, an object is commonly referred to as the *radiation dose.*

Let us consider (Fig. 9-1) a source of ionizing radiation, such as the focal spot on the target of an X-ray tube, a radioactive source emitting alpha or beta particles or gamma photons, a neutron source, etc. A material object with the shape of a cube (for purposes of simplification) is placed in the radiation field at a distance (d) from the source. A fraction of the radiation emitted by the source is delivered to the object through the side of the object facing the source. The amount of radiation energy (E_0) delivered to the object is called the *exposure dose.* Some of this energy is dissipated within the object; this fraction of E_0 is the *absorbed dose* (E_a).

A fraction of the radiation delivered to the object deposits only part of its energy within it. Due to one or more of several scattering processes occurring within the object (Compton scattering of X- or gamma photons, elastic or inelastic scattering of alpha or beta particles or neutrons, etc.), some of the particles or photons may emerge out of the object with residual energy (E_s). The residual energy of the scattered radiation is not absorbed by the object.

Still another fraction of the radiation delivered to the object does not deposit any of its energy within it, because the particles or photons do not interact at all with the matter of which the object is made. This radiation emerges from the object in the same direction as the incident radiation, and its energy may be referred to as transmitted energy (E_t). Because of the principle of conservation of energy, it must be

$$E_0 = E_a + E_s + E_t$$

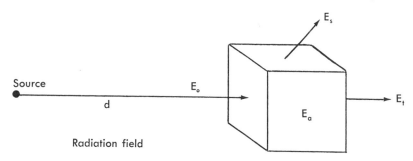

Fig. 9-1. Diagram to illustrate the concept of radiation dose. E_0, Exposure dose. E_a, Absorbed dose. E_s, Energy of scattered radiation. E_t, Transmitted energy. **d,** Distance of the object from the source.

The exposure dose depends on the following six factors:

1. If the radiation source is a radioactive source, the exposure dose is dependent on its activity, which is expressed in disintegrations per unit of time, or curies, as seen in a previous chapter. It is self-evident that, with all other factors being constant, a 2 Ci cobalt-60 source delivers to an object twice as much dose as a 1 Ci cobalt-60 source.* By doubling the activity of the source, the object receives twice as many beta particles and gamma photons per unit of time. If the source is an X-ray tube, the exposure dose is dependent on the intensity of the beam, which, as seen in Chapter 4, is affected by tube voltage and current and by filtration.

2. For radioactive sources, the exposure dose is dependent on the decay energy, and therefore on the particle or photon energy, and on the number of particles or photons emitted per disintegration. These characteristics are all graphically displayed by the decay schemes. Consequently, 1 Ci of different radionuclides delivers different amounts of radiation energy and therefore different exposure doses. If all other factors are constant, 1 Ci of cobalt 60 delivers to an object a higher gamma-ray dose than 1 Ci of cesium 137, because for cobalt 60 every disintegration produces two gamma photons, whereas for cesium 137 only 93.5% of the disintegrations produce one gamma photon. In addition, the cesium-137 gamma photon is less energetic than both cobalt-60 gammas (see pertinent decay schemes in Appendix III).

For X-ray sources, the exposure dose is dependent on the energy distribution in the X-ray beam, that is, on the quality of the beam (review the pertinent information in Chapter 4).

3. Because of the inverse square law, the exposure dose is inversely proportional to the square of the distance of the object from the source.

4. If the radiation emitted by the source is not collimated, the exposure dose depends also on the object's surface area exposed to radiation. It is obvious that the larger the area the more radiation energy is delivered to the object.

5. The nature of the absorber possibly present between source and object affects the exposure dose. Less radiation energy is delivered to the object when

---

*I hope that no student will ever be tempted to express radiation doses in curies! The curie is a unit of radioactivity and not of dose.

there is water between source and object than when there is air between them. Likewise, the gamma dose delivered to an object can be considerably reduced by placing thick lead shielding between the source and the object.

6. The exposure dose is directly proportional to the time of exposure.

In summary, the factors are activity or intensity of source, radiation energy, distance of object from source, object's surface area exposed, nature of absorber, and time of exposure.

An important practical application derives from the consideration of the factors affecting the exposure dose. In problems concerning radiation protection of personnel, one should remember that when other factors cannot be reduced the time and distance factors are quite influential in the reduction of the exposure dose. Thus, a nurse in care of a patient who has been given a therapeutic amount of a gamma-emitting radioisotope (several millicuries) can reduce the exposure dose to herself if she spends as little time as possible with the patient and most of all if she keeps the maximum possible distance from him.

In Fig. 9-1, it is evident that equal volumes (such as 1 cm^3) of different materials to which the same exposure dose has been delivered absorb different fractions of the exposure dose, because of the different densities of the materials. This fraction (that is, the absorbed dose) is small for 1 cm^3 of air, larger for 1 cm^3 of water or tissue, and even larger for 1 cm^3 of lead. What might not be so evident is that for the same incident dose equal weights (and therefore equal quantities of matter) of different materials do not necessarily absorb the same fraction of the exposure dose.

The dose absorbed by equal weights of different materials depends not only on the exposure dose, but also on the elemental composition of the irradiated material and on the radiation energy.

The elemental composition of the irradiated object determines its average (or "effective") atomic number. If we limit the discussion to X- and gamma radiation, the probability of the photoelectric effect (which results in total dissipation of the photon energy) increases with the atomic number of the absorbing material and decreases with the increase of the photon energy. Consequently, the fraction of an incident low-energy X- or gamma radiation that is absorbed by an object of high atomic number is relatively high.

With the increase of the X- or gamma ray energy, the Compton effect (which results only in partial dissipation of the photon energy) becomes more significant. Compton scattering does not depend on the atomic number of the absorbing material. Consequently, at higher photon energy, equal weights of materials with different atomic numbers may absorb the same fraction of the exposure dose.

In the next section these concepts are further elaborated and applied to the doses absorbed by different tissues.

The only fraction of the exposure dose responsible for the biologic effects in an organism, organ, or tissue is the absorbed dose, E_a.

The same radiation dose can be delivered to a system in a short time or a long time. The rate at which a dose is delivered is called the *dose rate* (that is, dose per unit of time). The magnitude of certain biologic effects depends not only on the total accumulated dose, but also on the dose rate, as discussed in later chapters.

Radiation dose units

The quantity *radiation dose*, which has been described in the previous section, must be accurately measured whenever an object is irradiated for any purpose. Therefore it is necessary to establish an appropriate and convenient dose unit.

For several years after the discovery of X-rays and radioactivity, radiologists were the only persons who were compelled to measure the radiation dose administered to their patients for diagnostic or therapeutic purposes. A unit of dose that was in use for some time in radiology was the "skin erythema dose," which was defined as the dose of X-rays necessary to cause a certain degree of erythema (reddening of the skin) within a specified time (1 to 3 weeks).* The vagueness of the unit and the impossibility of measuring exactly an X-ray dose with it are only too evident.

With the origin and development of other radiation sciences, notably radiation biology and radiation chemistry, and with the beginning of the atomic age, the problem of measuring doses of a wide variety of ionizing radiations interested a broader spectrum of scientists. The need was for a unit that could be defined less ambiguously, a unit based on some exactly measurable effect produced by radiation, such as ionization, energy dissipation, or induced chemical reactions.

Unfortunately, the history of radiation dose units has been marred by considerable confusion during the first 60 years of this century. Vague definitions, the selection of inappropriate units, the impossibility of using certain units for any type of ionizing radiation, the imperfect knowledge of some radiation interactions are all factors that have contributed to the confusion. At the present time, the situation is greatly improved, although some progress is still desirable.

The roentgen

The roentgen (named for the discoverer of X-rays) is a unit of exposure dose, defined in terms of number of ionizations produced in air by X- and gamma radiation.

It was first suggested by P. Villard in 1908 and then accepted and defined, rather ambiguously, 20 years later by the International Commission on Radiological Units (ICRU). It was redefined more precisely in 1937 (with slight modifications in 1962) as follows: "One roentgen is an exposure of X- or gamma radiation such that the associated corpuscular emission per 0.001293 g of air produces, in air, ions carrying 1 esu of quantity of electricity of either sign."

The abbreviation presently adopted for the roentgen is R (not r).

The definition of the roentgen needs explanation and comments. Let us consider (Fig. 9-2) a cube of dry air at $0°$ C and 760 mm of mercury pressure, with a volume of 1 cm³. The weight of this cube is therefore 0.001293 g; thus it contains a definite number of atoms. This volume of air is completely surrounded by air.

Let us imagine that we irradiate this air uniformly with X- or gamma rays for a certain period of time. Photoelectric and Compton interactions occur within and without the cube of air during irradiation. Of course, a large fraction of the incident radiation passes through it without interacting with any atom. The photoelectrons and the Compton electrons produced by the interactions constitute what, in the definition of the roentgen, is called "the associated corpuscular emission."

Before the electrons come to rest, they dissipate their kinetic energy in part

---

*Later, when the roentgen was adopted, it was shown that the erythema dose of gamma radiation from radium is about 1800 roentgens.

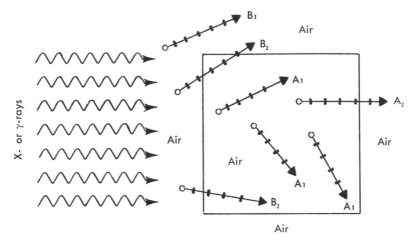

Fig. 9-2. Illustration of the roentgen unit. The square represents 1 cm³ of air. Straight arrows show the paths of photoelectrons and Compton electrons originating from interactions occurring at points indicated with small circles. Although the electron paths are really zigzag, they have been shown as straight lines for simplification purposes. The pairs of dots along the paths are ion pairs.

by producing ion pairs along their paths. Some electrons (type A in Fig. 9-2) originate from interactions that occur within the cube of air, and they may produce ion pairs exclusively within this volume of air (electrons labeled A_1 in Fig. 9-2), or partly within and partly without (A_2). Other electrons (type B) originate from interactions that occur outside of the 1 cm³ of air, and they may produce ion pairs exclusively outside of this volume (B_1), or partly outside and partly inside of it (B_2).

Now let us suppose that we can measure the total electric charge of *all* the positive ions and of *all* the negative ions produced by the electrons of type A *only.* The two total charges would be equal in magnitude, although of opposite signs. If the magnitude of the charges is 1 esu, then the quantity of X- or gamma radiation delivered is arbitrarily called 1 roentgen.

If the cube of air considered above were not surrounded by air, then the A_2 photoelectrons or Compton electrons would produce some ion pairs in a material different from air. The number of ion pairs produced by the same ionizing particle may be different in different materials. The definition of the roentgen specifies that ions should be produced by the associated corpuscular emission *in air.* This is why in our illustration we had to assume that the 1 cm³ volume of air is surrounded by air. The actual number of ion pairs produced by 1 R in 1 cm³ of air, in the conditions specified above, is 2.082×10^9, because the total electric charge of as many ions is 1 esu.

The average energy dissipated to produce 1 ion pair in air is 33.7 eV. Therefore the energy dissipated by 1 R of X- or gamma rays in air is about $33.7 \times 2.082 \times 10^9 = 7.01 \times 10^{10}$ eV/cm³, or 0.112 erg/cm³, or 86.9 erg/g.*

*These figures have been slightly changed several times during the past few years. The values adopted here are taken from Attix, F. H., and W. C. Roesch, editors. 1966–1968. Radiation dosimetry. Academic Press, Inc., New York.

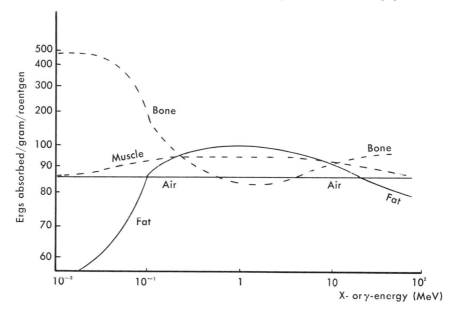

Fig. 9-3. Absorption of X- or gamma radiation energy in air, muscle, fat, and bone.

I cannot emphasize enough that the roentgen is a unit of exposure dose of X- or gamma radiation. Other types of ionizing radiations (alpha, beta, neutron radiations) cannot be measured in roentgens.

The rad

As stated above, the quantity of radiation responsible for biologic effects in the irradiated material is the absorbed dose. If 1 g of air is exposed to 1 R of 0.1 MeV gamma radiation, it absorbs a quantity of energy equal to 86.9 ergs. If instead 1 g of soft tissue is exposed to 1 R of the same type of radiation, it absorbs 95.1 ergs, and if 1 g of bone is exposed to 1 R of the same type of radiation, it absorbs about 175 ergs of energy. The quantity of energy absorbed by 1 g of material when it is exposed to 1 R of X- or gamma radiation depends on the effective atomic number of the material and on the radiation energy. Fig. 9-3 shows how this quantity of energy varies with radiation energy for air, muscle, fat, and bone. One can see that for air the amount of energy absorbed per gram per roentgen is not affected by the radiation energy, whereas it is considerably affected for bone and fat. The curve for water (not shown in the diagram) would not be very different from the curve for muscle and other types of soft tissues. The very high energy absorption in bone at low photon energies is justified by the relatively high effective atomic number of this tissue (13.8) as compared with that of water and soft tissue (7.5).*

Since the quantity of energy absorbed per gram per roentgen varies from one material to another, and since in most situations what we are really interested in is the dose absorbed by the irradiated object and not the exposure dose,

---

*The higher effective atomic number of bony tissue is due to the high content of calcium ($Z = 20$) and phosphorus ($Z = 15$) in this tissue.

a unit was introduced in 1954 by the ICRU for the measurement of absorbed dose. This unit is called the *rad* (an acronym for *radiation absorbed dose*). It has been defined as "the absorbed dose of any type of ionizing radiation that is accompanied by the liberation of 100 ergs of energy per gram of any absorbing material."

The rad has the considerable advantage of being usable with any type and energy of ionizing radiation, whereas the roentgen can be used only with X- and gamma radiation energy not exceeding 3 MeV.

We may say therefore that air absorbs 0.869 rads (that is, 86.9/100), soft tissue absorbs 0.951 rads, and bone absorbs 1.75 rads when they are exposed to 1 R of 0.1 MeV gamma radiation. Conversely, when air absorbs 1 rad of 0.1 MeV gamma radiation, it has been exposed to 1/0.869 or 1.15 R; when soft tissues absorb 1 rad of the same type of radiation, they have been exposed to 1/0.951 or 1.05 R; and when bone absorbs 1 rad of the same type of radiation, it has been exposed to 1/1.75 or 0.57 R. One can see that for water and soft tissues the exposure dose in roentgens and the absorbed dose in rads are almost equivalent numerically.

In many instances, the absorbed dose can be measured directly with a variety of dosimetric methods (described in several of the following pages). The absorbed dose can be calculated for any absorbing material if the exposure dose and mass absorption coefficient (μ/ρ, see p. 62) of the material are known, with the formula:

$$\text{Absorbed dose (in rads)} = \left[0.869 \frac{\mu/\rho \text{ of material}}{\mu/\rho \text{ of air}} \right] \times \text{R}$$

The term within brackets is known as the f factor; it has been calculated for different photon energies and for different types of tissues. Charts are published elsewhere,* with the f factor plotted as a function of HVL of the X- or gamma radiation used. The absorbed dose in rads is then obtained by multiplying the exposure dose in roentgens by the f factor.

The rem

In radiobiologic research, investigators have discovered and confirmed several times that the absorption of equal doses of different types of radiation by the same biologic system does not produce a given effect in the same degree. For instance, the frequency of chromosomal aberrations produced in *Tradescantia* pollen grains by a certain absorbed dose of 250 kV X-rays is twice as high as the frequency of aberrations produced by the same absorbed dose of ^{60}Co gamma rays. In other words, 250 kV X-rays are twice as effective as ^{60}Co gamma rays in the induction of chromosomal aberrations in *Tradescantia*. If a culture of mouse-ascites tumor cells has absorbed 150 rads of medium-energy X-rays, the percentage of abnormal anaphases among these cells 24 hours after irradiation is 30%. To obtain the same effect with neutron radiation, a much lower dose is needed. Neutron radiation is therefore more effective than X-rays in inducing anaphasic abnormalities in ascites tumor cells. Likewise, neutron radiation is 10 times more effective than X-rays in producing cataracts.

In order to take these important facts into account, the concept of *relative biologic effectiveness* (RBE) has been introduced in radiation biology. RBE can

---

*See, for instance, National Bureau of Standards Handbook 62. 1956.

be defined as a ratio:

$$RBE = \frac{\text{Dose in rads to produce a given effect with therapy X-rays}}{\text{Dose in rads to produce the same effect with radiation under investigation}}$$

One can see that the ratio is greater or lesser than unity depending on whether the radiation under investigation is more or less effective in producing the effect under study than 250 kV therapy X-rays, which are used as reference. Therefore the RBE may also be defined as the ratio of the biologic efficiency of the radiation used to the biologic efficiency of therapy X-rays in producing the same effect.

If the unit of absorbed dose (rad) is multiplied by the RBE of the radiation used, another unit of dose is obtained, and its name is *rem* (acronym for *r*oentgen *e*quivalent *m*an). Therefore we can write the following:

$$\text{Rads} \times \text{RBE} = \text{Rems}$$

The rem can thus be defined as the absorbed dose of any type of radiation that produces the same biologic effect as 1 rad of therapy X-rays. Consequently, if the RBE of neutron radiation for a specific effect is found to be 10 so that the same effect is obtained with 250 rads of therapy X-rays or 25 rads of neutron radiation, we say that 250 rems of either type of radiation are needed to produce the effect under consideration. With the introduction of the rem different absorbed doses of different types of radiation that produce the same biologic effect can be expressed with the same number. This fact is very useful in radiation biology and in problems concerning radiation protection of personnel. When a radiation dose is expressed in rems, we need not specify the type of radiation. For instance, if we state that 700 rems are sufficient to kill 50% of a population of rats in 30 days, we need not specify what type of radiation we are referring to; with this dose, any type of radiation produces the same effect. But if we state that 700 rads are sufficient to kill 50% of a population of rats in 30 days it is necessary to specify that we are talking about X- or gamma rays; in fact only about 70 rads of neutron radiation may be necessary to obtain the same effect (assuming an RBE of 10 for neutron radiation and for this specific effect).

Unfortunately, the conversion of rads to rems is many times difficult, because the RBE factor cannot be measured directly with any method or instrument. It can be only estimated by testing the effectiveness of the radiation of interest and the effectiveness of therapy X-rays in the production of the same biologic effect. Nevertheless, RBE values are, at times, vague and elusive, because the RBE may depend on a number of factors, such as (1) type of biologic effect being studied, (2) radiation type and energy, (3) organism, tissue, or cells under study, (4) presence or absence of oxygen during irradiation, (5) dose rate, etc.

Concerning radiation type and energy, there is an important property of the radiation used, which affects the RBE considerably. It is known as the *linear energy transfer* (LET). First introduced by R. Zirkle (1952), the concept of LET has proved to be very useful in radiation biology for the correct interpretation of many biologic effects. It is applied only to corpuscular radiation and is defined as the amount of energy (in keV) dissipated by an ionizing particle per micrometer of path:

$$LET = \text{keV}/\mu\text{m}$$

The LET also includes the energy dissipated along possible delta tracks, which are very common with alpha particles (see p. 91).

Most or all of the energy dissipated by a particle along its path is used in the production of ion pairs. It is evident then that the LET is roughly a function of the specific ionization of the particle.

Some approximate values of LET are given in the following table.*

Type of radiation	Approximate LET (keV/μm)
4 MeV gamma rays	0.3
^{3}H beta particles	5.5
7 MeV protons	10
Recoil protons from fission neutrons	45
0.6 MeV protons from $^{14}N(n, p)^{14}C$ reaction	65
Alpha particles from $^{10}B(n, \alpha)^7Li$ reaction	190
Fission fragments	4000–9000

One should bear in mind that in tissues an ionizing particle does not dissipate its entire energy along its path through the production of ion pairs only, for energy is also transferred by excitation to the medium. Excitation causes several biologic effects with a mechanism that is not always clearly understood.

Since the RBE depends on so many factors, RBE values are meaningful only when all those factors are specified. Values with the specification of the radiation type only should be interpreted, at most, as orders of magnitude, or as averages of values calculated for different biologic effects produced by the same type of radiation in different circumstances.

Some of these general RBE values are reported below:

Type of radiation	RBE
4 MeV gamma rays	≈ 0.7
Medium-energy X- and gamma rays	1
1 MeV beta particles	≈ 1
0.1 MeV beta particles	≈ 1.08
Beta particles of E < 30 keV	≈ 1.7
Thermal neutrons	≈ 4–5
Fast neutrons	≈ 10
Protons	≈ 10
Alpha particles	≈ 10–20

In summary, the following three radiation dose units have been introduced and defined in this section:

The roentgen is a unit of exposure and is defined in terms of ionizations produced in air by X- and gamma radiation.

The rad is a unit of absorbed dose and is defined in terms of energy deposited in any material by ionizing radiation of any type.

The rem may be considered as a unit of biologic dose and takes into account the relative biologic effectiveness of (any) radiation absorbed by a biologic system.

RADIATION DOSIMETRY

Radiation dosimetry is the measurement of the radiation dose delivered to, or absorbed by, an object. A dosimeter is any instrument or system that is

*From Storer, J. B., et al. 1957. Radiat. Res. **6:**188–288.

capable of measuring radiation dose, in roentgens or in rads, by providing a quantitative appraisal of a radiation effect produced within its sensitive volume.

The effect measured may be ionization in a gas (ionization chambers), the yield of a radiation-induced chemical reaction (chemical dosimeters and photographic emulsions), the heat that results from the degradation of the radiation energy (calorimeters), or atomic and molecular excitations (luminescent dosimeters).

The sensitive volume of the dosimeter in which the measurable effect takes place may be a gas (ionization chambers), a solution (chemical dosimeters), or a solid (photographic emulsions and luminescent dosimeters).

An *ideal* dosimeter should have the following characteristics:

1. Its response should be directly proportional to dose (linear response) so that if response is graphically plotted as a function of dose, a straight line should be obtained.

2. It should be useful over a wide dose range.

3. Its response should be independent of:
   a. Dose rate
   b. Certain radiation characteristics, such as type, energy, LET

4. The response should be precise, accurate, and reproducible.

5. The dosimeter should be simple and convenient to use and preferably portable.

Unfortunately, no single dosimeter possesses all the characteristics of an ideal dosimeter. For instance, chemical dosimeters have poor sensitivity in the dose range of a few hundred rads, some luminescent dosimeters are dose-rate and radiation-energy dependent, calorimeters and standard free-air chambers can be extremely precise and accurate, but because of the meticulous techniques involved and the complexity of the instruments, they cannot be used for routine dosimetric measurements.

A dosimeter is regarded as *absolute* if its construction or mechanism of operation is such as to give directly the dose absorbed in the sensitive volume after some appropriate calculations. Such a dosimeter does not have to be calibrated in a known radiation field. Absolute dosimeters, as they are described below, are the standard free-air ionization chamber, the Fricke chemical dosimeter, and the calorimeters. These dosimeters are used as standards for the calibration of other types of dosimeters.*

Description is given below only of those dosimeters presenting many of the characteristics of an ideal dosimeter. The emphasis is proportionate to their usefulness and popularity in radiobiologic work.

---

*Example: Suppose that we wish to calibrate a luminescent dosimeter. An absolute dosimeter is placed at an appropriate point in a radiation field, and the dose absorbed in a given time is measured and found to be, for example, 50 rads. The luminescent dosimeter is then placed at the same point for the same time, and its response (luminescence) is measured. The measured amount of luminescence corresponds then to 50 rads absorbed by the luminescent dosimeter. The same procedure is repeated for a different point of the same radiation field at which the dose is different. A curve of the amount of luminescence as a function of dose can thus be plotted. This curve is a straight line if the dose-response relationship is linear. For more accurate calibration, the procedure should be repeated for more than two points of the radiation field. In certain instances, the calibration of a nonabsolute dosimeter can be done with the use of an empirical formula.

Ionization chambers

Some dosimeters are ionization chambers operated on the same basic principles as the electroscopes and electrometers described in Chapter 7. The exposure dose is obtained by collecting and measuring the number of ion pairs produced by the "secondary corpuscular emission" generated by X- or gamma rays in a volume of air.

In the standard free-air ionization chamber (Fig. 9-4), an accurately measured volume of air (V) in a collimated beam of X- or gamma radiation is between two plates to which a high voltage is applied, so that one plate is positive and the other negative. The temperature and pressure of the air are also measured and controllable; therefore the weight of V can also be calculated.

To allow the secondary electrons generated in V to produce ion pairs exclusively in air (note definition of roentgen on p. 215), V is separated from the collecting plates by another volume of air (V_1). The distance (d) of the plates from V depends on the energy of the radiation used; the higher the energy the greater the distance, because the range of the secondary electrons generated by high-energy photons is longer than the range of the secondary electrons generated by low-energy photons. For very high energy radiation, dry pressurized air may be used, to keep the volume of the chamber within convenient size limits. In pressurized air, the range of the secondary electrons is proportionately reduced.

The ions produced in V and V_1 are collected on the charged plates, and thus an electric current is formed. This is a saturation current because the voltage applied to the plates is high enough (1500 to 2000 volts) to prevent any ion-pair recombination (see p. 125). A high-precision, very sensitive electrometer connected to the plates measures the cumulative charge of all the ions collected. If this charge is divided by the volume and corrected for standard temperature and pressure, the exposure dose (in roentgens) is obtained. Specifically, the exposure dose is given by

$$R = \frac{Q \times T \times 760}{v \times p \times 273}$$

wherein R is the exposure in roentgens, Q is the charge in esu, T is the temperature in Kelvin degrees, v is the sensitive volume of the chamber, in cm^3, and p is the pressure in the chamber, in mm of Hg. Since the standard free-air chamber meets all the requirements governing absolute determination of the roentgen, as this unit is defined, it must be considered as an absolute dosimeter, or primary standard. It is used by the National

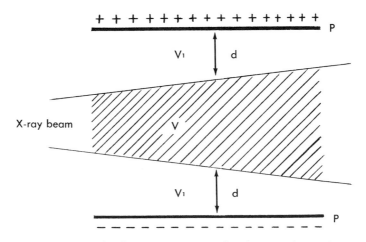

Fig. 9-4. Diagram of a free-air ionization chamber. **P**, Plates. (See text.)

Bureau of Standards and by some commercial manufacturers for the precise calibration of X-ray equipment and for the calibration of nonabsolute dosimeters.

For routine dosimetry of X- and gamma radiation, a condenser roentgen meter is more conveniently used. The most widely accepted model is the condenser R-Meter manufactured by the Victoreen Instrument Co.

This instrument (Fig. 9-5) consists of a DC power supply, an electrometer charger with its viewing microscope system, a lighting system for the microscope, all contained in a metal case, and an ionization chamber. Projecting from the top of the case are two operating controls and the viewing microscope. The ionization chamber is usually called a "thimble chamber" because the sensitive volume of air in which the ion pairs are produced is enclosed in a cavity with the shape of a thimble (Fig. 9-6). The thimble is made of an air-equivalent material (Bakelite or nylon), that is, a material with an effective atomic number identical to that of air. The size of the sensitive volume and the thickness of the thimble are such as to assure that all ion pairs are exclusively produced in the air volume and in the wall of the thimble. For this reason, different chambers are needed for different radiation-energy ranges.

Fig. 9-5. Condenser R-Meter with high- and medium-energy chambers. Another chamber is inserted in the socket. (Courtesy Victoreen Instrument Co., Cleveland, Ohio.)

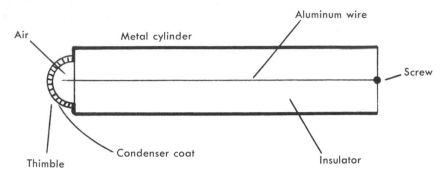

Fig. 9-6. Diagram of an ionization chamber for the Victoreen condenser R-Meter.

The thimble chamber is connected in parallel with a fixed capacitor contained in a metal cylinder, which forms one (grounded) electrode. The other electrode is a perfectly insulated aluminum wire that projects into the thimble at one end and is connected with a screw at the other end. The internal surface of the thimble is coated with a conductive material that is in electric contact with the metal cylinder. Only the thimble chamber is sensitive to radiation.

The instrument is used by inserting the ionization chamber in a socket of the metal case, and the capacitor is charged with a charger contained in the metal case by turning the control knobs. During this operation, the quartz fiber of the electrometer is viewed through the microscope against a scale calibrated in roentgens. The quartz fiber behaves exactly as in the Landsverk electroscope (see Chapter 7). When the image of the quartz fiber is on the zero position the capacitor is fully charged; that is, a certain potential difference is applied across the two electrodes. The chamber is then removed from the socket and placed in an X- or gamma field at the point where the exposure dose is to be measured. The ions formed in the thimble chamber are attracted to the oppositely charged electrodes of the chamber, thus reducing the charge held by the capacitor and the potential difference across its electrodes. When the chamber is reinserted into the socket of the instrument, the electrometer measures the reduced potential of the capacitor. The instrument has been calibrated by the manufacturer with a standard free-air chamber so that the reduced potential difference can be read directly as roentgens on the illuminated scale of the electrometer.

A wide selection of chambers is offered to cover different radiation energy and dose ranges. There are low-energy chambers for X-rays of 6 to 35 keV average energy, medium-energy chambers for the 30 to 400 keV range, and high-energy chambers for the 400 to 1300 keV range. Within each energy range, chambers for a variety of dose ranges are available (from 0.025 to 250 R), with rated accuracy varying from $\pm 10\%$ to $\pm 2\%$. Volumes of the thimble vary from 0.20 to 1450 cm^3.

The condenser R-Meter can be used for the measurement of total accumulated dose, as long as the dose does not exceed 250 R. When working with much higher doses, the instrument can be used to calibrate the radiation field at any distance from the source, by exposing the chamber in the field for a predetermined time interval compatible with the dose range of the chamber. If,

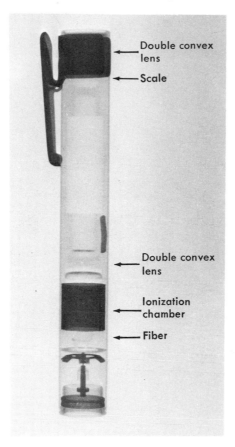

Fig. 9-7. X-ray picture of a pocket dosimeter.

for instance, the reading for a 1-minute exposure is 150 R at a given point of the field, and an exposure dose of 1500 R must be delivered to an object, the latter must be placed in the field at the same point for 10 minutes. In these conditions, the instrument is really used as a dose-rate meter. This procedure, of course, is valid only when the intensity of the source is constant during the exposure. Constant intensity is usually obtained with a gamma source, but not always with some models of X-ray machines.

The pocket dosimeter (Fig. 9-7) is an instrument used to monitor radiation exposures to personnel for radiation-protection purposes. It has the shape and size of a fountain pen and can be worn in a pocket. It incorporates, in compact form, all the components of a condenser roentgen meter (ionization chamber, capacitor, electrometer, scale, microscope), except the charger and the lighting system, which are contained in a separate box. The instrument gives a reading of the total accumulated dose and is available with various dose ranges (common ranges are 250 mR and 5 R full scale). These instruments are not suitable for accurate measurements of doses in a radiation field.

Other larger, yet portable, ionization chambers for monitoring purposes include the Cutie Pie and other similar instruments that give dose-rate readings in milliroentgens or roentgens per hour (Fig. 9-8). These instruments are very

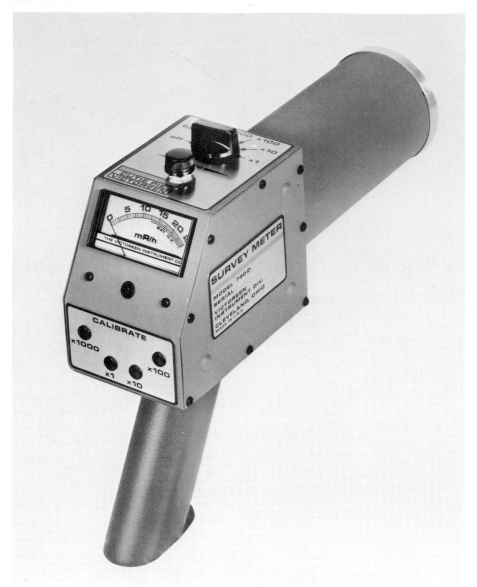

Fig. 9-8. Cutie Pie survey dose-rate meter. (Courtesy Victoreen Instrument Co., Cleveland, Ohio.)

useful and convenient for monitoring the radiation dose levels existing in proximity to radioactive materials, nuclear reactors, neutron generators, X-ray equipment, and other similar sources.

Calorimetric methods

Consider a measured volume of distilled water, thermally insulated so that no heat exchange is possible between the water and the surrounding environment. If the volume of water absorbs a dose of ionizing radiation and all the absorbed radiation energy is eventually degraded as heat, the temperature of the water rises, in proportion to the quantity of radiation energy added to the system. To raise the temperature of 1 ml. of distilled water by 1° C, an average of 1 calorie, or approximately 4.2×10^7 ergs of heat energy must be supplied. By measuring the temperature increase, one can determine the amount of radiation energy in ergs absorbed per milliliter of water and, with appropriate corrections, per gram of water. The value thus obtained can be easily converted to rads (note definition of rad on p. 218).

For instance, if the dose absorbed is 1000 rads, the amount of radiation energy absorbed is 1000×100 ergs per gram, or 10^5 ergs per gram. The temperature increase is

$$\frac{10^5 \text{ ergs}}{4.2 \times 10^7 \text{ ergs}} \approx 0.002° \text{ C}$$

Conversely, by measuring the temperature increase, the absorbed dose can be calculated. It is evident that a very sensitive thermistor or thermocouple would be needed to measure such small temperature changes.

A calorimetric dosimeter works basically on these principles. If the instrument is properly designed, it has the advantage of measuring energy deposition directly, and as such it should be classified as an absolute dosimeter. But it has also some disadvantages and limitations. Although very sensitive temperature-measuring devices are available for the determination of very small temperature changes, a calorimeter is limited to doses of no less than several hundred rads, since the quantity of heat deposited in a system by large radiation doses is relatively small. Furthermore, the fact is not always true that all the radiation energy deposited in a system is eventually converted to heat. If a radiation-induced chemical reaction takes place in the system and the reaction products persist after irradiation, part of the radiation energy is converted into chemical energy and therefore is not measured by the calorimeter. However, in some cases, radiation-induced decomposition products recombine so that the chemical composition of the system after irradiation is the same as before. All the radiation energy is then eventually converted into heat.

Calorimeters have the advantage over standard free-air chambers of measuring radiation energy deposited in any material by any type of ionizing radiation.

Calorimeters are not suitable for routine dosimetry. They are employed to measure the activity of radioactive sources, to determine the yield of radiation-induced chemical reactions (see Chapter 11), to calibrate the intensity of X-ray and electron beams, and to calibrate other dosimeters used as secondary standards.

Chemical dosimeters

The yield of some radiation-induced chemical reactions is directly proportional to the absorbed dose, and relatively independent of type of radiation, radiation energy, and dose rate. If the yield of the reaction products can be conveniently and accurately measured, the systems in which such radiation-induced chemical reactions take place can be used as chemical dosimeters. These dosimeters are described more appropriately in Chapter 11 after the chemical effects of ionizing radiations have been discussed.

Luminescent dosimeters

If certain solid materials are exposed to high-energy radiation (such as ultraviolet and ionizing radiation), they emit energy in the form of light. This phenomenon is called *luminescence.* If the emission of light occurs within a few nanoseconds after the absorption of the high-energy radiation, luminescence is more specifically known as *fluorescence,* and the materials showing such properties are called *fluors.* You may recall that scintillation detectors are based on this type of luminescence.

If the emission of light is delayed for more than 1 μsec, the phenomenon is more properly known as *phosphorescence,* and the materials showing such property are called *phosphors.*

Certain materials luminesce if irradiated with ultraviolet light. This phenomenon is known as *photoluminescence*.* If photoluminescence is enhanced by previous exposure of these phosphors to ionizing radiation, the phenomenon is known as *radiophotoluminescence*. Other materials luminesce if heated after exposure to ionizing radiation. This phenomenon is known as *radiothermoluminescence* (or, less properly, thermoluminescence), which is a thermally accelerated phosphorescence. Investigators have found that the amount of light emitted by radiophotoluminescent and radiothermoluminescent materials is dependent on the dose of ionizing radiation absorbed. This relationship justifies the use of these materials as dosimeters.

Radiophotoluminescent dosimeters

The materials used for radiophotoluminescent dosimeters are silver-activated phosphate glasses, with a silver content of about 4%. When these glasses are exposed to ionizing radiation and then excited with ultraviolet radiation (3650 A), they emit orange light with a peak intensity at about 6400 A. It seems that the luminescence centers are either Ag^{2+} ions (formed from Ag^+ ions by loss of an electron) or aggregates of Ag^0 atoms.

The glasses are available in a variety of shapes and sizes, like spheres, blocks, or miniature rods, small enough to be inserted into the tissues of animals and patients for the measurement of absorbed doses from internal emitters. Furthermore, high-Z and low-Z glasses are available (Z is the average atomic number).

The measurement of the dose absorbed by the dosimeter is made with an instrument, known as a *reader*, or *read-out system* (Fig. 9-9). The exposed dosimeter is irradiated with the ultraviolet light of a mercury vapor lamp. A special filter or monochromator allows the passage of near-ultraviolet light only (3650 A). The dosimeter emits orange light, which is made approximately monochromatic (about 6400 A) by another suitable filter. The intensity of the luminescence (proportional to the dose of ionizing radiation absorbed) is converted by a photomultiplier into an electric current, which is then amplified and read by a milliammeter calibrated in roentgens. These dosimeters must be calibrated, together with a particular reader, in a known radiation field with an absolute dosimeter by the manufacturer and periodically by the user.

The dose range in which the dose-response relationship is linear depends on several factors, such as the size of the dosimeter and its orientation in the ultraviolet beam. Miniature rods show a better dose-response linearity over the 10 to 10^3 rad range. Above the upper limit of this range the response is no longer linear with dose unless special precautions are taken. Some investigators have reported improved sensitivity of these dosimeters with special glasses that can measure doses well below 10 rads. In many types of phosphate glasses, the luminescence is not stable after the exposure to ionizing radiation. It builds up rapidly within a few hours after irradiation and then declines slowly over a long period of time. This phenomenon can be greatly attenuated if the exposed dosimeters are heated immediately after exposure. The temperature during exposure and reading has an effect on the response of these dosimeters.

Although these dosimeters do not show a dose-rate dependence, at least

---

*In this term the prefix "photo" refers to the radiation (that is, ultraviolet radiation) stimulating the luminescence, and not to the visible light emitted.

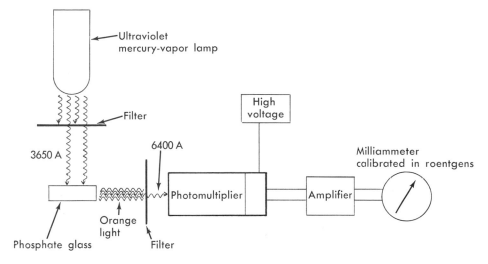

Fig. 9-9. Diagram of a reader for radiophotoluminescent dosimeters.

up to 10^8 rads/sec, they are considerably dependent on radiation energy. For gamma radiation they show a maximum response per roentgen at about 70 keV. For higher photon energies, the response declines steadily. A dependence on LET has also been shown.

Obviously radiophotoluminescent dosimeters are far from being ideal and versatile. Their serious shortcomings limit their usefulness to only a few particular situations in which they can be utilized routinely with appropriate calibration procedures.

Radiothermoluminescent dosimeters (TLD)

When a thermoluminescent phosphor is exposed to ionizing radiation, the ionization and excitation processes cause the trapping of electrons at sites of lattice imperfections. It is highly improbable that at room temperature the electrons will escape from the trapping centers and return to the ground state. At higher temperatures (well below the temperature of incandescence) sufficient energy is supplied to release them; when this release takes place, the phosphor luminesces, and the amount of light emitted is proportional to the number of trapped electrons and therefore to the dose of ionizing radiation absorbed.

Several thermoluminescent phosphors are available commercially with a light output large enough to qualify as useful and convenient dosimeters. Most of them are synthetic materials, obtained by growing crystals artificially with the addition of impurities as activators. Manganese-activated calcium sulfate ($CaSO_4$: Mn) is one of the oldest phosphors used in TLD. Manganese-activated calcium fluoride (CaF_2 : Mn) is superior to the natural calcium fluoride known as the mineral fluorite. Another widely used phosphor is lithium fluoride (LiF), and in particular the so-called TLD-100 developed by the Harshaw Chemical Company. The most recent addition to the family of the thermoluminescent phosphors is lithium borate; its characteristics are still under investigation.

Thermoluminescent phosphors are available in a wide variety of forms, shapes, and geometries, useful for different dosimetric problems, such as pellets, rods of various sizes, loose or encapsulated powders, single crystals, and ribbons.

It is interesting to note that many other materials exhibit thermoluminescent properties, although their light output is too weak to make them practical for dosimetry. For instance, a group of investigators of the Oak Ridge National Laboratory has attempted to use the rooftop tiles from Hiroshima and Nagasaki as thermoluminescent dosimeters to determine the radiation doses delivered at various points of those cities by the nuclear explosions.

The reader used in TLD is schematically shown in Fig. 9-10. The exposed dosimeter is placed in a pan and heated up at a variable and controllable rate (for example, 20° C/min) by means of an electric heater. When a certain temperature has been reached (different for different phosphors), which is, in any event, well below the temperature of incandescence, the phosphor begins to luminesce. As the temperature rises, the intensity of the light emitted is viewed by a photomultiplier, which converts light intensities into electric currents. A special filter blocks any infrared radiation emitted by the hot phosphor and prevents it from reaching the photocathode of the photomultiplier. After appropriate amplification, the output current of the photomultiplier is fed to a recorder, which is basically an X-Y plotter. A thermocouple in contact with the pan measures its temperature and feeds an electric signal to the same recorder, which plots a curve showing the variation of the thermoluminescent intensity with temperature. In an alternative mode, the recorder may plot thermo-

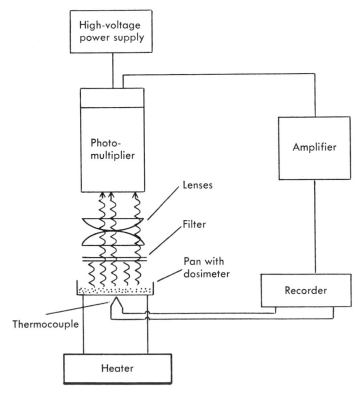

Fig. 9-10. Diagram of a reader for thermoluminescent dosimeters.

luminescent intensity as a function of time during which the temperature is increased. In either case, the curve is referred to as a *glow curve*. An examination of some glow curves (Fig. 9-11) shows that the peak luminescent intensity for different phosphors does not occur at the same temperature. It also shows that in certain instances there are several peaks, although only one of them is actually used for the dosimetric measurement. To measure the absorbed dose, one may use either the height of the main peak, or, more accurately, the total light emitted between the beginning and the end of the glow. This parameter is given by the area under the glow curve. Of course, the dosimeters must be calibrated together with a particular reader in a known radiation field.

Unlike radiophotoluminescent dosimeters, TL dosimeters are reusable after the reading and after the traps have been emptied by heating at sufficiently high temperatures. A complete reading system is shown in Fig. 9-12.

The characteristics of the three most commonly used phosphors in TLD are summarized in Table 9-1, in which differences in performance and usefulness are shown.

One advantage of LiF over the other two phosphors is the shorter wavelength of its luminescence light; this wavelength makes it easier to discriminate against infrared radiation by means of filters. It also shows the least energy dependence and its luminescence is quite stable. It is nearly tissue-equivalent (in terms of average Z), which makes it the phosphor of choice in several dosimetric problems encountered in radiation biology. Recently LiF has been found quite suitable and practical for radiation monitoring of personnel, as an alternative to film badges (see p. 234).

One distinct advantage of $CaSO_4$:Mn over the other two phosphors is its high sensitivity to exposures as low as a few microroentgens. It is difficult to think of any other dosimeter that can approach this degree of sensitivity. Also, the relatively low temperature at which the maximum luminescence occurs

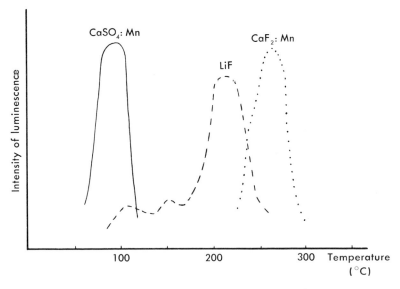

Fig. 9-11. Examples of glow curves for some irradiated thermoluminescent materials.

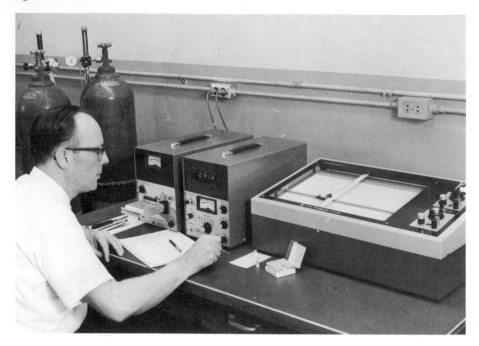

Fig. 9-12. TL analyzer with gas-flow apparatus, detector, picoammeter, and recorder. (Courtesy Harshaw Chemical Co., Solon, Ohio.)

Table 9-1. Characteristics of some thermoluminescent dosimeters

Characteristic	$CaSO_4 : Mn$	Synthetic $CaF_2 : Mn$	LiF (TLD-100)
Wavelength of luminescent light	5000 A	5000 A	4000 A
Temperature of maximum light emission	Single peak at 80° to 100°C	Single peak at 260°C	Main peak at 210°C
Useful dose range	A few μR to 10^4 R	2 mR to 2×10^5 R	0.3 R to 4×10^4 R
Dose-response relationship*	Nonlinear, at least above 10^4 R	Linear, over a wide range	Nonlinear, at least above 3×10^4 R
Dose-rate dependence	None, at least up to 10^{10} rads/sec		
Energy dependence at photon E < 0.1 MeV	Moderate	Severe	Negligible
Fading of luminescence after exposure	Severe	Moderate	Negligible
Tissue-equivalent	No	No	Yes

*Above the dose values indicated, saturation occurs and a plateau appears in the dose-response curve. Below the same value, linearity may not be rigorous.

allows for a small incandescent background signal emitted by the heated pan. The high sensitivity of this dosimeter is in part the result of the small background signal. The most serious drawback of $CaSO_4$ is the fading of luminescence after exposure, which necessitates significant corrections for fading losses in dosimetry applications.

Two of the most attractive advantages of CaF_2 : Mn over the other two phosphors are the broader useful dose range and a more linear dose-response relationship.

When thermoluminescent dosimetry was first suggested in the early 1950s, many investigators were rather skeptical about it, because of the imperfections inherent in the phosphors and in the read-out systems available at that time. At the present, through the close collaboration of science and technology, the state of the art has reached a point at which TLD has become the method of choice for dosimetric measurements in many situations. Improved phosphors and the precision, versatility, and accuracy of the read-out systems make TLD very reliable. What is most attractive in this type of dosimetry is the wide useful dose ranges and the absolute lack of dependence of the response upon dose rate. These advantages are not found together in ionization chambers or chemical dosimeters. Perhaps what keeps TLD methods from becoming more popular than they are is the high price of the readers. But progress is likely to be made in this respect, too.

Thermoluminescent dosimeters can be made extremely small and can be used as nearly point dosimeters. They find application in personnel dosimetry and recently in radiation therapy, thus replacing the more traditional ionization chambers. Judging from the present state of the art, we may safely say that the future of TLD looks more and more promising.

Methods utilizing radiation-induced changes of optical absorbance

Some materials (especially glasses and plastics) change their color upon absorption of high doses of ionizing radiation. This results in a change of their optical absorption spectrum. If the change of absorbance at a particular wavelength is fairly proportional to the absorbed dose over a wide range, the material may be used as a dosimeter.

A variety of materials can be used for this purpose. The same silver-activated glass used in radiophotoluminescent dosimetry shows an increase of absorbance in the ultraviolet and violet regions of the spectrum, when it is irradiated (Fig. 9-13). Cobalt-activated borosilicate glasses, clear plastics, such as methacrylates, polystyrene, and cellulose acetate, can also be useful in this type of dosimetry. Dyed gels, waxes, plastics, and glasses have been used with considerable success.

The procedure is to study first the absorption spectra of an irradiated and unirradiated sample in the ultraviolet region and, if necessary, in the visible region. A suitable wavelength is chosen so that the difference between the absorbances of the two samples is large. If possible, the wavelength chosen should not correspond with a steep slope in the spectra. The dosimeter is then calibrated in a known radiation field and the increase in absorbance per 1000 R per millimeter of dosimeter thickness at the selected wavelength is determined. This is the parameter of merit necessary to measure doses in unknown radiation fields with the same dosimeter.

With clear glasses and plastics, measurements are made in the ultraviolet region; with dyed systems, there is the distinct advantage that absorbances are measured in the visible region and therefore with easier techniques. An ordinary spectrophotometer is sufficient to measure the absorbances.

Silver- or cobalt-activated glasses, 1 to 3 mm thick, are useful in the dose range 10^3 to 2×10^6 R; others are useful in the 10^5 to 10^9 R range.

Plastics present the advantage of being nearly tissue-equivalent; their range is about

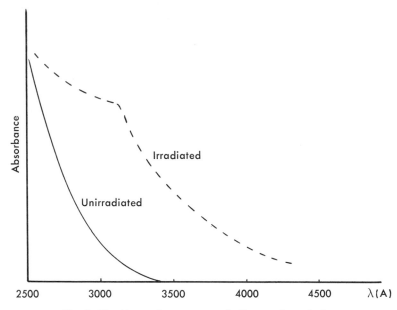

Fig. 9-13. Absorption spectra of silver-activated glass.

0.1 to 1 megarad. In addition, most of them show no fading of absorbance after irradiation, as some glasses do.

Most of the dosimeters of this group have shown dose-rate independence.

Film badges

Photographic emulsions have been widely used in radiation dosimetry. They lend themselves to the determination of dosage over a wide dose range, from a few millirads to several hundred rads, with the added advantage of being practically independent of dose rate. The degree of blackening that appears after the emulsion has been developed is proportional to the dose absorbed and is measured with a densitometer. The reading is compared with the reading of an identical emulsion exposed to a known radiation field.

Photographic films suitable for dosimetry are available with different sensitivities; some can measure doses from a few millirads to 10 rads, others from 500 to 10,000 rads. To make the response of the film as independent as possible of radiation energy, special absorbers between the radiation source and the film can be used.

The most common use of photographic films as dosimeters is in the so-called film badges for radiation monitoring of persons who are routinely exposed to radiation sources such as radioactive materials, X-ray machines, nuclear reactors, or particle accelerators. If these badges are properly designed, they can not only give a measure of the dose accumulated by the wearer in a specified period of time, but can also give an indication of the type of radiation to which he has been exposed, so that the proper precautions can be taken in the future. At least for this reason, film badges are more reliable and useful than pocket dosimeters for radiation protection of personnel.

The badge used by one film badge service (Gardray of the Picker X-ray Corporation—Fig. 9-14) consists of a plastic or metallic frame for the support

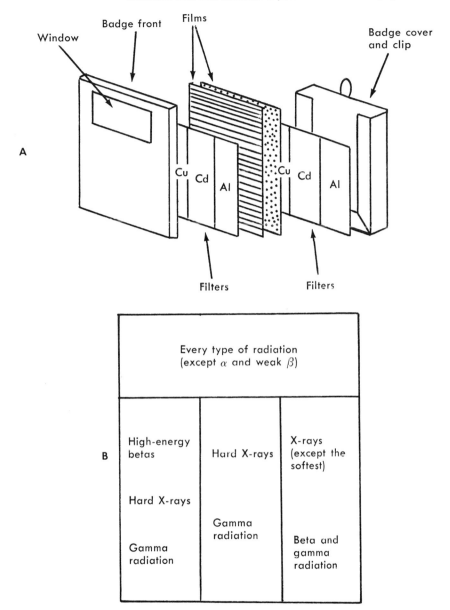

Fig. 9-14. A, Disassembled film badge. **B,** Diagram showing the types of radiation responsible for the blackening of the different areas of the film.

of two rectangular films of different sensitivity; one can record doses as small as 1 mrad and the other is sensitive only to large doses up to 300 rads. The addition of the second film is useful for the accurate determination of large doses of radiation to which the wearer of the badge may be accidentally exposed. The two films are wrapped in a thin, lighttight paper, on which the name and number of the wearer are imprinted.

To provide information on the radiation quality to which the wearer has been exposed, a trio of paired metal filters is lodged inside each badge, in the

Table 9-2. Characteristics of dosimeters

Dosimeters	Sensitive volume	Parameter directly measured	"Absoluteness"	Useful dose range
Standard free-air chambers	Air	Total electric charge of ion pairs	Yes	Variable
Condenser roentgen meters	Air	Capacitor discharge produced by ion pairs	No	From a few mR up to 250 R
Pocket dosimeters	Air	Capacitor discharge produced by ion pairs	No	Up to a few R
Survey dose-rate meters	Air	Electric current produced by ion pairs	No	mR/hr to hundreds of R/hr
Calorimeters	Any material	Radiation dissipated as heat	Yes	From a few hundred R up
Fricke chemical dosimeter	Solution	Yield of $Fe^{2+} \longrightarrow Fe^{3+}$ reaction	Yes	4000 to 40,000 rads
Ceric sulfate dosimeter	Solution	Yield of $Ce^{4+} \longrightarrow Ce^{3+}$ reaction	Yes	40,000 to 10^8 rads
Photoluminescent dosimeters	Solid phosphor	Intensity of luminescence	No	10 to 10^3 R
Thermoluminescent dosimeters	Solid phosphor	Intensity of luminescence	No	From a few μR to 10^5 R
Optical absorbance dosimeters	Glass or plastic	Change of optical absorbance	No	10^3 to 10^9 R
Film badges	Photographic emulsion	Blackening of emulsion	No	From a few mR to hundreds of R

Energy independence	Dose-rate independence	Accuracy in the useful dose range	Special uses	Convenience for routine dosimetry
Excellent up to 3 MeV	Good	Excellent	Calibration of other dosimeters	Poor
Excellent up to 1.3 MeV	Good up to 120 R/sec	±2% to ±10%	X-ray dosimetry	Good
Fair	Fair	Fair	Radiation monitoring of personnel	Excellent
Fair	———	Fair	Radiation monitoring	Excellent
Excellent	Excellent	Excellent	Calibration of other dosimeters, etc.	Poor
Good	Excellent	Good	Calibration of other dosimeters	Excellent
Good	Excellent	Good	———	Excellent
Poor	Excellent	Good	———	Fair
Below 0.1 MeV varies with phosphor	Excellent	Good	———	Good
Variable	Good	Good	With pulsed electron beams	Good
Poor	Good	±10% to ±50%	Radiation monitoring of personnel	Excellent

front and back of the film packet. A 1 mm aluminum filter stops the softest X-rays and weak beta particles, so that any blackening on the film in correspondence to this filter is caused by exposure to types of radiation other than those stopped by the filter. A 0.25 mm copper filter stops medium-energy beta particles and soft X-rays. A 1 mm cadmium filter allows only the passage of hard X-rays and gamma rays. One area of the film (in correspondence to the open window) allows the passage of any type of radiation, except alpha particles and very weak beta particles (which are absorbed by the paper wrapping the films). The doses received with these types of radiation are not intended to be measured with the film badge; they present no health hazard, because they are absorbed by the most superficial layers of the epidermis.

As a result, the blackening of the different areas of the film is due to the types of radiation as indicated in Fig. 9-14, *B*. The density of each area after development of the film allows the estimation of the dose due to each type of radiation.

A control badge is also used and kept on a film badge board or stored in a reasonably cool place away from any source of radiation. It is processed in the same manner as the badges worn by the personnel. It is intended to indicate exposures common to all badges during storage periods; its reading is therefore subtracted from those of the badges worn by the personnel.

Film badges may be obtained from some commercial firms. Used badges are sent back periodically (usually every month) to the issuing firm for processing and reading. The badge wearer receives an exposure report with the information about the type and dose of radiation to which he has been exposed for the period, as well as summaries of doses accumulated during previous periods. The accuracy of these dosimeters varies between 10% and 50% depending on the doses and radiation energy.

In accordance with the recommendations of the National Council for Radiation Protection, film badges should be used by any person (student, scientist, technician, etc.) who works routinely (not just occasionally) with sources of ionizing radiations such as X-ray machines, particle accelerators, nuclear reactors, and radioactive materials in millicurie quantities or larger.

From the preceding review of dosimetric methods and techniques, we see that the ideal dosimeter does not exist yet and that no single dosimeter can be used reliably for the solution of any dosimetric problem. However, the variety of instruments and techniques available at the present time is such that for every particular problem almost always a suitable and satisfactory method can be found.

Table 9-2 summarizes some characteristics of all the dosimetric methods and instruments described in this textbook. This information should help the student in the selection of the appropriate method or instrument when he is confronted with a particular dosimetric problem. Obviously, in a general summary of this kind, figures and ratings must be interpreted as approximate and relative. For more detailed information about specific methods, this text and the more technical literature suggested at the end of this chapter may be consulted.

EXPOSURE OF THE HUMAN POPULATION TO IONIZING RADIATIONS

The human body, or any organism, can be exposed to ionizing radiation in different ways. Radiation can be delivered to the whole body or only to some organs. We refer to these two types of irradiation as *whole-body exposure* (WBE) and *partial exposure* respectively. The same dose may produce entirely different biologic effects if delivered to the whole body or to only one organ. For example, 500 R of X-rays WBE delivered in a very short time may kill a man, whereas the same dose, delivered to the hand only, may cause at most a severe erythema.

A large dose delivered in a short time (that is, with a high dose rate) is an *acute* exposure, whereas the same dose delivered over a long period of time is a *chronic* exposure. There is no clear-cut criterion to determine when an exposure is to be defined as acute and when as chronic. The exposure of an animal to 500 R of X-rays within 1 minute is clearly an acute exposure, whereas a dose of 5 R delivered over 1 year is definitely chronic. Outside of extremes like these, it is difficult to classify certain "borderline" types of exposure; often the criterion must be subjective or at least relative to the biologic effect under investigation.

With the possible exception of some genetic effects, many biologic effects are strikingly dependent on dose rate. For instance, survival of a man irradiated with 500 R of X-rays is much less probable if the dose is from acute WBE than if it is chronic or fractionated (for example, 100 R a day, for 5 days).

Chronic exposure of the general human population

The human population is chronically exposed to a dose of ionizing radiations, coming from a variety of sources. This chronic dose may differ widely from one individual to another, depending on a number of factors, such as geographic area, dietary habits, and mainly whether the individual is in proximity to certain radiation sources. With regard to the last factor, the chronic exposure (*occupational exposure)* of individuals who work with X-ray machines, fairly large amounts of radioisotopes, particle accelerators, nuclear reactors, etc., may be considerably larger than that for the general population. In this section chronic occupational exposures are not considered.

Several sources contribute to the total chronic exposure of the general population. Some are natural sources, existing in the external environment or present in the body itself. Others are classified as artificial sources, such as the diagnostic X-rays, to which a large segment of the human population exposes itself.

Internal sources

The internal sources (interal emitters) are radionuclides that are taken up from the external environment mainly with food and water. Some of them have always been in the body, simply because they have always existed in the environment, at least since the origin of the human species. Others are the heritage of the atomic era, because they are artificial radionuclides released into the environment by nuclear explosions through fallout. From the atmosphere they have reached the human body through normal food chains.

Among the internal emitters, which have always been in the body, the most important is certainly ^{40}K. Natural potassium consists of the three isotopes ^{39}K, ^{40}K, and ^{41}K, of which ^{40}K is radioactive; its natural abundance is 0.0118% and its half-life is about 1.3×10^9 years. It decays by beta emission or by EC followed by a 1.46 MeV gamma ray (see decay scheme in Appendix III).

Potassium is classified as an essential element for practically every living protoplasm and organism. In the human body, this element is found in every cell, but especially in muscles, and its concentration is higher within the cells than in the extracellular fluids (the opposite is true for sodium). Since an organism cannot discriminate in favor of or against any isotope of a chemical element, all three isotopes of potassium are picked up from the environment, and consequently wherever potassium occurs in the body, there ^{40}K will be found with the same natural abundance as in the environment. According to a table published by the National Bureau of Standards the body of an adult 70 kg man ("standard man") contains 140 g of potassium, and therefore 1.65 $\times$ 10^{-2} g of ^{40}K. This corresponds to a body burden* of about 0.1 μCi of ^{40}K. This activity is larger than the activity of any other internal gamma emitter. The radiation dose to the body is about 20 mrems/year and is also greater than that of any other single internal emitter, not only because of the high activity of ^{40}K but also because of the high energy of its gamma ray. Obviously, because of its origin, the level of this internal emitter in the body will remain constant in natural conditions.

In 1955, C. E. Miller and L. D. Marinelli were measuring the ^{40}K levels in some human subjects with a whole-body counter at the Argonne National Laboratory. They discovered that the gamma spectra thus obtained showed a sharp peak in correspondence to 0.662 MeV and therefore that the bodies of their subjects contained ^{137}Cs. Shortly after, studies performed in other laboratories confirmed the same finding. Radiocesium, which was known to be one of the fission products occurring in fallout from nuclear explosions, had found its way into the human body and had to be considered as one of the internal emitters for the general population since the advent of the nuclear age.

Later it was clearly understood that radiocesium enters the body with food and especially with meat and milk. In studies on Norwegians, it was found that northern Norwegians who consume large quantities of reindeer meat and milk had levels of radiocesium in their bodies several times higher than other residents of the same country. The reason was that reindeer feed on lichens, and since these plants grow as thin covers on the soil of northern Norway, reindeer must graze over large areas to obtain sufficient food. Obviously, radiocesium from fallout had been deposited on the surface of the lichens and from there it was being concentrated in the bodies of reindeer. Other investigations have shown that most of the radiocesium that is ingested by cattle is the portion deposited on the foliage of the plants normally eaten by those animals. The radiocesium absorbed by the soil does not seem to enter the food chain readily, because it is not easily absorbed by the root systems.

Similar studies conducted among the Eskimos in Alaska have shown that Eskimos who eat almost exclusively caribou meat have much higher levels of radiocesium in their bodies than those who have a more diversified diet. The caribou, like the reindeer, feeds heavily on lichens.

Several laboratories have been monitoring, with whole-body counters, radiocesium body burdens in different populations of the northern hemisphere over a period of about 10 years. The data are presented in graphic form in Fig. 9-15, in which the body burden for this radionuclide is expressed, accord-

---

*In radiation biology the term "body burden" is often used to denote the amount, in microcuries, of a radionuclide in the body.

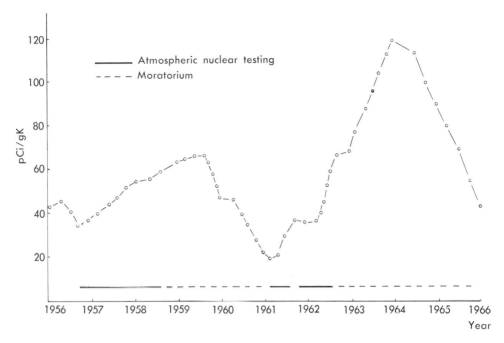

Fig. 9-15. Whole-body cesium-137 body burdens in man from 1956 to 1966 (the values plotted are averages of data from several laboratories).

ing to the common usage, in picocuries per gram of K in the body. The values plotted are the averages of the data issued by many laboratories. The curve shows that in the period of 1956 to 1966 there have been two maxima—in 1959 with 70 pCi/gK and in 1964 with 120 pCi/gK. The relationship of the maxima with periods of intensive atmospheric nuclear testing is evident; notice that the second maximum was reached about 1½ years after the end of the 1961–1962 testing, a consequence of the long permanence of the worldwide fallout in the atmosphere before some of its products enter the food web. During the summer of 1964, according to one report, the average radiocesium whole body burden in one population reached 1.28 μCi, with one individual of the same population containing 3 μCi! During the same year the average total body burden for all the populations surveyed was about 15 to 20 nCi (much less than the radio potassium body burden). It is estimated that the average total dose received from internally deposited ^{137}Cs in the United States during the period 1956 to 1961 was about 8 mrems/year. If the present moratorium of atmospheric nuclear tests continues, the rapid decline of radiocesium body burden that began after the 1964 peak should continue steadily.

Modern whole-body counters (see p. 184) are sensitive enough to detect and measure the gamma radiation emitted by radiocesium and radiopotassium in the human body. Indeed, these nuclides are the only internal gamma emitters usually detected by those instruments in normal subjects. An example of spectra taken with two normal subjects is shown in Fig. 9-16. The photopeaks of the two radionuclides in question are clearly visible in the expected energy bands.

The spectra show that subject *A* contains more radiopotassium than subject *B*, although its total body weight is smaller. The explanation of this difference

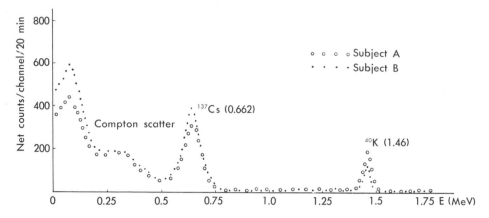

Fig. 9-16. Gamma spectra of the bodies of two normal human subjects taken with the low-level whole-body counter of the Medical Division of the Oak Ridge Associated Universities (200 channels = 0 to 2 MeV).

is to be found in the slightly different distribution of K in various tissues and organs of the body. Muscle, liver, and brain are very rich in K, whereas in adipose tissue and bones the K level is much lower. Consequently a heavy-boned or obese individual would contain somewhat less K per kilogram of body weight than a light boned or thin one. When the spectra of Fig. 9-16 were taken, subject *B* appeared definitely more adipose and heavy boned than did subject *A*.

On the other hand, the spectra show that subject *A* contains less radiocesium than does *B*. It is known that cow's milk is the principal dietary source of ^{137}Cs. Subject *B* did admit that he uses more milk and other dairy products (notably ice cream) than subject *A*.

The possibility of measuring the amount of body ^{40}K with a whole-body counter leads to the possibility of determining the total amount of K in the body. The method consists of comparing the intensity of the photopeak obtained from a known amount of K placed in a phantom (of the same size, shape, and density of the human body) with the intensity of the photopeak obtained from a human subject. Although the method is still being refined, it is more satisfactory and less tedious than other methods. The method of ^{40}K gamma measurement is both nondestructive and nontraumatic.

Like ^{40}K, radium is another internal emitter that has always been part of the total body burden of the human population. This element is an intermediate product in several natural decay chains. In particular, ^{226}Ra is a member of the ^{238}U decay chain. It is an alpha emitter with a half-life of 1620 years. In nature it is always found together with the many daughter nuclei, of which the most long-lived are ^{210}Pb (the half-life is 22 years) and ^{210}Po (the half-life is 138 days). The end of this chain is the stable ^{206}Pb. Radium is chemically similar to calcium, which is an element absorbed by the body and mainly utilized for the ossification of the bones. Radium follows calcium in the food chain leading to man and, like calcium, concentrates in the skeleton (bone seeker). It enters the body through water and a variety of foods. All natural waters contain small amounts of this element, although in some instances the level of radium in public water supplies may be very high (as in Joliet, Illinois). The average radium body burden for the adult general population is estimated as being 1.6×10^{-4} μCi, mostly concentrated in bones. The dose to the osteocytes can be as high as 38 mrem/year, because it is produced by the alpha particles (RBE $\approx$ 10), whereas the dose to the bone marrow is mainly due to the beta radiation of the daughter nuclei and is estimated as being about 0.6 mrem/year.

Like ^{137}Cs, ^{90}Sr is another internal source of ionizing radiation that has become part of the total body burden of the general population since the beginning of the nuclear era. This radionuclide is a beta-emitting, long-lived (the half-life is 28 years) fission product and, as such, is a regular component of worldwide fallout. It is slowly deposited on the surface of the soil and plants over a long period of time after a nuclear explosion.

Strontium is chemically similar to calcium, and therefore it moves with calcium along the various paths of the food chains. The path from the atmosphere to the human body is schematically as follows:

The amount of ^{90}Sr deposited on the surface of plants has fluctuated with the intensity of nuclear testing, whereas the amount present in the soil has changed less during the last 20 years.

^{90}Sr reaches man primarily through his consumption of dairy products and foods of plant origin. With regard to the latter source, the amount of radionuclide ingested can be sharply reduced if plant foods are thoroughly washed and the outer layers discarded before consumption. This practice should be very effective especially during, and immediately after, periods of intensive nuclear testing. Like radium and calcium, strontium is a bone seeker, and of course is readily assimilated in the bones and teeth of young individuals better than in adults. Once ^{90}Sr is concentrated in bones, it is eliminated very slowly, and this fact, together with the long half-life, justifies the concern about the internal hazard created by this radionuclide.

Research projects have been sponsored by various agencies to measure the amounts of radiostrontium deposited in soil and on plants over several years and to monitor the amounts present in milk, and even in human bones and teeth (Project Sunshine). Amounts of ^{90}Sr are usually measured in strontium units (formerly Sunshine units). The *strontium unit* is defined as 1 pCi of ^{90}Sr per gram of calcium.

These studies have produced a wealth of valuable data concerning the interrelations between nuclear tests and the distribution of ^{90}Sr in the various links of the food chain from atmosphere to man. But they have also shown that along the chain there are several mechanisms in effect that discriminate against strontium in favor of calcium. In other words, although calcium and strontium are chemically similar, in the passage of the two elements from one link to the next, the fraction of calcium absorbed is larger than the fraction of strontium absorbed.

This discriminating process is expressed quantitatively with the *discrimination factor* (DF), which is defined as the ratio of ^{90}Sr to Ca in one link of the chain divided by the same ratio in a successive link, such as

$$\frac{^{90}\text{Sr/Ca in milk}}{^{90}\text{Sr/Ca in bone}}$$

which can be abbreviated as

$$\text{DF(milk} \longrightarrow \text{bone)}$$

Although the DF(soil → plants) is difficult to evaluate, it seems certain that it is always greater than 1, which means that the ^{90}Sr/Ca ratio in plants is generally less than in the soil in which they grow. The experimental evidence also indicates that the values of DF(plants → milk) and DF(milk → bone) may be between 2 and 4. The physiologic processes operative in the animal and human body and responsible for the high DF values are mainly the differential intestinal absorption (the intestine absorbs a greater percentage of calcium present in the diet than of strontium) and differential urinary excretion, as a result of more efficient tubular reabsorption of calcium. As a consequence of the action of these discriminating mechanisms, the ^{90}Sr/Ca ratio in cow's milk is much lower than the same ratio in the cow's feed, and the ratio in human bone is lower than the same ratio in milk.

Thus a chain of discriminating processes reduces the deposition of radiostrontium in human bone. The discrimination factor from the beginning to the end of the food chain, that is, DF(plants → human bone), is then the product of the partial discrimination factors between any link and the next as follows:

$$\text{DF(soil} \longrightarrow \text{bone)} = \text{DF(soil} \longrightarrow \text{plants)} \times \text{DF(plants} \longrightarrow \text{animal)}$$
$$\times \text{DF(animal} \longrightarrow \text{milk)} \times \text{DF(milk} \longrightarrow \text{bone)}$$

In 1955, the DF(soil → bone) for children was found to be about 12.

The ^{90}Sr body burden has been highly variable during the last 20 years in conjunction with periods of intensive nuclear testing and with periods of moratorium. Radioassay studies have indicated that in 1959 the average skeletal burden for most of the general population was 0.3 strontium units with peaks of 2.1 units for 1-year-old children. In 1960, the largest body burdens (3.4 units) were found in 2-year-old children; the average skeletal burden for adults was 0.5 units, which is equivalent to a total ^{90}Sr body burden of 525 pCi, because the standard man contains 1050 g of calcium in his body.

Since ^{90}Sr is a pure beta emitter, the radiation exposure is limited to the osteocytes and bone marrow. Obviously the dose absorbed depends on the body burden. Calculations have shown that the dose to osteocytes is 2.7 mrem/year/strontium unit and the dose to bone marrow is 0.7 mrem/year/strontium unit. Thus in children with a skeletal burden of 3 strontium units, the osteocytes receive a dose of about 8 mrem/year and the bone marrow receives a dose of about 2 mrem/year.

It is still too early to assess precisely the potential long-term hazard resulting from the absorption of these doses. What is feared from ^{90}Sr, with its long half-life and slow turnover rate in bones is leukemia, bone necrosis, and bone cancer. An optimistic opinion is that the body burdens observed so far are well below the limits regarded as acceptable for the general population and well below those which might be expected to cause a detectable increase in the frequency of bone tumors.

In previous chapters, I have shown that ^{14}C originates in the atmosphere through a nuclear reaction between the cosmic radiation neutrons and the atmospheric nitrogen. After the advent of nuclear weapons, ^{14}C has been produced also by radioactivation of the same nitrogen by fission neutrons.

Carbon, and therefore also ^{14}C, exists in the environment (atmosphere, humus, and natural waters) mainly as carbon dioxide; in this form it is utilized by plants for photosynthesis and thus becomes incorporated in all organisms through the various food chains. As a consequence, ^{14}C contributes to the radiation exposure of the human population from internal sources.

The level of radiocarbon in the environment and organisms has been relatively constant at least for several thousand years before the beginning of the atomic era, because the new radiocarbon produced by cosmic rays would barely replace the amounts that disappeared through radioactive decay. The validity of the radiocarbon dating method is based on this important assumption. Nuclear explosions have disturbed this equilibrium by injecting significant amounts of additional radiocarbon into the environment and therefore into organisms. According to some estimates, ^{14}C concentration in the tropospheric carbon dioxide in the northern hemisphere had increased by 30% from 1955 to 1961.

At the present time, it seems that the concentration of radiocarbon in organisms is about 7.5 pCi/g of carbon. Since the total carbon content of the body of the standard man is 18% of the total body weight, the total ^{14}C body burden is about 0.1 μCi. Because ^{14}C is a pure, weak beta emitter, its radiation dose is much less than the dose from ^{40}K, which is present in the body with the same activity. The estimated dose from ^{14}C is only 1.6 mrem/year to the skeleton and 0.7 mrem/year to the gonads.

Unlike other internal emitters, ^{14}C does not concentrate in any particular organ, because carbon is an element essential in the composition of all organic molecules. The relatively rapid biologic turnover of this element does not result in a decrease of the radiocarbon body burden with time. The reason is that the level of this radioisotope in the environment remains constant at least over very long periods, due to its very long half-life (5730 years) and to the continuous production of it.

The presence of ^{14}C in the human body is a cause of concern, not only because it is a source of hazardous radiation, like all other internal emitters, but also for another reason. Many molecules (DNA, proteins, and enzymes) that perform certain vital functions in the organisms must contain one or more atoms of this carbon isotope in their structure. When these atoms decay, they become nitrogen atoms; this transmutation may be sufficient to inactivate a protein or enzyme molecule, or may be a source of mutations in the genetic material, which is made of DNA.

Tritium (^{3}H) should also be listed as an internal emitter, although the body burden is very small, with an estimated radiation dose of 1.8×10^{-3} mrem/year to soft tissue. It originates in the atmosphere through several interactions between cosmic rays and nitrogen. It is also produced and released into the atmosphere by thermonuclear explosions. Since tritium atoms can be found in any organic molecule (including DNA, proteins, and enzymes), its potential biologic hazard stems from the same reasons that are valid for ^{14}C.

External sources

Natural external sources of ionizing radiation, contributing to the chronic exposure dose of the general human population, are mainly cosmic rays and gamma emitters present in rocks, soil, and construction materials.

Measurements of this component of the chronic exposure dose made in several cities of the U.S.A. show that the mean annual doses vary greatly from a minimum of 73 mrads (New Haven, Connecticut) to a maximum of 197 mrads (Colorado Springs, Colorado).

The dose from cosmic rays depends very much on the altitude and latitude (see Chapter 6); it increases from 20-40 mrad/year at sea level to 300-450 rad/year at 20,000 ft. It also increases steadily from the equator to a latitude of 40° to 50° (northern and southern hemispheres). The typical dose from cosmic rays, at sea level and medium latitudes, can be taken as 50 mrem/year.

The gamma emitters present in rocks, soil, and construction materials that contribute to the chronic dose are mainly radium, uranium, and thorium, with their decay products, and potassium 40. Igneous rocks (such as granite) are more radioactive than sedimentary rocks (sandstones, shales, and limestones). If these materials do not contain unusually high levels of radioactivity, the radiation dose for the human population from these sources may be taken as 50 mrem/year. However, researchers have discovered during the past few years that there are certain inhabited areas of the world where people are exposed to considerably higher doses from these sources because of the presence of unusually radioactive mineral waters or radioactive minerals in the soil. The water of some mineral springs (in Austria, Japan, etc.) may contain radium concentrations a million times greater than the normal concentrations found in ordinary public water supplies. For many years these waters have been used for their allegedly therapeutic powers! In certain areas of Brazil and India there are towns and villages where the gamma radiation dose levels can be as high as 12 rads/year. These towns and villages are built on soil containing monazite (a mineral with a high content of thorium).

The human population living in culturally and medically advanced countries may be exposed to another source of ionizing radiation—diagnostic X-rays. This is an artificial external source to which individuals are exposed either several times a year or never during any particular year, depending mostly on their health and dental problems. The determination of an average dose for the total population from these sources would really have little meaning, since the exposure dose varies widely from one individual to another. Furthermore, unlike other sources considered so far, diagnostic X-ray doses for radiographic or fluoroscopic examinations are dependent on too many factors. Nevertheless, by gathering data on the frequency of examinations, on the organs of the body irradiated, on the dose delivered with each examination, and for as many individuals as possible, investigators could estimate that in the United States the average annual dose absorbed by the gonads from medical X-rays is about 140 mrads per person, for individuals under age 30. The skeletal dose is probably one half that value. However, similar studies made for other countries have led to quite different values.

Table 9-3. Sources of chronic exposure

	Internal sources								External sources		
	Natural				Radioactive fallout				Natural		Artificial
	^{40}K	Ra and daughter nuclei	^{14}C	^3H	^{14}C	^3H	^{137}Cs	^{90}Sr	Cosmic rays	Local gamma rays	Diagnostic X-rays, etc.
Body burden	0.1 μCi	160 pCi	—	—	0.1 μCi	Very small	5 nCi	525 pCi	—	—	—
Gonadal dose (mrem/year)*	20	0.8	—	—	0.7	2×10^{-3}	4	—	50	50	Highly variable
Skeletal dose (mrem/year)*	15	38	—	—	1.6	—	4	1-2	50	50	Highly variable

*Except for diagnostic X-rays and other artificial external sources of radiation, the gonadal dose from all sources is about 125 mrem/year and the skeletal dose from all sources is about 160 mrem/year.

In summary, the general human population is chronically irradiated by various sources of ionizing radiations, some external to the body, others located within the body itself. (All internal sources ultimately are of environmental origin.) Some of them have always existed, both in the environment and in the body; others began to exist with the advent of the atomic age. The presence of the latter in the human body is one price mankind has to pay for the exploitation of nuclear energy for warfare purposes. One internal source (^{14}C) did exist before the beginning of the atomic age, but its level has increased significantly as a consequence of fission and thermonuclear explosions. The radiation doses received from some sources (cosmic rays, potassium 40, and radium) vary little from one individual to another; the doses received from other sources (local gamma rays from soil, strontium 90, cesium 137, and medical X-rays) may vary widely with geographic areas. Nevertheless, determination of the average yearly doses from all the different sources responsible for chronic irradiation is useful. All the sources may be summarized and organized as shown in Table 9-3 (the average body burdens and dose values are valid at least for the standard man living in the northern hemisphere during the late 1960s).

Acute exposure—the concept of lethal dose

Large segments of the human population would be subject to acute whole-body exposure only in the event of a nuclear war. This type of exposure has already happened twice in 1945 in Japan.

Isolated or small groups of individuals have been exposed to large acute whole-body doses as a result of accidents, which are described in Chapter 15, along with the biologic damage caused by these exposures.

If the acute whole-body exposure dose is large enough, death may follow within a number of days, which depends on the dose. Early in the history of radiation biology investigators discovered that the minimum lethal dose varies from species to species, according to their radiosensitivity. Mammalian species are among the most radiosensitive—a problem discussed further in Chapter 15.

However, the same dose may be lethal for one individual but not for another of the same species, or at least the same dose may kill one individual within a few days and another much later. The reason is to be found in the high degree of variability existing within any species. Because of this fact, lethal dose values can be determined only on a statistical basis. Although the minimum lethal dose for a particular individual cannot be predicted, one can find experimentally a minimum dose that will kill a given percentage of a large population within a given time. The *median lethal dose* (LD_{50}) is the minimum dose sufficient to kill 50% of a population within a specified time. For most mammalian species all deaths directly attributable to acute whole-body exposure occur in less than 30 days. For this reason, the radiosensitivity of mammalian species is usually expressed in terms of $LD_{50/30}$, which is therefore the minimum dose sufficient to kill 50% of a population within 30 days.

For organisms whose life-span is less than 30 days, the LD_{50} is based on observation periods of less than 30 days; the periods should be specified.

The median lethal dose for man has never been established with certainty. The early estimates, derived from studies of the Atomic Bomb Casualty Commission on the mortality at Hiroshima and Nagasaki, were based on incorrect conclusions concerning the doses to which the fatalities had been exposed, or on the belief that certain fatalities were the results of whole-body exposure, whereas only partial exposures were involved. Furthermore, for man a $LD_{50/60}$ would be more meaningful and would reflect the mortality response more properly because many individuals exposed to a lethal dose actually die between 30 and 60 days after exposure. According to the current most authoritative opinions, the $LD_{50/60}$ for man is below 450 rems, and probably around 300 to 350 rems. Notice that the dose is expressed in rems; this means that an acute whole-body absorbed dose of 300 to 350 rads of X-rays is needed to kill 50% of a large human population within 60 days (RBE for X-rays $= 1$), but only 30 to 35 rads of neutron radiation (assuming an RBE ≈ 10) is needed to obtain the same effect.

MAXIMUM PERMISSIBLE DOSES AND MAXIMUM PERMISSIBLE BODY BURDENS

With the accumulation of a large body of knowledge concerning the deleterious effects of ionizing radiations on the human population, the need has arisen of suggesting or establishing guide lines in regard to the maximum dose levels to which the human body may be safely exposed. The problem has been one of the most formidable and difficult to solve in radiation biology.

When radiologists were the only large group routinely exposed to an external source of radiation, a maximum safe dose was calculated and suggested by some investigators; it ranged widely between 0.04 and 2 R/day. This dose had been arrived at mainly through the analysis of detectable, short-term deleterious effects. It was felt that these effects would not result from the exposure doses recommended.

When, in the 1930s, investigators discovered that radionuclides introduced into the body in certain amounts may cause serious damage,* they urged that guidelines be established also in regard to the maximum safe body burdens of the various radionuclides.

Immediately before World War II, the National Council on Radiation Protection (NCRP) had recommended that the maximum dose for whole-body external radiation should be 0.1 R/day, and the maximum body burden of radium should be 0.1 μCi. Although these suggestions were based on a body of information incomplete in many respects, they were basically safe, and it was somewhat providential that they became available just before the wartime atomic energy program was started. With so many persons involved, probably the credit for the excellent safety record connected with that program goes to those recommendations.

Since then, the maximum permissible doses and the maximum permissible body burdens of internal emitters have been revised several times by the NCRP and the International Commission of Radiological Protection (ICRP), in concomitance with the advancement made in radiobiologic knowledge.

The general criteria used by the two committees in their recommendations take the following facts and considerations into account:

1. Acute doses of radiation may produce immediate or delayed effects. The recommendation of maximum exposure doses and maximum body burdens should take both types of effects into account.

2. The dose rate affects the ability of the body to repair radiation damage; distinction should be made, therefore, between acute and chronic exposures.

3. Delayed effects include genetic effects that are not confined to the individuals actually exposed but are transmitted to the future generations; the recommendations should protect not only the individual but the species as well.

4. For many deleterious effects, it is difficult to establish whether a threshold dose exists, that is, a dose below which the effect is not produced at all. It is probable that there is no completely harmless amount of radiation exposure. At the present, we believe that there may be no threshold dose for certain genetic effects and for the production of certain types of neoplasms.

5. The child, the infant, and the unborn infant are more radiosensitive than are adults.

6. Different organs of the body are also differently radiosensitive.

7. The recommendations should allow as much of the beneficial uses of ionizing radiation as possible, while they assure that man is not exposed to undue hazard.

8. A distinction should be made between the recommended maximum exposure levels for occupational workers and those for the general population. The limits for the general population should be much lower for the following reasons:

    a. It is reasonable to expect that an individual who has chosen to work in an environment where high levels of radiation may be present should accept some risk of a magnitude at least comparable with the risk involved in other types of jobs.

    b. Since many individuals may be exposed in utero in the population at large, and since these individuals are more radiosensitive, the problem of the exposure of the total population obviously should be treated more cautiously.

    c. The number of occupationally exposed individuals is much smaller than the number of individuals in the population at large. The magnitude of the genetic effects, in terms of damage to the species, is a function of the number of individuals exposed. From the genetic point of view, the higher doses permitted

---

*This is the case of several girls who worked in a New Jersey factory as watch-dial painters. The paint used contained radium. The girls would shape their fine brushes with their tongue, thus introducing considerable amounts of radium into their bodies over several years. Some of these girls died of bone cancer.

to a few individuals should not be a matter of concern, or at least could be an acceptable risk.

The maximum permissible doses and maximum permissible body burdens for internal emitters are discussed separately.

Maximum permissible doses

As defined by the NCRP, the maximum permissible dose (MPD) is the highest dose of ionizing radiation (from external or internal sources) that, in the light of present knowledge, is not expected to cause appreciable bodily injury to a person at any time during his lifetime. The guidelines concerning the MPD's, as suggested by the NCRP, distinguish between radiation workers and the general population, in compliance with one of the criteria discussed above (p. 249). The previous recommendations published in the National Bureau of Standards Handbook no. 59 (1954) have been recently revised and published in the NCRP Report no. 39 (1971), to which the reader is referred for details. They can be summarized as follows:

1. For radiation workers
   a. From whole-body exposure to penetrating radiation (X- and gamma rays, high-energy beta radiation, neutrons)
      MPD in any one year shall be 5 rems (but not more than 3 rems in any calendar quarter)
      Long-term accumulated MPD shall be $5 \times (N - 18)$ rems (N is the age of the worker in years)
   b. From exposure of skin (other than hands and forearms) to radiation of low penetrating power
      MPD in any one year shall be 15 rems
   c. From exposure of hands
      MPD in any one year shall be 75 rems (but not more than 25 rems in any calendar quarter)
   d. From exposure of forearms
      MPD in any one year shall be 30 rems (but not more than 10 rems in any calendar quarter)
   e. From exposure of any other organ or tissue
      MPD in any one year shall be 15 rems (but not more than 5 rems in any calendar quarter)
   f. From whole-body or abdominal exposure of expectant mothers (radiation workers)
      MPD during the entire gestation period should not exceed 0.5 rems
2. For the general population
   a. Dose limit for whole-body exposure of an individual not occupationally exposed shall be 0.5 rems in any one year
   b. Dose limit for whole-body exposure of the population of the United States as a whole shall be a yearly average of 0.17 rems per person.

For radiation workers the suggested MPD values differ, depending on the organs exposed. The values are low when the whole body is exposed, or, in case of partial exposure, when such radiosensitive organs like bone marrow, gonads, and lens of the eye are irradiated. Allowing higher doses for exposure of the skin seems reasonable, as long as the radiation does not penetrate beyond the skin. The less stringent limits for partial exposure (except exposure involving the radiosensitive organs) find justification in several of the criteria discussed above and in a number of considerations that will be elaborated in greater detail later in this text.

The accumulated MPD is a function of age above 18. Obviously, the NCRP does not allow anyone under age 18 to be a radiation worker. According to the recommendations, a 40-year-old radiation worker should not have accumulated more than 110 rems from whole-body exposure. If he has worked with radiation only occasionally during the 22-year period after the age of 18, he is still allowed only 5 rems per year from whole-body exposure, although his total accumulated dose may be much below 110 rems.

The MPD's for calendar quarters have been established to take into account the dependence of biologic damage on dose rate. The 5-rem yearly MPD from whole-body exposure should be distributed as uniformly as possible throughout the year. A radiation worker, therefore, is not allowed to receive the whole 5-rem MPD within a few days (high dose rate). If a radiation worker has accumulated 3 rems during one quarter, he should reduce his exposure to the radiation source during the next three quarters in such a way as to receive only 2 rems for the same period. Exception to this rule can be made only in special circumstances and when accurate exposure records indicate that the individual's total accumulated dose is well below the $5(N - 18)$ rem limit.

The rigorous limit set for expectant mothers who work with radiation is justified by the high degree of radiosensitivity of the unborn infant (see details in Chapter 14). The recommended limit minimizes the possibility of prenatal or neonatal deaths and of congenital birth malformations.

With regard to the non-occupationally exposed population, note that a distinction is made between the dose limit for individuals and the average dose limit per person. Apparently, the NCRP does not allow a yearly dose limit of 0.5 rems for *every single individual.* The average yearly dose limit of 0.17 rems is based on the assumption that a dose of 5 rems to the gonads accumulated over a period of 30 years (the average duration of the reproductive activity) does not cause significant genetic damage to the population.

All the recommended limits do not include the doses from natural background radiation or from medical and dental radiation to which radiation workers and individuals of the general population are or may be exposed.*

The Federal Radiation Council emphatically states that "there can be no single permissible or acceptable level of exposure without regard to the reason for permitting the exposure. It should be general practice to reduce exposure to radiation, and positive efforts should be carried out to fulfill the sense of the recommendations. It is basic that exposure to radiation should result from a real determination of its necessity."†

---

*The reader should be aware of the fact that the NCRP recommendations have not been unanimously accepted by all radiation scientists. Some have criticized the standards as being biased and lax. Some critics have argued, for instance, that thousands of additional cancer, leukemia, and genetic deaths would result every year in the United States if the general population were actually exposed to an average of 0.17 rems every year. For a review of the controversial issues surrounding the radiation standards and the standards-setting agencies, see for instance P. M. Boffey: Radiation standards: are the right people making decisions? Science **171**:780, 1971.
†From Radiation protection guidance for federal agencies. May 1960. U.S. Government Printing Office, Washington, D.C.

Maximum permissible body burdens for internal emitters

Radionuclides can be introduced into the body with contaminated water, food, or air (and therefore via the gastrointestinal system or lungs). It would be desirable to determine a safe maximum permissible body burden (MPB) for each radionuclide likely to be introduced into the body. MPB values have been recommended by the NCRP and ICRP for a large number of radionuclides. They are reported in detail in the National Bureau of Standards Handbook 69. The basic criterion used in establishing the MPB values is that the total internal amount of a radionuclide should be such that the radiation dose delivered by it to the total body or to specific organs should not exceed the maximum permissible doses over a 50-year period.

The dose delivered by an internally deposited radionuclide is a function of the following main factors:

1. The type, energy, LET, and other physical characteristics of the radiation emitted by the radionuclide in question. In a previous section of this chapter, I mentioned, for instance, that the dose absorbed by osteocytes from a relatively small concentration of radium in the bones can be as high as 38 rems/year, because of the high energy and LET of the alpha particles emitted by radium. Similar considerations can be extended to other radionuclides.

2. The length of time it takes for the activity of the radionuclide in question to disappear from the body. Evidently the total amount of radiation emitted (and therefore the total dose delivered to the body) is a function of the reduction rate of the activity in the body.

Fortunately, this reduction rate is not exclusively a function of the half-life of the radionuclide, but also of the rate of its elimination from the body (turnover rate).

Atoms and molecules that enter into the composition of the body are continuously replaced by new atoms and molecules of the same elements or compounds through the normal metabolic processes. This replacement is what is referred to as the "biologic turnover." The elimination takes place through urine, feces, exhaled air, and perspiration. The turnover rate varies greatly from one chemical element to another and is determined by a variety of factors, such as the nature of the chemical compound containing the element in question, solubility or insolubility of the compound, and metabolic pathways in which the compound is involved. The turnover rate *does not* depend on the isotopic form of the element. Consequently, two isotopes of the same element are subject to the same turnover rate.

In radiation biology it is convenient to express quantitatively the turnover rate with the *biologic half-life* (T_b), which is defined as the time required for the elimination of one half of the amount of a given chemical element (in any isotopic form). Therefore, the same factors that affect the turnover rate will affect the biologic half-life. The radioactive half-life of radionuclides may be specifically called *physical half-life* (T_p). Values of biologic half-lives for several radionuclides are reported in Appendix III.

For some elements, the biologic half-life may be different in different organs of the body. For instance, the T_b for phosphorus is only a few days in the liver but over a thousand days in the bones, because of the different metabolic fate of the same element in the two organs (in bone, phosphorus exists as calcium phosphate, which has a very slow turnover). However, even in these cases, it is useful to consider the total body T_b.

With the introduction of the concept of biologic half-life, we can more specifically state that the reduction rate of the activity of an internally deposited radionuclide is a function of its physical as well as biologic half-life.

Suppose that radionuclide A has a T_p of 50 years and a T_b of 25 years and that 10 μCi of A are introduced into the body. The time it takes for this activity to be reduced to 5 μCi is not 50 years but less, because the reduction of the activity is caused not only by the physical radioactive decay process but also by the elimination of part of A from the body. The influence of both the decay process and biologic turnover on the reduction rate also explains why the activity of A in the body is reduced to 5 μCi in less than 25 years. The actual time it takes for a given activity of a radionuclide in the body to be reduced to one half, as a consequence of the decay process and biologic turnover, is called the *effective half-life* (T_e) of the radionuclide. It can be shown that the three half-lives in question are correlated by the following equation:

$$\frac{1}{T_e} = \frac{1}{T_p} + \frac{1}{T_b} \tag{9-1}$$

From equation 9-1 we derive

$$T_e = \frac{T_p \times T_b}{T_p + T_b} \tag{9-2}$$

which allows the calculation of the effective half-life of a radionuclide when physical and biologic half-lives are known. In the above example, where it was assumed that T_p was 50 years and T_b was 25 years, the effective half-life was

$$\frac{50 \times 25}{50 + 25} \approx 16 \text{ years.}$$

It is evident from equation 9-2 that T_e is always less than T_p and T_b and that the larger the difference between T_b and T_p, the closer the T_e value is to T_b or T_p (whichever is smaller). For example, if T_p is 1000 days and T_b is 3 days, T_e, calculated with equation 9-2, is 2.99 days $\approx$ 3 days (approximately equal to T_b).

Unfortunately, T_b values are more difficult to determine than T_p values, because of the magnitude of the biologic experimentation necessary for the determination. The T_b values generally reported in the literature (and in this textbook) are often approximate estimates at best and may be changed by more accurate experiments. The T_b of several radionuclides is still unknown.

The MPB of a radionuclide is based, among other things, on its effective half-life calculated for the whole body or for the critical organ. The *critical organ* is the organ to which the greatest damage is done by the radiation dose delivered by an internal emitter. According to the National Bureau of Standards Handbook 69, an organ is classified as critical for the radiation emitted by an internally deposited radionuclide when it accumulates the greatest concentration of the radionuclide, when it is essential for the normal functions of the whole body, and when it has a high degree of radiosensitivity. For instance, the thyroid gland is the critical organ for any radioisotope of iodine, because it accumulates more than 50% of any amount of radioiodine absorbed by the body and consequently receives the highest radiation dose among all organs of the body. Furthermore, the thyroid, with its hormone, controls several essential functions (growth, morphogenesis, basal metabolic rate, etc.). Bone is another example of a critical organ for a variety of radionuclides (bone seekers, such as radium, radiocalcium, radiophosphorus, and radiostrontium).

Certain elements (and therefore their radioisotopes) do not accumulate in any specific organ or tissue but are more or less uniformly distributed throughout the body (such as hydrogen and sodium). In this case, if the radiation damage is the same for all the organs, the whole body is said to be the critical organ. MPB values and critical organs for several radionuclides are reported in Appendix III. Although the MPB values reported have been arrived at through elaborate calculations, the interested reader can verify, with little effort, that they are based mainly on radiation type and energy, effective half-life, and radiosensitivity of the critical organ (see, for instance, the MPB values for the strontium radioisotopes).

Since it is desirable to predict the significance of the environmental contamination without waiting until it has accumulated in the human body, the NCRP has determined and published the *maximum permissible concentrations* (MPC) of several radionuclides in air and water. These values are determined after calculating the "permissible daily intake" by ingestion or inhalation, which would not result in an accumulation in the body greater than the MPB.

The MPCs in air and water are reported in the National Bureau of Standards Handbook 69; they are applicable to occupational workers, with the assumption that there is no exposure to external radiation. When such exposure does exist, the MPC values must be reduced so that the total dose to the critical organ from external sources and internal radioactive contaminant does not exceed the MPD.

The MPCs in air and water for the general population are lower (about one tenth) than the corresponding values for radiation workers.

REFERENCES

Arena, V. 1969. A problem on potassium-40 for radiation biology courses. The Amer. Biol. Teacher **31** (6):379.

Attix, F. H., and W. C. Roesch, editors. 1966–1968. Radiation dosimetry. Academic Press, Inc., New York, 3 vols.

Brues, A. M., editor. 1959. Low-level irradiation. American Association for the Advancement of Science, Washington, D.C.

Cameron, J. R., N. Suntharalingan, and G. N. Kenney. 1968. Thermoluminescent dosimetry. University of Wisconsin Press, Madison, Wisc.

Eisenbud, M. 1963. Environmental radioactivity. McGraw-Hill Book Co., New York.

Federal Radiation Council. 1960. Background material for the development of radiation protection standards. Report no. 1, Washington, D.C.

Henry, H. F. 1969. Fundamentals of radiation protection. John Wiley & Sons, Inc., New York.

Holm, N. W., and R. J. Barry, editors. 1970. Manual on radiation dosimetry. Marcell Dekker, Inc., New York.

Meneely, G. R., editor. 1961. Radioactivity in man; whole body counting and the effects of internal gamma-emitting radioisotopes. Charles C Thomas, Publisher, Springfield, Ill.

Meneely, G., and S. M. Linde. 1965. Radioactivity in man; whole body counting and effects of internal gamma-emitting radioisotopes (second symposium). Charles C Thomas, Publisher, Springfield, Ill.

Morgan, K. Z., and J. E. Turner. 1967. Radiation protection: a textbook of health physics. John Wiley & Sons, Inc., New York.

National Commission on Radiation Protection and Measurement. 1954. Permissible dose from external sources of ionizing radiation. NBS Handbook no. 59, U.S. Department of Commerce.

National Commission on Radiation Protection and Measurement. 1959. Maximum permissible body burdens and maximum permissible concentrations of radionuclides in air and in water for occupation exposure. NBS Handbook no. 69, U.S. Department of Commerce.

National Council on Radiation Protection and Measurements. 1971. Basic radiation protection criteria. NCRP Report no. 39, Washington, D.C.

Rees, D. J. 1967. Health physics; principles of radiation protection. The M.I.T. Press, Cambridge, Mass.

Richmond, C. R., and J. E. Furchner. 1967. Cesium-137 body burdens in man; Jan. 1956 to Dec. 1966. Radiat. Res. **32**:538.

Spiers, F. W. 1968. Radioisotopes in the human body. Academic Press, Inc., New York.

U. S. Atomic Energy Commission. Division of Technical Information. 1969. Biological implications of the nuclear age. Springfield, Va.

CHAPTER 10

The use of radionuclides in biologic and medical sciences

THE RADIOTRACER METHOD
The rationale

Ornithologists interested in the geographic patterns of bird migrations have used the tagging method for several years. An animal of the species of interest is captured, tagged, and then released. The tag may consist of a light metal disk or band on which the date and location are engraved. When the tagged bird is captured later at some other location, one can tell where it came from and the time it took to cover the distance between the two locations. By repeated tagging procedures applied to large numbers of birds, the routes followed in their seasonal migrations can be mapped.

This tagging method is useful and scientifically valid only if certain conditions are fulfilled. An obvious condition for the usefulness of the method is, of course, that the tag should be firmly attached to the animal, so that the tag is not lost in transit or removed by the animal itself. Furthermore, the tag should not interfere with the parameters related to the natural migration of untagged birds (such as route, altitude of flight, and time required to cover a certain distance). If, for instance, the tag is too heavy in relation to the size of the bird, it could affect one or more of these parameters. Or if the tag consists of a pouch of some sort, containing food, one cannot rule out the possibility that the animal could stop along the way to eat that food; if this happens, the tag has altered the normal natural migratory pattern.

When in biologic research the need arises to tag a small animal such as an insect, or a cell such as a bacterium, or a chromosome, or an organic molecule, certainly methods less crude than those used to tag birds are required.

Biochemists in particular have felt the need for labeling certain organic compounds present in the foodstuffs ingested by an animal, in order to trace their metabolic fate within the animal body. Investigators of the photosynthetic process had been unable for many years to establish whether the free oxygen liberated by photosynthetic organisms originates from carbon dioxide or from water, or from both. If one could somehow label the oxygen of one of the two raw materials used for the process, it would be easy to solve the problem by simply examining the free oxygen liberated to see whether it is labeled.

255

Early in this century biochemists have sometimes solved the problem of molecule-labeling by attaching to the organic molecule of interest special recognizable radicals like the benzene ring. Knoop, Dakin, and others used phenyl fatty acids to gain information about fatty acid metabolism, and most of the knowledge achieved with this method is still valid today. However, a fatty acid labeled with a benzene ring must be regarded as unphysiologic, because phenyl fatty acids are not normally present in the animal body, and therefore one cannot be absolutely sure that the tag used does not interfere in any way with the metabolic processes under investigation.

If one or more atoms of an element present in the molecule of interest are replaced with atoms of a different isotope of the same element, the molecule remains unchanged in its chemical and biologic behavior. Consequently, the organism is unable to distinguish between two chemically identical molecules containing different isotopes of the same element. But, although the organism under study cannot make the distinction, the investigator can. In fact, with the introduction of the isotope in the molecule, the molecule has been "tagged," and the tag (and therefore its bearer) can be identified by taking advantage of either of the following two characteristic properties of the tag:

1. If the isotope used to label the molecule is stable, the labeled molecule will be heavier or lighter than the unlabeled one, depending on whether the labeling isotope is heavier or lighter than the isotope normally present in the molecule. The difference in mass can be detected with an instrument called the *mass spectrometer*, which can separate labeled from unlabeled molecules of the same species by means of strong magnetic fields.

Stable isotopes often used in biologic research to label molecules are deuterium, ^{13}C, ^{15}N, and ^{18}O. It was with ^{18}O that Ruben and Kamen in 1941 could definitely establish that the free oxygen liberated in photosynthesis comes entirely from water. In fact, green plants that are allowed to absorb $H_2^{18}O$ and $C^{16}O_2$ liberate $^{18}O_2$, whereas if they are allowed to absorb $H_2^{16}O$ and $C^{18}O_2$, they release normal $^{16}O_2$.

Among the many useful applications of ^{15}N, suffice it to mention the determination of the average life-span of the human red blood cells.

2. If the isotope used to label the molecule is a radioisotope (unstable and radioactive), the labeled molecules can be distinguished from the unlabeled ones on account of the ionizing radiation they emit when the radioactive atoms decay. The radiation can be detected by one or more of the instruments and methods described in Chapter 7. Since these instruments and methods are also capable of measuring the intensity of the radiation emitted, one can, in many instances, determine quantitatively the number or concentration of the labeled molecules.

When the label is a radioisotope, the tracing or labeling method is known as the *radiotracer method*, which is the tracer method considered in this book.

The general assumption of the radiotracer method (with a few exceptions discussed below) is that *radioactive atoms or molecules in an organism behave in the same identical manner as their stable counterparts, because they are chemically indistinguishable from them.* The method loses its validity and usefulness whenever there is sufficient reason to suspect that the above assumption is unwarranted.

In biologic investigation, the radiotracer method can be used for a number of reasons and in a variety of research situations that are typified by the following examples.

Whole small animals (such as insects) can be labeled with a suitable radionuclide for the study of their migratory habits or of some other behavior. The labeling could be done by letting the insects feed upon food containing the radioactive material. The choice of the labeling radioisotope is determined by several considerations, such as its half-life, its tolerance by the organism, the length of time during which it will presumably remain in the insect's body, etc. In such situations the labeling radionuclide can be used in a variety of chemical forms (as an organic or inorganic compound). Even whole unicellular organisms like bacteria and protozoa can be tagged with a radioisotope and the technical problems to solve would be the same as those mentioned for insects.

In certain research problems it might be necessary to label only some intracellular structures. In this case the radioisotope and its chemical form should be such as to be selectively incorporated by the cellular organelle that one wishes to label. If, for instance, chromosomes must be tagged, one should recall that these structures contain most of the DNA present in the cell. One precursor that is exclusively utilized by the cell for the synthesis of DNA is the nucleoside thymidine. To label chromosomes, therefore, one can make available to the cell thymidine labeled with ^{14}C or tritium (^{14}C-thymidine or tritiated thymidine). The labeled chromosomes can be identified and distinguished from the unlabeled ones by means of radioautographic techniques.

Liquids circulating in the body of an organism (lymph in plants, or blood in animals) can be labeled with a radionuclide if one wishes to study their circulatory pathways or the speed of their flow. This is another problem for which the chemical form of the label is not critical and a variety of radioisotopes can be used. The choice is limited only by considerations of half-life, type of radiation emitted, toxicity of the label, and a few other factors.

The radiotracer method can be used to find out whether the minerals absorbed by the root system of a plant travel mainly through the phloem or through the xylem and at what speed. The labeling tracer can be a radioisotope of any of the elements commonly absorbed by a plant from the soil (Ca, Na, P, K), and it should be in an inorganic form. If the rate of blood flow must be measured in a human subject, a solution of radiosodium salt can be injected into one area of his circulatory system; a gamma probe is then applied externally to another preselected area and the time it takes for radioactivity to appear in this area can give an indication of whether circulation is normal or hampered by some vasoconstriction.

The radiotracer method is most useful for investigating the metabolic fate of certain chemical elements normally ingested by animals and for revealing in what organic molecules they are incorporated and to what extent. Also, the method provides information on whether a chemical element is concentrated in any particular organ or not. After injection of radioiron (such as ^{59}Fe) into an anemic rat, one can show that a high percentage of the injected activity is located in his bone marrow and red blood cells and that most of the iron injected has been utilized for the synthesis of new hemoglobin. On the other hand, X-irradiated animals show less uptake of radioiron into their bone marrow, an indication that X-rays have depressed the hemopoietic function of this tissue. A high percentage of the radioiodine injected into a normal rat can be found in its thyroid after a few hours. A lower-than-normal activity is an indication of thyroid hypofunction. With the simultaneous use of radiocalcium and radiostrontium, one can show that although these two elements are chemically similar the animal body does not absorb them at the same rate from the foodstuffs, but discriminates against strontium (see p. 243).

The most sophisticated and elegant aspect of the radiotracer method consists of the fact that it provides a means of labeling an organic molecule at almost any point of interest. Most organic molecules of biologic importance are made of three or four chemical elements (C, H, O, and N). Only C and H, however, can be radiolabeled usefully (with ^{11}C or ^{14}C and tritium, respectively), because the longest-lived radioisotopes of O and N available at the present time have half-lives of only a few minutes. The use of labeled organic molecules has elucidated and solved many problems of intermediary metabolism and synthesis of protoplasmic compounds. For example, let us consider the simple organic molecule acetic acid.

$$\begin{array}{c} ^{2}CH_3 \\ | \\ ^{1}COOH \end{array}$$

If we are interested in following the metabolic fate of the acetate introduced into the animal body, we can label the molecule with ^{14}C. Either C-1 or C-2 or both can be labeled as follows:

$$\begin{array}{ccc} CH_3 & ^{14}CH_3 & ^{14}CH_3 \\ | & | & | \\ ^{14}COO^- & COO^- & ^{14}COO^- \\ \text{Acetate-1-}^{14}C & \text{Acetate-2-}^{14}C & \text{Acetate-1,2-}^{14}C \end{array}$$

The choice depends on the particular purpose of the experiment. One can show that some of the acetate injected into a rat is used for the synthesis of glycogen, cholesterol, and fatty acids, while some of it is oxidized through Krebs' cycle and appears as carbon dioxide.

The radiotracer method can be used in biologic research for many other purposes beside those just exemplified. The list of purposes just given is by no means exhaustive. The applicability of the method is in many cases limited only by the ingenuity of the investigator and by the inherent physical properties of the radioisotopes available.

Origin and development of the method

G. Hevesy, a young Hungarian chemist, was the first to use the radiotracer method in 1923. He was working at a time when the only radionuclides available were the naturally occurring members of the actinium, thorium, and uranium radioactive series. Artificial radioactivity, and therefore the possibility of radioactivating elements of lower atomic number, had not yet been discovered. Hevesy used ^{212}Pb (known at that time as thorium B), with a half-life of about 11 hours, to study the absorption and distribution of lead in the broad bean plant *(Vicia faba)*. Roots of the plants were placed in solutions of labeled lead nitrate. With a primitive electroscope, Hevesy determined the amounts of radiolead taken up by the roots, stem, and leaves of the plants exposed to the solution. At that time, a radiotracer was referred to as a "radioactive indicator."

After the discovery of artificial radioactivity by Joliot and Curie in the early 1930s, small amounts of radionuclides of low Z could be produced on a limited scale in some laboratories, and consequently the scope and possibilities of the radiotracer method were expanded considerably. Hevesy himself carried out some experiments with radiophosphorus as soon as radioisotopes of phosphorus became available. These experiments have historic importance because they were the first radiotracer experiments that used the radioisotope of an element normally present in organisms.

After the construction of the first cyclotron (in the late 1930s) a variety of artificial radionuclides were prepared in microcurie amounts. In the cyclotron, targets made of stable isotopes were bombarded mainly with accelerated deuterons; the nuclear reactions involved were of the (d,n) or (d,α) type. The amounts of radioisotopes available for biologic work became increasingly larger, although their production was quite expensive. The first radioisotope of carbon (^{11}C) was produced with the cyclotron in 1937 by bombarding boron targets with deuterons.

$$^{10}B \ (d, n) \ ^{11}C$$

The half-life of ^{11}C is only 21 minutes; consequently long-range tracer experiments with this radioisotope were practically impossible. In spite of this handicap, Ruben, Hassid, and Kamen used ^{11}C in their pioneer experiments to investigate the path of carbon in the photosynthetic process. When barley plants were allowed to assimilate $^{11}CO_2$, the investigators found that radiocarbon was incorporated into carbohydrates.

But the really big impetus to the spreading of the radiotracer method came from the discovery of fission and the construction of the first nuclear reactors in the early 1940s. The nuclear reactor is an intense source of neutrons with a wide energy range. The possibility of producing large amounts of radionuclides by bombarding stable isotopes with neutrons was immediately seen. Furthermore, a large number of artificial radionuclides are found among the fission products, from which they can be isolated and purified by chemical means. Radionuclides were produced in millicurie and curie amounts at reasonably low prices and thus became available even to modest laboratories. At the beginning, the only supplier in the United States was the Oak Ridge National Laboratory. Later with the proliferation of nuclear reactors everywhere in the country, the Atomic Energy Commission licensed several commercial concerns for the production and marketing of radionuclides, so that the delivery of radioactive materials has been greatly expedited. Today it is possible to deliver shipments of short-lived radionuclides everywhere in a few hours by air mail. Reactor-produced ^{14}C and tritium, the two radioisotopes most used in biologic work, became available in large quantities and at increasingly lower prices.

Although the nuclear reactor has made outstanding contributions to the application of the radiotracer method in biologic and medical sciences, the cyclotron still plays an important role in the production of several radionuclides. The cyclotron is more versatile than the reactor because it can employ several types of charged particles (deuterons, protons, and alpha particles) for the bombardment of the target material, whereas the reactor is exclusively a source of neutrons. Some widely used radionuclides, like ^{74}As, ^{125}I, ^{54}Mn, and ^{22}Na, can be produced only by bombarding stable elements with charged particles.

The production of artificial radionuclides made available as research tools in biologic and medical investigation is one of the most remarkable examples of the peaceful uses of atomic energy.

Advantages of the method

Information about certain metabolic pathways could not have been obtained with any other method than the isotopic tracer techniques. It is doubtful that the tremendous amount of knowledge we have today on the intermediate step reactions of the photosynthetic process could have been acquired without some sort of isotopic tracers. But even when the tracer method can be substituted with some other techniques, isotopic tracers and especially radiotracers clearly present unique advantages.

In the preisotopic era, to study carbohydrate metabolism in the animal body, one needed to work in unphysiologic conditions. The experimental animal had to be pancreatectomized or had to be injected with diabetes-inducing toxic drugs like alloxan. The same problems can be better and more easily investigated on intact normal animals by injecting into them labeled compounds, whose stable counterparts are often naturally present in the diet of the animal. As mentioned previously, the animal body treats the labeled compounds exactly as it treats its stable counterpart; thus the experiment is carried out under strictly physiologic conditions.

Especially when radioisotopes are used, the amounts of tracer can be extremely small, and yet they can be easily detected by the available radiation-detecting instruments and techniques. The radiotracer method is very sensitive indeed. As an example, one should consider that 1 mCi ($= 3.7 \times 10^7$ dps) of iodine 131 weighs about as little as 8×10^{-3} μg.* This weight is so extremely small that it cannot be measured directly by any instrument or technique, and yet, the radioactivity associated with this quantity of iodine 131 is so high that it would jam most radiation counters. Even the least sensitive G-M counter can easily measure activities as low as 0.1 μCi ($= 1/10,000$ of a millicurie); therefore it can detect the presence of an amount of radioiodine as low as 8×10^{-7} μg! No chemical analytical test, no matter how sensitive, can detect such a small amount of iodine or of any material. The sensitivity of the method is high even with longer-lived radionuclides, for which the weight of 1 mCi of material is larger. The weight of 1 mCi of carbon 14 is about 0.226 mg. But since a good liquid scintillation counter can detect and measure activities as low as 3.7 dps, one can see that the presence of as little as 0.226×10^{-7} mg of carbon 14 presents no detection problem.

Since the radiotracer reveals its presence and amount by means of the emitted radiation and its intensity, tedious separation or extraction of labeled compounds from organs or tissues and their quantitative determination are often unnecessary. The percentage, or even the absolute amount, of radio-iodine injected into an animal and taken up by the thyroid can be determined without the extraction of the active compound from the organ, but simply by measuring the intensity of the radiation emitted by the organ. For less accurate determinations one need not even remove the thyroid from the animal. Since all radioisotopes of iodine are gamma emitters, the quantitative determination of the amount taken up by the organ can be done with a gamma probe applied to the neck region. Moreover, with special instrumentation (scanners—see below), even the relative distribution of the radioactive material in the various regions of the organ can be determined and recorded with an external probe. Of course, separation and purification of the active material is required when the need arises of identifying the chemical compound in which the administered radioisotope has been incorporated through the metabolic processes of the organism.

The use of radioisotopes as tracers, when feasible, presents advantages also

---

*In fact, $A = N\lambda$, where A is the activity in dps, N the number of radioactive atoms present in the sample, and λ the disintegration constant. See Chapter 5.

$$N = \frac{A}{\dfrac{0.693}{T_{1/2} \text{ (sec)}}} = \frac{3.7 \times 10^7 \text{ dps}}{\dfrac{0.693}{6.9 \times 10^5 \text{ sec}}} = 3.7 \times 10^{13}$$

The weight of 3.7×10^{13} atoms of ^{131}I is given by

$$\frac{131 \times 3.7 \times 10^{13}}{6.2 \times 10^{23}} \approx 8 \times 10^{-9} \text{ g} \approx 8 \times 10^{-3} \text{ μg}$$

over the use of stable isotopes. Radiation-detecting instruments and methods are generally much easier to handle and less expensive than mass spectrometers. The preparation of the final samples for radioassay is usually much simpler than the preparation required for the determination of stable isotopes. The sensitivity of the methods for measuring stable isotopes does not compare favorably with the sensitivity of most radiation-detecting instruments and methods. Furthermore, isolation of compounds labeled with stable isotopes from tissues and tissue extracts is more often necessary than when radiotracers are used. In biologic work, the use of stable isotopes is resorted to only when no radio-isotopes are available with a half-life long enough to make an experiment feasible. As mentioned above, this is exactly the case for nitrogen and oxygen. Nevertheless, one should remember that stable isotopes have the following two important advantages over radioisotopes: they are permanent, and they have no radiation effects on biologic systems.

Obviously, the use of radioisotopes as tracers poses problems of radiation hazards and safety precautions and requires a good knowledge of radiation-detecting instrumentation. These disadvantages should be carefully pondered by anyone who wishes to employ radiotracer methods for the first time. Likewise, the fact should be stressed that radioisotopes are not magic tools or panaceas that can solve any research problem. Many problems can be solved equally well with simpler and less sophisticated methods and equipment. The investigator who is unfamiliar with radionuclides should decide, after careful considerations, whether the use of radiotracers in his particular research situation is fully justified.

Experimental design

Any experiment using radioisotopes as tracers must be carefully planned in advance. The complexity of the problems to solve during the planning stage depends very much on the complexity of the experiment and on its purpose. Some of the problems that should be given careful attention are briefly illustrated in this section.

Once the investigator has decided which chemical element he wants to trace (a decision that is determined by the very basic nature of the experiment), he has to choose a suitable radioisotope of the element of interest. At the present time, with very few exceptions, there are several radioisotopes available for every element. The choice of the radioisotope in any particular experiment is determined by several factors.

In the first place, the physical half-life of the isotope should be taken into consideration. It is obvious that, in long-range experiments of the duration of several days, radioisotopes with half-lives of only a few hours cannot be used. Thus, if the element of interest is sodium, there is a choice between ^{22}Na and ^{24}Na, both gamma emitters. In a long-range experiment only the former can be used ($T = 2.6$ years), because the half-life of the latter is only 15 hours. In experiments of short duration it may be preferable, at times, to use short-lived isotopes. This consideration is especially necessary when the amount of radio-activity in the organism must be reduced to a minimum in order to avoid possible undesirable radiation effects or whenever radioisotopes are used with human subjects for diagnostic purposes. In these cases, a short-lived radioisotope is particularly advisable if the biologic half-life of the element is long. The effective half-life is reduced with a radioisotope of short physical half-life (see p. 253).

The choice of the isotope is also determined by the type and energy of the radiation emitted. In some experiments, it may be advisable to use beta emitters, in others gamma emitters. The choice is often dictated by the type of radiation-detecting instrumentation available. Without a good liquid scintillation counter, it is practically impossible to make accurate quantitative determinations of tritium or ^{55}Fe (see decay schemes in Appendix III). If, however, the intracellular location of labeled organic compound is to be determined by means of radioautographic techniques, a tritium label is preferable to a ^{14}C label because, as seen in a previous chapter, the very low energy of the tritium beta particles results in radioautograms of very good resolution.

When radioisotopes are not used as labels for organic compounds, one often needs to know their chemical form. With few exceptions, radionuclides are not available as free elements, but in some form of inorganic compounds (salts, acids, etc.). A radiotracer should be administered to an organism in the form of a compound that is not unnatural, harmful, or toxic.

In the case of labeled organic compounds, the choice of the compound is determined by the particular purpose of the experiment; it is impossible to give any general guidelines because of the wide variety of experiments that are possible with labeled organic molecules. However, the mere choice of the organic compound may not be sufficient; often the specific location of the label within the molecule deserves consideration. From this point of view, organic compounds may be labeled as follows:

1. *Totally labeled*, if all the atoms of an element present in the molecule are labeled. For example (* = label):

$$*CH_3 - *COOH$$

2. *Partially labeled*, if only one or some of the atoms of an element are labeled. The location of the label must be known, for example:

$$*CH_3 - COOH \quad or \quad CH_3 - *COOH$$

3. *Doubly labeled*, if atoms of two different elements present in the molecule are labeled, for example:

$$C*H_3 - *COOH$$

Again, the choice of the location of the label is determined by the particular nature of the experiment. As a general rule, the location of the label is important whenever one knows or suspects that the labeled compound administered is used by the organism for the biosynthesis of other organic compounds after some of its chemical bonds are broken. This is exactly the case for acetate; a decarboxylase may remove from the molecule —COO—, which is liberated as carbon dioxide, while the rest of the molecule may be used for the synthesis of glycogen, cholesterol, fatty acids, etc. On the other hand, thymidine is utilized as such for the synthesis of DNA; if the purpose of the experiment is to obtain the incorporation of this nucleoside in DNA, ^{14}C-thymidine may be used with the label attached to any point of the molecule.

In the analogy of bird tagging previously used, it was noted that if the tag is lost in transit the tagging loses all its usefulness. The tag must be firmly attached to the body of the animal. Similarly, the label in an organic molecule must be firmly attached to the molecule, or at least to that part of the molecule of interest to the investigator. This problem is particularly acute in experiments involving the use of tritiated compounds. Investigators have repeatedly observed that tritium bound in groups like —COOH, —OH, —NH$_2$, and —NH— is rapidly lost because it is readily exchanged with hydrogen ions, which are practically present everywhere in an organism. Therefore compounds with the tritium label in those configurations have little value for biologic studies.

How much of a labeled compound should be used in a radiotracer experiment? This question is perhaps the most important and critical that the investigator has to face in the planning stage of the experiment. In this context, the expression "how much" must be understood both in the sense of "how much activity" (expressed in millicuries or microcuries) and in the sense of "how much weight" (expressed in grams or fractions thereof). The basic principle to remember is the obvious fact that once the labeled compound is introduced into an organism it usually becomes diluted and mixed with similar or identical unlabeled compounds and also with other substances containing the stable isotope of the element under study. This fact is true especially when the labeled compound distributes itself uniformly throughout the body of the organism, but, to a lesser extent, it is also true when the compound is selectively concentrated in a particular organ or tissue. The consequence of this dilution is, of course, that the activity of the samples removed for radioassay is considerably lower, gram for gram, than the activity of the material introduced initially into the organism. Therefore, the level of activity to be injected into an animal should be such that the activity of the samples to be radioassayed is high enough to be accurately and reliably determined with the instrumentation available. The fact is immediately obvious then that the amount of activity to use should be proportionate to the weight of the animal.

But an excessive dose may not be advisable for a number of reasons. In the first place, radioisotopes and especially labeled organic molecules are quite expensive, and in many instances the cost of the radiotracer experiment must be kept within reasonable limits. Secondly, with too much activity introduced into the body of an organism, it is quite possible to reach a threshold above which radiation effects (see below) may be significant and may interfere with the natural metabolic processes of the organism and thus distort the experimental results. Thirdly, in many instances higher activity means also greater weight of the compound injected; in most experiments the weight of the material used should be as small as possible. Fortunately, activity is not always directly proportional to weight; in fact, it is possible to have 1 μCi of activity in 1 μg or in 1 ton of a labeled glucose, depending on whether some or all the molecules of the sample are labeled and on the number of labels per molecules. The term *specific activity* (SA, or sp. act.) is used to express the relation between activity of a sample and its weight. Specific activity can be indicated in a number of ways, such as μCi (or mCi, or Ci) per mg (or g, or millimole [mM], or mole) of compound. In biologic radiotracer experiments specific activity expresses also the number of dpm (or cpm)/g (or mg) of tissue or organ. Evidently then, one should use material with high specific activity to keep the activity high and the weight low. Some radioisotopes and labeled compounds can be obtained with any specific activity; for others the highest specific activity obtainable is so low as to make them practically useless in certain experimental situations. Because of the biologic dilution mentioned previously, the specific activity in the samples to be radioassayed at the end of the experiment is evidently lower than in the initial labeled compound. The specific activity of the labeled compound to administer to an animal is therefore determined by the desired minimum specific activity expected in the counting samples.

Sources of error

Like many other research methods, the radiotracer method, too, has its pitfalls. Abnormal and distorted results can be attributed to one of several possible causes, which the investigator should either avoid or at least be aware of. Some of the most common sources of error are briefly illustrated in this section.

Chemical impurities

Chemical impurities are traces of unwanted chemical species different from the one that the investigator intends to use. If the impurities are metal ions like Hg, Co, etc., they can interfere with the particular metabolic reaction under study; it is well known that those ions can inactivate certain enzymes. At other times the impurity is a chemical species only slightly different from the wanted one; it may even be labeled with the same radioisotope. A preparation of ^{131}I-labeled iodide may contain traces of ^{131}I-labeled iodate as impurity. A preparation of ^{32}P-labeled phosphate may contain some ^{32}P-labeled phosphite as impurity. Abnormal results caused by impurities of this type have often been reported in the literature.

The chemical purity of a radioisotope or labeled compound should never be assumed a priori. If impurities exist, they should be known as stated by the manufacturer. Purification of the preparation may be necessary when required by the nature of the experiment.

Radiochemical impurities

A radioisotope or labeled compound is radiochemically impure if it is mixed with traces of an unwanted radionuclide of a different chemical species.

Some of these impurities are created by the very process used for the preparation of the radioisotope, as illustrated by the following examples:

^{42}K is produced by neutron bombardment of a potassium target in a reactor. The nuclear reaction responsible for its production is

$$^{41}\text{K (n, }\gamma\text{) }^{42}\text{K}$$

^{41}K is present in natural potassium with an abundance of 6.88%. If the potassium target contains 1% of sodium as a contaminant, ^{24}Na also is produced through the reaction

$$^{23}\text{Na (n, }\gamma\text{) }^{24}\text{Na}$$

The radiopotassium obtained contains 13% of radiosodium as a radiochemical impurity. The reason for the high percentage of radiosodium as compared with the percentage of sodium initially present in the target is that the natural abundance of ^{23}Na is 100% (versus 6.88% for ^{41}K).

^{45}Ca can be produced by neutron bombardment of ^{44}Ca, which is present in natural calcium with an abundance of about 2%:

$$^{44}\text{Ca (n, }\gamma\text{) }^{45}\text{Ca}$$

If $CaCO_3$, containing as little as 0.01% of calcium phosphate as impurity, is exposed to the neutron flux, the radiocalcium obtained contains as much as 5% of ^{32}P as a radiochemical impurity. ^{32}P is formed by the neutron bombardment of the stable isotope ^{31}P, which is the only isotope present in natural phosphorus (100% natural abundance).

In cases like those exemplified above it is not difficult usually to remove the radiochemical impurity by chemical purification.

In other cases the radiochemical impurity is generated continuously by the very decay process of the radionuclide. This happens when the daughter nucleus is itself unstable. Thus the radionuclide ^{90}Y is always present in a ^{90}Sr preparation (see decay scheme of ^{90}Sr). Removal of the yttrium impurity obviously does not prevent its regeneration.

Isotope effect

At the beginning of this chapter I have stated that the validity of the radiotracer method is based on the general assumption that radioisotopes behave chemically like the stable isotopes of the same element. In certain instances, however, this assumption is not correct; researchers have observed several times that the atomic mass of an isotope may affect its chemical behavior. The different behavior is attributed to what can be called *isotope effect*.

Of course, concerning radioisotopes, a marked isotope effect is seen only when the ratio of the mass of the radioisotope to the mass of the most abundant isotope present in the natural element is significantly different from 1. This is the case for elements of low atomic number, and in radiotracer methodology the problem is particularly acute for carbon and hydrogen. For carbon, the ratio of the ^{14}C atomic mass to the ^{12}C mass (the most abundant isotope of natural carbon) is approximately 14:12. For hydrogen, the ratio of the ^3H ($= T$) atomic mass to the ^1H mass is approximately 3:1. In organic molecules labeled with ^{14}C or T the isotope effect stems from the fact that the stability of a chemical bond increases with the increase in the masses of the isotopes kept together by the bond. Consequently, the bond $^{14}C—^{12}C$ is more stable than the bond $^{12}C—^{12}C$, and the bond $T—^1H$ is much more stable than the bond $^1H—^1H$. In the case of hydrogen, since the mass ratio is very high, T and ^1H can hardly be considered as the same chemical element for many practical purposes. Likewise, H_2O and T_2O differ in many chemical properties and can hardly be considered as the same substance.

The isotope effect can be observed, for instance, in the rate of chemical reactions. A labeled compound reacts more slowly than the unlabeled compound of the same species, if the mass of the label is greater than the mass of the isotope normally found in the unlabeled compound; thus the higher the mass ratio, the lower the reaction rate. For instance, with T-labeled molecules, the reaction rate is 73% lower than with labeled molecules, with ^{14}C, the rate is 7% lower, and with ^{131}I, the rate is 1.5% lower. In the configuration of malonic acid-1-^{14}C, as shown below:

$$\begin{array}{c} COOH \\ | \\ CH_2 \\ | \\ ^{14}COOH \end{array}$$

one C—C bond ($^{14}C—^{12}C$) is stronger than the other. If this compound is

acted upon by a decarboxylase, the enzyme attacks the weaker bond preferentially, so that the released carbon dioxide is enriched with $^{12}CO_2$ and the
acetic acid is enriched with labeled acetic acid, as follows:

$$
\begin{array}{c}
\underset{}{COOH} \\
\overset{}{\underset{}{|}} \\
CH_2 \\
| \\
{}^*COOH
\end{array}
\longrightarrow CH_3-{}^*COOH + CO_2
$$

Consequently, the level of radioactivity of the carbon dioxide collected does
not reflect the level of activity of the malonic acid used as substrate.

Urease attacks unlabeled urea 10% faster than urea labeled with ^{14}C:

$$
\begin{array}{c}
NH_2 \\
\diagdown \\
{}^{12}CO \\
\diagup \\
NH_2
\end{array}
\xrightarrow{+H_2O} 2NH_3 + {}^{12}CO_2
$$

10% faster than

$$
\begin{array}{c}
NH_2 \\
\diagdown \\
{}^{14}CO \\
\diagup \\
NH_2
\end{array}
\xrightarrow{+H_2O} 2NH_3 + {}^{14}CO_2
$$

Another isotope effect can be seen in the introduction of optical activity by a
tritium label. Normal butane ($CH_3-CH_2-CH_2-CH_3$) is optically inactive,
because none of its carbon atoms is asymmetric. But the substitution with tritium of one of the hydrogen atoms attached to C-2 makes the molecule optically
active, because that carbon atom is now asymmetric:

$$
\begin{array}{ccccccccc}
 & H & & T & & H & & H & \\
 & | & & | & & | & & | & \\
H- & C & - & C & - & C & - & C & -H \\
 & | & & | & & | & & | & \\
 & H & & H & & H & & H &
\end{array}
$$

The slightly or markedly different chemical behavior caused by the label is
reflected by a different behavior of the labeled molecule in metabolic processes.
In studies of photosynthesis with barley seedlings the observation has been made
that the assimilation of $^{14}CO_2$ is about 17% slower than the assimilation of unlabeled carbon dioxide. Therefore, plants tend to discriminate against the
heavier carbon isotope. However, shells of snails grown in a balanced aquarium
containing a known percentage of ^{14}C-bicarbonate in the water have the tendency to discriminate against the lighter carbon isotope.

Isotope effects may be expected in experiments designed to study diffusion
processes through membranes (for example, plasma membranes). In general,
heavy isotopes diffuse through membranes more slowly than light isotopes of
the same elements. The difference in diffusion rate is appreciable between
labeled and unlabeled carbon compounds.

In practice, the possibility of isotope effects should be considered (and sometimes anticipated) in radiotracer experiments when the tracers used are tritium,
^{14}C, and a few other radioisotopes of elements of low atomic number. As the
atomic number increases, isotope effects become more and more negligible and

undetectable; as such they may be ignored, because they do not invalidate the experimental results.

Interesting to note here is that although the isotope effect is usually an unwanted "nuisance" it can also be exploited as a useful tool for the investigation of the mechanism of enzymatic reactions.*

The radiation effect

The isotope effect stems from the fact that the radioisotope used as a tracer has a mass different from that of its stable counterpart. The mass difference results in its different chemical behavior. But a radioisotope is different from a stable isotope of the same element in another property also; it is radioactive and therefore is a radiation emitter. Because biologic systems are affected by ionizing radiation, a radioisotope used as a tracer in a biologic experiment may show a biologic behavior different from that of its stable counterpart. The different biologic behavior may affect and distort the results of the experiment, and this distortion is what is referred to as the *radiation effect*. Thus a radiation effect may also invalidate the general assumption that is the basis of the biologic radiotracer method. In fact the radiation damage caused by a radioisotope introduced into an organism may be such as to impair or derail metabolic processes or alter the physiologic response of certain organs or tissues.

The magnitude of a radiation effect that may be observed in conjunction with the use of a radiotracer depends on such factors as dose of the tracer, type and energy of the radiation emitted, and duration of the experiment, and on whether the tracer is uniformly distributed throughout the body of the experimental organism or is concentrated only in one of its organs. In one study with rats the administration of as little as 0.045 μCi of ^{131}I per gram of body weight has been observed to cause definite impairment of thyroid function. For instance, the thyroid loses its ability to enlarge in response to a low iodine stimulus. Evidently, the use of large doses of ^{131}I in thyroid-uptake studies may invalidate the experimental results.

The high-energy beta particles emitted by ^{32}P incorporated in the chromosomal DNA may be sufficient to cause mutations. In this case, the very decay process with the formation of a daughter nucleus of a different chemical species (^{32}S) may also be the cause of molecular alteration and thus of biologic damage (see the transmutation effect in the appendix of Chapter 11).

Fortunately, in radiotracer experiments the radiation effect is important only inasmuch as it affects, and interferes with, the experimental results. In many of these experiments, radiation effects are nonexistent because the tracer doses generally used are much lower than the threshold doses required to induce radiation effects and because of the short duration of the experiments.

A radiation effect may be exerted on the radioactive compound itself, when the latter is stored on the shelf for prolonged periods of time. Tritium- or ^{14}C-labeled organic compounds may undergo decomposition, during storage, at a faster rate than their unlabeled counterparts. This is known as *radiation decomposition*, and it is significant in materials of high specific activity. The beta particles emitted by tritium and ^{14}C have enough energy to break chemical bonds in the surrounding molecules (self-radiolysis). ^{14}C-methanol, with a specific activity, of 8 mCi/mmole, is destroyed at the rate of 19% in 2 years. ^{14}C-choline with a SA of 1.8 mCi/mmole is destroyed at the rate of 63% per

---

*See examples in Wolf, G. 1964. Isotopes in biology. Academic Press, Inc., New York. Pp. 48–51.

year. For T-labeled compounds, it has been estimated that with a SA of 1 mCi/mg the radiation dose absorbed per day by the compound is about 3×10^5 rads! Obviously the phenomenon of radiation decomposition introduces chemical impurities into the preparation of a labeled compound.

Radiation decomposition can be minimized to some extent by storing the labeled compound in solution, or adsorbed to filter paper or mixed with glass beads. The solvent, the paper, or the beads absorb a large proportion of the beta radiation before it can do any harm to the labeled molecules.

Production of radionuclides and molecule-labeling techniques

As previously mentioned, in this atomic age artificial radioisotopes of practically any element can be produced by bombarding targets of stable isotopes with neutrons or charged particles. Several types of nuclear reactions (see p. 113) are used for their production. For large-scale production the nuclear reactor is used as a source of neutrons and particle accelerators (cyclotrons) as sources of charged particles.

When a radioisotope is produced by means of a nuclear reaction that leads to a change in atomic number, like (n, p), (d, n), and (n, α) reactions, it is an element different from the target element and it can be easily separated by chemical means from the precursor element, which has not reacted, and thus can be obtained free of radiochemical impurities. If no stable isotope (carrier) of the same element is added, the radioisotope preparation is said to be carrier free (CF). For example, carrier-free phosphorus 32 is a preparation of which all phosphorus atoms are ^{32}P. When a radionuclide is carrier free, it has the maximum specific activity possible. However, for many chemical manipulations one needs to add a carrier even to those radioisotopes that can be produced CF. When a radioisotope is produced by means of a nuclear reaction that does not involve a change in atomic number, like (n, γ) and (d, p) reactions, it is the same element as the target element. Only a limited fraction of the target material is actually converted to the desired radioisotope, which therefore is mixed with a stable isotope or isotopes of the same element. The product is not CF and usually has a relatively low specific activity.

Some radionuclides can be produced by one of several nuclear reactions; the choice of the reaction is determined by the specific activity desired. Two examples will illustrate this concept:

Zinc 65 may be obtained in a reactor by neutron irradiation of natural zinc, through a (n, γ) reaction. The target isotope is ^{64}Zn, with a natural abundance of only 49%. Consequently, the final product is ^{65}Zn mixed with other (stable) isotopes of the same element, and therefore with a low specific activity. The separation of the radioisotope from the stable isotopes is not possible by chemical means. However, with a cyclotron, ^{65}Zn may be obtained from a copper target by the reaction ^{65}Cu (d, 2n) ^{65}Zn. It can be easily separated from the residual copper, and if the target did not contain any zinc as impurity, the radionuclide can be obtained with high specific activity and even carrier free.

Sulfur 35 can be produced by any of the following reactions:

1. ^{34}S (n, γ) ^{35}S
2. ^{34}S (d, p) ^{35}S
3. ^{37}Cl (d, α) ^{35}S
4. ^{35}Cl (n, p) ^{35}S

The specific activity obtainable with reactions 1 and 2 is very poor because these reactions do not involve a change of atomic number and furthermore the natural abundance of ^{34}S in sulfur is only 4.22%. Reaction 3 is preferable to the first two, because the radionuclide obtained can be easily separated from the residual chlorine. But since the natural abundance of the target isotope is not very high (24.4%), the yield of the product is rather poor, although its specific activity may be high. The reaction of choice is no. 4, because the target is the most abundant isotope (75.5%) of chlorine; both yield and specific activity are the highest possible.

In certain instances it is possible to obtain a radionuclide with very high specific activity by means of a (n, γ) reaction in spite of the fact that the reaction does not involve a change of atomic number. If ethyl iodide is bombarded with neutrons, ^{128}I is produced $(^{127}I[n, \gamma]^{128}I)$. The gamma photon emitted in the reaction has enough energy (8 MeV) to cause an appreciable recoil of the newly formed ^{128}I, with the consequence that the chemical bond uniting the iodine atom to the carbon atom is broken and the iodine is liberated. The free radioiodine is thus easily separated by chemical means from the residual ethyl iodide. This very interesting type of reaction is known as *recoil* or *Szilard-Chalmers reaction* and is extremely important in radiation chemistry.

Many radionuclides are obtained as fission products in nuclear reactors; they can be purified chemically and be prepared carrier free.

Of the two radionuclides most often used in biologic work, ^{14}C and tritium, the first was discovered by Ruben and Kamen in 1940 and has replaced almost completely the previously available radioisotope of carbon, ^{11}C, mainly because of its much longer half-life (5730 years). It is produced in the nuclear reactor through the ^{14}N (n, p) ^{14}C reaction. The yield is high, because the natural abundance of ^{14}N is about 99.6%; the maximum specific activity obtainable is 2 Ci/g of carbon. The nitrogen compounds used as targets are NH_4NO_3, Be_3N_2, or AlN. The radionuclide is stored in the form of soluble or insoluble carbonate (Na_2CO_3 or $BaCO_3$), from which $^{14}CO_2$ can be easily obtained by treatment with an acid.

Tritium was discovered in 1939, when deuterium was bombarded with deuterons in a cyclotron. Its half-life is 12.6 years and therefore it is suitable as radiotracer in biologic experiments. Several nuclear reactions could lead to the production of this radionuclide, but for the production of large amounts, the ^6Li (n, α) ^3H reaction is used in nuclear reactors. The natural abundance of the target isotope is low (7.4%) but its cross section for thermal neutrons is high. The Li compounds used as targets are LiF or LiOH; the tritium produced is recovered as free gas and purified. It can be obtained carrier free (SA is 2.86 Ci/cm^3 of gas). It can be stored as gas or as tritiated water (^3H_2O).

In many tracer experiments radioisotopes can be used as free elements or as simple inorganic compounds, that is, in the same chemical forms as they are produced by the reactor or accelerator. In other experiments it is necessary to incorporate them in organic molecules, that is, to use them as labels. Organic molecules can be and have been labeled with a large number of radionuclides and notably with ^{14}C, T, ^{131}I, ^{32}P, and ^{35}S. In this section, ^{14}C- and T-labeling methods are briefly discussed.

^{14}C-labeling

Organic compounds labeled with ^{14}C may be obtained either by chemical synthesis or by biosynthesis.

Chemical synthesis. With the classic methods of synthetic organic chemistry, a large number of organic compounds can be synthesized in the laboratory from other organic precursors or from a carbon compound as simple as carbon dioxide. If the precursor is labeled with ^{14}C, the final product will also be labeled with ^{14}C. The choice of the chemical reaction is determined by the yield of the desired labeled compound, by the location of the label, and by other minor factors.

A general survey of the enormous number of synthetic methods available to prepare ^{14}C-labeled organic compounds cannot be given here. The interested reader is advised to consult the excellent review by Catch, cited at the end of this chapter. Another useful thing the student can do is to glance through the catalog of some leading commercial suppliers of labeled organic compounds to get an idea of the tremendous variety of labeled compounds that are available. At this point only one example can be given; the sequence of reactions employed for the production of acetic acid-2-^{14}C from $^{14}CO_2$ (*C = labeled carbon):

$$Ba*CO_3 \xrightarrow{+H_2SO_4} *CO_2 \xrightarrow{+LiAlH_4} \underset{\text{(Lithium aluminum methoxide)}}{LiAl(O*CH_3)_4} \xrightarrow{\text{Tetrahydro-furfuryl alcohol}}$$

$$\underset{\text{(Methanol)}}{*CH_3OH} \xrightarrow{+HI} \underset{\text{(Methyl iodide)}}{*CH_3I} \xrightarrow{+KCN} \underset{\text{(Acetonitrile)}}{*CH_3CN} \xrightarrow{+2H_2O} \underset{\text{(Acetate-2-}^{14}C)}{*CH_3COOH}$$

In general, any organic compound that can be synthesized in the laboratory can also be labeled with ^{14}C. Commercial firms have a large number of labeled compounds in stock; but they also offer a custom service for the synthesis of any partially or totally labeled compound that is not regularly stocked, in accordance with the specifications of the investigators.

Biosynthesis. For organic compounds that are not possible or convenient to label in the laboratory, an easy and inexpensive method is to let organisms label them from labeled precursors. Organisms will synthesize a variety of simple or complex organic compounds through their normal metabolic processes. The desired compounds are extracted from the organisms and purified by chemical means.

If $^{14}CO_2$ is made available to photosynthetic organisms, they will synthesize a variety of labeled organic compounds using $^{14}CO_2$ as a precursor. The facilities needed for the growth of plants that synthesize labeled compounds are called with the very descriptive name of "isotope farms." Famous are those of the Argonne National Laboratory. The green unicellular alga *Chlorella* is grown on a large scale in the presence of ^{14}C carbonates or bicarbonates. This alga is a producer of large amounts of labeled carbohydrates (monosaccharides, disaccharides, and polysaccharides) as well as of many amino acids. Nonaquatic plants of certain species are grown for the production of labeled drugs in sealed greenhouses and supplied with labeled carbon dioxide; labeled digitoxin can thus be obtained from *Digitalis purpurea* grown in these conditions.

Molds and bacteria are used for the labeling of vitamins and antibiotics; labeled penicillin can be obtained in this manner from *Penicillium notatum.*

Laboratory animals are utilized for the labeling of certain proteins (such as serum albumin), enzymes, and fatty acids. They are fed or injected with labeled precursors (amino acids, carbohydrates, etc.) synthesized by plants. Even isolated organs or tissue slices can be profitably used for the same purpose. Slices of beef adrenal cortex incubated in ^{14}C acetate have been used to synthesize labeled cholesterol and fatty acids.

The advantages of the biosynthetic methods over the laboratory synthesis are that organisms can label very complex and unusual organic materials that cannot be synthesized artificially. An added advantage is seen in the synthesis of optically active substances. It is well known that when one of these substances is synthesized in the laboratory a racemic mixture of both stereoisomers is obtained (although it is possible in certain cases to isolate one isomer from the other). On the contrary, organisms synthesize only one of the two isomers; for instance, only L-aminoacids are metabolized. Evidently, when an L-aminoacid is needed with a label, one can more conveniently obtain it by biosynthesis. The tedious work of separating it from a racemic mixture is thus avoided.

On the other hand, one can seldom obtain by biosynthesis a substance with the label in a specific predetermined location in the molecule. The compounds may be either totally labeled or may consist of a mixture of partially labeled molecules with labels at different locations. In many radiotracer experiments, of course, this is not objectionable.

Whether an organic compound should be labeled by chemical synthesis or biosynthesis is determined by several factors such as cost, yield, specific activity, total amount needed, difficulty of purification, and molecular location of the label.

Tritium labeling

The classical methods of organic synthesis could also be used to label organic molecules with tritium, starting from tritiated precursors and even from the simplest T compound—T_2O. In fact, these methods are used whenever they are convenient and feasible. But for tritium some other additional unique methods that are faster, less expensive, and less laborious are available. They are reduction of unsaturated precursors, catalyzed exchange, and the Wilzbach method.

Reduction of unsaturated precursors with tritium gas. If a precursor of the compound that one wishes to obtain labeled with tritium has a double bond between two carbons, it can be saturated by exposing it to tritium gas, as in the following reaction:

$$-\overset{|}{C}=\overset{|}{C}- \ + \ T_2 \ \longrightarrow \ -\overset{|}{\underset{T}{C}}-\overset{|}{\underset{T}{C}}-$$

Metallic platinum or palladium are the catalysts normally used for this reaction.

Catalyzed exchange in aqueous media. This method consists simply of heating together in a sealed tube the compound to be labeled, a hydrogen-transfer catalyst, and a tritiated solvent, which may be water, acetic acid, or trifluoracetic acid. An exchange takes place between the ^1H of the compound and the tritium of the solvent. This method of course, can be used only with compounds that remain stable at the high temperatures (about 120° C) required for the reaction.

The Wilzbach method (or gas exposure method). Unlike the preceding two methods, the Wilzbach method can be used to label any organic compound, even of unknown molecular structure, and the labeling is accomplished without a long chain of step reactions. It is used whenever other methods are unsuitable for the particular compound one wishes to label.

First introduced by Wilzbach in 1957, the method consists of the exposure of the compound in solution or in finely powdered form to several curies of tritium gas in a reaction vessel for several days or weeks at room temperature. The energy from the decay of tritium activates the molecules of the compound and dissociates the C-H bonds. The stable hydrogen of the compound is thus replaced by tritium, which therefore serves the twofold function of activating and replacing.

To have an idea of the activating energy involved in this process, one should recall that 1 cm^3 of carrier-free tritium corresponds to an activity of 2.86 Ci. The radiation dose absorbed by 1 g of the substance mixed with 1 cm^3 of tritium is about 30,000 rads/hour. Expressed in another way, the energy released by tritium is approximately 1.8×10^{19} eV/Ci of tritium per day. This energy is capable not only to dissociate C-H bonds but also C-C bonds of the compound to be labeled. The consequence is that the method is plagued by the formation of tritiated degradation products; this situation necessitates purification of the labeled compound.

The labeling yield is a function of exposure time and tritium activity; the fraction of tritium incorporated in the compound per day decreases as the molecular weight and the complexity of the compound increase. The labeling yield is good for carbohydrates, steroids, and aromatic hydrocarbons, but not so good for polypeptides and aliphatic hydrocarbons.

The exposure time can be shortened if additional activation energy is supplied from an external source. This energy may be supplied with gamma radiation (from cobalt 60), ultraviolet radiation, microwaves, or electric discharges. Unfortunately, this additional energy produces also a greater degree of destruction of the compound to be labeled.

One disadvantage of the Wilzbach method is that molecules are randomly labeled, so that it is not possible to obtain a partially labeled compound with the label in a predetermined molecular location.

Since carrier-free tritium is readily available, tritiated compounds may be obtained with higher specific activity (up to 13 Ci/mmole) than are ^{14}C-labeled compounds (only a few mCi/mmole).

• • •

In this section we have seen that a large number of organic molecules can be conveniently and successfully radiolabeled with ^{14}C or tritium. In certain situations the investigator may have no choice between the two labels because of the particular purpose of the experiment; in others, he may. Which criteria should determine his choice?

The factors determining the choice are intimately correlated with the nature of, and the techniques used in, the experiment. Before choosing the label the investigator should give a great deal of thought to the following facts:

1. The beta energy of ^{14}C ($E_{max} = 156$ keV) is considerably higher than the beta energy of tritium ($E_{max} = 18$ keV). Consequently, a good detecting or counting efficiency for tritium can be obtained only with liquid scintillation counters and radioautographic techniques, whereas good counting efficiencies for ^{14}C can be obtained also with other counting instruments (proportional or external scintillation counters). On the other hand, as repeatedly stated else-

where in this book, when the labeled compound is to be detected radioautographically, tritium gives a better resolution.

2. The half-life of ^{14}C is much longer than that of tritium. Correction for decay is never necessary for ^{14}C; it may be necessary for tritium in long-range experiments.

3. C-C bonds are stronger than C-H bonds. Consequently, the ^{14}C label is more firmly attached to a molecule than the tritium label. The latter, as previously stated, can be more or less easily exchanged with ^{1}H in biologic systems.

4. Higher specific activity can be obtained for tritiated compounds than for ^{14}C-labeled compounds.

5. In general, T labeling is easier than ^{14}C labeling.

6. Because of the difference in beta energy, tritium is radiobiologically less harmful. The radiation hazard should be considered in relation to the organism to which the labeled compound is administered, as well as to the personnel who must handle the labeled material.

7. Generally speaking, tritiated compounds are less expensive than ^{14}C compounds.

SOME EXAMPLES OF THE APPLICATION OF THE METHOD IN BIOLOGIC RESEARCH

There is no question that the radiotracer method has had a tremendous impact on practically every aspect of biologic investigation since its origin. One cannot overemphasize the significant contribution given by the method to the explosive growth of the life sciences during these past 20 years. There are very few areas of biology in which research problems cannot be usefully tackled with the radiotracer method. However, the use of the radionuclides has been particularly fertile in cytology and cell physiology, cytogenetics, molecular genetics, biochemistry, plant and animal physiology, microbial physiology, virology, ecology, and marine biology. Most of the glamorous breakthroughs in such vital problems like the biologic role of nucleic acids, the nature of the gene, the control of protein synthesis, the genetic code, have been made possible by radiotracers used in concomitance with other modern research tools. Every year, thousands of papers reporting on research done with the radiotracer method appear in the biologic literature.

A summary of all the contributions of the method to biologic investigation would obviously be an impossible task. All that can be done in an introductory textbook of this nature is to present two typical examples of research problems to which the radiotracer method has been successfully applied.

Studies on thyroid metabolism

Our knowledge of the biochemistry and physiology of the thyroid hormone has made considerable progress since Baumann first discovered (in 1895) that a high percentage of the body iodine content is concentrated in the relatively small thyroid gland. Much of what we know today has been learned from experimental animals injected with radioiodine, and especially with iodine 131.

Experiments have shown that about 30% of an injected dose of radioiodine is quickly accumulated in the thyroid, 10% is distributed to the rest of the body, and 60% is excreted with the urine. The uptake on the part of the thyroid begins immediately after injection and reaches a peak within a period of time that may vary in different species. In humans the maximum concentration is reached about 24 hours after injection. Gamma radiation from the organ, however, may be detected as early as a few minutes after injection.

The metabolic fate of the ingested iodine and its utilization in the synthesis of the thyroid hormones are schematically summarized in Fig. 10-1.

In natural conditions, iodine enters into the body with food and/or water in the form of inorganic iodide. After intestinal absorption the iodide is found in the blood plasma, from which it is taken up by the thyroid. In the follicles the iodide ions are oxidized (by a peroxidase) to atomic iodine, which is directly used to iodinate the amino acid tyrosine. The products of this reaction are mono-iodotyrosine and diiodotyrosine. What is known as thyroid hormone (or thyro-hormone) is actually a mixture of two different, but similar, chemical species with identical physiologic properties, which are thyroxin and triiodothyronine. The former is synthesized through the reaction between two molecules of diiodotyrosine; the latter results from the reaction of one molecule of mono-iodotyrosine and one of diiodotyrosine.

The two hormones are released into the general circulation immediately if their level in the bloodstream is low; otherwise they are stored within the colloid substance of the follicles after they are combined with a globulin, in the form of thyroglobulin and triiodothyronine globulin.

Radioiodine experiments show that a hyperactive gland (hyperthyroidism) can concentrate as much as 70% of an injected dose. The percentage is much lower than normal with hypofunctional glands (hypothyroidism).

The same experiments also show that the pituitary thyrotropic hormone (TSH) accelerates and stimulates the iodine uptake and the synthesis of the hormones, as well as their release into circulation. On the other hand, certain drugs have an inhibitory action on thyroid function, because they block one or another of the step reactions leading to the synthesis of the hormones. Thus the thyroid is prevented by potassium thiocyanate (KSCN) from taking up iodide from the bloodstream; whereas thiourea, thiouracil, and propylthiouracil inhibit the oxidation of iodide to atomic iodine. In both cases, a drastic depression in the thyroid utilization of radioiodine can be observed.

After injection of the radioisotope, the activity of the gland may be assayed with or without dissection of the animal. In the first case, the thyroid is removed and radioassayed by measuring either the intensity of its beta radiation (with a G-M or proportional counter) or the intensity of its gamma radiation (with a NaI(Tl) scintillation detector). The specific activity of the thyroid may also be compared with that of other organs.

Reliable data, however, may also be obtained by external counting. The following is a typical experiment that was carried out by use of this method:

Three dogs were used. Dog no. 1 received 10 USP units of TSH per day for 3 days; dog no. 2 was treated with 600 mg of propylthiouracil every day for 3 days; dog no. 3 did not receive any treatment and served as a control. After 3 days, the three animals were injected with 20 μCi of Na^{131}I solution, while an identical dose of the solution was kept in a test tube and used as a standard. The thyroid uptake was measured by external counting with a scintillation detector and a spectrometer. The scintillation probe was at a distance of a few inches from the neck of the animals (Fig. 10-2). Each animal was counted twice, once with, and once without, a B filter (consisting of a lead plate) interposed between the probe and the neck region. The counts collected without the filter represented the activity of the entire neck region; the counts collected with the filter

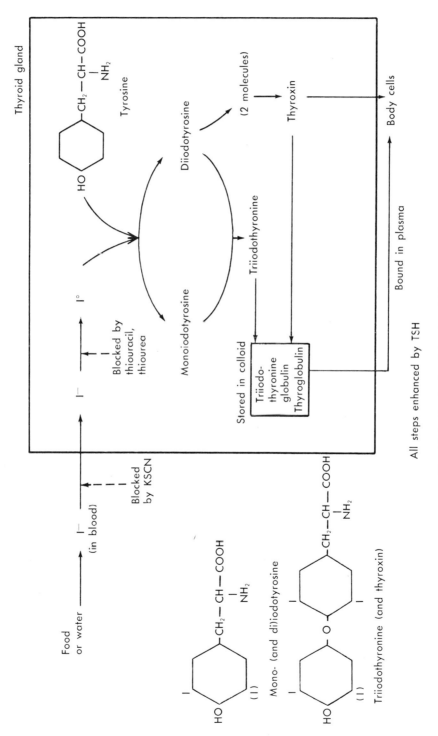

Fig. 10-1. Synthesis of the thyroid hormones.

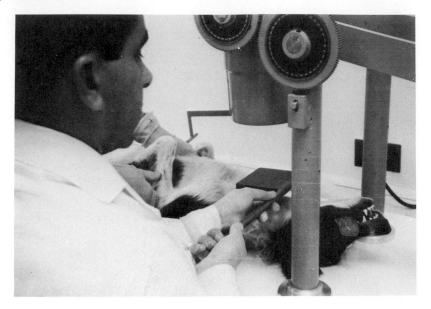

Fig. 10-2. Scintillation detector used in a radioiodine-uptake experiment.

represented extrathyroidal activity. The thyroidal activity was given therefore by the difference between the two counts.

The standard dose was also assayed in a plastic phantom, a device made of Lucite that reproduces some physical properties of the neck. The standard was also counted twice, once with and once without the filter.

The radioiodine uptake for each dog was calculated as follows:

$$\% \text{ uptake} = \frac{\text{Net cpm of thyroid}}{\text{Net cpm of standard}} \times 100$$

The pertinent data obtained are as follows:

	A	B	C	D
Dog 1	174,757	10,551	82,100	$\dfrac{82,100}{207,000} = 39\%$
Dog 2	17,937	5149	6400	$\dfrac{6400}{207,000} = 3\%$
Dog 3	126,070	8448	58,800	$\dfrac{58,800}{207,000} = 28\%$
Standard	439,167	25,209	207,000	———

A is the activity of the neck region (or standard) measured without the filter, in counts/2 min.

B is the activity of the neck region (or standard) measured with the filter (that is, extrathyroidal activity), in counts/2 min.

C is the thyroid (or standard) activity, in cpm, that is, $(A - B)/2$.

D is the percent uptake of radioiodine injected.

The results clearly show that the TSH enhances the radioiodine uptake and that propylthiouracil inhibits it. Other uses of radioiodine for thyroid studies in humans are mentioned in another section of this chapter.

Path of carbon in photosynthesis

Until 1940 nothing was known about the possible intermediate products formed in photosynthesis through the long chain of reactions leading from carbon dioxide, which is the raw material, to the carbohydrates. During the past century, various hypotheses had been suggested by some (Liebig, Baeyer) about the possible identity of the intermediate products; but none of them had any experimental evidence in its behalf.

The radiocarbon tracer method in photosynthetic studies was introduced for the first time by Ruben, Hassid, and Kamen in 1939. In the early experiments ^{11}C was used to label carbon dioxide. A few years later, after the discovery of ^{14}C, this radioisotope replaced the former one completely because of its much longer half-life and because it can be produced in large quantities.

In the late 1940s, work on the path of carbon in photosynthesis was continued by Calvin, Bassham, Benson, and others in the laboratories of the University of California at Berkeley. Their experiments showed that, besides carbohydrates, other organic compounds, like fats and amino acids, are the end products of the photosynthetic assimilation of carbon and that the only precursor common to all the final products is 3-phosphoglyceric acid (PGA), the first stable product of carbon assimilation.

The rationale of the tracer method as applied to photosynthesis is simple. A photosynthetic organism is allowed to abosrb $^{14}CO_2$ from the environment in the presence of light. Any radioactive organic compound extracted later from the organism must have been produced from radiocarbon dioxide. If the exposure to light is long, a large number of such compounds will be detected, and the longer the exposure the more complex they will be. Hopefully, if the exposure is kept very short, and the organism is immediately killed afterward, one should be able to stop the process of carbon assimilation at one of the early stages and thus isolate some of the early intermediate products.

Both unicellular algae (*Chlorella, Scenedesmus*) and leaves of higher plants have been used in the experiments. Algae are grown in large numbers in special culture vessels. An experiment begins by allowing radiocarbon (in the form of CO_2 or HCO^-) into a culture while a bright light is turned on. The exposure may vary from 5 to 30 seconds. At the end of the exposure period the algae are immediately killed by injecting into the culture vessel an appropriate amount of hot absolute ethanol so that after dilution with the culture water the alcohol concentration is reduced to 80%. The mixture is allowed to cool, and then it is centrifuged to remove all the insoluble materials. The supernatant is an alcoholic solution of several organic compounds extracted from the algal cells.

Leaves removed from a plant are placed in a chamber, where they are exposed to $^{14}CO_2$, which is liberated by the reaction of $Ba^{14}CO_3$ with an acid. The chamber is exposed to light for a predetermined period of time. After the excess carbon dioxide is flushed out of the chamber and fixed in barium hydroxide, the leaves are immediately immersed into boiling 80% ethanol, broken mechanically, and allowed to undergo extraction for 30 minutes. Leaf debris and other insoluble materials are removed from the alcoholic solution by centrifugation.

Lipids are removed from the alcoholic extract by mixing the latter with petroleum ether several times. After separation of the ether layers from the alcoholic layer by centrifugation, the ether layers are either discarded or saved for the analysis of the lipid components.

The alcoholic extract is concentrated to a very small volume by vacuum evaporation and then analyzed by two-dimensional paper chromatography. The chromatogram is developed along one dimension with a saturated aqueous solution of phenol and along

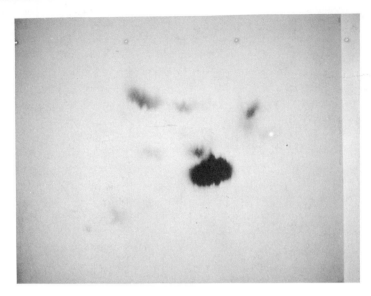

Fig. 10-3. Radiochromatogram from a photosynthesis experiment performed by students with bean leaves exposed to $^{14}CO_2$ and light for 10 minutes.

the other dimension with an aqueous solution of 1-butanol and propionic acid. When the chromatogram is dry, it is radioautographed by exposing it to a sheet of No-screen X-ray film. The black spots appearing on the radioautogram correspond to spots of radioactive organic compounds on the chromatogram (Fig. 10-3). With several methods, many radioactive compounds have been identified, such as amino acids, PGA, pyruvic acid, sucrose, and a variety of sugar phosphates, including some with 5 and 7 carbons (ribulose diphosphate and sedoheptulose phosphate). However, much fewer labeled compounds appear on the chromatogram if the exposure to light is reduced to 5 seconds, an indication that the process of carbon assimilation has been blocked at one of the early intermediate reactions. PGA is the predominant labeled compound even on chromatograms obtained in 5-second experiments. The fact that PGA is labeled only in its carboxyl group leads to the conclusion that it is synthesized by combining carbon dioxide with some other preexisting carbon compound. For more details, see the appropriate literature on the chemical mechanism of photosynthesis.

CLINICAL USES OF RADIONUCLIDES

The ever-increasing use of radionuclides for medical purposes has justified the creation of a new branch of the medical sciences, called "nuclear medicine."

Radium was the first radioactive material ever used for the treatment of certain malignant disorders, long before the first artificial radionuclides became available. It was encapsulated in needles or pellets that were applied on the skin, or introduced inside of body cavities. However, its cost has been, and still is, extremely prohibitive.

One of the first man-made radionuclides, phosphorus 32, was used for the therapy of some blood diseases as early as the 1930s. Since then, with production of an increasingly wider variety of radionuclides by reactors and accelerators, several radionuclides have been found suitable for use in the medical sciences, and consequently, during the past two decades, medicine has made exceptional advances uniquely dependent on radionuclides and undoubtedly will make many more in the future.

Radionuclides are used in medicine for the two following different purposes: (1) They are used for the diagnosis of certain disorders involving specific organs or the whole body. In diagnosis, with one or two exceptions, radionuclides are used as tracers according to the same rationale and principles illustrated on pp. 255 to 258. As radiotracers, they are introduced into the patient's body. The radiation they emit signals their presence, specific location, and concentration. (2) Radionuclides are used for the therapy of certain diseases of a neoplastic nature. As a therapeutic tool, a radionuclide is used essentially as a radiation source. If the therapy is properly planned, it is hoped that the ionizing radiation emitted will either destroy or depress the proliferative activity of malignant tissues, with little or no damage to the normal tissues. In therapy a radionuclide may be used either as an external or as an internal source.

When a radionuclide is ingested by the patient or injected into his bloodstream, it becomes an integral component of the body's chemical makeup, and its removal from the body by excretion is regulated exclusively by its chemical and physiologic properties and behavior. For how long it remains in the body, or how rapidly it is eliminated from the body is determined largely by its turnover rate or biologic half-life. Consequently, the total radiation dose delivered to the body and the hazard connected with it depend on the effective half-life of the radionuclide and on the type and energy of the radiation emitted. Any diagnostic technique involving the use of a radionuclide should be such as to reduce the radiation damage to the body to a minimum. Consequently, the amount of radioactivity introduced into the patient's body should be the minimum compatible with the sensitivity of the detecting instruments and with the duration of the study. If a diagnostic test is to be completed within a few hours, the use of radioisotopes of short physical half-life is preferable; for this reason, ^{124}I and ^{132}I are used in place of ^{131}I in some thyroid-uptake tests; ^{57}Co and ^{58}Co are preferable to ^{60}Co in procedures involving the use of radiocobalt-labeled vitamin B_{12} for the diagnosis of pernicious anemia. Radionuclides of long effective half-lives, like ^{90}Sr and ^{226}Ra, are never used as tracers for diagnostic purposes.

Any therapeutic technique involving the use of a radionuclide should be planned in such a way as to maximize the radiation damage to the abnormal tissues, while minimizing the damage to the surrounding normal tissues. Accordingly, when a radionuclide is injected into a patient's body for therapeutic purposes, the choice of the radionuclide and of the dose to be used is determined by this principle.

Unnecessary exposure of a patient to ionizing radiation is undesirable, and the benefits to the patient from a therapeutic or diagnostic procedure using radionuclides should be measured against both the potential and direct biologic effects of the irradiation. It should also be noted that, with a few exceptions, a clinical procedure involving the use of radioactive materials may be but one of several alternatives available to the physician in dealing with a specific clinical problem; radionuclides should be used for clinical purposes only when no other procedure can produce the same or better results. As an example, the radiotherapeutic treatment of a tumor should be considered only when surgery or chemotherapy is contraindicated.

If judiciously used by competent physicians specifically trained in radiation physics and nuclear medicine, radionuclides can be invaluable tools for the solution of several clinical problems. The number of radioisotope units or laboratories of nuclear medicine is increasing every day in many medical centers and medium-sized hospitals in the United States, along with opportunities for training of physicians and technologists and with societies and journals devoted to the progress of nuclear medicine. In the following sections, examples are given of how radionuclides are used in medical diagnosis and therapy.

Medical diagnosis

The chemical form of the radionuclides used in diagnosis may vary depending on the nature of the test. For certain studies they may be introduced into the patient's body in very simple inorganic forms such as NaI or NaCl when one desires that the radionuclide be diluted in the whole blood. In other tests, one needs to use radionuclides as labels for special inorganic or organic molecules (such as proteins, vitamins, and dyes) that are selectively absorbed and concentrated by the organ, tissues, or cells under study.

The radiotracers used in diagnosis must almost always be gamma emitters, because their detection, the determination of their location, and the measurement of their concentration are made with detectors placed outside of the body. Pure beta emitters would not emit radiation of sufficient penetrating power to be detected through several millimeters or centimeters of tissue. However, in certain blood tests, the activity of radiolabeled blood samples is measured in vitro. In this case, a pure beta emitter could be used, although gamma emitters are preferred because sample-preparation techniques for gamma counting are simpler.

The radioassay of tagged blood samples (blood cells, plasma, or whole blood) in vitro is done with a regular gamma scintillation counter, consisting of a scintillation detector with a NaI(Tl) crystal (of the well type), of a pulse-height analyzer, and of a scaler. The activity is usually measured in the window corresponding to the most intense photopeak of the radionuclide, so that the background count rate is reduced to a minimum.

When the activity of internal organs or tissues is measured, special scintillation probes are required. By means of properly designed collimation, these probes can detect only the radiation coming from very small areas, with the exclusion of the radiation possibly emitted by neighboring areas. Nevertheless, in certain situations accurate measurements can be made only with the use of special filters and phantoms (see experiment on thyroid-iodine uptake on pp. 274 to 276). This instrumentation is adequate when one is interested in the activity of an organ as a whole.

In certain diagnostic tests, a measurement of the overall activity of an organ that has absorbed the radiotracer is not sufficient. What is required is a knowledge of its differential distribution within the organ. This problem was solved for the first time in 1950 by B. Cassen with an instrument called the *scintiscanner* (Fig. 10-4). With this instrument, the area of the body overlying the organ of interest is scanned point by point with a narrowly collimated scintillation probe, which can "see" the radiation coming from areas as small as a few square millimeters. The probe is mounted on a motor-driven carriage that moves at a preselected speed. The probe moves back and forth along different lines over the area of the patient's body, thus scanning the radioactivity of the underlying or-

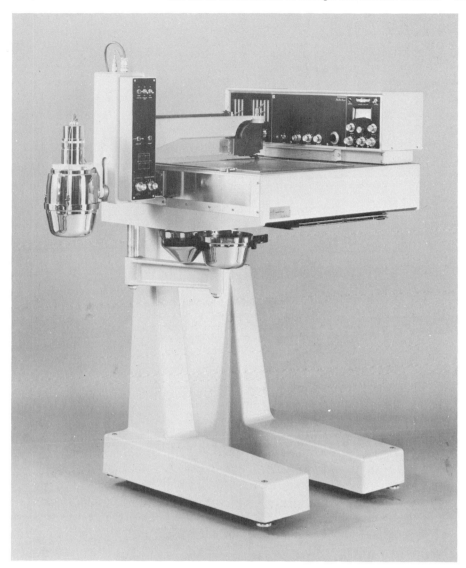

Fig. 10-4. Radioisotope scanner. This instrument is equipped with both a photorecording and a dot-recording system. The detector is supplied with a 3" × 2" NaI(Tl) crystal and three interchangeable collimators. (Courtesy Nuclear-Chicago, a subsidiary of G. D. Searle and Co., Des Plaines, Ill.)

gan. The probe is coupled mechanically with a readout device that records as a visual display the distribution of the organ's activity. The record is a map of the organ showing the differential distribution of the activity within the organ itself and is called a *scintiscan*, or *scintigram*.

The recording device is mounted on another carriage that is synchronized with the probe-moving carriage and therefore moves back and forth with it. Two methods of recording and mapping the activity of an organ are commonly employed—dot recording and photorecording. The dot recording employs a special electrosensitive paper called Teledeltos. This paper contains a black

Fig. 10-5. Dot scan of a thyroid gland. A cold nodule is visible in the upper left lobe, which appears on the right side of the scan.

material that is coated with a white chalklike substance. The recorder is essentially a stylus that transmits an electric impulse to the paper; when this happens, the stylus burns away the white coat and a black dot appears on the paper. Each impulse corresponds to a preset number of counts detected by the probe. On a *dot scan*, a variation in radioactivity distribution in the organ is represented by a variation in dot density (Fig. 10-5). The photorecording method employs a glow tube that moves above a photographic film and duplicates the motion of the detector over the patient. One type of photorecording device produces a flash of light from the glow tube for each detected count, whereas another type increases the brightness of the glow tube when the count rate increases. In either case a region of increased activity in the organ results in an increased film exposure. When the film is developed and fixed, a variation of radioactivity distribution is represented by a variation of darkening on the film. The scintiscan obtained with this method is more properly called a *photoscan* (Figs. 10-8 and 10-10).

Since the construction of the first scintiscanner, the scanning instrumentation has made considerable progress. Improvements have been directed toward higher degrees of automation and thus toward speedier operation, better resolution (which is obtained with improved collimation of the probe), contrast enhancement on the scintiscan, and easier and more intelligible visual display. One should note that scintiscanning techniques have the same purpose as radioautographic techniques. In photoscanning, the sophisticated instrumentation is necessitated by the fact that the photographic film cannot be placed in direct contact with the organ, as in radioautography. A photoscan can be considered as a "long-distance" radioautogram of an inaccessible organ.

The scintiscanning of an organ is often done as a substitute for radiography. In fact, in certain situations the normal X-ray radiographic techniques fail to supply a satisfactory image of the organ because of insufficient contrast. If the organ is allowed to concentrate a suitable radionuclide, a scintiscan can provide valuable information about its size, shape, and exact location.

Blood studies

In medical practice, one frequently needs to determine a patient's blood volume, such as before surgery or administration of a blood additive. The blood-volume determination is done very simply with the use of a radiolabeled compound utilizing the isotope dilution principle.

The principle is quite simple. Let us assume that we have a large mass of liquid (such as water), whose volume for some reason cannot be measured directly. A carefully measured volume (v) of a radionuclide solution is assayed for relative activity and then mixed with the unknown volume (V) of the liquid. After thorough mixing, an aliquot of the diluted solution exactly equal to v is withdrawn and its activity assayed again with the same instrument and in the identical geometric conditions as those used for the assay of the undiluted sample. If A_u is the activity of the undiluted sample and A_d the activity of the diluted sample, activities and volumes are related as follows:

$$\frac{A_d}{A_u} = \frac{v}{V}$$

and therefore

$$V = \frac{(A_u)\ (v)}{A_d}$$

For the determination of the blood volume, one method consists in the injection into the bloodstream of ^{131}I-labeled human-serum albumin (^{131}I-IHSA). About 10 ml of ^{131}I-IHSA solution of suitable concentration is radioassayed and then injected into the patient's antecubital vein. After one allows sufficient time for mixing (10 to 15 minutes), a 10 ml sample of blood is withdrawn and radioassayed. The degree to which the label has become diluted gives a measure of the blood volume.

With the same isotope-dilution principle, it is possible to measure the total red cell volume. Measurement may be done by tagging a measured volume of the patient's red cells. Red cells placed in a solution of sodium chromate ($Na_2Cr O_4$) bind this salt to their hemoglobin. If the chromate is labeled with ^{51}Cr (sodium radiochromate), the red cells will be radiolabeled. About 20 ml of blood is withdrawn from the patient and incubated, in the presence of an anticoagulant, with 75 μCi of sodium radiochromate for about 20 minutes at 37° C, or for 40 minutes at room temperature. During the incubation the mixture is agitated at regular intervals. After incubation, ascorbic acid is added to the mixture to reduce any excess of radiochromate that has not been absorbed by the red cells. The activity of the blood sample is measured, and then the sample itself is reinjected into the patient's body. After a few minutes, a new blood sample is taken and its activity measured. The total volume of red blood cells can be calculated with the use of the dilution principle.

If either total blood volume or total red cell volume is known, the total plasma volume can be calculated by measuring the hematocrit. The determination of total plasma volume can also be done directly with radiochromium trichloride, which binds itself almost exclusively to plasma proteins, rather than to red cells. For this procedure the tagging material is directly injected into the bloodstream.

Other radiotracers are used to measure the blood supply to extremities and to measure the rate at which certain substances are passed to the tissues. If a ^{24}NaCl solution is injected into an arm vein and a scintillation probe is placed against one of the feet, the recorded activity begins to rise as radiosodium reaches the foot and pervades its tissues. In a normal subject the activity rises evenly and begins to level off within 30 to 40 minutes. The increase is the result of interchange of radiosodium between blood and interstitial fluids. Equilibrium is established when the number of radiosodium atoms leaving the blood vessels is equal to the number returning to the circulation from the fluids. In many patients with a peripheral vascular disorder the equilibrium value is lower and is reached more gradually, showing an impairment in this supply system.

^{24}NaCl is also used to detect suspected constrictions of blood vessels (Fig. 10-6). After the injection of the tracer, the patterns of blood flow are compared simultaneously in the two legs with scintillation probes. This method is also valuable for measuring the improvement in vascular functions after therapeutic procedures. It can also help in deciding on the site of amputation if surgery becomes necessary.

The passage of blood through the heart may also be studied with the aid of ^{24}NaCl. Measurement is done by use of collimated scintillation probes placed at appropriate locations above different sections of the organ. The variation of the activity can also be recorded.

^{131}I-IHSA has been adapted for measuring quantitatively the cardiac output. An important advantage of this method is the fact that the albumin remains in circulation for comparatively long periods of time, as contrasted with freely diffusible inorganic sodium ions.

Several other radionuclides are used for other types of blood studies. ^{59}Fe (or ^{51}Cr) is valuable for red cell survival studies. The same radioisotope of iron

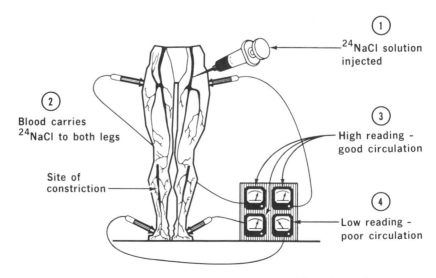

Fig. 10-6. Use of ^{24}NaCl to detect suspected constrictions of blood vessels. (Courtesy AEC.)

is used for the study of iron metabolism, erythropoiesis, plasma iron disappearance, red cell iron-turnover rate and for the diagnosis of some types of anemias.

Thyroid studies

Before radioisotopes of iodine became available in large quantities, techniques designed to test the thyroid function were based on the measurement of the basal metabolic rate (BMR). These techniques, however, were tedious, long, and uncomfortable for the patient, and their diagnostic accuracy was rather poor. Procedures utilizing radioiodine have yielded a greater accuracy. Furthermore, whereas the BMR method provides but one single measure, radioiodine methods can measure a number of parameters on which to base a diagnosis. Also, these procedures are more comfortable and convenient for the patient.

The simplest test for the study of thyroid function is the thyroid-uptake test; its purpose is to ascertain whether the gland is normal, hyperactive, or hypoactive in taking up iodine from the bloodstream and in synthesizing the hormone. The rationale and procedures used are practically identical to those discussed on p. 274 in connection with a typical experiment with dogs.

The radioisotope most commonly used is ^{131}I. Tracer doses of Na^{131}I in the range of 5 to 50 μCi are given orally to the patient, either in solution ("radioactive cocktail") or in capsules. An identical dose is set aside to be used as a standard. External radioassay of the thyroid is commonly done 24 hours later, although experience in some laboratories has shown that 6- and 8-hour uptake measurements can be equally valuable. Measurements of the activities are done with a gamma scintillation probe and with the use of appropriate filters and phantom necks.

For a normal thyroid the range of iodine uptake is about 10% to 40% in 24 hours. Uptakes significantly above or below this range are indicative of hyperthyroid or hypothyroid condition, respectively (Graves' disease, myxedema, etc.). As a double check, many investigators find it useful to combine an early measurement (after 1, 2, 4, and 8 hours) with a later measurement (24 hours) (Fig. 10-7).

Less simple studies of thyroid function involve the scintiscanning of the organ. This technique is used to estimate the exact location, size, or shape of

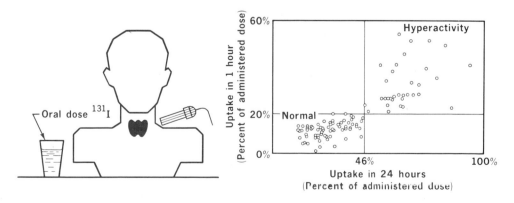

Fig. 10-7. Screening text for hyperthyroidism. (Courtesy AEC.)

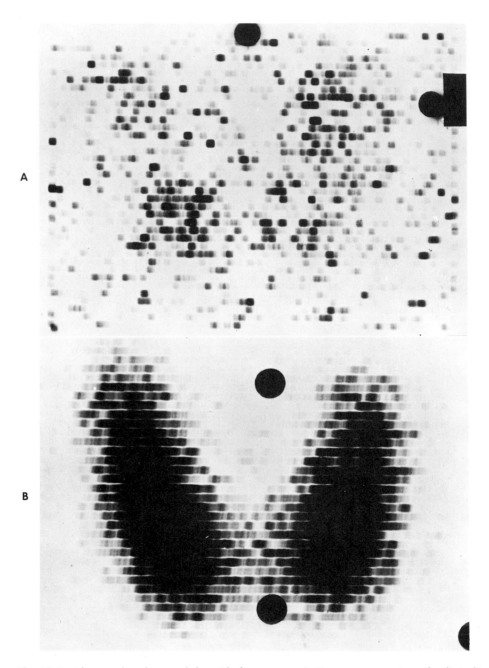

Fig. 10-8. Abnormal and normal thyroid photoscans. **A,** Poor concentration of radioiodine, associated with hypothyroidism. **B,** Same gland after the patient was placed on therapy; note the remarkable improvement in the scan picture. **C,** Photoscan of an enlarged multi-nodular thyroid. (From Early, P. J., M. A. Razzak, and D. B. Sodee. 1969. Textbook of nuclear medicine technology. The C. V. Mosby Co., St. Louis.)

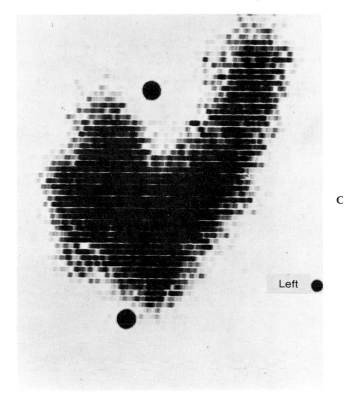

C

Left

Fig. 10-8, cont'd. For legend see opposite page.

the gland or to find out whether all its parts are equally functional or not. Scintiscans of an abnormal thyroid may reveal areas of greater or lesser radioactivity than the rest of the gland. Some of these areas are called nodules, and they can be felt by palpation. The scintiscan will reveal whether a nodule is "hot" (very active) or "cold" (inactive) (Figs. 10-5 and 10-8). Cold nodules may be either centers of neoplastic malignancy or adenomas, cysts or hemorrhagic areas. Hot nodules are almost always benign.

In certain cases, other radioisotopes of iodine are used for thyroid-scanning tests. ^{125}I has a half-life (60 days) longer than ^{131}I and therefore a longer shelf life. In addition, it emits a much less energetic gamma radiation, which in certain conditions may improve resolution and thus yield better photoscans. For short-term uptake studies (without scanning) ^{132}I may be preferable because of its very short half-life (2.3 hours) and its much more energetic gamma radiation. With this radioisotope the radiation dose to the whole body is reduced by a factor of 15. This reduction is particularly important when the patient is a pregnant woman or a child.

Another radionuclide is becoming increasingly popular for scanning tests of several organs, including the thyroid. This is technetium 99m, which is the daughter of molybdenum 99, a fission product. ^{99m}Tc (a half-life of 6 hours) decays to ^{99}Tc by isomeric transition, with the emission of low-energy gamma radiation (see decay scheme in Appendix III). ^{99}Tc, with a very long half-life, decays to the stable ^{99}Ru by pure beta emission.

Because of its extremely short half-life, ^{99m}Tc is not purchased as such but is produced in the same laboratory where it is used, from its parent radionuclide ^{99}Mo (a half-life of 67 hours). The device that generates ^{99m}Tc is known as a "radioisotope cow," and it consists essentially of an ion-exchange resin on which the parent nuclide is adsorbed. The resin is contained in a glass cylinder with its lower end tapering off in the form of a tube. As the parent nuclide decays, ^{99m}Tc is generated. If the daughter nuclide is left mixed with the remaining parent nuclide, a radioactive equilibrium whereby parent and daughter decay at the same rate will be reached. The daughter may be separated from the parent by chemical means, that is, by eluting the column with a suitable solvent ("milking"). After the cow has been milked, continued decay of the parent increases the amount of the daughter until an equilibrium condition is reached again. For best yields, the cow is milked once a day, every morning, to obtain the daily supply of ^{99m}Tc.

Similar cows are also used in laboratories of nuclear medicine to obtain other short-lived radionuclides from parents of longer half-lives. The generation of iodine 132 mentioned above from its parent tellurium 132 (a half-life of 77 hours) is another typical example.

Technetium, in the form of pertechnetate, is used by the body in a manner similar to iodine; the thyroid gland concentrates a large percentage of an injected dose in a matter of minutes. Scintiscanning of the thyroid with ^{99m}Tc has been very successful and rewarding. In many respects, this radionuclide presents several advantages over some of the radioiodine isotopes. For instance, its very short half-life and its complete lack of primary-particle radiation makes the radiation exposure of the patient exceedingly low (as low as 1/1,000 of that received from a ^{131}I scan).

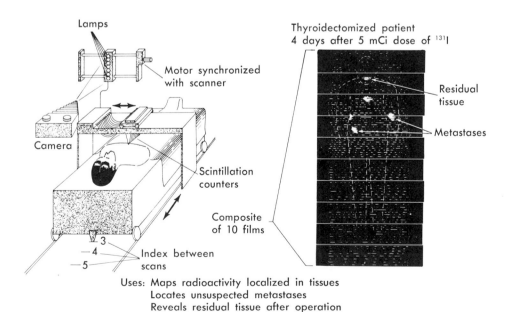

Fig. 10-9. Whole-body scanning for the location of malignant thyroid metastases. (Courtesy AEC.)

Iodine 131 has also been used successfully for the location in the body of malignant metastases of cancerous thyroid tissue. Cancers are known to inherit some of the physiologic characteristics of the normal tissues they originate from. Those of the thyroid gland are sometimes able to concentrate iodine and even to manufacture the hormone. When a thyroid cancer metastasizes to other parts of the body, the offshoots may continue to grow with thyroid characteristics. After injection of ^{131}I, a large proportion of these offshoots may show enough activity to be detected externally with the scintiscanning method. A whole-body scanner is used for this purpose (Fig. 10-9). To reduce the scanning time to a minimum, the scanner uses 8 or 10 collimated scintillation probes that scan a wide area of the patient's body at the same time. The scintiscan is often superimposed to, and matched with, a photographic picture of the patient's body to facilitate the location of the metastatic offshoots. Notice that the scan of Fig. 10-9 has been taken 4 days after the administration of 5 mCi of radioiodine. This technique has sometimes revealed the presence of metastases after the surgical removal of a cancerous thyroid.

Location of brain tumors

In the central nervous system of a normal adult a mechanism exists whereby certain metabolites present in the bloodstream are prevented from passing into the intracellular space of neurons and other elements of the nervous tissue. This mechanism, when in reference to the brain, is known as the "blood-brain barrier." However, the barrier breaks down in those areas of the brain affected by certain disorders, including malignant tumors. Such breakdown results in the incorporation, on the part of the neoplastic cells, of compounds that other-

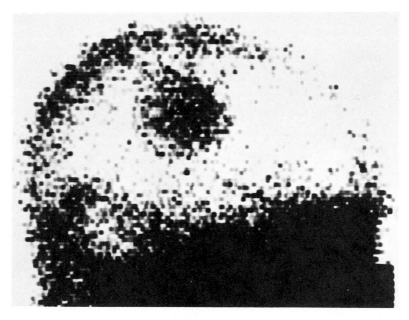

Fig. 10-10. Photoscan of an abnormal brain, showing a right parietal astrocytoma. The tracer used was technetium 99m. Right lateral view. (From Early, P. J., M. A. Razzak, and D. B. Sodee. 1969. Textbook of nuclear medicine technology. The C. V. Mosby Co., St. Louis.)

wise would remain within the bloodstream. If these substances are labeled with a radionuclide, they offer the possibility of localizing brain tumors with the scintiscanning technique.

^{131}I-IHSA is one tracer routinely used for the location of brain tumors. It seems that this albumin is incorporated by the malignant cells through a pinocytotic mechanism. After injection, the patient's head is scanned from several angles; this technique allows a more precise pinpointing of the location of the tumor and facilitates the work of the surgeon, if surgical removal of the tumor is indicated.

Technetium 99m is also profitably used for the scanning of brain tumors (Fig. 10-10).

Diagnosis of pernicious anemia

Pernicious anemia is prevented by vitamin B_{12}, which is ingested with the foodstuffs, and by the intrinsic factor of Castle, which is secreted by the gastric mucosa. Both are absorbed by the intestine and brought to the liver via hepatic portal circulation, where they are transformed into an antianemic factor. The intestinal absorption of the vitamin occurs only in the presence of the intrinsic factor. Failure of the gastric mucosa to produce the intrinsic factor results in the excretion of the vitamin with the feces and therefore in a low level of the vitamin in the blood and urine.

Vitamin B_{12} contains cobalt; therefore it can be labeled with radiocobalt (^{60}Co, or better ^{57}Co or ^{58}Co, which have a considerably shorter half-life). In the Schilling test, a 0.5 μCi dose of radiocobalt-labeled vitamin is administered orally to the patient; his urine is collected for 24 hours and radioassayed with a well scintillation counter along with a standard. A 24-hour urinary excretion of less than 7% of ingested dose is suggestive of pernicious anemia. If the same test is repeated after administration of the intrinsic factor, the urinary excretion of the labeled vitamin will increase in a patient affected by pernicious anemia, but not in patients affected by certain other disorders of intestinal absorption.

Liver function tests

For several years, a dye known as rose bengal has been used in testing liver functions. This procedure has recently been improved by labeling the dye with ^{131}I.

When rose bengal is injected into a vein, the liver removes it from the bloodstream and transfers it to the intestine with the bile to be excreted. The rate of disappearance of the dye from the bloodstream is therefore a measure of hepatic activity, blood flow in the liver, and patency of the biliary ducts. Changes in any of these functions cause a discernible change in the way the dye is taken up and/or excreted by the liver.

The tests with labeled rose bengal consist of injecting the dye intravenously and determining (1) clearance of the dye from the blood, (2) excretion of the dye into the intestine, and (3) uptake of the dye by the liver. The first two functions are assayed with scintillation probes immediately after the administration of the dye: one probe placed over the side of the head or thigh records the clearance of the dye from the bloodstream; a second probe is placed over the liver, and a third over the left side of the abdomen to record the passage of the dye into the intestine. The uptake of the dye by the liver is studied with the scanning of the organ. A hepatoscintiscan may reveal cancerous areas. Tests with labeled rose bengal may provide information on a variety of hepatic disorders, such as cirrhosis, obstructive jaundice, hepatitis, and ascites.

Neutron-activation analysis

Neutron-activation analysis (see p. 108) has recently been employed as a diagnostic technique for the quantitative determination of certain chemical elements present in the human body. One of the first facilities for this type of studies has been installed at the University of Washington's School of Medicine in Seattle. The procedure has been used on selected patients (affected by osteoporosis or other bone disorders) for the determination of their calcium content.

The patient stands on a turntable where he is exposed to a neutron source for a few seconds, once on his front surface and once on the back. The neutron source is a beryllium target that is struck by deuterons accelerated by a cyclotron. The stable ^{48}Ca isotope present in the bones is radioactivated to the short-lived, high-energy, gamma-emitting ^{49}Ca radioisotope. Immediately after the exposure, the patient is radioassayed with a multichannel pulse-height analyzer connected to a battery of gamma scintillation detectors.

It seems that in certain pathologic situations, neutron activation analysis is the only practical and nondestructive method to acquire information on changes of level of body calcium. On the other hand, some claim that the neutron radiation dose delivered to the patient is not higher than the dose delivered in several X-ray diagnostic procedures. The same method has also been used occasionally for the determination of the iodine content in the thyroid and for the quantitative analysis of sodium and potassium in muscles and of trace elements.

Gamma radiography

For quite some time the fact has been recognized that a truly portable radiographic X-ray unit would be very useful, because it could be carried by the doctor to a patient's bedside, to the scene of an accident, to a battlefield, or to locations where electric power is not available. Conventional X-ray equipment is too bulky and obviously not suitable to certain emergency applications.

Gamma-emitting radionuclides have solved this problem. A gamma emitter can be used as a source of penetrating radiation, and therefore it can make gamma radiograms. One such unit utilizes thulium 170 as a radiation source (with an 84 keV monochromatic gamma ray) and weighs only a few pounds. The thulium source is kept inside a lead shield; a shutter-release cable can be pressed to move it momentarily over an open port in the shielding. At a distance of 15 to 30 inches from the object, this device can produce an acceptable radiogram in a few seconds. Another radionuclide more recently used for the same purpose with success is iodine 125 (emitter of 35 keV monochromatic gamma radiation). The less-penetrating radiation is sufficient to radiograph arms and legs.

One advantage of these portable units over the conventional X-ray machines is that the gamma source can be made as small as 0.1 mm in diameter, virtually a point source, with the result of better sharpness of the image on the radiogram. The target of an X-ray tube cannot be made so small without the addition to the equipment of very efficient cooling systems.

In these gamma units only one or a few curies of activity are needed. If these devices are combined with rapid developing photographic films (similar to the sensitive papers used in Polaroid cameras), a physician is completely freed from dependence upon a film-processing laboratory. The film holder incorporates a self-contained processing system. A capsule contains film-developing chemicals that can be released and spread over the film without removing it from the holder. A finished picture is obtained in about 1 minute. Such facilities can be extremely valuable, for example, in an army field hospital, where sometimes neither X-ray equipment nor a darkroom is available.

Medical therapy

As already mentioned above, in therapy radionuclides are used as sources of beta or gamma radiation for the destruction of certain abnormal tissues or for the depression of the activity of certain normal tissues. Radionuclides can be used in several ways for these purposes. In teletherapy, large activities (hundreds or thousands of curies) of a gamma emitter are used as a sealed source in a suitable collimated device placed externally to the patient's body. In brachytherapy, a small sealed gamma source of much lower activity is placed in contact with the skin for the destruction of some superficial tumors, or is implanted within the body between organs or tissues (interstitial implants) or within body cavities (intracavitary insertion). In these cases, the radiation dose is terminated by simply removing the source.

In other procedures, radionuclides are introduced into the body as unsealed sources with activities of several millicuries. One type of intracavitary therapy consists of introducing the active material in colloidal form (radiocolloid) into a body cavity (pleural, peritoneal, etc.). The colloid state keeps the material confined within the space into which it has been introduced and prevents it from diffusing into the bloodstream or into surrounding organs or tissues. Or the active material is introduced into the bloodstream in a soluble chemical form, either directly by injection or via digestive system by ingestion. In this case, the radioactive material is geared with the metabolic processes of the organism, and after the body concentrates it in certain tissues, there it will serve as a source of localized irradiation. As in the case of many diagnostic procedures, two chief mechanisms account for the biochemical concentration of chemical elements in certain tissues—selective absorption of an element needed by the tissue for its specific function and differential turnover due to the increased utilization of an element in the more rapid metabolism of a particular tissue. When a radionuclide is used as an unsealed source, it obviously cannot be removed from the body artificially, and therefore the total radiation dose delivered to the body is exclusively determined by its physical characteristics, and in particular by its effective half-life.

The therapeutic methods utilizing radionuclides can be summarized as follows:

Therapeutic methods
    As external sources—sealed sources
        Teletherapy
        Superficial (brachytherapy)
    As internal sources
        Interstitial implants (brachytherapy)—sealed sources
        Intracavitary insertion
            Brachytherapy—sealed sources
            Radiocolloids—unsealed sources
        Introduction into bloodstream

Gamma-ray therapy is a direct outgrowth of X-ray therapy, with a number of advantages. Certain radionuclides have shown a superiority over X-ray equipment, because the gamma beam emitted is monoenergetic or at least consists of a few discrete energies. You may recall from Chapter 4 that a wide range of photon energies is present in an X-ray beam and much of the beam must be filtered out to eliminate the portion most harmful to the tissues lying above those that must be irradiated. To obtain a useful beam with radiation as penetrating as that from cobalt 60, very special and expensive X-ray equipment in the 2-million-volt range is required.

A cobalt-60 teletherapy unit consists basically of a small pellet of the radioactive metal placed in a superficial cavity of a lead ball (Fig. 10-11). By revolving the ball, the source can be moved in correspondence with a passageway or aperture from which the radiation beam can be directed toward the patient. The unit is "turned off" when the source is moved to a point on the opposite side of the aperture; in this position the source is surrounded by heavy shielding material on all sides, and no beam emerges from the aperture. This basic unit must be supported in such a way that the beam can be directed as desired. Further details of control are added, such as positioning controls, timing mechanisms, and even electronic computers to regulate the movement throughout a treatment.

Teletherapy units fall into two classes, depending on the source intensity and therefore the amount of shielding needed. The hectocurie units, containing only a few hundred curies of cobalt 60, can be light enough to be moved from place to place. The kilocurie units, equipped with sources of thousands of curies, are permanent installations.

In a rotational therapy unit the gamma source swings entirely about the pa-

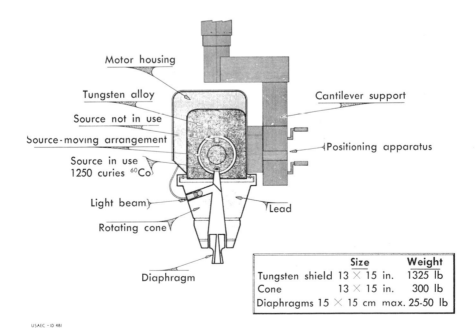

	Size	Weight
Tungsten shield	13 × 15 in.	1325 lb
Cone	13 × 15 in.	300 lb
Diaphragms	15 × 15 cm max.	25-50 lb

USAEC -ID 481

Fig. 10-11. Gamma-ray teletherapy unit. (Courtesy AEC.)

tient, with the gamma-ray beam continually focused onto a tumor. In this way a heavy dose can be delivered to a small volume without overdosing any one part of the skin or healthy tissue.

In some units, cesium 137 sources are used, whenever a gamma-ray beam of less penetrating power is satisfactory, as for the treatment of shallow tumors (recall that the gamma energy for ^{137}Cs is 662 keV versus 1.17 and 1.33 MeV for ^{60}Co). For these units less shielding material is needed than for cobalt-60 units. In addition, the much longer half-life of cesium 137 makes possible a less frequent changing of the source.

Phosphorus 32 was one of the first artificial radionuclides to be used as a topical therapeutic agent. Blotting paper and later plastic impregnated with the radionuclide have been employed to treat a variety of surface lesions (brachytherapy). A high dose of beta radiation can be delivered to the superficial layer with almost no radiation delivered to the underlying tissue. However, the short half-life is a decided disadvantage, as new applicators must be processed for each treatment. The technique does have definite advantages over soft X-rays in treating certain shallow widespread lesions.

Perhaps the greatest use of beta applicators is in the treatment of diseases of the eye, particularly of the cornea and epibulbar regions. As a result of the limited penetrating power of beta radiation, little of the dose is delivered to the interior of the structures. The sources are mounted in special applicators to allow an intimate contact with the diseased area. The most useful radionuclide for ophthalmic applicators is strontium 90 (Fig. 10-12).

The insertion of radium needles and radon seeds into cancer tissue has been a long-established method of treatment. More recently, cobalt-60 wire has been the material most used for this purpose, because it can be obtained with much higher specific activity and it is less expensive than radium. The wire is encased in an inert sheath, which absorbs the beta radiation and is unaffected

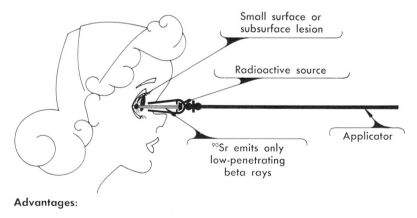

Small surface or subsurface lesion

Radioactive source

^{90}Sr emits only low-penetrating beta rays

Applicator

Advantages:

No extraneous gamma radiation

Removal of benign tumors without surgery

Readily adaptable to therapy of postoperative lesions

USAEC-ID-200

Fig. 10-12. Strontium-90 ophthalmic applicator for treating small lesions. (Courtesy AEC.)

by body chemicals. The encased wires are cut to the desired lengths and can be bent to a shape appropriate for the tissues to be treated.

Iridium 192 and tantalum 182 are used in the same manner, although they have much shorter half-lives than cobalt 60 does. The very short-lived gold 198 is also used in the form of wire or beads.

Cobalt-60 needles or rods of special configurations have been used to treat neoplasms such as carcinoma of the cervix.

Ceramic beads made of yttrium-90 oxide have been successfully used for the destruction of the pituitary gland. The implantation of the beads directly in the gland is an operation that presents less difficulty and less risk than the surgical removal of the organ and less damage to surrounding tissues.

Intracavitary injection of radiocolloids is used routinely for the palliative therapy of certain malignancies that result in the accumulation of large quantities of excess fluid (effusions) in the pleural or peritoneal cavity from their linings. At one time, frequent surgical drainage was the only useful remedy, and of course this was both uncomfortable and hazardous for the patient. Radiocolloids, being in liquid form, are well suited for placement in body cavities to irradiate their linings. Beta radiation can reach the tissues without being absorbed by the liquid; the effect is that of a thin radioactive coating. Moreover, the colloidal particles are rapidly deposited upon the surface of the lining and are engulfed by its cells. Colloidal gold 198 is commonly used for injection into the pleural and peritoneal cavities. More recently, colloidal ^{32}P chromic phosphate has been used for the same purpose. For this type of therapy, phosphorus 32 has the advantage over gold 198 of being a pure beta emitter; hence the total radiation dose delivered is confined to the area to be treated. It is agreed that about 50% of the patients with malignant effusions receive definite benefit from this treatment in the form of increased comfort, although there is no evidence of complete destruction of the neoplastic tissue. The general opinion is that this technique is indicated for patients in whom the neoplasm is growing slowly and the accumulation of fluid is the most significant symptom.

The selective absorption of iodine on the part of the thyroid gland is exploited for the therapeutic treatment of certain disorders of that gland with radioiodine (^{131}I). The use of this selective action to introduce a sufficiently destructive amount of radioiodine (several millicuries) into the gland is now considered a highly practical form of therapy for hyperthyroidism, including certain disorders, like Graves' disease and thyrotoxicosis, which do not respond to other types of treatment. Radioiodine therapy is especially useful when surgery would be too dangerous or when thyroid-inhibiting drugs are unsuitable. Attempts to treat thyroid carcinoma with the same technique have been only partially successful.

Radioiodine has also been used in the treatment of severe angina pectoris and congestive heart failure whenever these conditions do not respond to conventional medical therapy. By depressing the activity of the thyroid and consequently the basal metabolic rate, the work of the heart is reduced and the damaged blood vessels or muscles are more able to meet the circulatory requirements of the patient.

Although phosphorus is necessary for all living cells, nevertheless bone, hemopoietic tissues, and rapidly growing tissues tend to concentrate this element.

The importance of this effect in the treatment of certain diseases of the blood-forming system with phosphorus 32 is not definitely established, but the fact is that some of these diseases respond to total body irradiation, however it is administered. Polycythemia vera, consisting of the production of excessive numbers of red blood cells by the bone marrow, has been the most successfully treated disease of the hemopoietic system. The beta radiation from phosphorus 32 does not really cure the disease but reduces red cell formation nearly to normal levels, with considerable relief of the symptoms associated with the disease and with a significant increase of the patient's life expectancy. Radio-phosphorus has also been used to treat some forms of chronic leukemia, with beneficial results in many cases.

Finally, an interesting therapeutic method known as *neutron-capture therapy* has been under investigation in recent years at the Medical Division of the Brookhaven National Laboratory. The method is still in an experimental stage, but some of the results have been rather encouraging.

This therapy is used for the treatment of glioblastoma multiforme, an inoperable and extremely malignant form of brain tumor that is fatal, with an expectancy of only 1 year of life for the patient. The neoplasm projects roots into normal tissues so that its surgical removal becomes impossible.

If certain boron compounds are injected into the patient's bloodstream, they will be readily absorbed by the neoplastic tissue, because of the breakdown of the blood-brain barrier (see above), but not by the normal brain cells. Natural boron is made of two isotopes — ^{10}B and ^{11}B. The former has a high absorption cross section for slow neutrons. When ^{10}B is bombarded with neutrons, it is converted to ^{7}Li with the emission of alpha particles, as follows:

$$^{10}B \ (n, \alpha) \ ^{7}Li$$

In the neutron-capture therapy of glioblastoma multiforme, a boron compound is first injected into the patient's bloodstream. After a few minutes for allowance of boron absorption by the tumor cells, the patient's head is neutron irradiated for a few minutes with a beam emerging from a thermal column (see Chapter 8) of a research reactor. The therapeutic effect is attributed to the alpha particles generated in the nuclear reaction; with their very short range and high ionization density, they dissipate their energy entirely within the malignant tissue and hopefully destroy it, with practically no harm to the neighboring normal cells.

For a more successful and less hazardous therapy, attempts are being made to synthesize new boron compounds that have the greatest possible degree of selective absorption by the tumor cells. Since the technique has been used only in recent years on a limited number of patients, it is probably too early to establish whether or not a case of glioblastoma multiforme has ever been permanently cured.

REFERENCES

GENERAL

Blahd, W., F. Bauer, and B. Cassen. 1958. The practice of nuclear medicine. Charles C Thomas, Publisher, Springfield, Ill.

Broda, E. 1960. Radioactive isotopes in biochemistry. Elsevier Publishing Company, Amsterdam.

Chase, G. D., and J. I., Rabinowitz. 1967. Principles of radioisotope methodology. Burgess Publishing Co., Minneapolis.

Cloutier, R. J., C. L. Edwards, W. S. Snyder, and E. B. Anderson, editors. 1970. Medical radionuclides: radiation dose and effects. Symposium Series, no. 20. U. S. Atomic Energy Commission. Division of Technical Information. Springfield, Va.

Comar, C. L. 1955. Radioisotopes in biology and agriculture. McGraw-Hill Book Co., New York.

Deeley, T. J., and C. A. P. Wood, editors. 1967. Modern trends in radiotherapy. Butterworth & Co. (Publishers) Ltd., London.

Early, P. J., M. A. Razzak, and D. B. Sodee. 1969. Textbook of nuclear medicine technology. The C. V. Mosby Co., St. Louis.

Evans, E. A. 1969. A guide to tritium labeling services. Amersham/Searle Co., Des Plaines, Ill.

Feinendegen, L. E. 1967. Tritium-labeled molecules in biology and medicine. Academic Press, Inc., New York.

Fields, T., and L. Seeds, editors. 1961. Clinical use of radioisotopes. A manual of technique. The Year Book Medical Publishers, Inc., Chicago.

Francis, G. E., W. Mulligan, and A. Wormall. 1959. Isotopic tracers. The Athlone Press, London.

Hevesy, G. 1948. Radioactive indicators. John Wiley & Sons, Inc. (Interscience Publishers), New York.

Kamen, M. D. 1957. Isotopic tracers in biology. Academic Press, Inc., New York.

Kamen, M. D. 1964. A tracer experiment. Holt, Rinehart & Winston, Inc., New York.

King, E. R., and T. G. Mitchell. 1961. A manual of nuclear medicine. Charles C Thomas, Publisher, Springfield, Ill.

Owen, C. 1959. Diagnostic radioisotopes. Charles C Thomas, Publisher, Springfield, Ill.

Quimby, E. H., S. Feitelberg, and W. Gross. 1970. Radioactive nuclides in medicine and biology. Basic physics and instrumentation. Lea & Febiger, Philadelphia.

Raaen, V., G. Ropp, and H. Raaen. 1968. Carbon-14. McGraw-Hill Book Co., New York.

Silver, S. 1962. Radioactive isotopes in medicine and biology. Lea & Febiger, Philadelphia.

Vaeth, J. M., editor. 1968. Frontiers in radiation therapy and oncology. S. Karger AG, Basel.

Wagner, H. N. 1968. Principles of nuclear medicine. W. B. Saunders Co., Philadelphia.

Wang, C. H., and D. Willis. 1965. Radiotracer methodology in biological science. Prentice-Hall, Inc., Englewood Cliffs, N.J.

Wolf, G. 1964. Isotopes in biology. Academic Press, Inc., New York.

LITERATURE CITED

Bassham, J. A., and M. Calvin. 1957. The path of carbon in photosynthesis. Prentice-Hall, Inc., Englewood Cliffs, N. J.

Catch, J. R. 1961. Carbon-14 compounds. Butterworth & Co. (Publishers) Ltd., London.

CHAPTER 11

Action of ionizing radiation on molecules

GENERAL CONSIDERATIONS

One of the fundamental tenets in radiation biology is that any visible, radiation-induced injury to an organism at cellular or higher level always begins with a molecular damage. This statement means that such observable effects as chromosome breakage, destruction of entire cells, cancer, and radiation sickness are caused by radiation-induced changes of one or more types of important molecules present in the living protoplasm.

From a chemical point of view, living protoplasm is a complex system consisting of water (usually more than 50% of the whole protoplasmic mass) and a variety of small or large molecules with diverse biologic functions. Some of them are found in high concentrations, so that very high radiation doses are needed to destroy or inactivate a significant percentage of them. Others are fewer in number and play such an important role in the life of the protoplasm that even the inactivation of a few of them results in serious damage to certain vital activities of the cell. Molecules of this type are, for instance, nucleic acids, ATP, enzymes, proteins with special functions (such as myosin and hemoglobin), etc.

Radiation-induced chemical changes can be conveniently studied in nonliving systems, such as dry materials and simple or complex solutions. Although this study is primarily the concern of radiation chemistry, it has considerably contributed to a better understanding of the mechanisms responsible for the production of biologic damage.

A fraction of the radiation energy deposited in a chemical system is used to bring about chemical changes. The energy that is not so used is dissipated as heat, and its effect is to raise the temperature of the irradiated material. From a biologic point of view this energy is completely harmless because, as pointed out in a previous chapter, thousands of rads are needed to raise the temperature of the material by $1°C$.

A chemical change may be induced by radiation through ionization or excitation of molecules. When a molecule is ionized, with the loss of one or more electrons, it becomes highly unstable because its electronic configuration is unstable. Its fate is determined by a variety of circumstances; for instance, it may recapture the missing electron(s) or, more commonly, some of its chemical bonds will be broken, with the formation of fragments, some of which might still have an unstable electronic configuration. Free radicals are examples of these highly unstable and tremendously reactive molecular fragments. Free radicals have a very short lifetime (of the order of 10^{-11} second), because they quickly react chemically with other molecules that have not been "hit" directly by the ionizing particle or photon.

However, ionizing radiation may simply excite a molecule, that is, it may raise some of its electrons to higher energy levels. The energy needed to produce excitation is considerably less than the energy needed to ionize a molecule. Nevertheless, excited molecules are also in an unstable state. In certain cases, no change occurs in the excited molecule, because the electrons return to the ground energy level spontaneously; energy is thus given off in the form of light (luminescence). Otherwise the loss of the excess energy may result in the dissociation of the excited molecule, which obviously leads to chemical change, or the excited molecule may transfer its excess energy to another molecule by collision or some other mechanism. Excitation may thus be transmitted to another molecule.

It is not clear how extensively radiation-induced molecular excitation contributes to chemical changes and, in biologic systems, to visible damage. However, what is certain is that even nonionizing radiation like ultraviolet can induce chemical changes in certain systems (partial hydrolysis of sucrose, partial inactivation of certain enzymes, and alteration of DNA are well-known examples) and also produce irreparable damage to cells (for example, bacteria). If we consider that it takes the absorption of only 4 eV to break a C—C bond and that this is exactly the average photon energy of ordinary ultraviolet radiation, then the absorption of an ultraviolet photon by a molecule may possibly lead to an excitation whereby the excess energy involved is sufficient to break one of its chemical bonds. On the other hand, it is well known that for obtaining a molecular change of the same type and magnitude, the amount of ultraviolet radiation energy that must be absorbed by the system is at least a thousand times greater than the amount of ionizing radiation energy. Apparently, ionizations are much more effective than excitations in bringing about molecular changes.

It follows from the above considerations that the change of a molecule may be induced by ionizing radiation through direct or indirect action. The change takes place by direct action if the molecule becomes ionized or excited when it is directly hit by an ionizing particle or photon passing through or near it. Indirect action is observed when the molecule of interest does not absorb the radiation energy directly, but receives it through transfer from another molecule (usually of a different chemical species) or is acted upon by free radicals. Any chemical change observable in a pure system made of one chemical species (for example, a dry preparation of a protein in a purified form) evidently must be brought about exclusively by direct action. But in a system like a solution, wherein there are at least two chemical species (the solvent and the solute) in extremely different concentrations, a chemical change in the solute may be brought about by direct or indirect action. Because the solvent molecules are much more abundant, conceivably more solvent than solute molecules absorb radiation energy directly. Therefore more solute molecules are possibly changed by indirect action than by direct action. Since the protoplasm is essentially an aqueous solution, the problem of the determination of the relative contribution of the two types of action in inducing chemical changes in a solute is of considerable interest in radiation biology and is discussed below in greater detail.

The effectiveness of ionizing radiation in inducing chemical change can be appraised quantitatively, if the dose absorbed by the chemical system and the yield of the reaction product can be measured. In chemistry several methods are known for the determination of the yield of a chemical reaction. The most convenient way to express the yield of a radiation-induced chemical change is by means of the *G-value*, which is defined as the number of molecules changed for each 100 eV of radiation energy absorbed. If the reaction leads only to one chemical species (so that A always becomes B, and not B or C), the G-value is identical to the number of molecules produced for each 100 eV of energy absorbed. When the G-value is determined by measuring the number of molecules produced, one needs to ascertain that no other types of molecules are produced or may be produced as a result of the same reaction.

G-values are chiefly dependent on the nature and the conformation of the molecule that undergoes chemical change and also on the characteristics of the ionizing radiation (LET, photon energy, ionizing power, etc.). They may range from 0.01 or less to about 20. When a G-value is reported, one must specify whether it has been determined for the reaction product or for the molecule that has been changed. Determination is done by indicating the reaction product or the molecule altered (preceded by a minus sign) in parentheses as a subscript of the letter G. For instance, if a radiation-induced reaction consists of the oxidation of the ferrous ion to the ferric ion, the G-value may be expressed as

$$G_{(Fe^{3+})} \qquad \text{or} \qquad G_{(-Fe^{2+})}$$

It is useful to envision the whole action of ionizing radiation on a biologic system as proceeding through four consecutive stages. In the first stage (physical stage), the ionizing particle or photon produces along its track a number of ionized or excited molecules (activated molecules), which may be called "primary products." In the second stage (physicochemical stage), the extremely unstable primary products undergo secondary reactions leading mostly to the formation of highly reactive free radicals. In the third stage (chemical stage), free radicals react with one another or with surrounding molecules, thus producing new types of molecules that did not exist in the system before irradiation. Some of these newly formed molecules are quite stable, others are not and proceed to react with one another. After some time, a chemical equilibrium is attained in the system. In the fourth stage (biologic stage), the cellular protoplasm responds to the newly formed chemical products with some alterations that may be microscopically visible. In advanced organisms, the cellular injury may spread to higher organizational levels (tissues, organs, organ systems, or even to the whole organism).

The first three stages last a very small fraction of a second, whereas the biologic response is a slow process, which may last days or even years. A biologic effect may become visible after a long "latent period" (fourth stage). Why the death of an animal should follow as late as 30 days after the molecular damage has been done, or why leukemia should manifest itself several years afterwards, is a fascinating problem that can be solved only in part with the knowledge we have about the complex physiology of higher organsims. There are certain facets of this problem that are still elusive and difficult to explain.

An intriguing paradox is that although it takes doses of thousands of rads to inactivate in vitro significant percentages of such biologically important molecules as enzymes, proteins, and nucleic acids, even when they are in solution, only a few rads are sufficient to cause many visible biologic damages and a dose of only 500 rems may be fatal to many mammals. It would seem that the organism as a whole is much more radiosensitive than the molecules that enter into its composition. Perhaps, in an integrated organism the molecular damage is somehow modified and amplified by metabolic processes or by the different levels of organization, so that the biologic damage may seem out of proportion when compared with the initial molecular damage.

What is the nature of this "amplification" of the molecular damage by metabolic processes?

If an enzyme has been inactivated by radiation, through direct or indirect action, the damage will not be limited to the enzyme alone, as it happens when the enzyme is irradiated in vitro. The inactive or altered enzyme molecule will fail to catalyze its specific reaction, for example, the synthesis of a protein. The protein will be missing, and if it was designed to perform a certain particular function, its absence will interfere with the function. Note that if we assume that only one enzyme molecule located at a particular point in the cell was inactivated, not one but several molecules of the protein would be missing. Thus there is an amplification of the damage. Furthermore, if the suppressed function was causally correlated with other functions, they, too, will suffer damage of some type. The damage continues to be amplified, although it may take a long time (the latent period) before the end result of the whole process becomes detectable. Thus the injury may initially involve only a few enzyme molecules in a biologic system, and therefore it may take relatively small radiation doses to obtain biologic damage of considerable magnitude. Other examples, such as the initial radiation-induced formation of a "wrong" molecule, could equally well illustrate the same considerations. The very existence of problems of this type proves how much more there is to be learned about the obscure mechanisms responsible for biologic damage.

RADIOLYSIS OF WATER

Since water is the most abundant compound existing on earth and many chemical systems are aqueous solutions, it is natural that the attention of radiation chemists is particularly focused on the effects of ionizing radiation on water. Radiation biologists are also interested in the same problem, inasmuch as protoplasm is essentially an aqueous system. As pointed out above, molecules of biologic importance might be changed by radiation not directly but by their interaction with products derived from the direct action of radiation on the solvent molecules.

Much is known today about the chemical changes induced by radiation in water molecules. Indeed, this is one field of radiation chemistry in which our knowledge is most advanced. After the correction of initial errors and confusions, at present we are able to describe the process in considerable detail. The complex of chemical changes occurring in irradiated water is known as *radiolysis of water*.

As early as 1901, P. Curie noticed that free hydrogen and oxygen were evolved from aqueous solutions of radium salts, and this phenomenon was later attributed to the action of the radium alpha particles on the water molecules, which would result in their decomposition. However, also noticed by others was that the volumetric ratio of the two gases was not $2:1$ (as one would have expected from the equation $2H_2O \rightarrow 2H_2 + O_2$), but somewhat higher. The anomaly was explained when researchers discovered that irradiated water contains hydrogen peroxide; they thus surmised that some of the oxygen, instead of being released as a gas, would react with some water molecules and oxidize them.

Apart from the liberation of free hydrogen and oxygen, irradiated water appeared to be highly reactive, because it could easily induce chemical changes in certain compounds dissolved in it. It did not seem likely that the chemical reactivity could be attributed solely to its hydrogen peroxide content (which is known to be reactive). Irradiated water was referred to as "activated water."

The problem of defining the composition of activated water was intensively studied by H. Fricke, O. Risse, D. Lea, and others, using a variety of radiation sources of all types. They saw that X-irradiated air-free water does not show any development of hydrogen gas or formation of hydrogen peroxide, and yet it becomes "activated." We know now that ionizing radiation induces in water the formation of highly reactive chemical species, with the nature of ions and/or free radicals, as primary or secondary products; both primary and secondary products are involved in mutual interactions, or in reactions with other molecules of water or with molecules of solute that might be present in the system. The probabilities of some secondary reactions depend on a variety of factors, such as the specific ionization of the radiation used and the presence or absence of oxygen in the system.

A key role in irradiated water is played by two very reactive free radicals—the hydrogen radical and the hydroxyl radical. The fact has been firmly established that every water molecule acted upon directly by an ionizing particle or photon yields one hydrogen radical and one hydroxyl radical.

The G-value for the number of water molecules decomposed, $G_{(-H_2O)}$, has been accurately determined several times for oxygen-free water and found to range approximately between 3.9 and 4.5, depending on the radiation energy and LET.

The nature of the primary and secondary products and the possible mechanism of the radiolytic process are discussed below.

Free radicals

A free radical is a free (not combined) atom, molecule, or atomic group carrying an unpaired or odd electron. The "oddity" refers not only to the obvious fact that the total number of electrons in the radical is an odd number, but also to an odd spin. Orbital electrons not only revolve around atomic nuclei, but also spin around their own axes. The spin may be either clockwise or counterclockwise. In an atom or molecule with an even number of electrons, spins are paired; that is, for every electron spinning clockwise there is another spinning counterclockwise. This state of affairs is associated with a high degree of chemical stability, regardless of whether the atom or molecule is electrically neutral or charged (ionized). In an atom or molecule with an odd number of electrons there is one electron spinning in one direction for which there is no other electron spinning in the opposite direction; this is an unpaired electron. This situation is associated with a high degree of chemical reactivity, regardless of whether the atom or molecule is electrically neutral or charged.

To illustrate these concepts, let us consider the following example: The hydroxyl ion (OH^-), which is present in every aqueous solution, can be represented with the following electronic notation:

$$:\ddot{\underset{..}{O}}:H$$

The oxygen atom is surrounded by 2 electrons in the K-shell (not shown in the notation) and by 8 electrons in the L-shell. Two of the 8 electrons are shared with the hydrogen atom, thus forming a relatively strong covalent chemical bond. One of the two shared electrons belongs to the hydrogen atom. The whole system carries an even number of electrons (10); therefore the spins are paired and the system should be chemically stable, although it is a negative ion because it contains a total of 9 protons versus 10 electrons. In fact, our experience tells us that the hydroxyl ion is a very stable chemical species.

If the hydroxyl ion loses one of its L-electrons, its electronic configuration may be represented with the following notation:

$$:\ddot{\underset{..}{O}}\cdot H$$

This atomic group possesses a total number of 9 electrons, and therefore an unpaired electron is presented. It is a highly reactive free radical and is called a "hydroxyl radical." It happens to be electrically neutral because the total number of protons (9) is equal to the total number of electrons. Because of its high chemical reactivity, the lifetime of this free radical is very short: soon after its formation it will be involved in some chemical reaction and will disappear as such.[*] Other free radicals may have an ionic nature because the total number of electrons may be greater or lesser than the total number of protons. Examples are given in Table 11-1.

The short notation of an electrically neutral free radical carries a dot as a superscript of the chemical formula, for example, $OH^{\cdot}$. Ions that are also free radicals have no special short notation; their nature of free radicals can only be deduced from the electronic notation, for example:

$$H_2O^+ = H:\ddot{\underset{..}{O}}\cdot H$$

This ion is a free radical, which originates from a molecule of water that has lost one of its outer electrons.

For the student's convenience, all chemical species (neutral molecules, ions, and free radicals) that are present or formed in irradiated water are listed in Table 11-1.

The hydrogen radical is nothing but a free, electrically neutral atom of hydrogen. The extremely high reactivity of monoatomic hydrogen is well known. The H_2O^- radical is a water molecule that has captured an electron, whereas the hydronium ion is a molecule of water that has captured a proton (a hydrogen ion). The hydroperoxy ion and the hydroperoxy radical both originate from hydrogen peroxide through the loss of a proton, or a proton and an electron, respectively.

---

[*]The detection and analysis of free radicals in a system can be done with a technique called electron spin resonance spectroscopy; the description of the technique is outside the scope of this textbook.

Table 11-1. Chemical species present in irradiated water

Electronic configuration	Symbol or formula	Name	Total number of protons (p) and electrons (e)
H:Ö:H	H_2O	Water	10p + 10e
H	H^+	Hydrogen ion	1p + 0e
:Ö:H	OH^-	Hydroxyl ion	9p + 10e
H·	$H^{\bullet}$	Hydrogen radical	1p + 1e
:Ö·H	$OH^{\bullet}$	Hydroxyl radical	9p + 9e
H:Ö·H	H_2O^+	(Charged free radical)	10p + 9e
H:Ö:H·	H_2O^-	(Charged free radical)	10p + 11e
H:Ö:H H	H_3O^+	Hydronium ion	11p + 10e
H:Ö:Ö:H	H_2O_2	Hydrogen peroxide	18p + 18e
:Ö:Ö:H	HO_2^-	Hydroperoxy ion	17p + 18e
·Ö:Ö:H	$HO_2^{\bullet}$	Hydroperoxy radical	17p + 17e
H:H	H_2	Molecular hydrogen	2p + 2e
:Ö:Ö:	O_2	Molecular oxygen	16p + 16e

The mechanism

As pointed out above, the irradiation of pure water yields free hydrogen, oxygen, and hydrogen peroxide as stable molecular products. The actual yield for each product depends on the nature of the ionizing radiation used, on the experimental conditions present during irradiation, and on the degree of purity of water.

The final products are formed as a result of several possible types of reactions, in which the most important role is played by the highly reactive hydrogen and hydroxyl radicals. According to a generally accepted theory, these two free radicals originate not directly from the irradiated water molecule, but from a ion pair (H_2O^+ and H_2O^-).

The ionizing particle or photon ejects an electron from a water molecule, and thus a positive ion is formed:

$$H_2O \longrightarrow H_2O^+ + e^- \tag{11-1}$$

The electron released does not have enough energy to travel very far from the point of origin and ionize other water molecules. It is captured by another water molecule, which thus becomes a negative ion:

$$e^- + H_2O \longrightarrow H_2O^- \tag{11-2}$$

Both ions are also free radicals. The positive ion dissociates into a hydroxyl radical and a hydrogen ion:

$$H_2O^+ \longrightarrow OH^{\textbf{·}} + H^+ \tag{11-3}$$

Or it reacts with a molecule of water, with the formation of a hydroxyl radical and a hydronium ion:

$$H_2O^+ \xrightarrow{+H_2O} OH^{\textbf{·}} + H_3O^+ \tag{11-4}$$

The negative ion dissociates into a hydrogen radical and a hydroxyl ion:

$$H_2O^- \longrightarrow H^{\textbf{·}} + OH^- \tag{11-5}$$

According to this view the hydrogen radical originates from the initial negative ion and the hydroxyl radical from the initial positive ion.

According to another point of view, the two radicals may be formed by a different mechanism. The electron ejected in reaction 11-1 would lose its energy so rapidly that it could not escape from the electrostatic attraction field of the positive ion left behind. After traveling a distance of only 20 A, it would be attracted back to the positive ion. The result would be an excited (*) water molecule. The excitation energy would be used to dissociate the molecule into a hydrogen and a hydroxyl radical:

$$H_2O^+ + e^- \longrightarrow H_2O^* \longrightarrow H^{\textbf{·}} + OH^{\textbf{·}} \tag{11-6}$$

The occurrence of this alternate process leading to the formation of the two radicals has been questioned on several grounds; in the best of the hypotheses, it is thought to occur rather infrequently.

In pure water (absence of any solute), the hydroxyl radicals may combine in pairs, with the formation of hydrogen peroxide:

$$OH^{\textbf{·}} + OH^{\textbf{·}} \longrightarrow H_2O_2 \tag{11-7}$$

and the hydrogen radicals may combine in pairs with the formation of molecular hydrogen:

$$H^{\textbf{·}} + H^{\textbf{·}} \longrightarrow H_2 \tag{11-8}$$

The hydroxyl radical may also originate from the reaction of a hydrogen radical with a molecule of water. This reaction is another source of molecular hydrogen:

$$H^{\textbf{·}} + H_2O \longrightarrow OH^{\textbf{·}} + H_2 \tag{11-9}$$

Another possible reaction is the combination of a hydrogen radical with a hydroxyl radical, with formation of water:

$$H^{\textbf{·}} + OH^{\textbf{·}} \longrightarrow H_2O \tag{11-10}$$

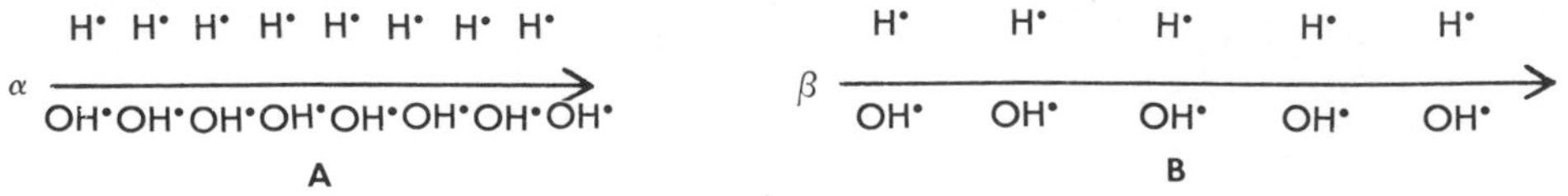

Fig. 11-1. Formation of hydrogen and hydroxyl radicals with radiation of high, **A**, and low, **B**, LET.

Reactions 11-7 and 11-8 are more probable than reaction 11-10 when ionizing particles of high LET (such as alpha particles) are used. In this case, because of the high specific ionization, the distances between hydrogen and hydroxyl radicals are greater than the distances between radicals of the same species (Fig. 11-1, A). Combination of radicals of the same species are therefore more probable. The opposite is true with radiations of low LET (Fig. 11-1, B). Reaction 11-10 is therefore more probable. Since reactions 11-7 and 11-8 yield molecular hydrogen and hydrogen peroxide, the theory explains why no appreciable amounts of these two molecular products can be detected when degassed water is irradiated with X-rays (low LET).

If the irradiated water is aerated (that is, contains free oxygen in solution), the yield of hydrogen peroxide is generally higher because of the formation of another free radical—the hydroperoxy radical. In the presence of free oxygen, the hydrogen radical reacts with it, thus forming the hydroperoxy radical:

$$H^\bullet + O_2 \longrightarrow HO_2^\bullet \tag{11-11}$$

Hydrogen peroxide can be formed directly from the hydroperoxy radical, when the latter reacts with another hydrogen radical:

$$HO_2^\bullet + H^\bullet \longrightarrow H_2O_2 \tag{11-12}$$

or through the intermediate formation of the hydroperoxy ion, if oxidizable ions or molecules (electron donors) are available in the system as solutes:

$$\begin{aligned} HO_2^\bullet + e^- &\longrightarrow HO_2^- \\ HO_2^- + H^+ &\longrightarrow H_2O_2 \end{aligned} \tag{11-13}$$

However, other processes are responsible for the destruction of the hydrogen peroxide produced by the preceding reactions. Both the hydrogen and the hydroxyl radical can destroy hydrogen peroxide:

$$H^\bullet + H_2O_2 \longrightarrow OH^\bullet + H_2O \tag{11-14}$$

and

$$OH^\bullet + H_2O_2 \longrightarrow HO_2^\bullet + H_2O \tag{11-15}$$

Notice that the two reactions add more hydroxyl and hydroperoxy radicals to the system. The amount of hydrogen peroxide appearing in irradiated water is therefore the net result of an equilibrium between two antagonistic types of reactions.

The hydroperoxy radical is also responsible for the formation of free oxygen through the following reactions:

$$HO_2^\bullet + HO_2^\bullet \longrightarrow H_2O_2 + O_2 \tag{11-16}$$

$$HO_2 + OH^\bullet \longrightarrow H_2O + O_2 \tag{11-17}$$

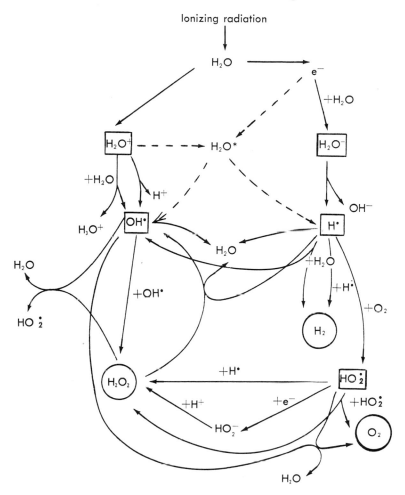

Fig. 11-2. Radiolysis of water. H_2O^* means an excited water molecule.

All the preceding reactions are schematically summarized in Fig. 11-2. From the analysis of the reactions one can infer that the mechanisms involved in the radiolysis of water, and the intermediate and final products deriving therefrom, are dependent mainly on the following two factors—the type of radiation used and the absence or presence of free oxygen in solution.

If inorganic or organic materials are dissolved in the irradiated water, some of the free radicals and final products will attack the solute molecules with reactions that depend on their degree of reactivity, nature of the solute, and pH of the solution.

The hydroxyl radical has a high electron affinity because by gaining an electron it becomes a hydroxyl ion, which is chemically stable. Consequently, it is a powerful oxidizing agent. For instance, it can oxidize the ferrous ion:

$$Fe^{++} + OH^{\bullet} \longrightarrow Fe^{+++} + OH^-$$

The hydrogen radical has a strong reducing power (even stronger than the oxidizing power of the hydroxyl radical), because if it loses its electron it becomes a hydrogen ion, which is more stable. Therefore, one would expect that the oxidations brought about by the hydroxyl radical in an irradiated solution should be "undone" by the hydrogen radical. However, the fact is that the net result of the irradiation of solutions is generally oxidation. Part of the explanation is found in reactions 11-7 and 11-8: some of the hydrogen radicals are eliminated by the formation of the inert molecular hydrogen (reaction 11-8), whereas the combination of hydroxyl radicals leads to a product that also has oxidizing properties (reaction 11-7). In the presence of oxygen, another oxidizing agent, the hydroperoxy radical, becomes available and thus also contributes to the overall oxidation of the solute.

Not much is known about the properties of the hydroperoxy radical in solution, except that it is a less powerful oxidizing agent than the hydroxyl radical. When it oxidizes a solute molecule, it becomes a hydroperoxy ion:

$$HO_2^{\cdot} + e^- \longrightarrow HO_2^-$$

The oxidizing power of one of the molecular products formed in the radiolysis of water, hydrogen peroxide, is well known in inorganic chemistry. This product also contributes significantly to the oxidation of the solute.

EFFECTS ON MOLECULES IN AQUEOUS SOLUTIONS

When the aqueous solution of a material is irradiated, chemical changes occur in the solvent molecules, through the formation of free radicals, as well as in the solute molecules. The latter may be changed either by direct action or indirectly when they are acted upon by the free radicals and other active products originating in the solvent. Hydrogen and hydroxyl radicals may attack organic molecules and convert them to free radicals. The hydroxyl radical may remove hydrogen from an organic molecule and change it into an active free radical:

$$RH + OH^{\cdot} \longrightarrow R^{\cdot} + H_2O$$

Free organic radicals of different species may react with one another and combine, thus giving origin to new chemical species.

Observations on simple solutions with one solute

When the solute molecules are large (enzymes, proteins, and nucleic acids), most of the radiation energy conceivably is absorbed by the solvent and chemical changes observed in the solute are mainly brought about via indirect action. Obviously, the reason is that the total number of solvent molecules is very large in relation to the number of solute molecules. It would be interesting to determine the relative contribution of the two mechanisms (direct and indirect) to the chemical alteration of the solute.

Let us assume that an enzyme solution is irradiated at different solute concentrations with a constant radiation dose and that the inactivation of the enzyme caused by radiation can be assayed in terms of number of molecules inactivated by the dose used. Let us plot (Fig. 11-3) *absolute number* of inactivated molecules as a function of enzyme concentration, to obtain the *concentration-response* relationship. A straight, horizontal line is an indication of indirect action, because in this case the number of enzyme molecules inactivated is exclusively dependent on the number of free radicals produced by radiolysis of water. This number should be constant for a constant absorbed dose. However, at very low enzyme

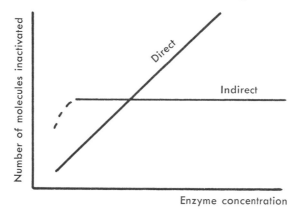

Fig. 11-3. Effect of enzyme concentration on the number of enzyme molecules inactivated.

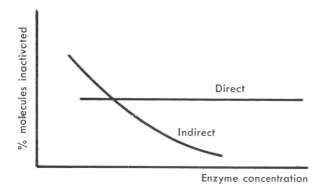

Fig. 11-4. Effect of enzyme concentration on the percentage of enzyme molecules inactivated.

concentrations, the absolute number of molecules inactivated decreases, because some of the water radicals react with one another before they encounter the very diluted solute molecules. A certain critical minimum concentration exists, therefore, for which all the free radicals available react with solute molecules. If the action on the solute is direct, the absolute number of enzyme molecules inactivated increases with concentration. In this case, the more molecules are present along the paths of the ionizing particles or photons, the more of them are "hit" and inactivated.

The concentration-response relationship may be represented graphically (Fig. 11-4) by plotting *the percentage of molecules* inactivated (rather than absolute number) as a function of solute concentration. If the action is indirect, the curve falls off as the concentration increases. If instead the action is direct, then the percentage of molecules inactivated remains constant with increasing concentration.

The behavior of the concentration-response curve can, therefore, be used as a test to determine whether the action of radiation on a solute is direct or indirect. Actually, it is unlikely that the chemical change in the solute will be brought about by one mechanism with the exclusion of the other. In most cases, both mechanisms are responsible for the change, and therefore the concentration-response curves are never exactly like those in Figs. 11-3 and 11-4, wherein we have hypothetically assumed the existence of one mechanism with the exclusion of the other. From the analysis of the actual curve obtained one should always be able to determine which of the two mechanisms prevails.

This test (often called the "dilution test") is decisive and very useful for in vitro systems. Unfortunately, it cannot be applied to biologic systems because, obviously, one cannot vary the concentration of the complex solutions that make up the protoplasm of a cell. However, certain primitive organisms and biologic structures (bacteria, spores, and seeds) can be dehydrated and hydrated to a certain extent, with little effect on their viability. It would seem that in these instances the dilution test could be used with some success in biologic systems.

Another useful test to determine the contribution of the indirect and direct mechanisms in the radiation-induced changes of a solute molecule is the freezing test. It consists of comparing the magnitude of the chemical change obtained when the system is irradiated in the frozen state with the change obtained when it is irradiated in the liquid state. If the magnitude is the same, it is an indication of direct action; if it is different, an indirect action must be mainly responsible for the change. The validity of the test is based on the fact that in a frozen solution the mobility of the water-free radicals is greatly hindered; consequently, the probability of their encounter with solute molecules is sharply reduced, and if the mechanism is indirect, the effectiveness of the radiation

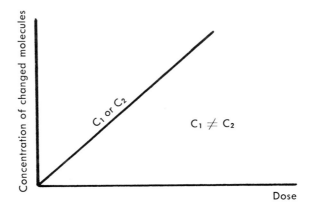

Fig. 11-5. Indirect action and dose-response curve. The curve is not affected by the solute concentration.

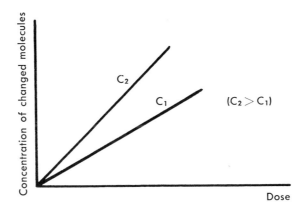

Fig. 11-6. Direct action and dose-response curves. The slope of the curves is affected by the solute concentration.

dose used should be lower. The freezing test can be applied to those biologic systems that can be frozen and thawed without any significant harmful effects (some bacterial species).

There is a wealth of evidence showing that enzymes in solutions are mainly inactivated by indirect action. For instance, the dose of ^{60}Co gamma rays needed to inactivate 37% of yeast invertase is about 6 megarads if the enzyme is irradiated in solution, and about 11 megarads if it is irradiated when dry. The same is true for many proteins.

The type of relationship that exists between the chemical change and the radiation dose (*dose-response relationship*) also depends on whether the action on the solute is direct or indirect. When the indirect action predominates, there is a linear relationship between dose and response, which is independent of the concentration, as long as the latter is above a critical value. If the concentration of changed molecules is plotted as a function of dose (Fig. 11-5), a straight line whose slope is dependent on the nature of the solute, on the characteristics of the radiation used, but not on the solute concentration, is obtained. This is a consequence of the fact (see above) that with an indirect action the number of molecules changed with a given dose is independent of concentration.

If the action is direct, the dose-response relationship is still represented graphically by a straight line; but its slope depends on the concentration used (Fig. 11-6). The slope increases with increasing concentration since in this case the number of molecules changed with a given dose increases with concentration.

Observations on solutions with two solutes

When a solution contains two solutes, one can sometimes observe a competition between the two solutes for the water-free radicals. This competition may result because although the two solutes may be present with the same molar concentrations one of them is more likely to be attacked by the water-free radicals than the other is.

A classical experiment was performed by W. M. Dale (1942). A protein (D-amino acid oxidase) and a dinucleotide (alloxazine adenine nucleotide) were present together in a solution with approximately equal molar concentrations (2×10^{-7} M). The reactivity of the dinucleotide with the water-free radicals is 2.5 times greater than the reactivity of the protein, so that with an equal number of collisions between radicals and solute molecules, more dinucleotide molecules than protein molecules should be changed. However, the molecular size of the protein is much larger than that of the dinucleotide; as a consequence, the probability of collision of the water-free radicals with the protein molecules is greater. In fact, one can show that more protein molecules than dinucleotide molecules are changed, in spite of the equal concentrations of the two solutes. The solutes are competing for the same radicals, but since the reactions take place preferentially between radicals and protein, the protein protects the dinucleotide. A similar type of protection is even more evident with other solutes, as one can see in Chapter 16.

Action of radiation on solutions of ferrous sulfate—chemical dosimetry

In 1927 Fricke and Morse reported that ferrous sulfate in aqueous solution is oxidized to ferric sulfate by ionizing radiation, and they suggested that this reaction may be used for dosimetric purposes.

It is believed that in aerated solutions the ferrous ion is oxidized through the following reactions:

$$Fe^{++} + OH^{\bullet} \longrightarrow Fe^{+++} + OH^{-}$$

$$Fe^{++} + HO_2^{\bullet} \longrightarrow Fe^{+++} + HO_2^{-}$$

$$Fe^{++} + H_2O_2 \longrightarrow Fe^{+++} + OH^{-} + OH^{\bullet}$$

One can see that the oxidizing agents are the hydroxyl radical, the hydroperoxy radical, and hydrogen peroxide.

If certain conditions concerning the composition of the solution are met, a ferrous sulfate solution may be used as a chemical dosimeter because the yield of the chemical change is linearly proportional to absorbed dose over a wide dose range and is independent of type of radiation, radiation energy, dose rate, and temperature.

The ferrous salt commonly used in the ferrous sulfate dosimeter (Fricke dosimeter) is actually ferrous ammonium sulfate, $Fe(NH_4)_2(SO_4)_2 \cdot 6H_2O$, with a concentration of about 10^{-3} M, although the concentration is not critical, since the yield of the ferric ion obtained does not seem to depend much on it. NaCl with a 10^{-3} M concentration is added to suppress the enhanced oxidation of the ferrous ion due to organic impurities that are invariably present in normal distilled water. For accurate results, it is advisable to use glass-bidistilled water in the preparation of the solution.

The solvent originally suggested by Fricke was 0.8 N sulfuric acid, and this solvent is used for most applications. However, some investigators point out that for X-rays of low energy (100 kVp or less) 0.1 N sulfuric acid presents several advantages.

The $G_{(Fe^{3+})}$-value has been measured many times for different types of radiation with ionometric and calorimetric methods by several investigators. For X- and gamma rays of different photon energies the G-value varies between 14 and 16. Some recently obtained values are shown in Table 11-2.

Several methods can be used to measure quantitatively the concentration of ferric ion present in the irradiated dosimetric solution. The most convenient and rapid method is the determination of the optical absorbance of the irradiated solution in the ultraviolet region with a spectrophotometer. The ferric ion shows a strong absorption in the ultraviolet region, whereas the absorption of the ferrous ion is practically negligible in the 200 to 350 nm range.

Fig. 11-7 shows the ultraviolet absorption spectra of ferric ammonium sulfate solutions (in 0.1 N sulfuric acid) at different concentrations, within the wavelength range of 200 to 315 nm. Also shown is the absorption spectrum of the ferrous ion in the unirradiated Fricke solution. The absorbance of the ferrous ion is negligible at least for the longer wavelengths. The absorbance of the ferric ion shows a minimum at 275 nm and two peaks—one at about 305 nm and a higher one at about 225 nm. It is customary to measure the absorbance of the ferric ion at 305 nm. The absorbancy* at this wavelength has been determined by many authors; a widely accepted value is 2196 at a temperature of 25° C with 0.8 N sulfuric acid used as solvent. When 0.1 N sufuric acid is used, the absorbancy is slightly higher (see specific values in Table 11-2). The absorbancy at 305 nm is quite temperature sensitive. Some authors (Scharf and Lee) have advised measuring the ferric ion concentration at the 225 nm peak

---

*Formerly "molar extinction coefficient." Absorbancy is, in effect, the absorbance of a 1 M solution, 1 cm thick. It is dependent on the nature of the solute, wavelength, and temperature.

Table 11-2 Constants for the calculation of dose with the Fricke dosimeter

Solvent	Density (ρ)	G-value			Absorbancy (ϵ)	
		110 kVp X-rays*	^{137}Cs gamma rays†	^{60}Co gamma rays‡	at 305 nm	at 225 nm
0.1 N SA	1.002	14.7	15.3	15.3	2187 (at 23.7° C) +0.6%/° C	4542 (at 25° C) +0.1%/° C
0.8 N SA	1.024	14.7	15.3	15.3	2174 (at 23.7° C) +0.7%/° C	4565 (at 25° C) +0.1%/° C

*With 0.65 mm Al filter. Mean energy = 33 keV.
†Energy = 662 keV.
‡Mean energy = 1.23 MeV.

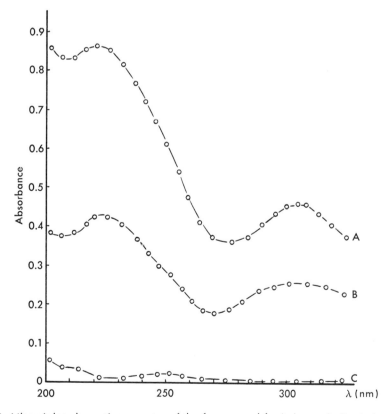

Fig. 11-7. Ultraviolet absorption spectra of the ferrous and ferric ions. **A,** Ferric ion solution with a concentration greater than 10^{-4} M. **B,** Ferric ion solution with a concentration equal to 10^{-4} M. **C,** Spectrum of the unirradiated Fricke dosimetric solution (ferrous ion). Reference is the solvent (0.1 N sulfuric acid). All spectra taken with the Beckman DU-2 spectrophotometer.

and have pointed out several advantages in favor of this procedure. The absorbancy at this wavelength is greater than the one at 305 nm (about double); this greater absorbancy means that one can accurately measure smaller concentrations of the ion, which would result in increased sensitivity of the dosimetric method. The absorbancy at 225 nm is much less dependent on temperature and also less dependent on the normality of the sulfuric acid used as solvent. However, in the opinion of many users of the Fricke dosimeter, the advantages are largely offset by certain disadvantages, and mainly by the well-known difficulty of accurately measuring ultraviolet absorption at short wavelengths, even with the best spectrophotometers.

For the determination of the absorbed radiation dose, a sample of the dosimetric solution is placed in the radiation field. Soon after irradiation, the absorbance of the sample is measured with a spectrophotometer at 305 nm (or 225 nm), with a sample of unirradiated solution as reference. The absorbed radiation dose is then calculated with an equation that is derived as follows:

If 1 cm cuvettes are used for the determination of the optical absorbance, then we have the equation*:

$$c_{Fe^{+++}} \text{ (moles/liter)} = \frac{A}{\epsilon}$$

c is the concentration, A is the absorbance, and ϵ is the absorbancy at the wavelength used.

$$c_{Fe^{+++}} \text{ (moles/ml)} = \frac{A}{10^3 \epsilon}$$

$$c_{Fe^{+++}} \text{ (moles/g)} = \frac{A}{10^3 \epsilon \rho}$$

ρ is the density of the solution.

$$\text{Number of } Fe^{+++} \text{ ions/g of solution} = \frac{A}{10^3 \epsilon \rho} \times \text{Avogadro's number} = \frac{A \times 6.023 \times 10^{20}}{\epsilon \rho}$$

Since G ions are formed per 100 eV absorbed, we have the equation:

$$E_{abs} \text{ (in eV/g)} = \frac{\dfrac{A \times 6.023 \times 10^{20}}{\epsilon \rho}}{G \times 10^{-2}} = \frac{A \times 6.023 \times 10^{20}}{\epsilon \rho G \times 10^{-2}}$$

$1 \text{ eV} = 1.602 \times 10^{-12}$ ergs. Therefore

$$E_{abs} \text{ (in ergs/g)} = \frac{A \times 6.023 \times 10^{20}}{\epsilon \rho G \times 10^{-2}} \times 1.602 \times 10^{-12}$$

Since 1 rad = 100 ergs/g, then

$$\text{Dose (in rads)} = \frac{A \times 6.023 \times 10^8 \times 1.602}{\epsilon \rho G \times 10^{-2} \times 10^2} = \frac{9.65 \times 10^8 \times A}{\epsilon \rho G}$$

The correction for temperature is made by dividing the dose obtained with the preceding equation by

$$1 + C(t_2 - t_1)$$

---

*Concentration of solute, absorbance, absorbancy, and thickness of solution are related by Beer's Law: $A = \epsilon c l$.

where C is the temperature coefficient, t_2 is the temperature at which A is measured, and t_1 is the temperature at which ϵ was determined. Therefore:

$$\text{Corrected dose} = \frac{9.65 \times 10^8 \times A}{\epsilon \rho G \; [1 + C(t_2 - t_1)]}$$

When the 305 nm peak is used for the determination of absorbance, the highest dose that an air-saturated dosimetric solution can register accurately is about 40,000 rads. This upper limit is set by the exhaustion of oxygen; beyond this limit the yield of ferric ion falls off and is no longer proportional to the dose. The maximum measurable dose is four times higher if oxygen, rather than air, is dissolved in the solution, or if the solution is continuously aerated during irradiation. The lower limit is the dose that produces sufficient oxidation to be accurately measured. With use of a 1 cm cuvette, this is about 4000 rads or less, depending on the reliability of the spectrophotometer in measuring absorbances below 0.100. Lower doses can be measured by using longer cells.

The Fricke dosimeter presents several advantages over other types of chemical dosimeters. Its response is independent of dose rate up to about 10^7 to 10^8 rads/sec. It is also independent of the ferrous ion concentration between 5×10^{-2} M and 10^{-4} M, and of sulfuric acid normality between 0.1 N and 1.5 N. Consequently, when the solution is made up these concentrations do not have to be accurately measured. The yield of the dosimeter is not significantly affected by the temperature of the solution during irradiation between $0° C$ and $65° C$.

The Fricke dosimeter shares, with free-air ionization chambers and calorimeters, the prerogative of being an absolute dosimeter (see Chapter 9), because it measures directly the radiation energy absorbed.

The ceric ion (Ce^{4+}) in aqueous acid solution is reduced to cerous ion (Ce^{3+}) by ionizing radiation. It is believed that the hydrogen radical, the hydroperoxy radical, and hydrogen peroxide are responsible for this reduction. This radiation-induced reaction meets several requirements necessary for chemical dosimetry for doses higher than the upper limits of usefulness of the Fricke dosimeter.

A solution of ceric sulfate (or ceric ammonium sulfate) in 0.8 N sulfuric acid may be used as a chemical dosimeter. However, it must be calibrated against a Fricke dosimeter because it is extremely sensitive to organic impurities. The useful dose range is between 4×10^4 and 10^8 rads. The ceric sulfate concentration used varies with the dose range for which the dosimeter is employed, from 0.001 N for the 0.10 to 0.30 megarad range and up to 0.05 N for the 6 to 15 megarad range. The upper dose limit is set by the solubility of ceric sulfate. The $G_{(Ce^{3+})}$-value for the most commonly used types of radiation varies within a very narrow range, that is, from 2.5 to 2.9. The solution is very sensitive to visible light and ultraviolet radiation; therefore it must be stored in the dark.

The concentration of cerous ion produced is conveniently measured spectrophotometrically. The ceric ion shows a maximum absorbance in the ultraviolet at 320 nm, with an absorbancy value of 5610, whereas the absorbancy of the cerous ion at the same wavelength is only 2.7. Unlike what happens with the Fricke dosimeter, in the ceric dosimeter the optical absorbance decreases with the radiation dose absorbed. The dose is calculated with an equation similar to that used for the Fricke dosimeter by using the appropriate G and ϵ values. The temperature coefficient is small, so that no correction for temperature is necessary.

The response of the ceric sulfate dosimeter is also independent of dose rate and of the presence or absence of oxygen. The usefulness of this dosimeter is justified by its capability of measuring very high doses, well above those that can be covered by the Fricke dosimeter, although it does present a few drawbacks, as evidenced from the brief description given above.

Other chemical dosimetric systems are also available, although none of them is as reliable as the ferrous and ceric dosimeters. Detailed descriptions may be found in more specialized literature.

Radiation scientists are interested in chemical dosimeters because they are inexpensive, easy to use, do not require any special instrumentation, and are capable of measuring total accumulated doses of thousands of rads.

For radiobiologic use, the ferrous and ceric dosimeters present at least the following four distinct advantages:

1. Their G-values have been accurately determined as function of LET and chemical characteristics of the solutions.

2. They can be used in containers that can mimic nearly any biologic object.

3. They are almost water equivalent, so that correction to tissue equivalence can easily be made.

4. They measure average absorbed doses directly.

The ceric dosimeter, in addition, shows a response that is almost independent of the type of radiation, so that it is particularly useful in the dosimetry of mixed neutron-gamma fields.

Unfortunately, in mammalian radiation biology in which doses lower than 1500 rads are commonly used, these dosimeters cannot be employed because of their low sensitivity. Various techniques have been tried to increase their sensitivity with different degrees of success.

ACTION OF IONIZING RADIATION ON BIOLOGIC MACROMOLECULES

The radiobiologist is particularly interested in the lesions produced by ionizing radiations in certain chemical compounds of high molecular weight because some of them play important roles in metabolic activities and in inheritance. The injury of these macromolecules is more likely to be the cause of biologic damage than the injury of smaller and more abundant molecules. The macromolecules of biologic importance are those of proteins, enzymes, nucleic acids, and certain polysaccharides (such as glycogen, starch, and cellulose). Radiation-induced damage of these molecules can be caused by both direct and indirect action. In certain instances, the two mechanisms may cause different types of damage to the same molecule.

Small alterations in biologic macromolecules can be easily detected because they are accompanied by changes in some of their physical or physicochemical

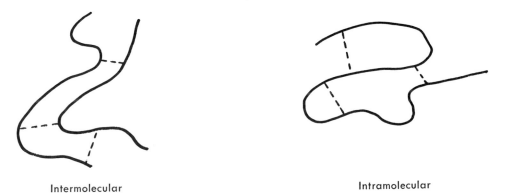

Intermolecular   Intramolecular

Fig. 11-8. Intermolecular and intramolecular cross-linking.

properties, which can be readily measured with appropriate methods and instrumentation. In fact, the appraisal of the radiation-induced damage is done by observing the alterations of these properties. The properties that are most frequently altered are solubility, viscosity, ultraviolet or visible absorption spectra, molecular weight, sedimentation rate, and electrophoretic behavior. Sometimes, the careful analysis of the alteration of these properties may lead to the discovery of the molecular lesion or injury; other times it does not, because the relationship between the property and the molecular composition or configuration is obscure. Most of the uncertainty still associated with this type of radiobiologic research finds an explanation in this lack of knowledge.

Nonetheless, from an extensive study of the radiation-induced changes in the physicochemical properties of biologic polymers, one can deduce that ionizing radiation can produce the following general types of alterations in their molecules:

1. Degradation, which is the result of the disruption of the chain(s) of basic units, may be indicated by a reduced molecular weight.

2. Simple alteration of secondary, tertiary, or quaternary structure occurs through disruption of secondary bonds (hydrogen bonds, disulfide bridges, etc.), but the main chains of basic units remain intact.

3. Intermolecular or intramolecular cross-linking is an effect caused by the creation of secondary bonds between two molecules, (intermolecular cross-linking), or between two points of the same molecule, which were not linked before irradiation (intramolecular cross-linking) (Fig. 11-8). Cross-linking may result in reduction of solubility, reduction of viscosity, and gelification.

The specific effects of ionizing radiations on the most important biologic polymers are briefly described below.

Effects on proteins

The direct and indirect mechanisms may be responsible for the alteration of different physicochemical properties in proteins or for the alteration of the same properties but in different degrees, as evidenced by the fact that if a protein is irradiated in a dry state the observable effects may be different from those induced by the irradiation of the same protein in solution. Furthermore, G-values are higher when the direct mechanism is prevalently involved; direct action is thus much more efficient than indirect action. This fact is probably due to the high molecular weight and therefore to the large size of the protein molecules.

Often denaturation is a readily detectable effect. If a fresh, shelled chicken egg is irradiated with 600,000 rads, the thicker component of the egg white (a solution of albumin) disappears. In many instances, denaturation is thought to be caused by the disruption of secondary bonds (disulfide or hydrogen bonds) with consequent alterations of the steric configuration of the molecule. Alterations in the primary structure are not necessarily involved.

Extensive studies on bovine serum albumin showed that irradiation of this protein results in an increase of the sedimentation constant, but not in alteration of the molecular weight. Therefore no degradation occurs. The fact that the irradiated serum albumin is more reactive is attributed to the radiation-induced opening or unfolding of the polypeptide chain(s), with the exposure of disulfide

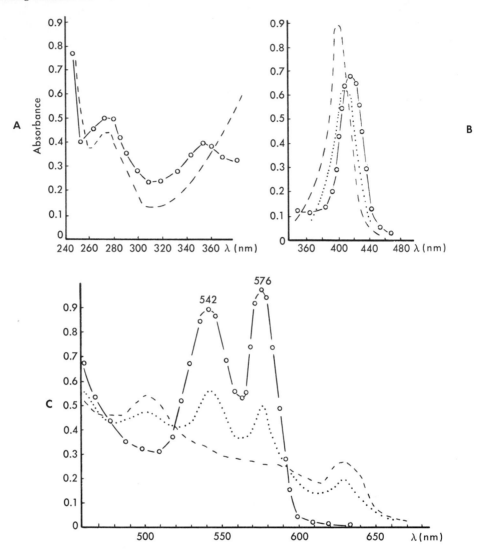

Fig. 11-9. Absorption spectra of unirradiated *(circles)* and irradiated *(dots)* oxyhemoglobin, and methemoglobin *(dashes)* in solution. **A,** Ultraviolet, **B,** Soret band. **C,** Visible. *Concentrations in **A** are one third and in **B** one ninth of the concentrations used in **C**. Concentrations of oxyhemoglobin and methemoglobin for the same region are identical.* Ultraviolet spectra taken with a Beckman DB spectrophotometer; spectra in the Soret and visible bands taken with a Cary-15 recording spectrophotometer. (From V. Arena. Unpublished observations.)

bridges, which thus become more vulnerable to the attack by oxidizing and reducing agents.

The effects of radiation on oxyhemoglobin (HbO_2) in vitro can be conveniently studied spectrophotometrically, that is, by observing the alterations produced in the ultraviolet and visible absorption spectra of this molecule. The first studies on X-irradiation of HbO_2 solutions in vitro were published by Fricke and Petersen in 1927. Using the primitive photometers available at that time, they noticed the conversion of HbO_2 to methemoglobin. (MetHb). This form of hemoglobin contains trivalent iron, has no ability to bind O_2 loosely, and therefore cannot be used as respiratory pigment in the body.

Fig. 11-9 shows the absorption spectra of HbO_2 and MetHb in the ultraviolet and visible regions. The absorption in the far violet is shown separately because the absorbancy at these wavelengths is much higher than in other regions. The strong absorption band in the far violet is known as the Soret band and is caused by the heme; in fact, the absorption spectra of free hemes and of their several derivatives, after separation from the globin moiety, still show the same band.

In the visible region of the spectrum (Fig. 11-9, *C*) HbO_2 shows two narrow absorption peaks at 576 and 542 nm, whereas MetHb shows two broad absorption maxima at 630 and 500 nm. The Soret band (Fig. 11-9, *B*) of HbO_2 has a peak absorption at 412 nm and is less intense than the Soret band of MetHb, whose maximum absorption is shifted to 405 nm (at least in slightly acid solutions). The ultraviolet absorption (Fig. 11-9, *A*) shows a peak at 276 nm; this peak is characteristic of practically all proteins and is attributed to the presence of aromatic amino acids (phenylalanine, tyrosine, tryptophan). However, the ultraviolet absorption of HbO_2 is more intense than that of MetHb, at least in the 250 to 350 nm range. Apparently the ultraviolet absorption of the globin moiety is somehow affected when the ferrous ion is oxidized to ferric ion in the $HbO_2 \rightarrow$ MetHb conversion.

The absorption spectra of an X-irradiated solution of HbO_2 in the visible and Soret regions are also shown in Fig. 11-9. The shape of the spectrum in the visible region indicates that irradiation has caused a partial transformation of HbO_2 to MetHb. A significant transformation may be observed with doses as low as 20,000 to 30,000 rads. With higher doses, the spectrum of the irradiated solution would approach even more closely the spectrum of a pure MetHb solution.

The Soret band of the irradiated solution shows a shift of the absorption peak toward shorter wavelengths, another indication of MetHb formation; but the lower intensity of the band seems to suggest that other molecular changes have been caused by radiation, such as a partial disruption of the porphyrin ring of the heme (as proposed by Barron and Johnson).

How X-irradiation affects the ultraviolet absorption of hemoglobin is controversial. Some investigators have observed a decrease of absorption (Barron and Johnson); others have observed an increase, which on the other hand, has been noted also for human serum albumin. As Fricke and Petersen had already suspected, it seems that X-irradiation of HbO_2 alters simultaneously the state of iron, the porphyrin ring, and the globin moiety.

The results of some irradiation experiments have elucidated the nature of some chemical changes produced by radiation in proteins. A release of ammonia and hydrogen sulfide can be observed when solutions of free amino acids are irradiated. Ammonia originates from deamination, that is, from the destruction of the amino groups, and hydrogen sulfide originates from the attack of the sulfhydryl radical of cisteine. There is evidence also that the imidazole ring of histidine is easily attacked. Release of ammonia and hydrogen sulfide has been observed also upon irradiation of native proteins. Apparently the two gases originate from the free amino and sulfhydryl groups that are found along the polypeptide chains. Indeed, these two radicals are considered to be the most radiosensitive sites in amino acids and proteins.

An attack on the peptide linkage and its disruption has been observed in some proteins (gelatin, ribonuclease) irradiated in solution, in the presence of oxygen. The disruption results in the formation of amide and carbonyl groups. The reaction proceeds as follows:

$$-CO-NH-\underset{\underset{R}{|}}{\overset{|}{C}}H \xrightarrow{+O_2+H_2O} -CO-NH_2 + O=\underset{\underset{R}{|}}{\overset{|}{C}} + H_2O_2$$

$$\text{(Amide)} \qquad \text{(Carbonyl)}$$

The resulting products are quite different from those obtained by hydrolysis of the peptide bond.

Effects on enzymes

The most important biochemical property of an enzyme is its catalytic activity, which can always be more or less easily assayed quantitatively when the enzyme is placed in presence of its specific substrate in appropriate conditions (of temperature, pH, etc.). The fact has been known for a long time that this enzymatic property is affected by ionizing radiation. Enzymes may be inactivated in solution or in the dry state. Since inactivation always involves some type of molecular damage to the enzyme, its quantitative appraisal is a means of assessment of the damage.

The results of early experiments of radiation-induced inactivation of enzymes have little value, because crude enzyme preparations were used; later the observation was made that impurities may have a considerable protective effect on the enzyme molecule. More reliable data are obtained if solutions of purified enzymes are used.

The percent of enzyme inactivated in solution decreases with the increase of concentration, while the G-value remains constant—an indication that in solution the indirect mechanism predominates (Figs. 11-3 and 11-4). The dose required to inactivate 63% of a pepsin solution with a concentration of 10 mg/ml is about 2.5 million rads, whereas to obtain the same percent inactivation of a solution with a concentration of 0.1 mg/ml, only about 40,000 rads are needed.

The freezing test (see p. 310) also indicates that in solution enzymes are mainly inactivated by indirect action. In fact, a frozen solution of pepsin (10 mg/ml concentration) shows only a 5% to 10% inactivation with 2.5 million rads. This percentage rises sharply around 0° C. However, as it happens for all proteins, the direct action is more efficient than the indirect action.

For some enzymes, inactivation is not affected by the presence or absence of oxygen, a possible indication that the hydroxyl radicals, rather than the hydroperoxy radicals and hydrogen peroxide, may be responsible for the molecular damage.

Inactivation can be partially or completely prevented, if the enzyme is irradiated in the presence of its substrate. In accordance with the modern views concerning the nature of the contact established between enzyme and substrate, this finding could be logically interpreted as protection of the active site of the enzyme by the substrate molecule. However, the results of some recent experiments do not seem to confirm this interpretation. A more general "radical scavenger" mechanism would account for the protective effect (Winstead and Gass).

Several years ago, it was commonly thought that enzymes containing free sulfhydryl groups in their molecule were considerably more radiosensitive than other enzymes. Examples of sulfhydryl enzymes are succinic dehydrogenase, yeast hexokinase, ATP-ase. Nonsulfhydryl enzymes are carboxypeptidase, catalase, RNA-ase, cytochrome oxidase, lactic dehydrogenase, and trypsin. Barron and others in a number of papers claimed that the sulfhydryl enzymes could be inactivated with G-values as high as 3 (versus values of 0.05 to 0.3 for nonsulfhydryl enzymes), that the G-value would increase considerably in presence of oxygen, and that the enzymes could be reactivated, after irradiation, with sulfhydryl compounds. In their opinion, radiation was responsible for the destruction of the sulfhydryl group, on which the catalytic activity depended. Several investigators have attempted to replicate Barron's results with no success. They have been unable to obtain the same high G-values claimed by Barron with sulfhydryl enzymes. At the present, it is generally believed that sulfhydryl enzymes are no more radiosensitive than other enzymes (Winstead).

Of certain interest (although difficult to explain) is the fact that very low doses (100 rads or less) may enhance the catalytic activity of papain (a proteolytic enzyme); larger doses, however, will inactivate the same enzyme. The same effect has been reported for actomyosin; 2.5 krads may enhance the enzymatic activity of this protein, whereas higher doses will inhibit it. This occurrence might be a case of a radiation-induced creation of an active site that did not exist before, or a case of increased reactivity of a previously existing active site. Obviously, more satisfactory answers can be obtained only when the configuration and the active sites of these enzymes are known in detail.

For the correct interpretation of some results of radiation-induced inactivation of enzymes, one should bear in mind that the integrity of the whole molecule of the enzyme is not an essential condition for the preservation of the catalytic activity. Certain points of the molecule may be drastically altered without a reduction in enzymatic activity, because they are "inert" in terms of activity, whereas other points (active sites) are so critical that even a slight alteration of their steric configuration, not necessarily accompanied by a chemical change, is sufficient to make them inactive and incapable of combining with the substrate molecule. Ionizing radiation energy may be deposited at random in different points of the large molecule of an enzyme. Because of this fact two molecules of the same enzyme may absorb the same number of electron volts of energy, and yet one may be inactivated and the other may not.

Effects on nucleic acids

Most of the investigators' attention has been focused on the damage induced by radiation on DNA. This damage has been studied with various techniques on DNA irradiated in vitro (in solution or in the dry state) or in biologic systems. Several types of damage have been observed.

Irradiated solutions of DNA clearly show a decrease of viscosity that is linearly proportional to dose. Viscosity is a physical property that can be most readily assayed quantitatively in these solutions. The decrease of viscosity is believed to be caused by either of the following two types of alterations:

1. The long double helix of the unirradiated DNA molecule may be straight or broadly coiled, but in either case it shows a certain degree of "stiffness." Radiation reduces this stiffness, and consequently the molecule is free to take up a more compact configuration by further folding up around itself (Fig. 11-10). Thus it occupies a smaller volume, and the solution shows lower viscosity.

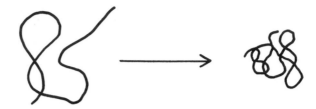

Fig. 11-10. The folding-up of a stiff long molecule. The length remains the same.

2. The decrease of viscosity may also be accounted for by a process of molecular degradation that would occur if both strands of the helix are broken less than 5 nucleotides apart *(double-strand break)*. The result is a fragmentation of the helix into smaller pieces (main-chain scission). The shorter fragments would show lower viscosity in solution. It seems that a double-strand break may take place especially when the DNA molecule is traversed by an ionizing particle of high LET.

When DNA-nucleoproteins are irradiated, the effect on viscosity is reduced; apparently the protein moiety has a protective action on the DNA helix.

One-strand breaks are more probable than double-strand breaks. Rejoining usually occurs between the broken ends of the strand; in the presence of oxygen, rejoining may be prevented by the formation of peroxides on one of the broken ends.

Strangely enough, unlike what happens with ultraviolet radiation, the hydrogen bonds of the double helix do not seem to be particularly sensitive to ionizing radiation, which, on the other hand, easily disrupts the same type of bond in globular proteins (see above). It is thought that in proteins the interchain or intrachain hydrogen bonds are not necessarily associated with the most stable molecular configuration, so that when they are broken the molecule may rapidly change its configuration (for example, by unfolding). The same would not be true for the DNA double strand.

Damage of DNA may manifest itself as *intermolecular cross-linking*, if irradiation occurs in the absence of oxygen. The initial event that leads to this type of cross-linking is a single-strand break (Fig. 11-11). One of the two ends of the break is chemically reactive and can join with another similar end of another molecule where a single-strand break also occurred; a cross-link is thus formed between the two molecules. Cross-linking results in an increase of the average molecular weight. It is thought that in the presence of oxygen cross-linking is prevented because the reactive end of a break is peroxidized and incapable of joining with another similar end.

Other experiments show that chemical alterations may occur in the bases; possibly some of the single-strand breaks may be caused by these alterations, which may consist in deamination or more drastic destruction of the bases. The pyrimidine bases are more radiosensitive than the purine bases; thymine is the most sensitive of all. In X- or gamma-irradiated solutions of DNA, in the presence of oxygen, the bases are apparently destroyed, with a G-value between 1 and 2, which is considerably greater than the same value observed for the destruction of the sugar.

Recently the radiosensitivity of several strains of phage DNA has been studied (Blok et al.). In this material what is assayed is the biologic activity of the nucleic acid, that

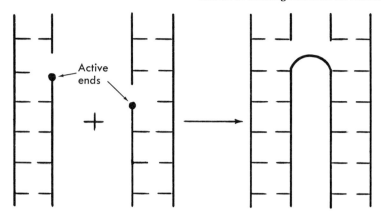

Fig. 11-11. Cross-linking in DNA.

is, its ability to produce more bacteriophage particles in the host's protoplasm. The DNA isolated from the *Escherichia coli* phage φX174 is particularly suitable because it is still biologically active after prolonged purification. This DNA is single stranded and circular (like the DNA of several other viruses) and every chain break leads to inactivation. On the other hand, the molecular weight is accurately known, and therefore breaks in the strand can be readily detected.

Experiments designed to assess the radiation-induced damage of DNA in more complex biologic systems (from bacteria to mammalian cells) have established that the nuclear and extranuclear DNA is subject to the same alterations observed in vitro. Several types of molecular alterations may be the cause of genetic mutations (see Chapter 13). The damage to the parental strands leads to faulty replication of the molecule at the damaged site, unless the site is fully repaired prior to replication. The numerically more abundant single-strand breaks are efficiently repaired (presumably by enzymatic processes). Double-strand breaks appear to be nonrepairable, except perhaps in extremely radio-resistant microorganisms. It is common opinion that DNA damage is the chief contributing factor to radiation injury leading to cell death.

Other experiments with simple cell systems and whole mammals indicate that irradiation may either delay or inhibit DNA synthesis, depending on the dose and on a variety of other circumstances. Reduction of DNA synthesis in rats may be demonstrated after irradiation with only a few hundred rads (see, for instance, Nygaard and Pottinger). These experiments are often coupled with radiotracer methods (for example, the extent of DNA synthesis is assessed by determining the amount of tritiated thymidine taken up by the cell nucleus).

Effects on polysaccharides

Work on the radiation-induced alteration of polysaccharide molecules has not been particularly abundant. However, the few reports available seem to show that the general effect on these molecules is degradation or depolymerization.

In studies on starch, Desrosier and others have used a very convenient spectrophotometric technique. Natural starch consists of two different types of molecules—amylose and amylopectin. Amylose is made of long unbranched chains of glucose (about 10,000 units); amylopectin is more complex because it consists of a branched system of chains of glucose units. The addition of an iodine solution to starch produces a characteristic blue color (iodine test). The color is attributed to the formation of an amylose-iodine complex. The absorption spectrum of an iodinated starch solution shows an absorption peak at 660 nm. The absorbance decreases if the starch molecules are depolymerized.

Using this method, one can conclude from irradiation experiments that radiation induces depolymerization in starch and increases its susceptibility to enzymatic hydrolysis. Depolymerization is also indicated by a reduced viscosity and by the appearance of reducing sugars (maltose and glucose).

Studies on the glycogen content of unirradiated and irradiated beef muscle indicate that the content decreases with increasing radiation dose from 0.44 mg/g of muscle with 0 rads, to 0.13 mg/g with 8 megarads. This effect is also thought to be caused by molecular degradation.

Irradiated solutions of cellulose show a decreased viscosity, liberation of glucose molecules and increased susceptibility to enzymatic hydrolysis. The consequences of these events in a cellulose structure (for example, cell wall) are decreased mechanical strength and reduced resistance to bacterial attack.

Even from the brief summary given above, the reader can easily notice that our understanding of the effect of ionizing radiations on macromolecules of biologic importance is still hazy and incomplete. This type of research is still in its infancy, as evidenced, among other things, by the contradictory results obtained by different investigators in certain instances. Probably the causes of this unpleasant state of affairs are to be sought in the lack of more appropriate methodology and in the incomplete knowledge we still have about the structure of these important molecules. Hopefully more significant progress may be made in the near future.

APPENDIX: The "transmutation effect" in labeled molecules and its radiobiologic implications

The molecular chemical changes previously discussed are induced by the direct or indirect action of an ionizing particle or photon whose source is external to the target molecules. If a molecule is labeled with a radioisotope, it will undergo a chemical change as a result of the radioactive decay of its label. Let us consider a DNA molecule in which one of the atoms of phosphorus has been replaced by ^{32}P:

$$
\begin{array}{ccccc}
 & & & \diagup \mathrm{OH} & \\
-\mathrm{Sugar}-\mathrm{O}-{}^{32}\mathrm{P}-\mathrm{O}-\mathrm{Sugar}- \\
 & | & \parallel & & | \\
 & \mathrm{Base} & \mathrm{O} & & \mathrm{Base}
\end{array}
$$

When the $^{32}_{15}P$ atom decays, it emits a negatron and a neutrino; the daughter atom is no longer phosphorus but sulfur ($^{32}_{16}S$). The change of the atomic number is accompanied by a rearrangement in the orbital electrons of the atom. Until this rearrangement takes place, the daughter atom is in an excited chemical state; it is also a positive ion because of the excess of one positive nuclear charge as a consequence of the increase of Z by one unit.

The momentum of the atom before the decay event is zero. Since the momentum of the beta particle is different from zero and total momentum must be conserved, the vector sum of the momentum of the emitted particle and of the residual atom must also be zero. Some kinetic energy is thus imparted to the daughter nucleus, which recoils in a direction opposite to that of the beta particle. The recoil energy of the daughter nucleus in beta emission is not as large as in alpha emission (see p. 90), wherein the momentum of the emitted particle is considerably larger; nevertheless, it is considerably different from zero and depends on the E_{max} of the beta particle emitted. The nuclear recoil energy in the decay of ^{32}P is 78 eV (beta $E_{max} = 1.71$ MeV); in the decay of ^{14}C, it is only 7 eV (beta $E_{max} = 0.156$ MeV), etc.

If the radioactive label is an atom that decays by electron capture (such as ^{125}I or ^{55}Fe), the decay results in a vacancy in one of the inner electronic shells, which is filled by an outer electron. This electronic transition is accompanied by a release of energy. The vacancies in the outer shells of the daughter atom may be filled by electrons from neighboring atoms in the molecule.

All the events accompanying the decay of the labeling atom (recoil of the daughter atom, electronic rearrangements, etc.) may involve enough energy to affect the chemical bonds that link the daughter atom with the neighboring atoms. These bonds may even be broken, with consequent disruption of the molecule. Note that if we consider only the recoil energy of the daughter atom of ^{32}P (78 eV) this energy is well above average bond-energy values.

A disruption of the molecule is a chemical change. But even when no chemical bonds are broken, the fact that the daughter atom is an atom of a different element implies a chemical change for the molecule. This type of chemical change that is not caused by radiation but by the change of atomic number, excitation, and recoil accompanying the decay process that occurs within a molecule is called the *transmutation effect*.

The chemical changes attributable to transmutation effects have been investigated by radiation chemists in molecules like ^{3}HH and $^{14}CO_2$. In the first case, calculations show that no disruption of the chemical bond occurs in 93% of the decays of ^{3}H; the resulting product is $(^{3}HeH)^+$. For $^{14}CO_2$, the chemical bonds remain intact in 81% of the decays; the resulting product is $(NO_2)^+$.

If the labeled molecule is a biologic macromolecule incorporated in a virus or bacterial cell, etc., the transmutation may result in a biologic effect. Several experiments have been conducted to investigate the biologic effects produced by the radioactive decay occurring in nucleic acid molecules labeled with ^{32}P, ^{14}C, or ^{3}H. In many instances one can distinguish between radiation-induced and transmutation-induced biologic effects. The organisms used have ranged from the simplest phages to *Drosophila*, mammals, and higher plants. The effects most intensively studied are change of viability in phages, genetic mutations, chromosomal aberrations, transforming power of bacterial DNA, and cell death. ^{32}P has been used more than any other radionuclide to label the molecule of interest because it may be obtained incorporated in a fairly large variety of organic molecules with a very high specific activity, and because the transmutation effect can be easily identified, due to the short half-life of this radioisotope.

The typical technique for these experiments consists of allowing the organism to incorporate a labeled precursor that is known to be used for the synthesis of the molecule of interest. For instance, if thymidine (including ^{14}C- or ^{3}H-labeled thymidine) is made available to an organism, it will be used for the synthesis of DNA.

In a classic experiment, A. D. Hershey and others showed that ^{32}P-labeled phages kept under refrigeration exhibit a progressively decreasing viability. The decrease is exponential and roughly proportional to the decrease of the decay rate of ^{32}P, which would indicate that a single disintegration event can activate an entire phage. With appropriate calculations, Hershey estimated that only a small percentage of the observed lethalities could be caused by the beta radiation emitted by the decaying atoms. Most of them were the result of transmutation effects in the DNA of the phage. In fact, unlabeled phages suspended in a solution of ^{32}P providing the same radiation dose as in the first experiment showed a much better viability. An experiment of this type is often called, with a highly descriptive name, a "suicide" experiment, because the origin of inactivation or death is within the very molecular makeup of the organism affected.

If unlabeled phages are allowed to infect labeled bacterial cells, their viability is not significantly affected; an additional evidence that "murder" (caused by radiation from a source external to the DNA molecule) contributes to the lethality of these phages much less than "suicide."

A description of other interesting suicide experiments can be found in the review by Krisch and Zelle, cited at the end of this chapter.

Suicide experiments are often useful to identify the type of molecule whose alteration is suspected to induce a biologic effect. On the other hand, the possibility that biologic effects may be caused by the radioactive transmutations occurring in labeled molecules of biologic importance should always be kept in mind when radiotracer methods are used in biologic research (see p. 267).

REFERENCES

GENERAL

Allen, A. O. 1961. The radiation chemistry of water and aqueous solutions. D. Van Nostrand Co., Inc., Princeton, N.J.

Altman, K. I., G. B. Gerber, and S. Okada. 1970. Radiation biochemistry. Vol. 1: Cells. Academic Press, Inc., New York.

Burton, M., and J. L. Magee, editors. 1969-1970. Advances in radiation chemistry. John Wiley & Sons, Inc., New York. 2 vol.

Desrosier, N. W., and H. M. Rosenstock. 1960. Radiation technology in food, agriculture and industry. Avi Publishing Co., Westport, Conn.

Errera, M., and A. Forssberg. 1961. Mechanisms in radiobiology. Academic Press, Inc., New York. Vol. 1, Chapter 2.

Haissinsky, M. 1957. La chimie nucléaire et ses applications. Masson & Cie. Editeurs, Paris.

Haissinsky, M., editor. 1961. The chemical and biological action of radiation. Academic Press, Inc., New York.

International Atomic Energy Agency. 1962. Biological effects of ionizing radiation at the molecular level. Vienna.

International Atomic Energy Agency. 1967. Solid state and chemical radiation dosimetry in medicine and biology. Vienna.

Lea, D. A. 1962. Action of radiation on living cells. Cambridge University Press, New York.

Krisch, R. E., and M. R. Zelle. 1969. Biological effects of radioactive decay; the role of the transmutation effect. In Augenstein, L. G. and M. Zelle, editors. Advances in radiation biology. Academic Press, Inc., New York. Vol. 3, pp. 177-211.

Kuzin, A. M. 1964. Radiation biochemistry. (English translation.) Daniel Davey & Co., Inc., New York.

O'Donnel, J. H., and D. F. Sangster. 1970. Principles of radiation chemistry. Elsevier Publishing Co., Amersterdam.

Pryor, W. A. Aug. 1970. Free radicals in biological systems. Sci. Amer. **223**:70.

Spinks, J. W. T., and R. J. Woods. 1964. An introduction to radiation chemistry. John Wiley & Sons, Inc., New York.

Vereshchinskii, I. V., and A. K. Pikaev. 1965. Introduction to radiation chemistry. (English translation.) Daniel Davey & Co., Inc., New York.

LITERATURE CITED

Barron, E. S. G., and P. Johnson. 1956. Studies on the mechanism of action of ionizing radiation. XV: X-irradiation of oxyhemoglobin and related compounds. Radiat. Res. **5**:290.

Blok, J., L. H. Lutjens, and A. L. Roos. 1967. The radiosensitivity of bacteriophage DNA in aqueous solution. Radiat. Res. **30**:468.

Dale, W. M. 1942. Effects of X-rays on the conjugated protein D-aminoacid-oxidase. Biochem. J. **36**:80.

Fricke, N., and S. Morse. 1927. The chemical action of roentgen rays on diluted ferrous sulphate solutions as a measure of dose. Amer. J. Roentgen. **18**:426.

Fricke, H., and B. W. Petersen. 1927. Action of roentgen rays on solutions of oxyhemoglobin in water. Amer. J. Roentgen. **17**:611.

Haskill, J. S., and J. W. Hunt. 1967. Radiation damage to crystalline ribonuclease. Radiat. Res. **32**:287.

Nygaard, O. F., and R. L. Pottinger. 1959 and 1960. Effect of X-irradiation on DNA metabolism in various tissues of the rat. Radiat. Res. **10**:462; **12**:120, **12**:131.

Scharf, K., and R. M. Lee. 1962. Investigation of the spectrophotometric method of measuring the ferric ion yield in the ferrous sulphate dosimeter. Radiat. Res. **16**:115.

Weiss, J. 1952. Chemical dosimetry using ferrous and ceric sulfates. Nucleonics, **10**(7):28.

Winstead, J. A. 1967. The role of sulfhydryl groups in radiation sensitivity and chemical protection in enzymes. Radiat. Res. **30**:832.

Winstead, J. A., and A. E. Gass. 1967. Studies on the mechanism of substrate protection of enolase and lactic dehydrogenase against ionizing radiation. Radiat. Res. **30**:208.

CHAPTER 12

Effects of ionizing radiations at cellular level

In the preceding chapter we have seen that the biologic effects of ionizing radiations are the consequences of events that occur at molecular level in three stages of very short duration. During the much longer fourth stage (the biologic stage) the organism responds to the molecular damage with functional and structural alterations. The structural alterations may be visible only at microscopic level or may manifest themselves also macroscopically. The time of appearance of the biologic effects after irradiation depends on several factors, but in particular on the total accumulated dose and dose rate.

The biologic damage manifests itself first at the cellular level, that is, in the very structural and functional building blocks of practically every living organism. The extent of the damage depends largely on the total dose absorbed by the cell. It may range from damage to chromosomes only (chromosome aberrations), which is visible with doses as low as a few rads, to the derangement of certain metabolic processes leading possibly to the formation of neoplastic (cancerous) cells, to the alteration of the cell cycle, and even to cell death.

With the exception of chromosomic damage, gross morphologic damage to cellular structures, such as cytoplasmic organelles, becomes apparent only after irradiation with massive doses. A cell may fail to reproduce or may even die without showing extensive structural alterations—another indication of the important role biochemical lesions may play in the induction of the final biologic damage.

In higher multicellular organisms, depending on the type of cells that have suffered the radiation damage, the biologic effects produced at cellular level may be transferred to higher levels of organization (tissues, organ systems, and the whole organism), because a complex multicellular organism is an integrated system in which all cells, tissues, and organs are mutually and intimately correlated and coordinated. For example, bone marrow damage induced with local irradiation of this tissue may cause anemia and therefore impairment of the cellular respiratory processes in all other tissues of the body.

For the study of the radiation-induced biologic effects at cellular level the use of systems made of cells that are relatively independent of one another would be desirable, so that any damage to any cell of the system can be safely attributed directly to radiation. In a biologic system whose cells are morphologically and physiologically interrelated, the response given by one cell may be evoked directly by radiation or simply by the damage induced in the neighboring cells.

The simplest independent cell systems that can be used for this type of investigation are bacterial cultures grown on solid media or in liquid nutrient broths. These cultures can be easily handled with the usual bacteriologic techniques; large numbers of identical cells may be obtained in a short time in relatively small volumes, and each cell of a population is to a great extent independent of other cells. Yeasts are also a convenient material for the same reasons. Fungal or fern spores can be obtained in large numbers. Although spores must be considered as dormant cells, investigations have shown that radiation affects their germinating power. Pollen grains and immature anthers of *Tradescantia* and *Trillium* have been often used for the assay of radiation-induced chromosomic aberrations at various stages of gametogenesis. Likewise, immature and mature animal gametes are used to investigate the effects on viability. Mature insect and vertebrate eggs are particularly suitable for many types of experiments because of their relatively large size. Among protozoa, several species of amebas have been favorite material for different irradiation experiments. Single amebas may be handled with microsurgical techniques in studies of the differential radiosensitivity of the nucleus and cytoplasm. Independent systems of mammalian cells are, for instance, suspensions of red blood cells. Although these cells are atypical in many respects, they are a useful experimental material for the study of radiation-induced effects on plasma-membrane permeability.

A greater wealth of radiobiologic information can be gathered from studies of other types of mammalian cells in active division and with a nucleus clearly distinguishable from the cytoplasm. (You may recall that in mammalian red cells such distinction does not exist.) To obtain populations of relatively independent cells, various techniques have been developed utilizing well-established tissue-culture cells from embryonic or adult tissues of humans or laboratory mammals. Fibroblasts and neuroblasts have been successfully grown in tissue cultures. HeLa cells (strain of human neoplastic cells) have been grown in vitro for several years in many laboratories and recently have been used as experimental material in radiobiologic studies by Puck, Marcus, Cieciura, and others. For a typical experiment, a measured volume of cell suspension of known concentration is inoculated into a petri dish, and some additional nutrient liquid medium is added. The single isolated cells soon attach themselves to the bottom and start dividing rapidly, thus forming colonies, or clones, which can be seen with the unaided eye and counted. In normal conditions the number of visible colonies should be equal to the number of cells initially inoculated into the dish. If the dish is irradiated immediately after the attachment of the cells to the bottom and then incubated for 10 to 14 days at 36° C, several types of biologic effects may be detected. Some of the initial cells may fail to divide; they either die or grow to enormous size. In either case the number of colonies will be less than the number of cells inoculated into the dish. In certain instances, the colonies may vary in size (that is, number of cells per colony) or in other interesting characteristics that can be easily analyzed microscopically. The same cloning technique has been widely employed by several other investigators with cells from the Chinese hamster. Techniques have also been developed to synchronize mammalian and human cells in vitro (see p. 356) so that all the cells of the same population go through the same stage of the cell cycle at the same time. Synchronization is especially desirable when the object of the study is the dependence of the biologic effects on the various phases of the cycle.

With a somewhat different technique (Till and McCulloch) cell colonies from single isolated cells may be grown in vivo, rather than in glassware with artificial nutrient media. If bone marrow cells removed from a normal mouse are counted and inoculated intravenously into another preirradiated mouse, gross nodules will develop in the recipient's spleen. These nodules are considered to be colonies, or clones, originating from the proliferation of some of the marrow cells injected. Some investigators have estimated that one nodule develops for every 10^4 cells inoculated. Recipient mice are sacrificed 10 or 11 days after inoculation and their spleens are examined. This in vivo culture method has the advantage that the inoculated cells not only multiply in the host but also differentiate. The irradiated recipient provides the receptacle, medium, and control of temperature, pH, and humidity, which are usually required for the in vitro cultivation of marrow cells. If the donor's marrow cells are irradiated in vitro before inoculation, the effects of radiation can be assessed by counting and examining the spleen nodules.

Recently, an interesting organism, widely used before for other types of studies, has been made the object of irradiation experiments by Rusch, Guttes, Nygaard, and others, with the intent of investigating radiation-induced effects on nuclear metabolism and division. This is the plasmodial slime mold *Physarum polycephalum* (Myxomycetes). During its vegetative stage (plasmodium) this mold grows as a gigantic syncytium, consisting of a large mass of protoplasm, which contains hundreds of nuclei exhibiting spontaneous synchrony of mitotic divisions. A method has also been developed to grow this organism in pure culture in a partially defined soluble medium. It may be grown either as a single large plasmodium on surface culture (on filter paper) or as a suspension of tiny plasmodia in submerged cultures. In certain conditions, the tiny plasmodia merge to form a single large plasmodium, which exhibits synchronous mitoses.

The plasmodium of *P. polycephalum* possesses other favorable characteristics that recommend it for radiobiologic studies. For instance, it can be cut into several pieces, which can be subjected to different experimental treatments, with a minimum disturbance to the protoplasmic activities. Some fragments may be irradiated with different doses or at different times (that is, in different phases of the mitotic cycle), while other fragments of the same plasmodium may serve as controls.

Although this organism cannot be considered, strictly speaking, to be a system of independent cells (because of its syncytial nature), it may be extremely useful when the purpose is the investigation of the effects of ionizing radiation on the nuclear activities and divisions, or on the physical and chemical characteristics of the protoplasm.

Obviously, mammalian radiation biologists, or any other investigator who is mainly interested in the radiation-induced biologic effects in the human organism, will see little value in the experimental results obtained from work with a mold. Indeed, any attempt to extrapolate the conclusions derived from work with one organism to other (quite different) organisms should always be made with extreme caution.

Although systems consisting of populations of independent cells, especially if synchronized, are a desirable material in many research situations, much valuable information on the effects at cellular level has also been obtained with other multicellular systems. Extensive studies have been conducted on radiation-induced chromosome aberrations in the cells of *Vicia faba* (broad bean) root tips and in the buds of several other plant species. In these materials cells are mitotically very active (although the cell divisions are not synchronous); when the chromosomes are few and large, beautiful microscopic preparations may be obtained with the usual staining techniques.

RELATION BETWEEN RADIOSENSITIVITY AND CELL TYPE—THE LAW OF BERGONIÉ AND TRIBONDEAU

A basic problem in cellular radiation biology is the following: Are cells of different organisms, or of different tissues and organs of the same organism, equally radiosensitive? Is there any difference in sensitivity between very specialized, differentiated cells and embryonic tissues that have not yet reached their final stage of differentiation?

Only a few years after the discovery of X-rays, the first radiation therapists had observed with interest and surprise that X-rays may kill neoplastic cells with no apparent damage to the surrounding healthy tissue or even to the tissue that has been invaded by the tumor. In 1906, J. Bergonié and L. Tribondeau echoed this discovery in a famous paper, which is one of the most important contributions in the history of radiation biology. In their paper, the authors also observe that X-rays may discriminate not only between healthy and cancerous tissues, but also between different healthy tissues of the same organ or organism. In fact, they report that in X-irradiated rat testis only germinal cells are destroyed, whereas the interstitial tissue remains unimpaired. The authors conclude that germinal cells are more radiosensitive than interstitial cells are, because they have a greater reproductive activity. More generally, X-rays are more effective on cells that are in dividing activity, on those cells that have "a longer dividing future ahead," and on those cells whose "morphology and function are least fixed." Since cancerous cells have a greater reproductive activity and are less differentiated than are the surrounding healthy tissues, they are even more vulnerable to the killing action of X-rays.

The authors also pose the vexing question of why the same X-rays that are capable of destroying malignant tumors can also induce in the negligent radiation therapist tumors identical to those they can destroy. Their interpretation is that moderate doses not lethal to cells may transform them into "teratocytes," that is, into monster cells with atypical mitoses, or they may somehow "influence the further evolution" of the cells. Neoplastic cells are in fact monstrosities with this type of anomaly. Therefore according to the authors, treatment of tumors with large, massive radiation doses is preferable, rather than with small and fractionated ones. However, they recognize that radiation therapy is contraindicated when the histologic characteristics of the malignant tissues to be destroyed are similar to those of the surrounding healthy tissues with respect to reproductive activity; in these cases there would be no selectivity on the part of the X-rays.

In summary, Bergonié and Tribondeau are the first scientists to recognize formally that ionizing radiation like X-rays may discriminate between different types of cells of the same organism and that the basis of this discrimination is to be found in the degree of reproductive activity and in the differentiation of the cells irradiated. The so-called law of Bergonié and Tribondeau may be stated concisely as follows: *the radiosensitivity of cells is directly proportional to their reproductive activity and inversely proportional to their degree of differentiation.*

A few words of clarification might be in order at this point. If we limit our consideration to the development of a high organism like a mammal, we find that the organism begins its life as a zygote, that is, as a cell being unspecialized physiologically and morphologically but having tremendous reproductive (that is, mitotic) power. In fact, immediately after fertilization the zygote divides several times, thus producing two, four, eight, etc., blastomeres. The blastomeres divide in synchrony and so rapidly that the overall size of the embryo at these early stages is not significantly larger than that of the original zygote. For some time the blastomeres even look identical, although later some of them will become neurons, others epithelial cells, etc. As the development proceeds, the divisions of the blastomeres slow down but not at the same rate in every embryonic region. Quite early, some embryonic cells undergo a process of specialization and differentiation, whereby their biochemical and structural organization becomes strikingly different from that of other types of cells and adapted to the performance of very specialized functions. It suffices to think, by way of example, of the evolution of neuroblasts into neurons, with their complex physiologic and structural organization. The formation of neurons is practically already completed before birth. The faster the cells of a lineage undergo the process of specialization and differentiation, the sooner they will lose their power of self-division and therefore their reproductive activity. The time when this loss occurs depends also on the degree of final differentiation. Since neurons are much more differentiated than epithelial tissues, they have lost the ability to divide long before birth, whereas epithelial cells retain the power of self-division permanently.

Unlike neurons, epithelial tissues are much less different from the embryonic tissues that gave origin to them. Epithelial cells are the final product of a process of differentiation that was not pushed too far, as it was in the case of neurons; their biochemical and structural organization is less differentiated and complex. Consequently, even in the adult organism they still retain one of the properties of the embryonic cells: the power of self-division.

If the law of Bergonié and Tribondeau is correct, one would expect a higher radiosensitivity in embryonic than in adult tissues and a higher radiosensitivity in less differentiated types of cells with a high reproductive activity (like epithelial cells and bone marrow cells) than in more specialized types (like neurons and muscular fibers) that have completed their process of differentiation and lost their power of self-division. In more than 60 years of radiobiologic research, this expectation has been found to be generally correct; for instance, if the degree of radiosensitivity at cellular level is reflected in the radiosensitivity of the whole organism, it is true that immature organisms, like embryos, fetuses, and larvae, are more sensitive to radiation than are fully developed organisms (see Chapter 14).

However, the results of many observations and experiments have also shown that the validity of the "law" is not as general as it was thought to be at the beginning, because of several exceptions, which have not yet been completely explained. Furthermore, it seems that additional factors, other than those recognized by Bergonié and Tribondeau, may determine the degree of radiosensitivity in certain types of cells and that the law finds a more rigorous application if it is used to compare the radiosensitivities of different types of cells within the same organism, or within relatively small taxonomic groups.

In the first place, when testing the radiosensitivity of a particular type of cell, one should select an appropriate criterion or type of damage and compare the sensitivities of different types of cells in relation to the same criterion. The easiest type of damage that can be used for this purpose is, of course, cell death, that is, the cessation of all cellular activities followed by disintegration of the cell itself. This criterion may always be applied, regardless of the types of cells under study, as long as other variables, like time required after irradiation for death to occur, etc., are taken in due consideration. The criterion of cessation of cell division can be used only if all the types of cells, whose radiosensitivities are to be compared, show the same degree of reproductive activity. If the frequency of chromosome aberrations is chosen as a criterion of radiosensitivity, the problem may become quite complex; for instance, it would seem necessary to irradiate all the cells of interest during the same phase of the cell cycle, because it is quite possible that chromosomes may not be equally sensitive at different phases of the cycle. To realize these experimental conditions, one might need to overcome serious technical difficulties.

Beside the criteria of sensitivity mentioned above, several others may be used. If we compare the radiosensitivities of two types of cells belonging to organisms of entirely different taxonomic groups, using cell death as a criterion of sensitivity, the limitations of the law become immediately apparent. An ameba cell and a mammalian lymphocyte are both free cells of comparable size and morphology. They also show some similar physiologic properties, such as mechanism of locomotion and phagocytosis. The main difference is that an ameba has a high reproductive activity, whereas a lymphocyte divides only exceptionally. One would also think that a lymphocyte is a more specialized type of cell than an ameba (although the problem of cell differentiation in protozoa is extremely thorny and controversial). On the basis of these considerations, one would predict that amebas should be either equally as, or more radiosensitive than, lymphocytes. Yet, the opposite is true. The fact is well known that a whole-body exposure of less than 100 R can reduce the lymphocyte count considerably, within a few hours, whereas about 100,000 R are required to kill 50% of a population of amebas.

But an exception to the law can be seen even when we compare the radiosensitivity of mammalian lymphocytes with that of the intestinal mucosa cells (epithelial cells). The latter show a considerable reproductive activity, whereas lymphocytes do not; yet the mucosa cells are more radioresistant than lymphocytes.

Recently, an interesting experiment has been performed to compare the radiosensitivity of dividing and nondividing cells of the same morphologic type. Haber and Rothstein, of the Oak Ridge National Laboratory, have investigated gamma ray–induced sensitivity to photodestruction of chlorophyll in chlorenchyma cells of tobacco leaves (Fig. 12-1). We know very well from plant morphology that leaves grow in size at the beginning mainly by cell division and later by simple increase in the size of their cells without division. Consequently, younger leaves exhibit more cells in mitosis than do older ones. If leaves of any age are exposed to light, a gradual loss (photodestruction) of chlorophyll takes place in the chlorenchyma cells. Bleaching of green tissue in higher plants is a general characteristic of senescence and a precursor of cell death. The rate of photodestruction of chlorophyll is enhanced if the leaves are exposed to ionizing radiation prior to their exposure to light. Disks punched from leaves of different ages were exposed to several different doses of gamma radiation and then exposed to light for 24 hours. The rate of photodestruction was found to be exclusively a function of radiation dose and not of the developmental stage of the cells exposed. Thus this experiment shows that cells of the same organ and of the same type, but with different rates of cell division, exhibit the same radiosensitivity—a fact certainly not in agreement with the law.

Less recent experiments show that cellular radiosensitivity may be related to factors other than degree of differentiation and reproductive activity. In several plant species radiosensitivity has been found to be related to nuclear volume and chromosome number. Sparrow (1961) has tested a number of *Chry-*

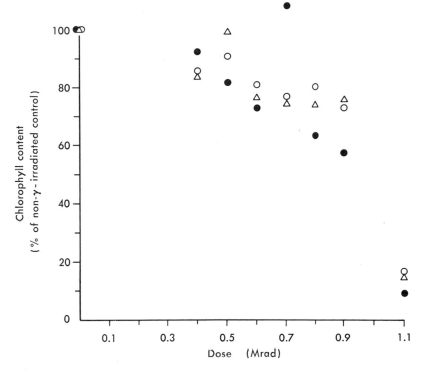

Fig. 12-1. Susceptibility to photodestruction of chlorophyll induced by gamma rays in tobacco leaves. For the three groups, disks were cut from growing leaves of the same plants. The disks were gamma-irradiated and then illuminated for 24 hours with white, incandescent light. Each point represents analysis of eight leaf disks. *Dots,* "Young" leaves (20 to 50 mm long). *Circles,* "Middle" leaves (80 to 110 mm long). *Triangles,* "Old" leaves (130 to 155 mm long) with no cell divisions occurring in the mesophyll. (Courtesy Haber, A. H., and B. E. Rothstein. 1969. Science **163:** 1338-1339. Copyright 1969 by the American Association for the Advancement of Science.)

santhemum species, all with the same number of chromosomes in the basic complement (n = 9), but with different ploidy (from 2n to 22n). The radiosensitivity decreases with increasing ploidy. In yeasts (*Saccharomyces cerevisiae*), from which varieties with different ploidy may be obtained, monoploid cells are more sensitive than diploid; but the sensitivity increases above the diploid stage, so that tetraploid cells are more sensitive than diploids and hexaploids are more sensitive than tetraploids. The interpretation of these results is made difficult by the conflict existing between the data obtained with yeasts and those obtained with *Chrysanthemum* and other species. Concerning the yeasts, a tentative explanation is that in cells of any ploidy lethality is caused by some type of chromosome damage, probably lethal mutations; monoploid cells may be killed by recessive lethals, whereas for diploids and cells of higher ploidy most of the lethal effects may be due to the induction of dominant lethal mutations. This not-too-convincing explanation is not shared by all the investigators who have obtained the same results with yeasts.

Sparrow (1965) and others have compared the radiosensitivities of several plant species with diploid chromosome numbers ranging from 6 to 136, by determining the acute exposure doses necessary to induce either death, a severe

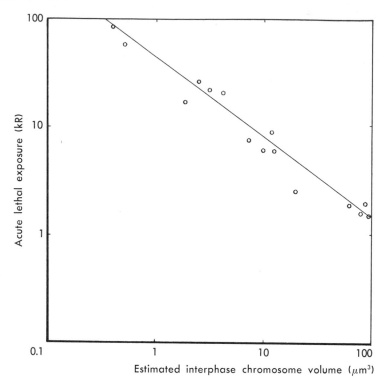

Fig. 12-2. Relationship between lethal exposure and interphase chromosome volume for 15 herbaceous species after acute irradiation. (Modified from Sparrow, A. H. 1963. Relationship between nuclear volumes, chromosome numbers, and relative radiosensitivities. Science **141**:164.)

growth inhibition, or a slight growth inhibition. They found that the radiosensitivity of a species is dependent on the average volume of the interphase nucleus (measured in root meristems) and on the chromosome number. If the chromosome number is the same in two species, the one with a larger nucleus is more radiosensitive. If two species have the same nuclear volume, the more radiosensitive is the one with fewer chromosomes. To unify the two variables (average interphase nuclear volume and chromosome number), the authors have calculated their ratio, which has been named *average interphase chromosome volume*. If the acute lethal exposure dose is plotted on log graph paper as a function of interphase chromosome volume (Fig. 12-2), the fact is then evident that radiosensitivity is directly proportional to interphase chromosome volume.

When the radiation energy absorbed per chromosome was calculated for every exposure required to obtain a specified effect (death or growth inhibition), it was found that all values are similar for all species studied. These values were about 3.7×10^6 eV for lethal effects, 2.3×10^6 eV for severe growth inhibition, and 1.07×10^6 eV for slight inhibition. The interpretation of these results seems to be that a relatively constant amount of energy must be absorbed per chromosome and therefore that a relatively constant amount of damage per chromosome is required for the development of a given effect.

The greater radioresistance of polyploids as compared with diploids (for example, in *Chrysanthemum*) may be the result of their reduced interphase chromosome volume (because the increase of chromosome number is not matched by a proportionate increase of nuclear volume) and not necessarily the result of genetic redundancy.

Although the experiments described above show that cellular radiosensitivity must be dependent on a large number of cellular characteristics, the law of Bergonié and Tribondeau is still a valid and useful generalization, in spite of its limitations and exceptions.

It is quite useful to compare the radiosensitivities of different types of mammalian cells for a better understanding of the acute radiation syndrome (see Chapter 15). The criterion of sensitivity used is mainly based on the time it takes after irradiation to observe the death or disappearance of the cells. The most important types of cells and tissues can be arranged in an arbitrary number of groups such that the radiosensitivity increases from the first to the last group, as follows:

1. Muscular fibers and neurons are considered to be the most radioresistant cellular elements of the mammalian body. They have achieved a high degree of differentiation and show no reproductive activity after birth.

2. Cells of certain connective tissues, such as fibrocytes and chondrocytes.

3. Red blood cells and cells of the endothelial lining of large vessels.

4. Epithelial cells of the pulmonary alveoli, granulocytes, osteocytes, and mature sperm cells.

5. Spermatocytes, spermatids, osteoblasts (bone forming cells).

6. Malpighian layer of the epidermis, most cells of the gastrointestinal mucosa, endothelial cells of capillary vessels, and myelocytes (precursors of granulocytes in the bone marrow).

7. The three most radiosensitive types of cells are the most primitive spermatogonia, which exhibit little differentiation and a high reproductive activity, the erythroblasts (precursors of red cells in the bone marrow), and mature lymphocytes. The extreme radiosensitivity of the lymphocytes has been discussed on p. 332.

This radiosensitivity scale for mammalian cells is based on data obtained from several experiments of irradiation of whole animals, as well as of cells in tissue cultures. It is also in good general agreement with the law of Bergonié and Tribondeau, with very few exceptions, which have not yet been explained.

DIFFERENTIAL RADIOSENSITIVITY OF THE NUCLEUS AND CYTOPLASM

Some of the experiments described in the preceding section already seem to suggest that in certain cases the cause of radiation-induced cellular damage might be a severe lesion occurring within the nucleus. Other experiments were designed to compare more directly the radiosensitivity of the nucleus with that of the cytoplasm. The conclusion that can be derived from most of them is that indeed the nucleus is far more radiosensitive than the cytoplasm. This greater sensitivity was to be expected because the nucleus is the controlling and directing center of all the vital activities of the cell and the carrier of the genetic material. Furthermore, the cytoplasm contains multiples of several structures or organelles (mitochondria, chloroplasts, ribosomes, etc.) that are the sites of cytoplasmic metabolic functions, so that the inactivation of only some of them does not result in a complete cessation of those functions. In contrast, the nucleus seldom possesses more than one representative of each of its component structures. This is especially true of monoploid nuclei. For example, even a slight damage in a small region of a chromosome may cause alteration of the DNA of that region, which in turn will upset the formation of messenger RNA and thus the synthesis in the cytoplasm of some important protein.

Some of the classic experiments designed to study the relative radiosensitivity of the nucleus and cytoplasm are described and discussed below.

Fern spores

One of the first experiments was performed by R. Zirkle with the spores of the fern *Pteris longifolia*. These spores have an eccentric nucleus; its location is marked by a spot, which can be easily recognized on the spore coat. The spores were laid on a flat surface with the nuclear spots facing downward and irradiated with a beam of polonium alpha particles from above. The experiment was designed so that the nucleus could fall in different segments of the alpha tracks and thus would absorb different amounts of radiation energy (recall that the LET varies along the track of the same alpha particle). By interposing suitable thin aluminum filters between the source and spores, Zirkle could also stop the alpha particles in the cytoplasm and prevent them from reaching the nucleus. He found that much lower doses are required to induce damage, like inhibition of germination and of chlorophyll development or cracking of the spore coat if the nucleus is irradiated rather than if only the cytoplasm is irradiated.

Habrobracon eggs

Similar results were obtained with the eggs of *Habrobracon juglandis* (a parasitic wasp) by R. C. von Borstel. These eggs, too, have a small and eccentric nucleus (Fig. 12-3). It is possible to irradiate with alpha particles the nucleus alone or the cytoplasm alone of these eggs. Such irradiation may be accomplished also with other types of radiation by using microbeams that can strike areas only a few micrometers in diameter. Von Borstel found that eggs will not hatch if only one alpha particle, on the average, hits the nucleus. However, if the nucleus was not included in the radiation field, calculations showed that about 17 million alpha particles are needed to kill 50% of the irradiated eggs.

If unfertilized eggs of *Habrobracon* are irradiated with 2,400 to 54,000 R of X-rays, the nucleus dies, but the eggs can still be fertilized with unirradiated sperms; they will develop into monoploid embryos whose chromosome complements are exclusively paternal (androgenesis). Apparently, if any damage had been done to the egg's cytoplasm, it was not sufficient to prevent the development of the egg. Furthermore, the irradiated cytoplasm does not kill the male pronucleus.

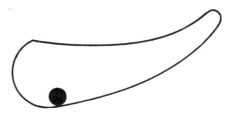

Fig. 12-3. *Habrobracon* egg.

If the same experiment is performed with eggs of the silkworm moth, the result is different: the irradiated cytoplasm of the egg inactivates the sperm nucleus. A possible explanation of this phenomenon may be found in the presence in the cytoplasm of radiation-induced peroxides, which would diffuse into the unirradiated nucleus and inactivate its chromatinic material.

Amebas

Amebas of practically all species are a very convenient material for this kind of study because they are relatively large and yet unicellular or syncytial organisms. Microsurgical techniques have been successfully used with these organisms by several investigators for a variety of research purposes.

The normal activities of an irradiated protoplasm can be restored by unirradiated protoplasm, as shown by the experiments of E. W. Daniels with the giant ameba *Pelomyxa.* This organism (known also with the name *Chaos*) is a large multinucleated ameba. If it is X-irradiated with a supralethal dose, it dies or at least its divisions slow down considerably. But if part of the protoplasm of an unirradiated ameba is allowed to fuse with an irradiated ameba, the latter will recover. The therapeutic properties of the donor protoplasm have been intensively investigated. By means of appropriate centrifugation one can obtain a donor protoplasm free of nuclei and other large cytoplasmic organelles, such as mitochondria. This protoplasm when allowed to fuse with an irradiated ameba still shows radiorestorative ability. This ability is not even destroyed by treatment of the donor protoplasm with nitrogen mustard, which is known to possess radiomimetic properties.* According to Daniels, the irreversible site of lethal X-ray damage in *Pelomyxa* appears to be intracytoplasmic rather than intranuclear. On the other hand, the nitrogen mustards react primarily with DNA, and with RNA to a lesser extent. If an ameba is treated with nitrogen mustard, its death is the result of nuclear damage. However, its cytoplasm does retain the therapeutic components of unirradiated cells and therefore can be successfully used to restore the normal activities of an X-irradiated ameba.

With more recent experiments, Daniels has shown that the active therapeutic ingredients are inactivated by 253.7 nm ultraviolet radiation. On the basis of this finding and electron microscopic studies, he believes that the restorative agents are polyribosomes or large subunits of them. When a lethally irradiated ameba receives unirradiated polysomes, it may synthesize proteins again, including enzymes needed for the repair of the damaged nuclear DNA. Damage to nuclear DNA in an untreated irradiated ameba cannot be repaired because of damage to the nuclear and cytoplasmic RNA-containing systems.

Cells in tissue cultures

In vitro cultivated newt heart cells and chick fibroblasts have been irradiated with microbeams of protons or alpha particles. If the nucleus alone is irradiated, a few protons are sufficient to produce typical abnormalities in the cells, whereas the irradiation of the cytoplasm alone is without effect even when thousands of protons pass through it. Cells in mitosis showed significant alterations only when chromosomes were directly hit by the microbeam.

---

*Radiomimetic substances are chemicals (usually alkylating agents) that induce in biologic systems all the end effects observed after treatment with ionizing radiations.

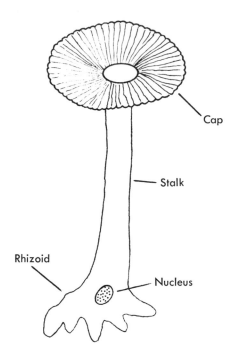

Fig. 12-4. *Acetabularia mediterranea*

Acetabularia

Acetabularia is a unicellular green alga that has been extensively used by Hämmerling and Brachet for the study of nucleocytoplasmic relationships and by others for the study of cytoplasmic RNA synthesis. The organism (Fig. 12-4) is about 3 to 4 cm long and consists of a rhizoid, where the nucleus is located, a slender stalk, and a cap, which in *A. mediterranea* looks like a disk with radial furrows. In the early stage of its development removal of the nucleus from the rhizoid is possible. The enucleated alga will survive for months, will retain the ability of synthesizing proteins, and will exhibit a limited morphologic differentiation. If nucleated and enucleated algae are irradiated in early stages of their development, the chance of survival for the enucleated algae is less than for the nucleated ones. Furthermore, when a fully developed and differentiated alga is irradiated, the cytoplasm farthest from the nucleus dies before that nearer to the nucleus. These results seem to indicate that the presence of the nucleus (although irradiated and presumably damaged) either limits the damage induced in the irradiated cytoplasm or accelerates its repair.

Amphibian eggs

The remarkable experiments performed by W. R. Duryee on amphibian eggs seem to suggest that irradiated cytoplasm has a considerable influence on the induction of nuclear damage. The nucleus of amphibian eggs directly taken from the ovary can be easily removed with the usual microsurgical techniques. Isolated nuclei are very radioresistant; the first lesions are detectable with exposure doses of several thousand roentgens. But when unirradiated isolated nuclei are placed back into the cytoplasm of the eggs, they show the expected high radiosensitivity observed when whole eggs are irradiated.

If an unirradiated nucleus is transplanted into an enucleated irradiated egg, it will exhibit the same alterations visible when it is irradiated with the cytoplasm.

If a very small amount of cytoplasm from an irradiated egg is transplanted into an unirradiated egg very close to the nucleus, the nucleus will show, a few hours after the transplantation, the same type of damage visible if the nucleus were irradiated. Similar results are obtained when the source of radiation is incorporated in the egg itself. ^{32}P has been used for this purpose.

Evidently, these experiments show that nuclear damage is significantly enhanced by cytoplasmic damage. Duryee believes that at least in amphibian eggs radiation-induced toxic products may diffuse from the cytoplasm into the nucleus

•   •   •

Looking back at the representative experiments described above, we see that those with fern spores, insect eggs, and cells in tissue cultures clearly indicate that the nucleus is by far more radiosensitive than the cytoplasm and that for equal doses the nuclear damage is more detrimental to the whole cell than cytoplasmic damage is. The other experiments suggest that repair of damaged cytoplasm is sufficient to reactivate the whole protoplasm even in the presence of irradiated nuclei, that the presence of an irradiated (damaged?) nucleus limits cytoplasmic damage (cytoplasm more radiosensitive than nucleus?), and that nuclear damage is enhanced by the presence of irradiated cytoplasm. Probably the way out of these contradictions may be the consideration of the complex and delicate interactions existing between nucleus and cytoplasm; these interactions may differ significantly in their nature and effects in different types of cells.

EFFECTS ON INTRACELLULAR STRUCTURES AND ORGANELLES

A very radiosensitive cell, like a mammalian lymphocyte, may be killed by exposure to relatively small doses of ionizing radiation, and yet it may show no microscopically visible morphologic alteration or damage in any of its structures. The reason is that death may be the result of (invisible) biochemical damage. To observe morphologic damage in most cellular structures, one must use massive doses, which may be well above the minimum dose required for cell death. There is one remarkable exception. Unlike other cellular organelles, chromosomes are very radiosensitive, even when the criterion used for the appraisal of their sensitivity is visible morphologic damage. In fact, after absorbing doses well below a dose lethal for the cell as a whole, chromosomes show changes in number and morphologic alterations known as *aberrations*.

In the following section radiation-induced chromosomal aberrations are discussed before the effects induced on other types of intracellular structures.

Chromosomal aberrations
Nature

In certain types of cells, irradiation may induce a clumping of the chromosomes, especially if the cells are irradiated as soon as the mitotic process begins. This phenomenon is known as "stickiness" and is thought to be caused by the alteration of some physicochemical properties of the chromosomic nucleoproteins. It is also induced, at times, by high doses of ultraviolet radiation. One consequence of the stickiness of the chromosomes may be seen at anaphase; the two sets of sister chromatids are hampered in their migration toward the two poles of the spindle by strands of chromosomic material persisting between them (anaphase bridges). However, there is evidence that the stickiness of the chromosomes is a temporary and reversible phenomenon.

A much more important effect induced by ionizing radiation in chromosomes is *breakage*. Chromosome breaks are gaps in the continuity of the chromosomic arms.* Depending on the dose, an irradiated cell may show breaks in one or more of its chromosomes. A single chromosome may be broken in one, two, or more points. Two breaks may occur in the same arm of a chromosome or in different arms on the opposite sides of the centromere. Obviously, the latter type of breakage is more probable with metacentric and submetacentric chromosomes.

The broken ends of a chromosome may recombine with each other, through a recovery process, thus reconstituting the chromosome exactly as it was before the break occurred. In the jargon of radiation cytogenetics, this restoration is better known as *restitution*. In this case, no lesions are detected in the chromosomes, nor is any alteration induced in their number and morphology. It is estimated that normally a high percentage of radiation-induced chromosome breaks do restitute.

If the breaks do not restitute, one of the following two events may occur:

1. Centric (containing the centromere) chromosomal fragments will survive and be transmitted to the future cell generations through the normal processes of chromosome replication and anaphasic separation of sister chromatids. However, they will be the carriers of a deletion. The acentric fragment(s) resulting from the break(s) will not be transmitted to both daughter cells, because the lack of a centromere makes it impossible for them to attach themselves to the spindle fibers, and consequently they are unable to migrate to the poles of the spindle at anaphase. They will either be lost in the process of cytokinesis, or they will remain in only one of the daughter cells. In this latter case, the fact seems highly probable that the acentric fragments will be lost some time during the subsequent cell divisions.

2. Broken ends may rejoin with one another, but in such a way as to produce new, abnormal chromosomic configurations. Two nonhomologous chromosomes may exchange the fragments resulting from breakage (reciprocal translocation), or the fragment of a chromosome may recombine with the other fragment of the same chromosome but after a rotation of 180 degrees (inversion), or the centric fragments of two chromosomes may combine together to form one chromosome with two centromeres (dicentric chromosome). Other similar abnormal recombinations are described below. All these rearrangements are referred to as *chromosomal aberrations*, and the abnormal chromosomes resulting from these aberrations may be referred to as *aberrant chromosomes*.

The consequence of the preceding considerations is that chromosome breaks are a necessary condition for the formation of aberrant chromosomes.

Radiation-induced chromosomal aberrations were first discovered and described by Karl Sax in 1938 in X-irradiated *Tradescantia* microspores. Since that time, hundreds of

---

*If necessary, see the morphology of chromosomes and pertinent nomenclature in any modern textbook of cytology.

reports have been published on this extraordinary effect induced by ionizing radiation, and a large volume of information has been accumulated concerning aberration mechanisms, dose-response and radiation type–response relationships. Radiation botanists have continued to use *Tradescantia*, but now other plant material has been found equally suitable for the study of chromosomal aberrations, such as *Trillium* microspores (A. H. Sparrow), because of their few (2n = 10) and large chromosomes, and *Vicia faba* root tips (S. Wolff). Animal material has consisted mainly of dipteran salivary cells with their giant chromosomes.

More recently, through the development of new techniques, the investigation has been extended to the aberrations induced in mammalian and human cells irradiated in vivo or in tissue cultures. The successful use of this material is of extreme interest in radiation biology.

Chromosomal aberrations can be induced during any phase of the cell cycle. including interphase. However, since in interphase (and to a lesser extent in prophase) chromosomes are not visible as individual and distinct units, aberrations can usually be detected and scored when cells are fixed and stained in metaphase and anaphase. Various techniques may be employed to facilitate this study, such as the treatment of the material with colchicine to increase the percentage of metaphasic cells, and with hypotonic solutions to induce the swelling of the nucleus and thus the spreading of the chromosomes. Aberrations induced during metaphase or anaphase are usually detectable during the subsequent mitotic division.

If an aberration occurs in a somatic cell and this cell still possesses reproductive activity (for example, epithelial cell and bone marrow cell), the aberrant chromosome(s) will be transmitted only to the cells deriving from it (somatic aberrations); therefore the aberration will be confined in, and die with, the individual. If an aberration occurs in an immature or mature gamete, the aberrant chromosome(s) will be transmitted to all the cells of the offspring originating from the damaged gamete (gametic aberrations).

The mechanism of chromosome breakage has been the object of animated discussion. How does radiation break chromosomes? Is a chromosome break the result of a direct hit on the part of an ionizing particle or photon, just like the snapping of a wire by a bullet? Or is it purely the morphologic response of a chromosome to a biochemical disturbance caused by radiation and therefore the result of very complex interactions of several different factors? The results of different experiments do not lead to the same conclusion in regard to this problem. The sensitivity of the chromosomes to breakage can be modified by previous exposure to ionizing radiation, or by previous or subsequent exposure to far red and infrared radiation. Chromosomal aberrations indistinguishable from those induced by ionizing radiation may also be caused by treatment with radiomimetic compounds or by virus infection. Chromosomes are not equally radiosensitive throughout the cell cycle (see pp. 348 to 349). Chromosomic lesions induced by irradiating metaphasic or anaphasic cells do not become visible until the following mitosis. All these facts would seem to suggest that a chromosome break is not as simple an event as the snapping of a wire by a bullet.

On the other hand, in one *Tradescantia* experiment investigators estimated that chromosome damage had occurred in every nucleus that had been hit by an alpha particle. Radiation of high LET is more efficient in inducing aberrations than radiation of low LET. These facts have been construed by some as evidence that the interpretation of chromosome breaks as results of radiation-induced biochemical disturbances is not entirely satisfactory.

We must admit that the breakage mechanism is still obscure. Perhaps it will be clarified when we acquire more information about chromosome structure and when we understand exactly what breaking of chromosomes entails in terms of their molecular composition.

Another question that has not been satisfactorily answered yet is whether certain regions of a chromosome are more susceptible to breakage than others. Although the experimental evidence shows that breaks may occur at any point along a chromosome, heterochromatic regions are probably somewhat more sensitive. Again, a better solution may be found for this problem when the structure of chromosomes at molecular level is better known.

Types of aberrations—genetic implications

Whether restitution of a broken chromosome occurs depends on several factors, and in particular on the morphologic state of the chromosome (which, in turn, depends on the stage of the cell cycle) at the time when the break occurs. It also depends on how long the gap remains open; the longer it remains open, the lesser the chance that restitution will occur. If the gap remains open for a long time, there is a good chance that the broken ends will join with other broken ends of the same chromosome or with other chromosomes that also suffered breakage (recombination). Which broken ends recombine is mainly determined by random encounters; for instance, if a fragment of chromosome *A* happens to be closer to a fragment of chromosome *B* than to the other fragment of chromosome *A*, recombination between the fragment of *A* with the fragment of *B* is more probable than restitution between the two fragments of *A*.

For some students it might be rather difficult to understand how chromosomal fragments can join or fuse spontaneously in restitution or recombination. The phenomenon might seem to be just as improbable as the spontaneous welding of two steel bars that are simply placed in contact by their ends. However, one should remember that fusion of broken ends occurs also in crossing over (the exchange of chromatid pieces between two homologous chromosomes), which normally takes place in the first meiotic prophase. Apparently there must be strong forces of cohesion between the broken ends of any two threads of chromatic material. This cohesion must be related with some physicochemical properties of the chromatic material itself. If such is the case, then spontaneous fusion of chromosomic fragments should not be any more difficult to understand than the spontaneous coalescence of two drops of mercury that are close enough together.

As already mentioned above, chromosomal aberrations are the results of recombination. Two general types of aberrations must be distinguished immediately—those of the chromosome type and those of the chromatid type (Fig. 12-5).

If a cell is irradiated at a time in its cycle when DNA has not yet been duplicated (early interphase), chromosomes behave as though they were single stranded, and any break produced in a chromosome at any point involves the whole thickness of the chromosome. When chromosome replication occurs later in interphase, the break also is duplicated in both sister chromatids at exactly the same point; a chromosome break does not interfere with chromosome replication, and both centric and acentric fragments replicate normally. Aberrations resulting from this type of break are known as *aberrations of the chromosome type*. Since this type of aberration affects both sister chromatids of a chromosome,

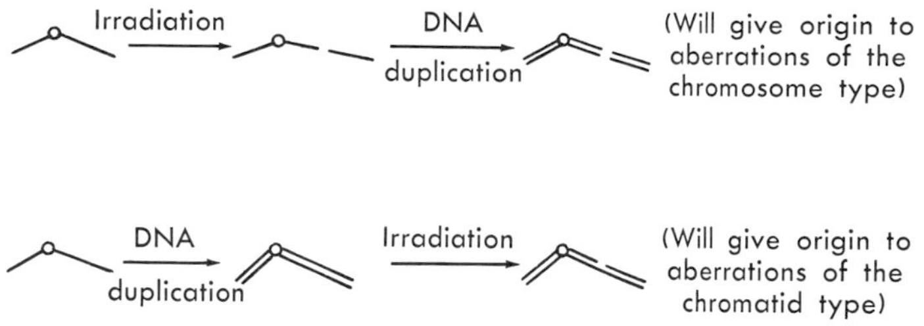

Fig. 12-5. Chromosome and chromatid types of aberrations

it is passed on to both daughter cells, because when the original cell divides, the two sister chromatids of every chromosome become the chromosomes of the two daughter cells respectively.

If a cell is irradiated after DNA duplication (late interphase), chromosomes behave as though each were made of two parallel strands and any break produced in a chromosome involves only one of the strands. When chromosome replication occurs later, the break is found in only one of the two sister chromatids. Aberrations resulting from this type of break are known as *aberrations of the chromatid type*, and they are passed on only to the daughter cell that happens to receive the chromatid where the break took place.

Aberrations of the chromosome type are readily distinguishable from those of the chromatid type with the use of several criteria. However, the abnormal chromosomic configurations resulting from recombination are basically the same in both classes of aberrations. The sequence of mechanical processes that lead to the formation of an aberrant chromosome is sometimes easy to trace back to the original break; at other times it is difficult. A chromosome may be aberrant and yet not detectable as such because its morphologic appearance is not different from that of the chromosome before the break took place (such is the case for some types of inversions, which are discussed below). Only through genetic analysis can certain aberrant chromosomes be identified as such. The following description is limited only to those aberrations leading to chromosomic configurations that can be readily detected as abnormal.

When one break occurs in one chromosome (Fig. 12-6, *A*), a centric and an acentric fragment are obtained. Assuming that no restitution or recombination takes place, the two fragments replicate normally. The centric fragments are able to move, in anaphase, to the poles of the spindle, but the daughter cells receive them with a terminal deletion. The acentric fragments are unable to migrate; they either disappear in cytokinesis, or they happen to be transmitted to only one of the daughter cells. In the course of subsequent cell divisions they will likely be lost.

An intrachromosomal recombination may occur with one break in one chromosome (Fig. 12-6, *B*). Sometime between chromosome replication and anaphase, recombination may take place between the broken ends of the chromatids of the centric fragment and between the broken ends of the chromatids of the acentric fragment. When the centromere splits, the centric fragment is made of two incomplete sister chromatids joined together by their broken ends. This abnormal configuration is called a *dicentric* (with two centromeres); it is also called an *isochromosome* because the joined chromatids originate

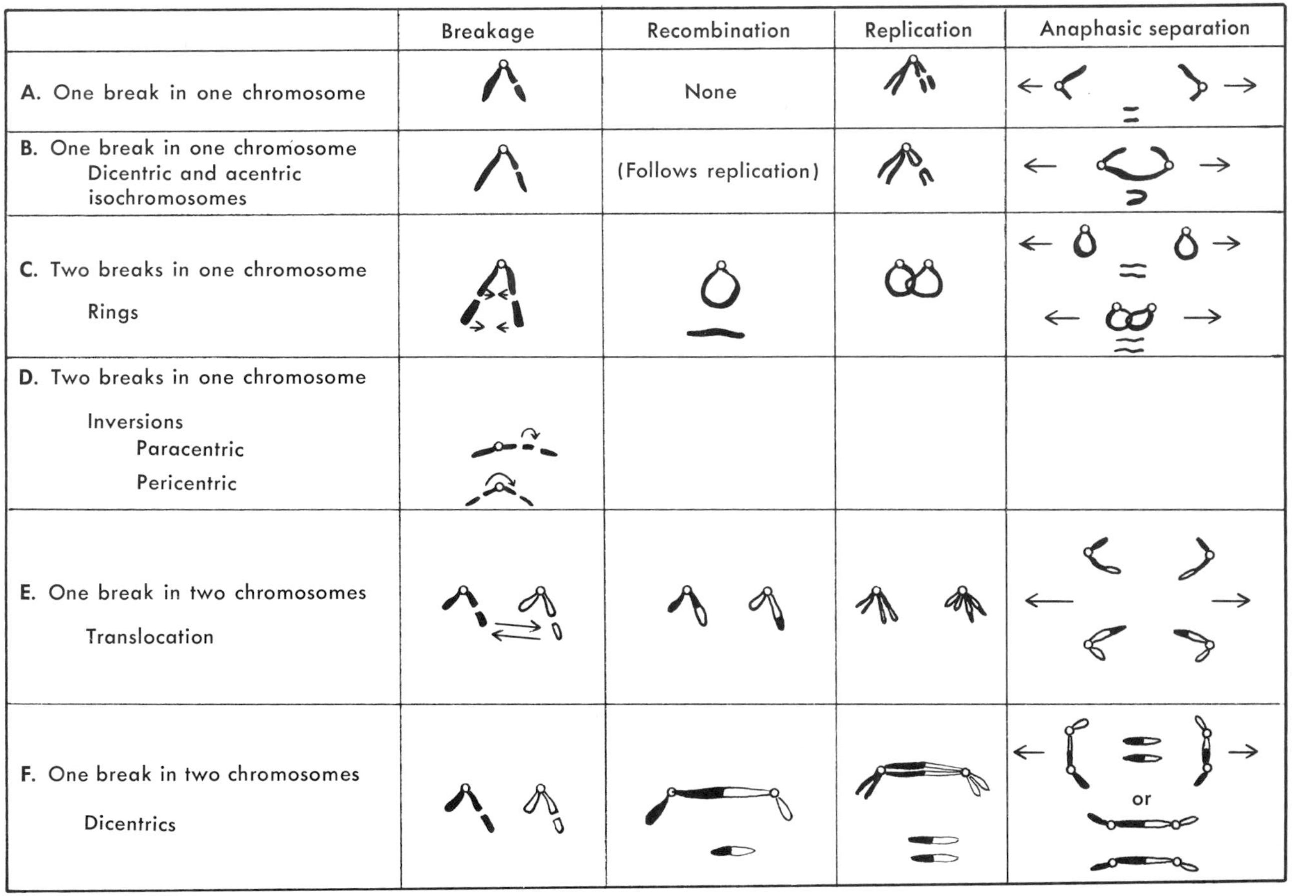

Fig. 12-6. Some chromosome aberrations.

from the replication of the same chromosome. At anaphase, as the two centromeres are pulled toward the opposite poles, an anaphase bridge can be seen. If the bridge is broken after considerable stretching, each chromatid is drawn into each of the daughter nuclei. If the bridge is not broken, the isochromosome is passed on to only one of the daughter cells or to neither. In this case, one or both cells are monosomic (containing only one member of a chromosome pair, with a total chromosome number of 2n-1).

The result of some aberrations may be an abnormal ring-shaped chromosome. Two breaks in the same chromosome are necessary to explain the formation of a ring (Fig. 12-6, C). The breaks must occur in both arms. The two broken ends of the centric fragment recombine and thus form a ring; the recombination of the two acentric fragments yields an abnormal chromosome with an interstitial deletion. The ring-shaped chromosome undergoes normal replication, and the two (ring-shaped) sister chromatids may separate normally at anaphase. But if the centric fragment happened to twist before recombination, the two sister chromatids will be interlocked and will not be able to separate without breaking open.

With two breaks in the same chromosome, an inversion may take place if a nonterminal fragment rotates 180 degrees before recombining with the other two fragments of the same chromosome. If the rotating fragment contains the centromere, the inversion is *pericentric;* if it does not contain the centromere, the inversion is *paracentric.* Replication and anaphase separation of the two aberrant sister chromatids are normal (Fig. 12-6, D).

Translocation may follow when two chromosomes suffer one break (Fig. 12-6, E). The acentric fragment of one chromosome recombines with the centric fragment of the other chromosome and vice versa. Or the centric fragments of the two chromosomes may recombine with each other by their broken ends, with the production of a dicentric (Fig. 12-6, F). The two acentric fragments may also recombine with each other. After replication the two dicentric chromatids may separate from each other or they may form anaphase bridges, which may or may not break, with the possible consequences already outlined above for Fig. 12-6, B.

Since chromosomes are the carriers of the genome, chromosomal aberrations evidently not only result in abnormalities of the mitotic process and in the transmission of abnormal chromosome complements to the daughter cells, but also in genetic damage. When a segment of a chromosome is missing in an aberrant chromosome, all the genes lined up on that segment are also missing. A daughter cell may thus inherit an incomplete genome, or the genome is complete, but certain genes are not carried by the proper chromosomes (as in the case of translocations), or they are not in the proper sequence (inversions). In this latter case, the genetic damage is a consequence of a "position effect"; the phenotypic expression of a gene may be altered when the gene is in the wrong position on a chromosome.

The extent of the total genetic damage transmitted with chromosomal aberrations depends on a variety of factors, such as the type of original cell in which the aberration was induced, number and types of aberrations transmitted to daughter cells, number and kind of genes deleted, translocated, or inverted, and most of all whether the aberration originated in a somatic or gametic cell.

In some cases an aberration may involve such serious genetic damage as to be lethal for one or both daughter cells. Radiation-induced cell death may be the direct effect of chromosomal aberrations produced in the irradiated cell. On the other hand, a minor deletion, even though present in all the cells of an individual (because it originated from a gametic aberration) may be compatible with an almost normal life if it is heterozygous, that is, carried only by one homologue of a chromosome pair.

Whatever the magnitude of the genetic damage may be, with regard to the human population, chromosomal aberrations should be a cause of great concern because they can be induced even by very small radiation doses, as discussed in the following section.

Dose-response relationships

Chromosome aberrations are known to occur also spontaneously, with a frequency that has been calculated for certain species. This frequency is quite low (a few aberrations per hundred cells scored), but too high to be accounted for by background radiation alone. Chromosome breaks and aberrations may simply be the consequence of casual derangement of the chromosomes from their normal behavior, quite apart from any external cause. Radiation-induced aberrations are simply superimposed above the normal burden of spontaneous aberrations; the two types of aberrations are not qualitatively different. In other words, what radiation really does is to increase the frequency of the chromosomal aberrations above the level of the spontaneous ones.

It has been recognized for a long time that chromosomal aberrations are one type of biologic effect that can be induced by doses as low as a few rads. Indeed, we feel that for this effect there probably is no threshold dose.

The shape of a dose-response curve may depend mainly on the following factors:
1. LET of the radiation used.
2. Dose rate, or dose fractionation.
3. Number of breaks required to account for an aberration. In this respect, the aberrations illustrated in Fig. 12-6, A and B, are referred to as one-break aberrations; the others are referred to as two-break aberrations.

The dose-response curve for one-break aberrations is linear and independent of dose rate (Fig. 12-7, A). The LET of the radiation used does not affect the linearity of the curve but only its slope. The linearity is rigorous enough to justify the use of radiation-induced one-break aberrations as a dosimetric method. Thus immature *Tradescantia* anthers (containing microspores) could be used as "biologic dosimeters"; and in fact they were used as such with some success in the nuclear tests at the Bikini atoll.

The linearity of the dose-response relationship for one-break aberrations has been interpreted as meaning that a single break is produced by the passage of a determinate number of ionizing particles or photons. It also means that the probability, p, that radiation will break a chromosome is proportional to the dose. The probability that radiation will produce two independent breaks is then $p \times p = p^2$; therefore this probability should be proportional to the square of the dose. In fact, for two-break aberrations the dose-response curve is not linear when radiation of low LET is used (Fig. 12-7, A). The curve is commonly called a "dose-square curve," although two-break aberrations do not increase

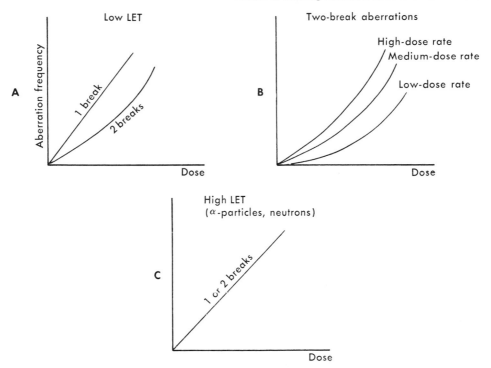

Fig. 12-7. Dose-response relationships for chromosomal aberrations.

exactly as the square of the dose, as expected, but as a power of the dose which is somewhere between 1 and 2.*

Variation of dose rate or fractionation of the total dose delivered affects the dose-response curve for two-break aberrations. In Fig. 12-7, *B* shows that for a given dose the frequency of two-break aberrations induced decreases with decreasing dose rate, because with low dose rates or fractionated doses there is a good chance that one of the two breaks will restitute; a two-break aberration thus becomes impossible.

With radiations of high LET, such as alpha particles and neutrons, the dose-response curve is linear for both one- and two-break aberrations (Fig. 12-7, *C*). The explanation is that probably these radiations lose enough energy per unit length of track in a nucleus to cause more than one chromosomal break. Although the multiple breaks may involve different chromosomes, they are produced by the same ionizing event and therefore they are not statistically independent. The consequence is that, for a given dose, the probabilities of one- or two-break aberrations are equal and both linearly proportional to the dose.

Neutrons are much more efficient than X-rays in inducing chromosome breaks. This difference seems to indicate that in order to produce one break, more than one ionization is necessary. In fact, if one ionization were all that is needed for a break, one would expect that densely ionizing radiations should be less efficient than X-rays, because many of the ionizations occurring close

---

*In *Drosophila,* translocations (two-break aberrations) induced in irradiated sperm increase as the 1.5 power of the dose.

together along the track of the particle would simply be wasted. According to some calculations, the energy dissipated by 15 to 20 ionizations is required for a chromosome break.

Chromosomal radiosensitivity in relation to the cell cycle

A review of the main events that occur during the cell cycle is given at this point for the reader's convenience.

In a population of actively dividing cells (a clone of paramecia, the cells of an epithelial tissue, etc.), each cell goes through a sequence of events that repeat themselves regularly and cyclically. These events involve mainly nuclear DNA, chromosomes, the nucleus as a whole, and the cytoplasm. The sequence is known as the *cell cycle*. At a certain point of the cycle the cell duplicates itself; each daughter cell will go through another cycle.

The cycle may be conveniently described by starting from the end of cell division. Each daughter cell passes through a so-called resting stage, which has been long known as "interphase," before it divides again. Sometime during the interphase the chromosomal DNA replicates itself, so that its total amount is doubled. This event occurs during that portion of interphase known as the S (synthesis) period. The beginning of the S period usually does not coincide with the beginning of interphase; it is preceded by a presynthetic period called G_1 (G means gap), which usually follows the telophase of the preceding cell division. The S period is followed by a postsynthetic period known as G_2 which, in turn, is followed by the prophase of the next division. Interphase, therefore, is made of three subphases in the following chronologic order — the G_1, S, and G_2 periods.

Except in meiosis, chromosome duplication, or longitudinal cleavage of each chromosome into two chromatids, takes place during interphase. The chronologic relationship between DNA duplication and chromosome duplication is not at all clear. There is no sound evidence available yet to believe that chromosomes duplicate only after DNA duplication is completed. The two duplications possibly occur at the same time during the S period, or the structural doubling of chromosomes may take place at G_1. Obviously one reason for these uncertainties is that in interphase chromosomes are not visible as discrete units. Certainly, chromosomes will appear double-stranded (that is, each made of two chromatids) during the next prophase.

Karyokinesis is that stage of the cell cycle accomplishing the division or duplication of the nucleus, that is, the exact partition of the original chromosome complement into the two daughter nuclei, so that each of them receives a complete set of chromosomes and therefore a complete genome.

After two nuclei have been obtained, the cytoplasm divides in two subequal or unequal parts by one of several mechanisms. This phase of the cell cycle is known as cytokinesis. Normally, both daughter cells, independently of each other, go through another cycle starting with G_1.

The cell cycle is schematically illustrated in Fig. 12-8. The relative duration of the different phases of the cycle varies with types of cells. For karyokinesis, anaphase is the shortest phase. Generally speaking, interphase lasts longer than the actual cell division, except in actively dividing cells like the blastomeres in the early embryonic stages. The relative durations of G_1, S, and G_2 also vary; for instance, in Ehrlich ascites tumor cells in tissue culture, G_1 lasts 6 to 8 hours, S 13 to 15 hours, and G_2 6 hours (mitosis lasts 0.75 hours). Other examples are given in the following sections.

In a clone (population of cells originating from the same ancestral cell) the cycles of the individual cellular members may be synchronous or asynchronous; if they are asynchronous, they may be synchronized, in certain instances, with various techniques.

By exposing *Tradescantia* pollen grains to 200 R of X-rays, P. C. Koller, in 1953, was the first to recognize that chromosomes are differently radiosensitive to breaks and aberrations at different stages of the cell cycle. He concluded that their radiosensitivity reaches a peak during the stage when their "division" occurs.

At the present, the results of experiments carried out with different mate-

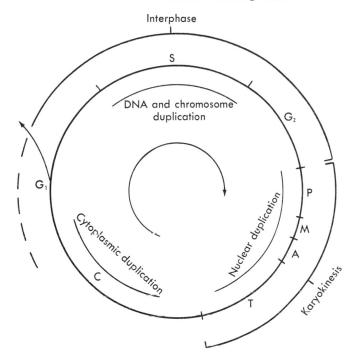

Fig. 12-8. The cell cycle. **P,** Prophase. **M,** Metaphase. **A,** Anaphase. **T,** Telophase. **G₁, S,** and **G₂** are subphases of interphase.

rials may be somewhat discordant, but nonetheless the general picture is that G_1 is a period of relatively low radiosensitivity for chromosomes. As the cell proceeds through S and G_2, the sensitivity increases and then drops sharply just before metaphase. Refined studies with *Tradescantia* microspores (whose cell cycle lasts several days) have shown that chromosome radiosensitivity undergoes significant fluctuations between the beginning and the end of the G_2 period. The relative radiosensitivity of chromosomes during karyokinesis is rather difficult to appraise, because any aberration resulting from breaks occurring during this stage can be detected only during the next cell division.

Different types of aberrations are preferentially induced at different phases of the cell cycle. In regard to the two general classes (chromosome type and chromatid type), aberrations of the first type evidently should be the only ones that can be induced in G_1, or in general before chromosome duplication. Chromatid type of aberrations are commonly induced in G_2, prophase, and metaphase. Different frequencies for inversions, translocations, rings, dicentrics, etc., at different stages of the cycle have also been found.

The change in chromosome sensitivity and the variation in frequency of different types of aberrations through the cell cycle have been attributed to a variety of factors and causes—to different amounts of oxygen available to the cell at different stages, to different amounts of water, and chiefly to different structural configurations and behavior of the chromosomes at different stages. Certainly their degree of spiralization, thickening, etc., should have considerable bearing on the susceptibility of chromosomes to breakage. The type of

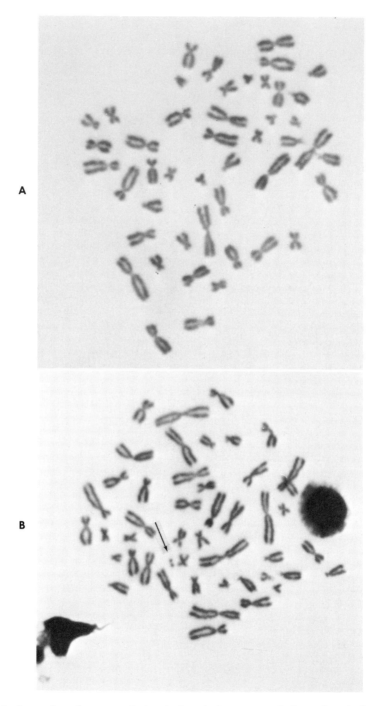

Fig. 12-9. Examples of some radiation-induced chromosomal aberrations in human peripheral leukocytes. **A,** Normal male complement. **B,** Complement with deletion. **C,** Complement with a dicentric chromosome. **D,** Complement with ring chromosome. Arrows indicate the aberrant chromosomes. (Courtesy Bender, M. A. Dec. 1967. An. Acad. Brasil. Ciencias **39:**77-93.)

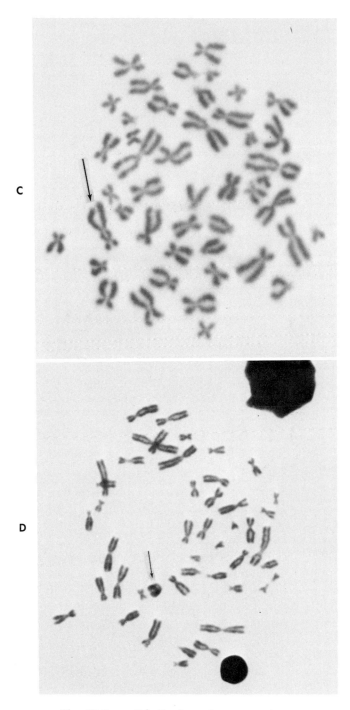

Fig. 12-9, cont'd. For legend see opposite page.

aberrations induced should reflect the spatial relations existing among chromosomes at the time of irradiation.

On the basis of the above considerations, investigation of radiation-induced chromosomal aberrations will become an invaluable research tool for the clarification of the structure of the chromosomes at different stages of the cell cycle, and in particular at interphase when they are not directly visible as such.

Chromosomal aberrations in irradiated humans

In the past the study of radiation-induced chromosomal aberrations in mammals has been made practically impossible by several technical difficulties, and notably by the high chromosome numbers involved, which make the observation and counting extremely unreliable. In more recent times, some special techniques that are actually offshoots of the more general tissue culture techniques have been developed; they have opened the way to a relatively easy analysis of the chromosomal aberrations induced in vivo by radiation in human subjects. One of these techniques is the widely used *human peripheral leukocyte culture method.*

A blood sample is obtained from the irradiated subject. Leukocytes are separated from the red cells and cultured in vitro with the usual techniques used for other types of mammalian cells. When they are in the body, leukocytes do not divide; they are in a permanent pre-S phase. The in vitro culture stimulates them to divide; if they are fixed and stained at metaphase, their chromosomes are microscopically visible. An analysis of the aberrations induced when the leukocytes were irradiated in vivo is thus possible. Of course, this analysis is limited to only one type of cell of the irradiated subject, whereas aberrations certainly occur in other tissues as well.

The same method may be employed for the analysis of gametic aberrations induced in an irradiated subject, by using blood samples obtained from his or her offspring. As already mentioned above, any gametic aberration that is compatible with the production of a mature and viable zygote should appear in all the somatic cells of the offspring. The study of this type of aberration could not be done easily on the gametes, because these cells, in mammals, are even more difficult to handle than somatic cells; furthermore, the analysis of human female gametes would not be practically feasible.

After the introduction of the peripheral leukocyte technique, numerous reports have appeared in the literature on chromosomal aberrations induced in the cells of irradiated human individuals. The following groups of subjects have been studied:

1. Patients subjected to either therapeutic or diagnostic partial or whole-body X- or gamma irradiation
2. Patients who had received exposures from incorporated radionuclides (such as radioiodine) and radium dial painters (see p. 249)
3. Marshallese islanders and Japanese fishermen exposed to fallout from a nuclear device exploded at Bikini (see p. 209)
4. Radiation workers
5. Hiroshima and Nagasaki survivors and their offspring
6. Victims of radiation accidents

In all the above subjects chromosomal aberrations attributable to radiation have generally been found (Fig. 12-9). Unfortunately, in most cases no correlation could be assessed between aberration incidence and dose, because the latter could not be established with certainty, with the exception of patients irradiated for therapeutic reasons. It has been suggested, however, that the measurement of the frequency of aberrations could be used as a "biologic dosimeter," especially in those cases when information on radiation dose cannot be obtained from other sources, as in radiation accidents. Although this type of dosimetry seems quite simple with the peripheral leukocyte method, it does present limitations and problems of calibration, mainly because the dose-response curves are not always linear for any type of aberration (see pp. 346 to 348). Further discussion of this biologic dosimetry can be found in Bender's review (1969).

Chromosome abnormalities have been often observed in patients affected by radiation-induced leukemias. Whether they are the direct cause of the neoplasm or simply a concomitant epiphenomenon is not known; although, of course, one chromosomal abnormality (the Philadelphia chromosome, carrier of a deletion in one of its arms) has been shown to be associated with chronic myeloid leukemia.

Radiation cytogenetics is still unable to assess accurately the genetic damage associated with radiation-induced chromosomal aberrations in the human population. We do not have sufficient evidence yet on the radiosensitivity of the human chromosomes in the germinal cells. Data from experimental animals should be extrapolated to humans with extreme caution. Although chromosomes of different mammalian species might be equally radiosensitive in relation to breakage, the dose-response relationship and the types of aberrations induced must differ among karyotypes of different species on the basis of the chromosome number and of the configuration of the individual chromosomes.*

Effects on other cellular structures

The plasma membrane is not simply a structure that protects and delimits the cellular protoplasm from the immediate surrounding environment. Its main function is to police the traffic of materials between protoplasm and environment. Diffusion of materials through the plasma membrane takes place by means of two different mechanisms: passive-transport phenomena (caused by concentration gradients or electric potential gradients), which require no expenditure of biologic energy, and active-transport phenomena, which consume energy. An active-transport mechanism accounts for the concentration of K^+ ions being constantly higher inside of cells than outside, and for the concentration of Na^+ ions in the interstitial fluids being higher than the concentration of the same ions in the protoplasm. Any excess of Na^+ ions that might penetrate into the cell is promptly expelled through the plasma membrane (the so-called sodium pump mechanism, which plays an important role in the transmission of a nerve impulse along the axon). Unlike artificial membranes, the plasma membrane is selectively permeable.

The first experiments designed to investigate the possible effects of ionizing radiations on plasma membranes date back 30 to 40 years. They have been conducted mainly with red blood cells and yeast cells. In general, radiation causes a definite increase of plasma-membrane permeability, although with moderate doses no visible morphologic damage can be detected. In more recent times, the radiotracer method has been applied to these studies. Using ^{42}K and ^{22}Na or ^{24}Na, several investigators have reported that K^+ ions leak out of the cell and Na^+ ions leak in after irradiation. Rabbit hearts exposed to 1000 R lose in the perfusion liquid four times more potassium than the controls. The sodium pump breaks down in the irradiated giant squid axon. Other

---

*For instance, one should not expect a high frequency of aberrant, ring-shaped chromosomes in a karyotype with a high percentage of acrocentric chromosomes.

experiments indicate that also the permeability to dyes is affected. All these results may be interpreted in the sense that ionizing radiation probably alters the molecular ultrastructure of the plasma membrane (see Danielli's model) or interferes with whatever biochemical mechanism is responsible for active-transport processes.

With massive doses, hemolysis of red cells has been observed and attributed to a possible morphologic damage of the plasma membrane. If red cells of *Amphiuma* (an amphibian) are irradiated, the effect is a total shrinkage of the cells first, followed by swelling and eventually by hemolysis. The same cells have been partially irradiated with microbeams of alpha particles. In this case shrinkage is limited to the irradiated point of the cell only. However, this localized shrinkage is followed by swelling of the entire cell and finally by hemolysis. No observable morphologic damage has been reported in mammalian red cells irradiated with equally large doses. This fact might be comprehensible with the consideration that mammalian red cells lack a definite nucleus.

The radiation-induced increased activity of many cellular enzymatic systems has been reported several times in the radiobiologic literature. One possible explanation of this strange phenomenon is that in normal healthy cells the topographic distribution of certain enzymes and of their specific substrates prevents those enzymes from being active. Enzymes and their substrates are sometimes localized in different protoplasmic compartments so that mutual contact (essential for enzymatic activity) does not occur. The barriers that prevent the contact may be the several membrane systems existing within the cell and especially in the cytoplasm (endoplasmic reticulum, Golgi system, etc.). The disruption of these barriers by any means may result in the contact of the enzymes with their substrates and therefore in a chemical reaction not observable in the intact cell. For example, the white or yellow flesh of the caps of certain mushrooms *(Russula, Boletus)* turns black or blue within seconds after being cut. Although some of these reactions may be triggered by the presence of free oxygen made available to the cut surface, we believe that other reactions are triggered by the injury of cellular structures caused by cutting. This type of cellular injury with disruption of intracellular membrane systems may be induced by ionizing radiation, as shown by Bacq and Herve with *Russula*. Irradiation of uncut caps of this mushroom with soft X-rays (7000 R) causes blackening in the flesh to a depth of 3 to 4 mm, whereas the blackening in the controls is much less extensive.

The lysosome is one cytoplasmic organelle wherein the separation of certain enzymes from their specific substrates is a well-known fact. Lysosomes are essentially small bags (smaller than mitochondria) filled with several enzymes, most of which have been identified. Some of them are proteolytic enzymes (cathepsins) and nucleases, which could hydrolyze several cytoplasmic proteins and nucleic acids if they were allowed to come in contact with them. In an intact and healthy cell, the enzymes are confined within the lysosomes and thus prevented from doing damage to the surrounding protoplasm. The opening of the lysosome and the release of its enzymes has been observed in many cases in concomitance with the death of the cell. The consequence is the autodestruction of the cell.*

---

*For this reason lysosomes have been nicknamed the "suicide bags" of the cell.

In view of the preceding facts, Bacq and Alexander have suggested that radiation-induced cell death may be the direct result of the triggering of the cell's autodestruction mechanism. Radiation would induce the morphologic breakdown of those membrane systems (notably lysosomes) that form a barrier between certain enzymes and those chemical components of the protoplasm that are their specific substrates. The radiation-induced breakdown of lysosomes would release their enzymes into the surrounding cytoplasm and cause the destruction of the cell. This point of view is known as the *enzyme-release hypothesis*. However, there is no direct evidence yet that the low radiation doses sometimes sufficient to cause cell death induce the breakdown of lysosomes. Experiments on isolated lysosomes show that very massive doses are required before any measurable leakage of enzymes from these organelles can be observed. Of course, it is possible that lysosomes are more radiosensitive when they are within the cell.

The electron-microscopic studies of A. Goldfeder on radiation-induced morphologic alterations of the cell's ultrastructure are worth mentioning because of their thoroughness and because of some important general conclusions, that they seem to justify. Two strains of transplanted mouse tumors, as well as normal lymph nodes, were exposed to X-rays with doses as high as 10,000 R with partial body exposure. The electron micrographs clearly show extensive structural changes in the irradiated cells, such as disruption of plasma membranes, dilatation of vesicles such as the cisternae of the endoplasmic reticulum, swelling of mitochondria, and even destruction of mitochondrial cristae. And yet, cells in which such extensive cytoplasmic damage is evident often show an intact nucleus and an intact nuclear envelope.

Based on her own observations integrated with those of other investigators, Goldfeder suggests that the radiosensitivity of different types of cells (with cell death as the criterion) is determined by a variety of cytoplasmic factors, beside the factors proposed in the original law of Bergonié and Tribondeau. The integrity of the cellular membrane systems seems to be a critical factor in cellular radiosensitivity, so that the extent of the radiation injury is determined by the extent of the breakdown of those systems. Death of cells irradiated in interphase is caused by damage to the plasma membrane. The differential radiosensitivity of different types of cells may be determined by the number of mitochondria present in the cytoplasm; the fewer the mitochondria, the greater the radiosensitivity. Thus, the radioresistance of cardiac cells may be explained by the large number of mitochondria they contain. Conversely, the scarcity of mitochondria may be at least a contributing factor to explain the extremely high radiosensitivity of lymphocytes, in spite of the fact that these cells do not normally divide. Conceivably, if a cell is provided with a large number of organelles of the same type and only some of them are inactivated by radiation, enough of them may be left intact, thus continuing their vital functions and reproducing new units for the replacement of the injured ones.

ACTION ON CELL DIVISION

The effect of ionizing radiations on cell division can be investigated properly on populations of cells with reproductive activity and with a fairly high mitotic rate. Early in the history of radiation biology researchers found that radiation interferes with the mitotic process either by retarding it, or by inhibiting it altogether permanently. This permanent inhibition may or may not be followed by cell death.

The experimental material most used for this type of study has consisted mainly of root-tip meristematic cells, pollen grains, fertilized eggs in early stages of cleavage, and, more recently, normal or neoplastic mammalian and human cells in tissue cultures (for example, human carcinoma HeLa cells, mouse L-cells, and Chinese hamster ovarian or lung cells). In the past, most of the experiments were carried out with populations of randomly dividing cells. However, the correct interpretation of the results was made difficult by the presence in the system of cells that were in all possible phases of the cell cycle, both at the time of irradiation and at the time of observation. Later, methods have been developed to synchronize most or all the cell divisions occurring in a population. These methods have been especially used with mammalian cells in tissue culture. The description of all the methods is beyond the scope of this book. It suffices to mention that one of them consists of treating the cells with hydroxyurea prior to irradiation; it has been shown that this drug kills the cells that happen to be in the S phase at the time of treatment without affecting the progression of the cells through the other phases of the cycle. By timing the treatment in accordance with the previously known duration of the cycle, one can obtain an accumulation of all the surviving cells at the G_1 phase and consequently a synchronization of their cycles. Unfortunately, the synchrony achieved with these artificial methods does not last for long. Experiments must, therefore, be carried out within a reasonably short time after synchronization.

The duration of the entire cell cycle (generation time) for synchronous or asynchronous mammalian cells in tissue culture varies with the type of cells although it is constant for constant conditions. Table 12-1 gives an idea of the duration of the cycle and of its phases for three types of cells that have been intensively used as experimental material during recent years.

When an actively dividing cell is irradiated, one or another of the following events may be observed, depending on the total exposure dose, dose rate, type of cell, and phase of the cell cycle at the time of irradiation:

1. The cell dies during irradiation.

2. Cell death occurs some time after irradiation, following a few or no mitoses.

3. The cell survives, but mitoses are permanently inhibited.

4. The cell survives and divides, but the first mitosis or mitoses after irradiation are delayed.

When a large population of cells is irradiated, different cells may be differently affected, especially if they are not synchronized. The events listed above are described and analyzed separately below.

Instant death. To kill mammalian cells while they are still in the radiation field, very high doses and dose rates are required (about 100,000 rads absorbed in a few minutes or less). Obviously this type of *instant death* must be caused by gross disruption of the morphologic structure and gross alterations of the

Table 21-1. Duration (in hours) of cell cycle for some mammalian cells*

	G_1	S	G_2	Mitosis	Total
HeLa cells	9	9	3	1	22
Mouse L cells	6	8	4	1	19
V79 Chinese hamster lung cells	$1\frac{1}{2}$	6	$1\frac{1}{2}$	1	10

*Data provided by W. Sinclair.

chemical machinery of the cell. Massive doses cause rapid coagulation of proteins and depolymerization of DNA. Evidently such damage is not even compatible with a survival of short duration.

Interphase death. With smaller doses, some time may pass before the intracellular damage is amplified to the point leading to cell death. If no mitoses occur between irradiation and death, the cell is said to die of *interphase death,* after the phase of the cell cycle when death occurs. This type of death is known also by other names, such as *nonmitotic death, nondivision death,* and *immediate death.* When cells continue dividing for some time after irradiation and before dying, their death is referred to as *delayed death, clonal death,* or *abortive colony formation.*

The mechanism of interphase death is somewhat obscure; it does not seem to be the consequence of the unsuccessful attempt of the cell to divide, because at times a few divisions may be observed before the cell dies. Furthermore, the interphase death of mammalian cells in tissue culture is comparable to that of lymphocytes exposed to relatively small doses; and lymphocytes are not actively dividing cells. Radiation may induce permanent inhibition of mitosis with survival of the cell (see below). Thus interphase death seems to be unrelated to mitosis.

The dependence of the probability of interphase death on the position of the cell in its cycle at the time of exposure may be conveniently studied with populations of synchronized mammalian cells in tissue culture. A known number of cells is irradiated in a petri dish. The colony-forming ability is taken as a criterion of survival. The number of macroscopically visible colonies formed some time after irradiation is equal to the number of irradiated cells that survived. If any cells survived without forming colonies (see below), they can also be identified with or without the help of a microscope.

Survival data (Sinclair, 1968) show that in Chinese hamster cells with short G_1 the radiosensitivity for interphase death is highest in mitosis and in G_2, less in G_1, and even less in the latter part of the S phase. In cells with a long G_1, there is also a resistant stage in the early part of this phase, whereas the latter part may be as sensitive as mitosis.

Some experiments with root-tip meristematic cells of *Vicia faba* (Hall et al.) have given results that closely resemble the survival patterns shown by hamster cells. The meristematic cells were partially synchronized by treating the root tips with hydroxyurea and exposed to X-rays (100 to 300 R) at various stages of their cycle. Radiosensitivity was appraised by use of growth inhibition of the primary root as a criterion; in fact, reduction of this growth reflects the incidence of cell death in the meristematic region. The meristematic cells were most sensitive during mitosis and resistant during DNA synthesis.

Reproductive death. When mammalian cells in tissue cultures are exposed to moderate doses, it may happen that at least some of them will not form colonies because their proliferating power has been permanently inhibited. Yet they will not die but will continue to metabolize and synthesize nucleic acids and proteins. This phenomenon is referred to as *reproductive death,* or *mitotic death.*

The most interesting contributions to the understanding of reproductive death have been made by T. Puck and others with HeLa strain cancer cells in tissue culture (Fig. 12-10), although even normal cells from a variety of organs behave in the same manner. Using the technique described above, they found that reproductive death may occur in certain cells with doses as low as 50 R. With

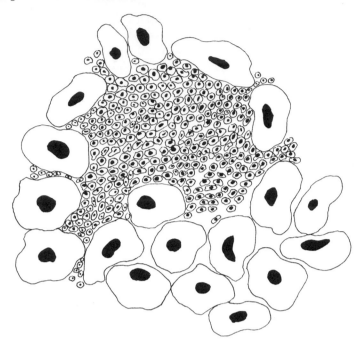

Fig. 12-10. Diagram of radiation-induced reproductive death in HeLa cells. A colony of a few cells was exposed to a dose of a few hundred roentgens. Most of the cells suffered reproductive death and are forming giants (the large peripheral cells). Small cells are the descendants of a cell of the original colony, which escaped reproductive death. They are of normal size.

500 R, only about 1% of the irradiated cells form large colonies comparable to the unirradiated controls. Some of the cells form colonies of normal size, but the cells of these colonies are considerably larger and fewer in number than the cells of the colonies formed by unirradiated cells. In this case, reproductive death has not followed irradiation immediately but after a moderate number of mitotic divisions. If reproductive death follows irradiation immediately, no colony is formed, but the irradiated cell will grow to huge size and thus become a giant cell visible with the naked eye (up to 1 mm in diameter). The metabolic normalcy of these reproductively dead giant cells has been tested in several ways; for instance, viruses have been able to replicate themselves within these cells, an indication that the metabolic machinery of the protoplasm was still in good working condition. Giants have also been observed by others among irradiated mouse leukemia cells in tissue culture. They may grow up to 30 times the normal volume. In Puck's opinion, cellular gigantism observed among his irradiated cells is the result of a chromosomic mutation.

Although the results of these experiments on human cells in tissue cultures are extremely interesting and fascinating and although they provide the best available information in human cellular radiobiology, they should be applied to in vivo situations with a great deal of caution. It is well known that cells in tissue cultures are made "abnormal" by the very fact that they are cultured in glassware, isolated from one another, and fed by artificial mixtures of nutrients, which often bear no resemblance to the interstitial fluids of the body. The reacquisition of the proliferating power by cells that in the body do not divide is in itself an indication of abnormality. The same may be said for the phenomenon of dedifferentiation, which so many times is observed in tissue cultures. The reciprocal interactions existing between cells living together side by side in the body, with their limitations of vital space and of nutrients are certainly absent in tissue cultures. Undoubtedly, the life in tissue culture is subject to many artifacts, and therefore we cannot be sure that the radiosensitivity and behavior of cells irradiated in vitro are the same as when they are irradiated in situ. Nevertheless, in spite of so many shortcomings inherent in tissue culture methods, there are several lines of evidence that seem to indicate that the study of the in vitro behavior of cells is useful for understanding at least some of the events observable in vivo.

In the radiotherapy of neoplastic tissues, both interphase and reproductive death play an important role.

Mitotic delay. Radiation-induced mitotic delay more than any other type of action on cell division has been and is being investigated by a large number of radiation biologists.

Depending on the dose used and type of cells studied, an irradiated cell may fail to divide at the expected time. This condition is a temporary mitotic inhibition from which the cell apparently recovers, because after a certain delay time cell divisions are resumed at normal rates.

The first studies on mitotic delay date back to the 1930s (Henshaw and Francis). The test material used consisted of single fertilized eggs of *Arbacia*. In 50% of these eggs the first cleavage is completed within 1 minute after fertilization. If either the sperms, the unfertilized eggs, or the zygotes are irradiated, the time elapsing between fertilization and completion of the first cleavage is prolonged (cleavage delay). The duration of the delay does not depend on whether the sperm or the egg is irradiated, but depends on the dose. After centrifugation an egg can be cut into two halves, one with the nucleus and the other without. Both halves can be fertilized; after fertilization they cleave normally. If the enucleated half egg is irradiated before fertilization, no cleavage delay is observed, as it is when the nucleated half egg is irradiated. These results would suggest that the site of damage responsible for the cleavage delay is in the nucleus and not in the cytoplasm.

A different method is used for the study of mitotic delay induced by radiation in a large population of cells with reproductive activity, such as cells in tissue cultures. An asynchronous, unirradiated culture of cells contains at any time some cells that are going through mitosis and others that are in interphase. The percentage of cells in mitosis in relation to the total number of cells examined is called the *mitotic index*. The mitotic index depends mainly on the relative duration of the mitotic stage, compared with the duration of the whole cell cycle; the shorter its relative duration, the lower the mitotic index. Of course, variations in environmental conditions (such as temperature and supply of nutrients) also affect the mitotic index of the same culture.

Shortly after the irradiation of an asynchronous culture, mitotic figures start disappearing from the cell population, as indicated by a drop of the mitotic index. To ascertain whether the drop is caused by radiation or not, one compares the mitotic index of the irradiated culture with that of a sham-irradiated culture used as control and examined at the same time. The following ratio may

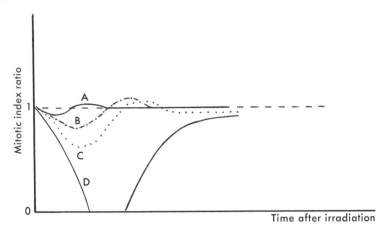

Fig. 12-11. Variation of mitotic activity after irradiation. Doses increase from **A** to **D.**

be referred to as the *mitotic index ratio* and is taken as an indication of the magnitude of the disappearance of mitoses due to irradiation:

$$\frac{\text{Mitotic index of irradiated culture}}{\text{Mitotic index of control}}$$

This method was used for the first time by Spear in his classical studies of the mitotic delay induced by different radiation doses in cultures of chick fibroblasts. The results of other experiments carried out with other types of cells have shown the same general patterns, which can be represented graphically by plotting the mitotic index ratio as a function of time elapsed after irradiation for the different doses used. A family of curves is thus obtained (Fig. 12-11).

Immediately after irradiation (time 0) the mitotic index ratio is usually 1 for all doses. But later it drops gradually until a certain minimum is reached, depending on the dose. The lower the minimum, the longer the time taken to reach it. Note that for very high doses (curve *D*) the minimum may be zero and remain at this value for some time; in this case no cells are going through division during this time. The downward bending of the curves indicates that radiation delays the division of some cells but not of others. Since the population is asynchronous, one may assume that certain stages of the cell cycle are more radiosensitive than others in regard to mitotic delay. The different minima for the different doses are an indication that the duration of the delay is a function of dose.

After a minimum has been reached, the mitotic index ratio climbs back to higher values, which may exceed 1 if the radiation dose was not too high. When the ratio is greater than 1, it means that more mitotic figures are visible in the irradiated culture than in the control. This rise of the curves is due to the fact that the cells whose division was delayed start dividing again at the same time as those not affected by irradiation. The divisions of both groups of cells are visible in the culture at the same time. This phenomenon is a sort of compensating mitotic wave. After reaching a peak, the mitotic index returns to the normal preexposure value and thus the mitotic index ratio becomes 1.

With very large doses (curve *D*) it is possible that not only no compensating wave is observed but also the mitotic index does not return to its normal value.

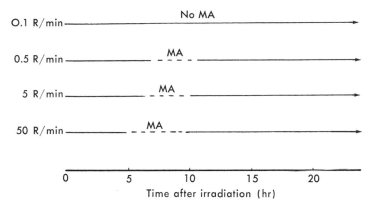

Fig. 12-12. Influence of dose rate on onset and duration of mitotic arrest **(MA)** in *Tradescantia* pollen grains induced by 200 R of X-rays.

The reason is that cells whose mitotic capacity has been inhibited permanently (reproductive death) are present in the culture.

A dependence of the onset time of the mitotic arrest and of its duration on dose rate was shown in *Tradescantia* pollen grains by Koller (Fig. 12-12). With a dose of 200 R of X-rays, the time elapsed between irradiation and onset of the mitotic arrest increases with decreasing dose rate. No arrest is seen with a dose rate of 0.1 R/min (that is, 200 R delivered in 2000 minutes). Also, the duration of the mitotic arrest seems to decrease with dose rate. These experiments would indicate that the radiation damage responsible for the inhibition of mitosis is transient (and therefore reversible and repairable); radiation delivered at a very low rate is either unable to induce the damage, or the damage induced is repaired so quickly that it is temporarily ineffective in inhibiting mitosis.

Studies with mammalian cells in culture have consistently shown that the sensitivity of these cells to mitotic delay depends on two parameters: (1) radiation dose and (2) stage of the cell cycle at the time of irradiation.

When a synchronous culture is irradiated at any stage of the cycle, the delay time is almost a linear function of the dose, as it had already been observed with *Arbacia* eggs (see above); but the slope of the curve is different for different stages. The delay is quite long for cells irradiated in the mitotic phase, is small for cells in G_1, and reaches a maximum for cells in S. With regard to G_2, differences in sensitivity have been found in different types of cells. In mouse L cells and human kidney cells the delay is longer for G_2 than for S.*

In Chinese hamster cells and in HeLa cells the sensitivity in late S phase is greatest for division delay and least for cell death. This fact seems to suggest that the cellular sites concerned with these two radiation-induced effects are different. Furthermore, one can presume that an irradiated S cell has a good chance of survival because the longer time interval elapsing before the beginning of the next division allows the cell to repair the radiation damage before such damage is expressed.

---

*Experiments with plasmodia of the slime mold *Physarum polycephalum* (see p. 329) have shown that in this organism the maximum delay of nuclear division is obtained when the plasmodia are irradiated during prophase or immediately after telophase, when DNA synthesis takes place (G_1 is practically nonexistent in this organism). Irradiation in G_2 results in a considerably shorter delay (Nygaard and Guttes, 1962).

The delay of cell division could be accounted for either by a uniform prolongation of the whole cell cycle or by the prolongation of only some of its stages, whereas the others proceed at their normal pace. The results of several experiments seem to indicate that, although cells irradiated in mitosis are quite sensitive to delay, the cell division occurring at the time of irradiation is not slowed down (provided that cells are irradiated at late prophase or later); the subsequent division is the one that is delayed. The best evidence available at the present time from HeLa and hamster cells is that the mitotic delay is mainly accounted for by a block in G_2, regardless of the stage of the cycle during which the cell is irradiated. For cells irradiated in the S stage, an additional contribution to the mitotic delay comes from the prolongation of S. This added factor would explain why these cells show the longest delay when irradiated in the S stage.

The old hypothesis that the direct cause of mitotic delay may be the inhibition of DNA synthesis is no longer tenable in the light of more recent studies. In fact, mitosis may be inhibited permanently (reproductive death) and yet the cells continue to synthesize DNA. There must be some cellular process (quite unrelated to DNA synthesis) directly responsible for the onset of mitosis; radiation would interfere with such a process. Investigation of its nature is still underway. According to a viewpoint based on recent studies with mouse leukemic cells (Doida and Okada, 1969), the temporary blockage of G_2 might involve interference of radiation with the synthesis of a protein or proteins necessary for the triggering of mitosis. That these proteins might be involved somehow in the formation of the mitotic spindle is a definite possibility.

Besides G_2 block and prolongation of the S phase, a third phenomenon has been observed as a contributing factor to mitotic delay in cells irradiated during prophase. Cells irradiated before a particular critical stage of prophase may not only fail to proceed through mitosis but may actually return to an earlier prophase substage. This radiation-induced *prophase reversion* has been reported for grasshopper neuroblasts and several types of avian and mammalian cells in culture by Carlson and reported by others for *Tradescantia* microspores and onion root tips. Using a hanging-drop culture technique and phase contrast microscopy, Carlson has been able to follow the mitotic behavior of single cells. The prophase of grasshopper neuroblasts may be subdivided into eight distinguishable stages, labeled by Carlson as very early, early, initial middle, intermediate middle, terminal middle, initial late, intermediate late, and terminal late. Each substage can be identified by certain microscopically observable features. Slides containing cells at different prophase substages were X-rayed with different doses (from 8 to 1024 R) and then examined microscopically at regular intervals. Cells irradiated at intermediate-late and terminal-late prophase with any dose are neither delayed nor reverted. Irradiation at earlier stages with low doses induces delay but not reversion. After 128 to 1024 R, nearly all cells irradiated before intermediate-late prophase show reversion. The cells may proceed through more advanced stages of prophase for some time after irradiation before stopping and reverting to earlier stages. The duration of this postirradiation "inertia" is inversely proportional to dose but largely independent of the particular stage irradiated. How far back in prophase the cell reverts also depends on dose; the higher the dose, the earlier the prophase stage reached at the end of the reversion process. Analogous results were obtained with chick fibroblasts and a variety of mammalian cells, except with HeLa cells.

For the discussion and interpretation of these results in relation to the possible biochemical mechanism responsible for prophase reversion, the reader is advised to consult Carlson's original reports. Suffice it here to say that in grasshopper neuroblasts DNA synthesis (and therefore the S phase) extends from middle telophase through very early prophase. Therefore G_2 coincides with some stages of prophase. Neuroblasts irradiated during these stages are actually irradiated during a phase that is described as G_2 in other types of cells, in which S is only a portion of interphase. Thus the block of mitosis in irradiated prophase neuroblasts is really a G_2 block.

Further studies are certainly desirable for a better clarification and understanding of prophase inertia and reversion.

SURVIVAL STUDIES—THE TARGET THEORY

The analysis of the correlation between accumulated dose and the surviving fraction of an irradiated population of organisms has contributed significantly to our understanding of the mechanism of action of ionizing radiations in producing certain effects.

A survival curve is obtained by irradiating a population with increasing doses and then plotting the surviving fraction as a function of dose. When the population consists of a suspension of certain bacteria, or of dried and crystallized viruses, and it is irradiated in certain conditions, the survival curve may have a shape as shown by Fig. 12-13, *a*. The same type of curve may also be obtained when a "population" of macromolecules (enzymes and proteins) is irradiated in the dry state. In this case, "survival" is indicated by lack of inactivation or of any other measurable chemical change.

An inspection of curve *a* shows immediately a very interesting fact: equal increments of dose do not kill equal numbers of organisms. If a certain dose, for example, 5000 rads, kills 50% of the original number of organisms, a dose twice as high (10,000 rads in our example) does not kill 100% of the original number but less. A closer analysis of the curve shows that each dose increment kills the same percentage of organisms surviving after the preceding dose. If dose 1 kills 50% of the population, an additional unit dose (total dose is two units) kills 50% of the population that has survived dose 1, that is, 75% of the original population. An additional unit dose (total dose is three units) kills 50% of the population that had survived dose 2, so that after dose 3 the surviving fraction of the original population is 12.5%, and so forth.

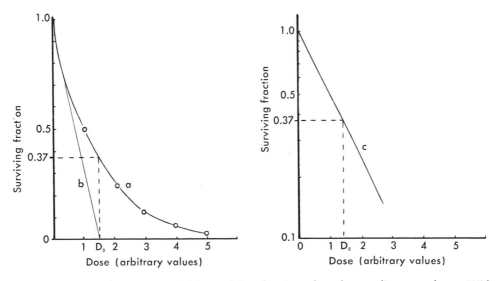

Fig 12-13. Survival curves. **a,** With surviving fraction plotted on a linear scale. **c,** With surviving fraction plotted on a logarithmic scale. **b,** Line extrapolated from low-dose portion of curve **a.** D_0 is the dose necessary to provide an average of one hit per target.

These observations prove that the dose-survival relationship is of an exponential nature; that is, the surviving population decreases at an exponential rate. An exponential rate denotes a constant fractional decrease of the number of organisms per unit dose absorbed. In fact, if the surviving fraction is plotted on a log scale (Fig. 12-13, c), a straight line is obtained.

The survival curve, therefore, fits the equation

$$S = e^{-kD} \tag{12-1}$$

where S is the fraction of the original population surviving after dose D, e is the base of the natural logarithms, and k is a proportionality constant that is the slope of curve c.

Equation 12-1 can also be written in the following log form*:

$$\ln S = -kD \tag{12-2}$$

A theory to explain the survival curve described above was suggested in the 1930s by J. A. Crowthers, N. W. Timoféeff-Ressovsky, and Max Delbruck and later developed by D. Lea. It is known as the *hit* or *target theory*. The elementary notion of the theory in its simplest form is that certain critical regions or sensitive sites of a discrete structure and limited dimensions exist in a biologic system like a bacterial cell or a virus particle. They are referred to as "targets." An ionizing event occurring in or very near the target is responsible for the observed effect (such as the death of a bacterium or the inactivation of a virus in our examples). The target may be a critical molecule, a cell organelle, or even a whole cell. The inactivating event is often called a "hit." If, for instance, the target of a bacterial cell is hit and the hit results in the inactivation of the target, the bacterium will die. A hit at any other point of the bacterial cell outside of the target will be ineffective. Thus, one hit is sufficient to kill a bacterium, if it involves the target; two hits are no better than one; and where there is no hit, there is no effect.

If this is indeed the mechanism of action, certain survival patterns are to be expected, and this can be illustrated with an analogy. Consider a system of boxes as in Fig. 12-14, symbolizing a certain number of targets (16). If eight balls are thrown randomly against the boxes, 50% of the boxes will be struck. This group of eight balls may symbolize a radiation dose (dose 1) that is sufficient to hit and inactivate 50% of a number of targets. If eight more balls (dose 2) are thrown randomly against the boxes, they will not necessarily strike all the remaining 50% of the boxes, although dose 2 per se is sufficient to strike 50% of all the boxes. After dose 2, one can see that some boxes have been hit twice, others once (either by dose 1 or dose 2), and others have not been hit at all. Each successive group of 8 balls (each successive dose increment) reduces the number of unhit boxes (the number of survivors) by one half of their previous number.

From the analogy one can see that if a number of targets is exposed to a dose that is sufficient to hit all of them, some targets will actually be hit once, others more than once, and still others not at all. How many targets will be hit once,

---

*Since the dose-survival curve described above is of the same nature as the time–radioactive decay relationship, the equations describing these relationships are also similar. A review of the mathematic treatment of the decay curve is in Chapter 5.

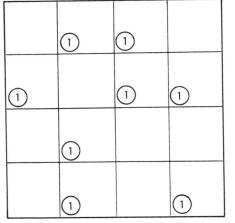

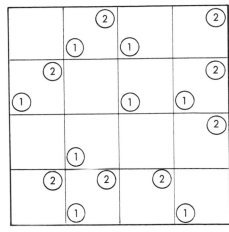

Dose 1   Dose 1 and dose 2

Fig. 12-14. Box analogy to illustrate the shape of the survival curves.

twice, three times, or not at all can be predicted with the Poisson formula, which is the mathematical expression of the Poisson distribution, as follows:

$$P_n(a) = \frac{e^{-a} a^n}{n!}$$

or (when $a = 1$)

$$P_n = \frac{e^{-1}}{n!} \qquad (12\text{-}3)$$

In our case the formula gives the probability (P) that a target will be hit n times when the average (a) number of hits per target is 1 (that is, when the number of hits is equal to the number of targets); e is the base of the natural logs.

Consequently:

37% of the targets will not be hit at all $\left(\frac{e^{-1}}{0!} = \frac{e^{-1}}{1} = 0.37 \right)$

37% of the targets will be hit once $\left(\frac{e^{-1}}{1!} = \frac{e^{-1}}{1} = 0.37 \right)$

18% of the targets will be hit twice $\left(\frac{e^{-1}}{2!} = \frac{e^{-1}}{2} \approx 0.18 \right)$

6% of the targets will be hit three times $\left(\frac{e^{-1}}{3!} = \frac{e^{-1}}{6} \approx 0.06 \right)$

etc.

Returning to the dose-survival relationship shown in Fig. 12-13, one can clearly understand that equal increments do not kill equal numbers of organisms (curve a) because as the dose increases some of the targets are hit more than once, and if one hit is sufficient to inactivate a target, additional hits to the same target are simply wasted. In other words, as the dose increases, the organisms that have not yet been hit are "protected" by the organisms that have already been inactivated. This is tantamount to saying that the inactivation efficiency of the radiation decreases with increasing dose.

With very small doses (whereby the number of hits is much smaller than the number of targets) so few targets are hit that the chance of hitting the same target twice is vanishingly small.* Consequently, at low doses (note upper tract of curve *a*) the dose-survival relationship is practically linear rather than exponential. If this linear region is extrapolated until it intersects the horizontal axis (curve *b*), the dose that can be read at the point of intersection (D_0) represents the dose that would be necessary to provide an average of one hit per target. If this dose were distributed uniformly and not at random, survival after dose D_0 would be 0%. The actual surviving fraction can be obtained graphically by drawing from D_0 a line perpendicular to the horizontal axis until it intercepts the curve *a*. From the point of intercept another line is drawn perpendicular to the vertical axis; the intersection occurs at the 0.37 value of the surviving fraction. Therefore, with a dose D_0 (or D_{37}, as it is sometimes designated) sufficient to produce one hit per target, 37% of the original population of organisms will still survive.

The D_0 dose varies for different biologic systems. It is often used as an index of radiosensitivity of the system under study. It is also mathematically related to the target size, so that the target size, in certain conditions, may be calculated with several methods. This calculation has as its purpose the hope of identifying the target with some known cell structure.

The target theory in its simplest form as described above is applicable whenever the survival curve has an exponential shape and when the effect is independent of dose rate. These conditions are fulfilled with simple biologic systems like populations of bacteria irradiated under certain conditions, dried and crystallized viruses, and even enzymes and other proteins irradiated in the dry state. In these systems it seems indeed that the radiation effect can be identified with a single event occurring in a sensitive volume and is not brought about by accumulation of injuries or by other indirect causes. Consequently, the target theory should be confined to systems wherein a direct action (see Chapter 11) is the sole or predominant mechanism of inactivation. Therefore, if the effect observed in a cell is found to be caused by changes in the surrounding tissues or in the cell's chemical environment as a result of the action of radiation-induced free radicals, the target theory is not applicable. In fact, in these cases the survival curve does not have the simple exponential shape, and a dependence of the effect on dose rate is likely to be observed, an indication that recovery processes are involved. For more complex organisms, the survival curves, in general, are not as simple as the exponential curves. This is an indication that the radiation effect is brought about by more complex mechanisms than postulated by the target theory.

Some survival curves that are not strictly exponential can nonetheless be interpreted in terms of a modified target theory. In Fig. 12-15, although the surviving fraction has been plotted logarithmically, the curves are straight lines only above a certain dose. Below this dose the shape is sigmoid. Survival curves of this type are obtained for instance with *Serratia marcescens* irradiated in presence of oxygen, with mammalian cells irradiated in tissue culture, or with normal mouse marrow cells irradiated in vitro and then cultured in the spleen of a recipient mouse (see the spleen nodule culture method on p. 329).

---

*For instance, if the number of hits is one tenth the number of targets, the probability of hitting a target twice is given by the equation $P_2(0.1) = \dfrac{e^{-0.1} \times 0.1^2}{2!} \approx 0.0045$.

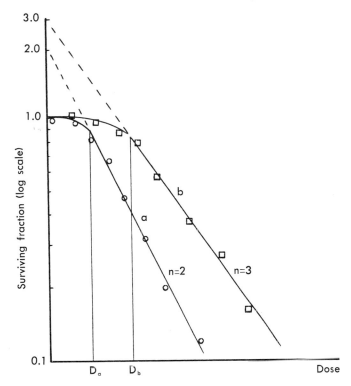

Fig. 12-15. Logarithmic plots of the dose-survival relationship for the multitarget and multihit models. **a** is the survival curve of an organism more radiosensitive than the organism whose survival curve is **b.**

These survival curves can be interpreted with either of the following two models:

1. The biologic system under study contains more than one target; all of them must be inactivated before the observable effect is produced. This model could be called the *multitarget* model. In this case, a sigmoid survival curve must be interpreted as meaning that low doses do not cause appreciable decrease of survival, because with low doses (few hits) and with a large number of organisms the probability that all the targets of the same organism will be inactivated is very low. The larger the number of targets in an organism, the higher the dose needed to detect an appreciable decrease of survival. Thus the survival curve *b* in Fig. 12-15 would indicate that the organism in question contains more targets than the organism whose survival curve is *a.*

At doses where the curve becomes exponential (D_a or D_b) many organisms will have sustained inactivation of almost all their targets; the probability that any further hits will inactivate the one or few targets still left intact is very high. As a consequence, above dose D_a or D_b the dose-survival relationship becomes exponential because the number of organisms left with only one target still intact is large. Therefore the relationship is of the same nature as for organisms with a single target.

Obviously the larger the number of targets present in an organism, the higher the dose required to obtain a given percentage of survival. Therefore D_0 increases with the number of targets.

If the exponential portion of a multitarget survival curve is extrapolated to 0 dose, it will intercept the vertical axis at a point above 1. The ordinate represented by this point is referred to as the *extrapolation number* (or *n* number). In Fig. 12-15, the extrapolation numbers are 2 and 3 for curves *a* and *b*, respectively. One can show mathematically that the extrapolation number may be interpreted as the average number of targets per organism, and therefore as the average number of targets that must be hit before the biologic effect is observed. When the effect measured is survival, extrapolation numbers for mammalian cells in culture vary between 2 and 20.

2. The biologic system under study contains a single target, which, however, must be hit more than once to be inactivated. Only when the target is inactivated, will the effect (death or other) be observed. This model could be called the *multihit* model.

A mathematical treatment of the multihit model would show that actually the dose-survival relationship is expressed by an equation that would not really fit the type of curves shown in Fig. 12-15. However, the differences between a multitarget curve and a multihit curve are so small that they could be detected only if the curves are plotted with extremely accurate data. For most purposes one can say that the type of curves shown in Fig. 12-15 fits either model.

The extrapolation number of a multihit curve may be interpreted as the average number of hits necessary to inactivate a target.

As pointed out above, the target theory is not of general application and the survival curves are not always as simple as those discussed here. Nevertheless, survival studies of systems for which the target theory with its modifications is valid have been extremely useful for the comprehension of the mechanisms of action of ionizing radiations in organisms. The analysis of survival curves provides a means to determine the number of targets (that is, the number of key molecules, cellular structures, etc.) that are responsible for the various functions and whose inactivation results in damage to the functions. Of course, knowledge of their number is not the same as knowledge of their identity. What these critical targets are is not known in many cases. But, for simple systems like those used above as examples, the analysis of survival curves, in conjunction with the calculation of target size and evidence derived from other experiments, may lead to the identification of the target(s). For different organisms and for different effects the target has been variously identified with the DNA molecule, a chromosome, the bacterial chromatid (in *E. coli*), the plasma membrane, or other well-known cellular structures.

REFERENCES

GENERAL

Bacq, Z. M. and P. Alexander. 1961. Fundamentals of radiobiology. Pergamon Press, Inc., New York. Chapter 10.

Bender, M. A. 1969. Human radiation cytogenetics. In Augenstein, L. G., R. Mason, and M. Zelle, editors. Advances in radiation biology. Academic Press, Inc., New York. Vol. 3, pp. 215-269.

Bond, V. P. and T. Sugahara, editors. 1970. Comparative cellular and species radiosensitivity. Proceedings of an international seminar. Kyoto, Japan. The Williams & Wilkins, Co., Baltimore.

Casarett, A. P. 1968. Radiation biology. Prentice-Hall, Inc., Englewood Cliffs, N. J. Chapter 7.

Cellular radiation biology. (A symposium on radiation effects in the cell and possible implications for cancer therapy.) 1965. The Williams & Wilkins Co., Baltimore.

Elkind, M. M., and G. F. Whitmore. 1967. The radiobiology of cultured mammalian cells. Gordon & Breach Science Publishers, Inc., New York.

Errera, M. 1959. Effects of radiations on cells. In Brachet, J. and A. E. Mirsky, editors. The cell. Academic Press, Inc., New York. Vol. I, pp. 695-733.

Harris, R. J. C., editor. 1961. The initial effects of ionizing radiations on cells. Academic Press, Inc., New York.

Powers, E. L. 1962. Considerations of survival curves and target theory. Phys. Med. Biol. **7:**3.

Wolff, S. 1961. Radiation genetics. In Errera, M. and A. Forssberg, editors. Mechanisms in radiobiology. Academic Press, Inc., New York. Vol. 1, Chapter 6.

Wolff, S., editor. 1963. Radiation-induced chromosome aberrations. Columbia University Press, New York.

LITERATURE CITED

Bergonié, J., and L. Tribondeau. 1906. De quelques résultats de la radiothérapie et essai de fixation d'une technique rationelle. C. R. Acad. Sci. (Paris) **143:**983. English translation in Radiat. Res. **11:**587, 1959.

Carlson, J. G. 1969. A detailed analysis of X-ray induced prophase delay and reversion of grasshopper neuroblasts in culture. Radiat. Res. **37:**1.

Carlson, J. G. 1969. X-ray induced prophase delay and reversion of selected cells in certain avian and mammalian tissues in culture. Radiat. Res. **37:**15.

Daniel, J. W., and H. P. Rusch. 1961. The pure culture of Physarum polycephalum on a partially defined soluble medium. J. Gen. Microbiol. **25:**47

Daniels, E. W. 1961. Recovery of reproductive function in supralethally X-irradiated amoebae following cytoplasmic microtransfer. In Progess in protozoology. Proc. 1st Int. Congress, of Protozoology, Czechoslovakian Academy of Sciences, Prague, p. 238.

Daniels, E. W. 1965. Effect of nitrogen mustard on the radiorestorative portion of amoeba protoplasm. Argonne National Laboratory Biological and Medical Division Annual Report. U. S. Department of Commerce, Springfield, Va. P. 198.

Daniels, E. W., and E. P. Breyer. 1970. Rescue of supralethally X-irradiated amoebae with nonirradiated cytoplasm. Radiat. Res. **41:**326.

Daniels, E. W., and H. H. Vogel. 1958. Protective component of nonirradiated protoplasm of amoebae. Proc. second Int. Conf. on Peaceful Uses of Atomic Energy. Geneva, Vol 3, p. 29.

Doida, Y. and S. Okada. 1969. Radiation-induced mitotic delay in cultured mammalian cells. Radiat. Res. **38:**513.

Duryee, W. R. 1949. The nature of radiation injury to amphibian cell nuclei. J. Nat. Cancer Inst. **10:**735.

Goldfeder. A. 1963. Cell structure and radiosensitivity. Trans. N. Y. Acad. Sci., Ser. II, **26**(2):215.

Haber, A. H., and B. E. Rothstein. 1969. Radiosensitivity and rate of cell division: "law of Bergonié and Tribondeau." Science **163:**1338.

Hall, E. J., J. M. Brown, and J. Cavanagh. 1968. Radiosensitivity and oxygen effect measured at different phases of the mitotic cycle using synchronously dividing cells of the root meristem of Vicia faba. Radiat. Res. **35:**622.

Henshaw, P. S., and D. S. Francis. 1936. Effects of X-rays on cleavage in Arbacia eggs. Biol. Bull. **70:**28.

Koller, P. C. 1953. Progress in biophysics. Pergamon Press, Inc., London.

Lea, D. E. 1962. Actions of radiations on living cells. Ed. 2. Cambridge University Press, Cambridge, England. Chapter 3.

McGrath, R. A., R. W. Williams, and R. B. Setlow. 1964. Increased ^{3}H-thymidine incorporation into DNA of irradiated slime mould. Int. J. Radiat. Biol. **8:**373.

Nygaard, O. F., and S. Guttes. 1962. Effects of ionizing radiation on a slime mould with synchronous mitosis. Int. J. Radiat. Biol. **5:**33.

Nygaard, O. F., S. Guttes, and H. P. Rusch. 1960. Nucleic acid metabolism in a slime mold with synchronous mitosis. Biochim. Biophys. Acta **38:**298.

Puck, T. T. April 1960. Radiation and the human cell. Sci. Amer. **202:**142.

Puck, T. T., and P. I. Marcus. 1956. Action of X-rays on mammalian cells. J. Exp. Med. **103:**653.

Sax, K. 1938. Induction by X-rays of chromosome aberrations in Tradescantia microspores. Genetics **23:**494.

Sax, K. 1950. The effects of X-rays on chromosome structure. J. Cell. Comp. Physiol. **35**(suppl. 1):71.

Sinclair, W. 1968. Cyclic responses in mammalian cells in vitro. Radiat. Res. **33:**620.

Sparrow, A. H. 1965. Relationship between chromosome volume and radiation sensitivity in plant cells. In Cellular radiation biology. The Williams & Wilkins Co., Baltimore. Pp. 199-222.

Sparrow, A. H., R. L. Cuany, J. P. Miksche, and L. A. Schairer. 1961. Some factors affecting the response of plants to acute and chronic radiation exposures. Radiat. Botany **1:**10.

Spear, F. G. 1953. Radiation and living cells. Chapman & Hall Ltd., London.

Till, J. E., and E. A. McCulloch. 1961. A direct measurement of the radiation sensitivity of normal mouse bone marrow cells. Radiat. Res. **14:**213.

Von Borstel, R. C., and R. W. Rodgers. 1957 and 1958. Alpha-particle bombardment of the Habrobracon egg. Radiat. Res. **7:**484; **8:**248.

Wolff, S. 1968. Chromosome aberrations and the cell cycle. Radiat. Res. **33:**609.

Zirkle, R. E. 1932. Some effects of alpha irradiation on plant cells. J. Cell. Comp. Physiol. **2:**251.

CHAPTER 13

Fundamentals of radiation genetics

GENERAL CONSIDERATIONS

In the preceding chapter (p. 345) we have seen that chromosomal aberrations induced by radiation in gametes may result in genetic damage in following generations. This type of genetic effect may be termed *intergenic*, because a chromosome break (which is the essential condition for the formation of an aberration) takes place between gene loci and does not affect the structure of the gene. The sudden appearance in a population of a new, inheritable phenotypic characteristic caused by a chromosomal aberration is a *chromosomic mutation*. But new phenotypic characteristics may suddenly appear without any chromosomal breaks or structural rearrangements; they are caused instead by changes in the very structure of the genes. These changes are known as *gene*, or *point, mutations*.

Not very much is known about the true mechanism of a point mutation. From what we know at the present time about the nature of the gene, a gene may conceivably mutate as a result of an error in the process of DNA duplication. The consequence may be either the complete absence of a purine or pyrimidine base at a certain point of the newly synthesized DNA strand (base deletion), or the presence of a wrong base in the new strand, that is, a base not complementary to the corresponding base of the preexisting strand (base change). It is also possible that a gene mutates at some time other than during DNA duplication under the influence of causes that may be known or unknown.

Alternatively, a point mutation may consist of the total loss of a gene, so that any phenotypic manifestation under its control simply fails to appear. Albinism in the human species and white eyes in *Drosophila* are believed to be caused by the absence of the gene responsible for skin pigmentation or eye color. Once a gene has mutated, it is replicated exactly with its new configuration in all the following DNA duplication processes, unless the mutated gene undergoes a new mutation in a different direction or reverts to its original configuration (back mutation), which is a very rare event indeed. For all practical purposes, a mutation can be therefore considered as an irreversible process.

Ionizing radiations are capable of inducing point mutations; as such, they are mutagenic. This is another type of genetic effect, which may also be termed *intragenic*. Apparently the ionizations produced by radiation have enough energy to break chemical bonds and to upset the molecular configuration of a gene. Although the phenotypic manifestations of intergenic and intragenic alterations may be similar, in this chapter we shall consider only the problems related with radiation-induced point mutations.

Historical background

Shortly after the birth of genetics, investigators recognized that gene mutations may occur spontaneously in nature. The word "spontaneously" does not necessarily mean that these mutations are always the result of mere chance; it means that these mutations could be the result of chance or could be produced by some external environmental agents existing in nature, which we are not always able to identify. Geneticists believe that the few or many alleles of the same gene found in the genetic pool of a species arose as a consequence of several spontaneous mutations of an original gene, which occurred during the long existence of the species.

However, if we examine the individuals of a limited number of generations in a species, looking for mutations, we reach the conclusion that the frequency of spontaneous mutations is very low indeed. To score a significant number of mutants, one should examine thousands of individuals of several generations. The task, of course, has been possible with *Drosophila*. Within a few years after *Drosophila* had been introduced in the genetics laboratory, and after thousands of flies had been examined for several generations, a few new phenotypic characteristics that were attributed to gene mutations were discovered. Perhaps one of the most famous examples is the appearance of a white-eyed male in a culture bottle in T. H. Morgan's laboratory.

Geneticists recognized that the study of gene mutations could pave the road to the understanding of the nature of the gene itself. Since the study of mutations was hampered by the extremely low frequency of spontaneous mutations, geneticists began seeking ways to increase this "sluggish" mutation rate. In 1911, Morgan and his group exposed fruit flies to radium radiation. Some wing mutations were obtained, but because of the genetic impurity of the material used, the investigators were reluctant to conclude that those mutations were radiation induced.

A frantic search for mutagenic agents continued incessantly in several laboratories during the second and third decades of this century. A variety of physical and chemical agents were tried (temperature, ultraviolet radiation, and a large number of chemical compounds). The results were either negative or at best inconclusive, mainly because of the inadequate methods used for the quantitative estimation of any increase of mutation frequency.

After developing the proper techniques (see p. 376), H. J. Muller exposed *Drosophila* males to X-rays and produced definite evidence that X-rays induce point mutations in sperm cells. His paper "Artificial Transmutation of the Gene," published in 1927, is one of the classics in the history of genetics. In this first paper, Muller already recognized the dangers deriving from X-irradiation of the germinal tissues and distinguished between the hazard connected with somatic effects and the implications of the long-term genetic effects of radiations.

The mutagenic effects of X-rays and other ionizing radiations were soon confirmed by other investigators working with other materials. In 1928, L. J. Stadler announced that X-rays are mutagenic for barley and maize seeds. In his second paper, he reported that the rate of induced mutations was independent of the temperature at the time of irradiation and that the relation between mutation rate and total dosage was linear.

Types of mutations

The only point mutations that can be inherited by the progeny of an individual are the *gametic* mutations, that is, those mutations occurring in the individual's mature or immature sex cells. But mutations may occur or be induced also in the genome of somatic cells; *somatic* mutations are obviously noninheritable and are extinguished with the death of the individual that carries them.

The phenotypic expression of a somatic mutation is detectable when the mutation occurs in a tissue of high mitotic activity (for example, epidermis and bone marrow) and in particular in embryonic or larval tissues. A brown-eyed child born from parents both with blue eyes (bb × bb) may be the result of a somatic mutation, taking place in the zygote or at some slightly later developmental stage, of the allele b (blue eyes) to B (brown eyes). Eyes of different colors in the same individual are also explained by a somatic mutation that must have occurred in the primordial cell of one eye. If the mutation occurs some time after the one-cell primordium, only part of the organ may be affected. In *Drosophila* the development of the eyes begins in the larval stages and is completed some time at the end of the pupal stage. J. T. Patterson exposed normal larvae of different ages to X-rays and examined the eyes of these flies in the adult stage. The larvae used were heterozygous Ww females and hemizygous WO males; their eyes were expected to develop with the red color of the wild type, if no mutation in the allele W occurred.* In a significant percentage of irradiated individuals, Patterson noticed red eyes of the wild type with white patches. The size of the patches was large in flies irradiated in early larval stages and small in flies irradiated at later stages. Apparently, the earlier the stage of irradiation the more ommatidia were affected by the mutation. The results show that the mutation was induced in one single primordial cell; when mutation was induced in an early primordial cell, it was transmitted to a large number of ommatidia.

Similarly, red dahlia plants irradiated chronically in the gamma field of the Brookhaven National Laboratory with 118 R/day have occasionally produced red inflorescences with few or many white flowers. This is another example of somatic mutation; it is induced by radiation at an early or late stage of the inflorescence primordia.

Some radiation biologists suspect that radiation-induced leukemia might be caused by some type of somatic mutation induced in the cells of the bone marrow or of the lymphoid tissues.

Since somatic mutations are restricted to the individual and are not transferable to the following generations, they do not alter the balance existing in the genetic pool of a species, and therefore are much less important than gametic mutations in terms of population genetics. In the remainder of this chapter we shall consider only gametic mutations.

A point mutation may be either dominant or recessive, depending on whether the mutated gene behaves as a dominant or as a recessive allele in a diploid chromosome complement. A dominant mutation manifests itself even when it is heterozygous. Therefore, if a dominant mutation occurs within a

---

*Recall that in *Drosophila* the gene for eye color is sex linked and that the allele w (white eyes) is recessive to the allele W (red eyes of the wild type).

gamete and this gamete is involved in fertilization, the phenotypic expression of the mutation will be observable in the individual that develops from the zygote. On the contrary, a recessive mutation may not express itself in the phenotype for a number of generations; it will be observable only when by mere chance it happens to be associated with the same allele in the same zygote. The reason is that recessive alleles can express themselves in the phenotype only when in homozygosity. An exception, of course, exists for sex-linked genes; recessive alleles express themselves phenotypically when hemizygous, that is, when carried by the heterogametic sex. Thus in *Drosophila*, in which the heterogametic sex is the male (XY), a recessive allele carried by the X chromosome is hemizygous in males and therefore will express itself.

The phenotypic expression of a mutation may be observed as a more or less gross morphologic abnormality *(visible mutation)*. Examples of visible mutations are curly wings in *Drosophila* and veinless, crumpled, or clipped wings in *Habrobracon*. Other times mutants are difficult to distinguish and identify, because the mutated gene affects either some minute morphologic traits, or some physiologic or biochemical characteristics.

Other mutations are not visible for an entirely different reason. A mutation may be of such a nature as to be incompatible with the life of the individual carrying it; the biologic function under the control of a gene may be so vital that if it is lost or altered the organism cannot live. The individual will die very early during its embryonic or larval development. Such mutations are called *lethal*. Of course, dominant lethal mutations are always phenotypically effective (in killing the carrier), whether they are homozygous, heterozygous, or hemizygous; whereas recessive lethals are phenotypically effective only when they are homozygous or hemizygous. Carriers of dominant lethals or of homozygous or hemizygous recessive lethals are never visible in a population.

Certain mutations, although not completely incompatible with life, may more or less handicap the life of their carriers. They are referred to as *semilethal* (or sublethal) and *detrimental;* and many times they are visible mutations in the sense explained above. In *Drosophila*, a semilethal mutation is defined as one that allows 0 to 10% of its carriers to survive; and a detrimental mutation is one that allows more than 10% but less than 100% of its carriers to survive.

Parenthetically, it should be noted that in radiation genetics the term "detrimental mutation" is often used in a broader sense than the one stated above; it is used as meaning "unfavorable" to the individual that carries it, because it decreases the individual's chance of survival in the competition with nonmutants of the same species. Unfortunately, most mutations, both spontaneous and radiation-induced, are unfavorable or harmful; they introduce a point of weakness into a species. Most mutants are eliminated by the natural selection pressure. However, as old mutations are wiped out, new genes arise by mutations, so that what is called the "genetic load" of a species is kept more or less constant.

In regard to gametic mutations, when a mutation occurs (or is induced) in an immature sex cell, it will be transferred to all mature gametes originating from it. Consequently, the number of gametes carrying a given mutation depends on the stage in gametogenesis when the mutation occurred. Thus a mutation induced in a spermatogonium is likely to be found in more sperms than a mutation induced in a primary spermatocyte. The larger the number of sperms carrying the same mutation, the higher the probability that the mutation will be perpetuated in the following generations.

Organisms studied

The study of radiation-induced mutations, their scoring, and, most of all, the quantitative correlation of their frequency with radiation parameters (such as dose, dose rate, and energy) are not equally easy with any organism that can be handled in a laboratory. However, appropriate techniques have been developed for some particularly favorable organisms.

Drosophila

The usefulness of this organism in many fields of genetic research is well known to any student who has taken an introductory course in genetics. Because of the prolificity of this species, its short life cycle, and the simplicity of the culture media, thousands of flies and several generations can be obtained in a few months. Much has been learned from *Drosophila* about the mutagenic effect of ionizing radiation. Some of the techniques used are mentioned and described on pp. 376 to 379.

Mammals

The desire for learning more about the genetic effects from species more closely related to man has stimulated the ingenuity of some investigators in the search of a mammalian species suitable for radiation genetic studies. So far, the mouse has been the best answer to the problem. Large colonies of mice (by the thousands) must be reared for this type of study. Because of a lesser prolificity and longer life cycle, studies with mice are more time consuming than studies with *Drosophila*. But this is the price science must pay for learning facts from an organism phylogenetically akin to man. The techniques used in mouse experiments are described on p. 382.

Neurospora

The bread mold *Neurospora*, and in particular the species *N. crassa*, has been one of the favorite organisms in genetic studies since the 1940s and has contributed to the knowledge of the mechanism of the phenotypic expression of the gene especially through the classical experiments of G. W. Beadle and E. L. Tatum. Some of these experiments involved the induction of mutations by means of X-rays and ultraviolet radiation.

Neurospora has a short life cycle (about 10 days) and can be propagated abundantly by means of asexual spores (conidia). With the exception of the zygote, the life cycle is entirely monoploid (in *N. crassa* n = 7), which is of considerable genetic advantage, because any induced mutation manifests itself immediately in the phenotype, regardless of whether it is dominant or recessive.

As it often happens in many protozoa and other fungi, there are two mating types in *Neurospora*, designated *a* and *A*, and therefore two types of mycelia. Conjugation occurs between a hypha of a type *A* mycelium and a hypha of a type *a* mycelium. The result of the fusion is a diploid zygote, which immediately undergoes meiosis, with the formation of four monoploid cells, which then divide once more by means of normal mitosis to yield a total of eight ascospores, four of each mating type. The ascospores originating from the same zygote remain enclosed in the same ascus and therefore do not mix with ascospores deriving from other zygotes. Furthermore, in the ascus they are aligned in a row in the same order in which they are produced through the three divisions. This is a significant advantage in many types of genetic experiments. When the ascospores germinate, they give origin to mycelia, which grow in size and reproduce asexually by means of conidia. A zygote is formed again when mycelia of opposite mating types happen to come in contact.

A mycelium of the wild type grows on a minimal medium consisting of water, a carbohydrate (sucrose), certain inorganic salts, and the vitamin biotin. With these materials, the organism can synthesize many other metabolites and especially amino acids and all other vitamins needed. Nutritional mutants are unable to synthesize one or more vitamins or amino acids. They grow only on a minimal medium to which the substance they are unable to synthesize has been added. For example the mutant "pantothenic-less" is unable to synthesize pantothenic acid (a vitamin) and therefore can grow only on a minimal medium to which pantothenic acid has been added.

When a large number of conidia of the wild type are X-irradiated, a significant percentage of mycelia originating from them will be nutritional mutants. The exact identity of the mutation induced can be ascertained by first growing the mycelia on a complete medium (containing all vitamins and amino acids that are known to be synthesized by the wild type) and then testing their ability to grow on a complete medium minus one of the vitamins or amino acids. If a mycelium does not grow on a complete medium that is deficient in pantothenic acid, it must be a "pantothenic-less" mutant.

Morphologic mutants also occur. The morphology of one mutant, for instance, is characterized by a buttonlike colonial growth on agar (similar to the typical colonies of most bacteria), whereas the wild type shows a weblike growth with long branching hyphae rising from the surface of the agar medium.

In summary, the main advantage of *Neurospora* in radiation genetics is the monoploidy of its vegetative stage; this monoploidy makes possible the immediate identification of any mutation induced (dominant or recessive) without the necessity of waiting for several generations, as is often the case for diploid organisms. For further details on the genetic importance of *Neurospora* see *General Genetics*, by Srb and Owen, cited on p. 392.

Bacteria

Bacteria have some advantages in common with *Neurospora*. The evidence from bacterial cytology and genetics is that although bacteria do not possess true chromosomes the genome is often organized in the form of a circular filament of DNA and that genes are in single doses. Therefore, bacteria are also monoploid and any mutation becomes visible in the colony arising from the mutant cell. In addition, the life cycle is much shorter than in *Neurospora*. Millions of organisms can thus be obtained from a few cells in a very short time — a definite advantage in mutation studies.

As in *Neurospora*, the phenotypic expression of many mutations in bacteria is of a biochemical nature; that is, some mutants cannot synthesize certain compounds that the nonmutants can. Other mutations affect certain morphologic and macroscopically visible characteristics of the colonies (color, shape, size, aspect, etc.). Still others affect the sensitivity or resistance to antibiotics, the ability or inability to produce certain antigens, and the capacity of being stained by particular dyes.

Identification of mutations can be done on the colonies formed by the irradiated cells on solid medium. It is a valid procedure because each colony is a clone, and therefore all the cells of the colony have exactly the same genotype as their ancestral cell.

Viruses

The phenotypic expression of mutations induced in the viral genetic material is best observed when the irradiated virus is allowed to infect its host cell; if a mutation was induced in the virus, the damage produced to the host by the infection or the reaction of the host to the infection is possibly different. With phages, a mutation can be revealed by studying the appearance, size, or shape of the area cleared on an agar plate inoculated with the host bacterium when the irradiated virus is added to the plate. A mutated plant virus might produce lesions of a different nature in the leaves of the host. Mutated animal viruses may evoke skin reactions different from those evoked by the nonmutated strain.

Many experiments have been performed on viruses with mutagens other than ionizing radiations (nitrous acid and mustard gas). These studies have contributed significantly to a better understanding of the molecular mechanism of mutation.

DROSOPHILA STUDIES
Methods

In *Drosophila*, the detection of radiation-induced mutations is often complicated by the diploidy of this organism, because if the mutation induced is recessive it might not appear in the offspring until it becomes homozygous.

Let us assume that a thousand sperms are irradiated and that in 20 of them a dominant mutation is induced. If the mutation is visible (not lethal) and all the irradiated sperms fertilize a thousand eggs, 20 mutants will be scored among the thousand flies of the first generation. If the dominant mutation is lethal, the individuals formed by the eggs fertilized by the sperms carrying the mutated gene will never reach the adult stage, and the number of flies of the first generation will be 980 instead of 1000. In either case the mutation and its frequency will be detectable immediately.

But if the mutation is recessive, it will not manifest itself in any of the flies of the first generation, because in the zygotes the mutated gene will be associated with a dominant allele. This fact is true, whether the mutated gene is autosomal or sex linked (that is, carried by the X chromosome). One should remember that in *Drosophila* (as well as in all other species with XY males) the father's X chromosome is transmitted only to the female progeny. The recessive mutation may be detectable only in one of the subsequent generations. In addition, the recessive mutation will manifest itself phenotypically only in some of the offspring born from siblings carrying the mutated gene in the heterozygous condition. However, a sex-linked recessive mutation will manifest itself in any male that happens to inherit the X chromosome carrying the mutation, because males are hemizygous for all sex-linked genes.

Since most mutations (spontaneous or induced) are recessive, more or less laborious and sophisticated techniques are required to detect radiation-induced mutations and to measure their frequency. Several methods have been developed for this purpose; they are designed to detect recessive mutations (lethal or visible) induced in the X chromosome or in certain autosomes.

The most ingenious and classical technique is the so-called ClB method, first developed and used by Muller in his pioneer experiments on the mutagenic effect of X-rays. This method is designed to detect sex-linked recessive lethals in *Drosophila* and to measure their frequency (Fig. 13-1).

A special *Drosophila* stock is needed, the females of which have one normal X chromosome and the other X chromosome marked by a crossing-over suppressor (C), a recessive lethal gene (l), and the dominant marker bar eye (B).

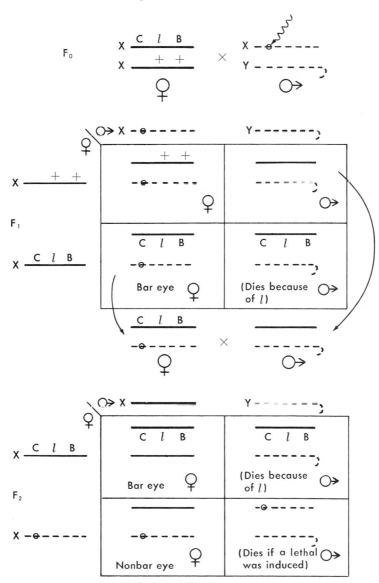

Fig. 13-1. The ClB method for the detection of sex-linked recessive lethal mutations in *Drosophila*. The small circle on the paternal X chromosome represents a radiation-induced recessive lethal.

C is an extensive inversion that is known to prevent crossing over between the ClB chromosome and its homologue, so that no recombination of sex-linked alleles is possible in these females. B and l are both in the inverted segment. These females are therefore heterozygous for C, l, and B and are bar eyed. Males of the wild type are irradiated and then mated with ClB females. In the F_1 offspring, the females receive one X chromosome from the mother and one (irradiated) from the father; 50% of them are bar eyed and carriers of l and C. Fifty percent of the F_1 males die because they receive the ClB X chromosome from their mother and the lethal gene expresses itself phenotypically by a lack of viability.

The F_1 bar-eyed females are then mated individually with normal males (one pair in each culture vial). The F_2 females are all viable; 50% of them are carriers of a ClB chromosome and therefore have bar eyes; the other 50% carry the X chromosome that was irradiated in the sperm cells of their grandfather. If a recessive lethal mutation was induced in this chromosome, it is heterozygous in these females, which would therefore be viable. Fifty percent of the F_2 males die because they are the carriers of the gene l in hemizygosity. The other 50%, which receive the irradiated chromosome from their mother, do not survive if this chromosome carries a recessive lethal mutation. In this last case, no males appear in the F_2. Consequently, if an F_2 culture vial contains no males, the mother of this offspring was the carrier of an irradiated X chromosome in which a recessive lethal mutation had been induced. The experimental procedure in the ClB technique is such that the number of F_2 cultures is equal to the number of irradiated sperms involved in fertilization in the F_0 cross (irradiated males × ClB females). Therefore, the percentage of F_2 cultures with no males is identical to the percentage of irradiated sperms used in fertilization that carry an induced recessive lethal mutation. Thus the great advantage of the ClB technique is that to detect the number of mutations induced all one need do is to count the maleless F_2 cultures.

Other techniques have been developed, by Muller and others, for the detection of recessive mutations in *Drosophila*. Since the inversion in the ClB chromosome does not completely suppress crossing over and therefore does not completely prevent the loss of a possible radiation-induced mutation in the irradiated chromosome, an improved technique, known as the M-5 technique, has been introduced. Its purpose is also the detection of recessive, sex-linked lethal mutations; two generations are also needed, as in the original ClB technique. In addition, the M-5 technique may also be used to detect mutations induced in irradiated eggs.

The "attached X" test is useful for the detection of sex-linked, recessive, lethal or visible mutations. A special stock of females with attached X chromosomes is needed for this test. The advantage is that only one cross is necessary for the completion of the test.

A method developed by Child can be used to detect lethal recessive mutations induced in the autosomes II and III. The method is laborious and requires three generations.

The frequency of spontaneous or induced mutations is expressed with figures that always need qualifications, as it is seen in the following examples.

When Muller applied the ClB technique to unirradiated males, he found that approximately two sperms out of a thousand used were carriers of a sex-linked, recessive lethal mutation. Consequently, the frequency of spontaneous recessive lethal mutations in any locus of the X chromosome in *Drosophila* is 0.002 or 0.2%. With an exposure of males to 1000 R of X-rays the frequency of recessive lethal mutations on the X chromosome was 0.03 (or 3%). Since there is some evidence that the *Drosophila* X chromosome contains about 300 gene loci in which lethals may be induced, with 1000 R the frequency of lethal recessive mutations at any one locus is 0.03/300, or 0.0001 (= 0.01%). The frequency of sex-linked recessive lethal mutations per locus per roentgen is 0.0001/1000 or 0.0000001; that is, there is one chance in 10 million that exposure of a single gene of the X chromosome to the dose of 1 R will result in a lethal recessive mutation.

The doubling dose is the radiation dose necessary to give an induced mutation frequency equal to the spontaneous mutation frequency. The estimation of the doubling dose is another useful way to appraise the mutagenic efficiency of radiation in different species.

Results

With various techniques several investigators have studied the mutagenic effects of ionizing radiations in *Drosophila,* their relation to dose, dose rate, and LET, and the nature of mutations induced. The results of their experiments are summarized below.

1. Although one would expect that the magnitude of the genetic damage produced should be a function of the radiation dose, the radiation-induced genetic damage is not represented by mutations more detrimental than those which occur spontaneously. From a qualitative point of view, radiation-induced mutations are of the same nature as the spontaneous mutations observed in natural unirradiated populations. What radiation does is to increase the mutation frequency above the frequency level of the spontaneous mutations. Furthermore, the mutations induced by radiation are not qualitatively different from those induced by other mutagenic agents. Radiation does not discriminate in favor or against any particular gene; consequently, at the present time and with our current knowledge, there is no way to direct the mutagenic effect of radiation toward any predetermined locus and not even to predict which gene will be affected. However, like spontaneous mutations, induced mutations occur more frequently in those genes that are less stable.

2. A statistically significant increase of frequency above the level of spontaneous mutations can be detected with doses as low as 25 R, as long as the number of individuals irradiated with such low doses is very large.

If the data of several experiments involving different types of induced mutations (dominant, recessive, sex-linked, autosomal, lethal, or nonlethal) are pooled together, the same generalized dose-response curve can be plotted (Fig. 13-2). The curve *A* shows that there is a direct linear relationship between total dose accumulated and mutation frequency, in the low dose range as well as in the high. Since experiments with doses below a certain minimum (about 25 R in *Drosophila*) are very laborious and tedious and surpass the human limits of an investigator's patience,* the curve must be extrapolated below the minimum dose used. The validity of this extrapolation is accepted by many radiation geneticists. If the extrapolation is done, the curve will not intercept the origin of the axes, but rather a point on the Y axis corresponding to the spontaneous

---

*In one study, an induced mutation frequency significantly above the spontaneous level has been reported for *Drosophila* with an X-ray dose as low as 5 R (see Glass H. B., and R. K. Ritterhoff. 1961. Science **133:**1366).

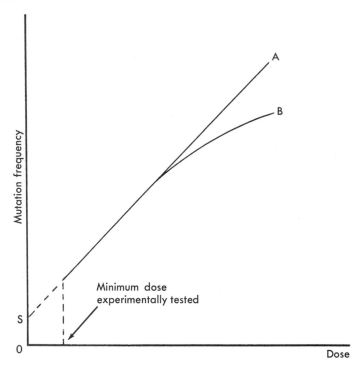

Fig. 13-2. Generalized relationship between mutation frequency and radiation dose in *Drosophila*. **S,** Frequency of spontaneous mutations. **A,** Curve obtained after correcting the experimental data for chromosomes carrying two or more lethals. **B,** Experimental curve.

mutation frequency. The consequence is that although with no dose only spontaneous mutations are observed, with any dose above 0 R (no matter how small) there is probably an increase in the mutation frequency. In other words, as far as the mutagenic effects of radiation are concerned, the common opinion is that there is no threshold dose, unlike what happens for several other types of biologic effects. The practical consequence of this finding is that probably there is no small radiation dose that is entirely harmless biologically.

The slope of the dose-response curve for sex-linked recessive lethals is such that the dose of 1000 R increases the frequency of these mutations from 0.2% (spontaneous level) to about 3%. Every additional 1000 R results in a frequency increment of 3%. However, at very high doses the curve plotted with the experimental data might bend slightly downward (curve *B* in Fig. 13-2). The downward bending is not necessarily an indication that at high doses the induced mutation yield falls off. The ClB technique does not allow the differentiation of X chromosomes carrying one lethal mutation from those carrying two or more. With high doses, the chance of more than one mutation on the same chromosome increases. The relationship between dose and number of point mutations induced is still linear. In Fig. 13-2 curve *A* is obtained by correcting the experimental data for chromosomes carrying two or more lethals.

The doubling dose in *Drosophila* is estimated to be somewhere between 30 and 100 R.

The straight linear relationship between dose and effect seems to suggest that a radiation-induced mutation is the product of a single "hit," that is, the mutation is produced by a single ionizing particle or photon. If an interaction of hits were necessary for the formation of a mutation, the yield would not be linearly proportional to the first power of dose, as it happens, for instance, in the dose-response relationship for the two-break aberrations (see Chapter 12).

3. That induced mutations are the result of single hits is further evidenced by the absence of a dose-rate effect in *Drosophila*. The experimental results obtained by several investigators show that a given accumulated dose produces the same yield, regardless of whether the dose is delivered at high rate, low rate, or fractionated. The dose rates have ranged between 0.001 and 2700 R/min, and yet no difference has been found in the yield. If mutation were the result of interaction between independent ionizing events, one would expect that with low dose rates a repair of the effects of earlier ionizations could take place before the total number of necessary ionizing events was completed. The absence of the dose-rate effect then shows that the mutagenic effect is not subject to the possibility of recovery (like so many other biologic effects induced at low dose rates) and that a radiation-induced mutation is an irreversible phenomenon (except, of course, for the rare possibility of a spontaneous back mutation). If radiation-induced mutations are irreversible with no possibility of a repair process, the *Drosophila* studies would indicate that the genetic damage is cumulative; the mutations induced by any successive radiation exposure are added to those induced by all the previous exposures, no matter how much the exposures are spread in time. The genome of an individual is like a dosimetric film badge; it keeps a full account of the total radiation dose received, whether emitted by luminescent dials of wrist watches, radioactive fallout, atomic explosions, or diagnostic X-rays, or whether delivered in small partial doses widely spread in time or in one single acute dose. Every irradiation, small or large, contributes to the total score; therefore any exposure is genetically harmful.

4. Many experiments have been carried out with *Drosophila* to test mutagenic effectiveness of radiations of different wavelength, particle energy, specific ionization density, and LET. The reports of different experiments seem to be contradictory; some results indicate that the yield of mutations increases with LET; others indicate that the yield is independent of LET. However, the methods and techniques used in some experiments are such that they do not distinguish between intragenic and intergenic effects produced. As pointed out in the preceding chapter, the yield of certain types of chromosomal aberrations may be dependent on LET. But in those experiments in which the true point mutations are not masked by some mutations that are intergenic effects, observers have noted that the yield of point mutations is independent of LET. High-energy X-rays show the same yield as X-rays with energy as low as 10 keV, although occasionally and in particular experimental conditions decrease in yield has been noted with very soft X-rays. But probably the decreased yield is to be attributed to the poor penetrating power of the radiation used and therefore to a decrease of the true dose delivered to the gametes.

Experiments by Timoféeff-Ressovsky and Zimmer suggest that neutron radiation is less effective in inducing sex-linked recessive lethal mutations than are X-, gamma, and beta radiations. These experiments have been criticized on the ground that neutron dosimetry was not so accurate as X-ray dosimetry when the experiments were carried out. However, Lea's authoritative opinion is that there seems to be little doubt that neutrons are less efficient than X-rays in inducing sex-linked recessive lethals in *Drosophila*.

MAMMALIAN STUDIES
Methods

The general principles of radiation genetics established by the middle 1940s were based mainly on experimental data from *Drosophila* and a few nonmammalian organisms. At that time, when the concern about the genetic hazard of ionizing radiation for the human species was quite widespread, there was no available information on radiation mutagenesis in any mammal. All that could be done was to "extrapolate" to mammals and man the conclusions reached through *Drosophila* experiments. Unfortunately, this was done also with other conclusions concerning *Drosophila* genetics (such as sex chromosomes and the genetic mechanism of sex determination). Later, geneticists were to regret the error of extrapolating to mammals and man the results obtained with a dipteran!

Immediately after World War II, an extensive program was started, especially under the auspices of the Atomic Energy Commission, to investigate the genetic effects of radiation in mammals, because it was felt that the information acquired from this research could be more validly used for an estimation of the genetic hazard of radiation to man. One project was started in 1947 at the Biology Division of the Oak Ridge National Laboratory under the direction of W. L. Russell. The mouse was selected as the most favorable experimental mammalian organism for this type of study. For this project, mice have been and are being reared in very large numbers; the Oak Ridge facilities are considered to be the largest "mouse factory" in the world.

The technique used by Russell for the detection of radiation-induced mutations in mice is relatively simple and involves only the examination of the first generation of the irradiated animals. It is based on a specific locus test; the mutations detected and scored are not those induced at any locus of any chromosome but only those induced at specific loci. A special strain of mice is needed that is homozygous for the following seven autosomal recessive genes: a (nonagouti), b (brown), c^{ch} (chinchilla), d (dilute), p (pink eye), s (piebald), and se (short ear). An animal of this strain will show all the seven phenotypic characteristics listed in parentheses. A normal male or female mouse (homozygous dominant for all seven genes and therefore with none of the above characteristics) is irradiated and then mated with a mouse of the homozygous recessive strain. If no mutation has been induced in any of the seven loci of the irradiated gametes, the progeny of this cross will appear normal, although it will be heterozygous for all seven genes. If a mutation has been induced in one or more of the seven loci, some of the offspring of the cross will not be normal for one or more of the seven characteristics, because it no longer carries one or more of the dominant alleles (A or B or C^{ch}, etc.). Obviously, in order to obtain valid and statistically significant frequency data, one must collect them from a very large number of crosses.

The lethal radiation dose for mice is much lower than it is for *Drosophila*. Thousands of roentgens cannot be used for mutagenic studies; the highest sublethal doses that can be used are only a few hundred roentgens.

Some of the results obtained in mouse experiments can be best explained in the light of some peculiar characteristics of mammalian gametogenesis. In the mature testis, spermatogonia are present in the seminiferous tubules throughout the reproductive life-span. While some spermatogonia produce spermatocytes and then spermatids and mature sperms, others continue to divide by regular mitosis. Some of the daughter cells become spermatocytes; others keep dividing, etc. Consequently, throughout the period of sexual maturity all stages of spermatogenesis (from spermatogonia to sperms) can be seen in the tubules. The production of mature spermatozoa from a spermatogonial cell takes about 5 weeks in mice (and almost twice as long in man). This knowledge has made possible the estimation of the time required for each stage of spermatogenesis and the planning of radiation experiments for observation of the response of any selected stage. For instance, if a male is irradiated and the mated 5 weeks later, the offspring will be produced by sperm cells that were irradiated at the spermatogonial stage. At the mating time, sperms irradiated in some postspermatogonial stage were no longer available for fertilization.

The effects of acute irradiation on male fertility are well known. Males irradiated with 300 R or more are fertile for a few weeks after irradiation, become temporarily sterile for some time depending on the dose, and then regain fertility for the remainder of their reproductive life. This type of response shows that some spermatogenetic stages are more radiosensitive than others; certain spermatogonial stages are extremely radiosensitive (with LD_{50} as low as 20 to 24 R of gamma rays).

In females, the primordial germ cells pass through all the oogonial divisions and reach the oocyte stage in the embryo; consequently, the ovary of a newborn female contains no oogonia. Soon after birth, the oocytes go through a resting stage, called "dictyate," in which they remain until a few hours before ovulation, which occurs in the secondary oocyte stage. Thus in an adult ovary only primary and secondary oocytes are present. When an adult female is irradiated and then mated, the offspring is always produced by egg cells that were irradiated in the primary or secondary oocyte stage, depending on the time interval elapsed between irradiation and mating.

In contrast to males, female mice may be permanently sterilized by low acute radiation exposures. According to the results of some experiments, all early oocytes are destroyed by 50 R, but the sensitivity decreases with increasing maturity of the follicle.

From a radiobiologic point of view, one should remember that the general patterns of human gametogenesis do not seem to be substantially different from those observed in the mouse.

Results

The first experiments were simply designed to study the dose-response relationship, that is, the relationship between total accumulated dose and the frequency of induced mutations at the seven loci under study. As in *Drosophila*, the relationship, in general, can be graphically represented by a straight line, although the usable dose range must be much narrower than the one that can be used for *Drosophila*. However, for mice the slope of the line is much steeper, because the average induced mutation rate for the seven loci is much higher than in *Drosophila*. After more appropriate experiments, researchers found that the average induced mutation rate per locus in the mouse is about 15 times the same rate in *Drosophila*. Furthermore, they showed that the mutation frequency for a given dose in the offspring produced by sperms irradiated at the spermatogonial stage does not depend on the time of the reproductive life-span when spermatogonia are irradiated. In fact, offspring conceived, for example, 18 months after irradiation of the male were as likely to bear a mutation as those conceived 3 months after irradiation. These findings on the dose-response relationships are themselves of considerable interest, because they show that two species as phylogenetically remote as a mammal and an insect respond quite differently to the mutagenic action of ionizing radiation.

But perhaps the most unexpected finding from the genetic work on mice done at Oak Ridge is that, unlike what happens with *Drosophila*, the dose rate does affect the mutation frequency. In preliminary experiments Russell observed that a given dose delivered at the rate of 90 R/min produced 3 to 4 times more mutations than those of the same dose delivered at the rate of 0.9 R/min. For a while he suspected that the effect was due to radiation quality rather than dose rate, because in the first experiments the acute dose had been delivered with X-rays and the chronic dose with ^{137}Cs gamma rays, for reasons of convenience. Further experiments with X-rays for the chronic dose and gamma rays for the acute dose ruled out the possibility of an influence of radiation quality and showed that the dose-rate effect was real.

However, the relationship between dose rate and mutation frequency is not the same in the two sexes (Fig. 13-3). For spermatogonia a true dose-rate effect exists from 90 R/min down to about 0.8 R/min. Below this value, a further decrease of dose rate to values as low as 0.001 R/min does not result in a decrease of mutation frequency. The results from females give a different picture. A dose-rate effect exists at least between 90 and 0.009 R/min. Furthermore, the mutation frequency at 90 R/min is higher in oocytes than in spermatogonia, whereas the opposite is true at a very low dose rate. The frequency of radiation-induced mutations in oocytes at 0.009 R/min with a total dose of 400 R is not significantly above the spontaneous mutation frequency. In summary,

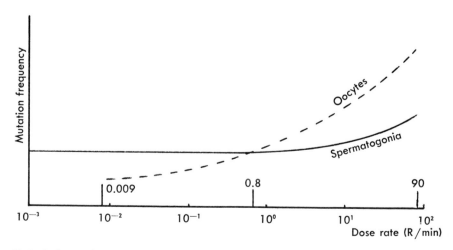

Fig. 13-3. Relationship between dose-rate and frequency of mutations induced in mouse spermatogonia and oocytes.

although the dose-rate effect exists in both sexes, it is more pronounced in females.

The discovery of a dose-rate effect for radiation mutagenesis in mammals, which is so much at variance with the results of the *Drosophila* experiments, has stimulated an intensive search for possible explanations and implications. With more recent experiments, Russell could rule out the possibility of cell selection, that is, the possibility that the pattern of cell damage, cell killing, or selective elimination of germ cells might be different at different dose rates, with the possible result of a different mutation frequency in the surviving cells. Some experiments already mentioned above rule out also the possibility that during a long chronic radiation exposure the radiosensitivity of the germ cells might vary with the stage of the cells. On the other hand, in *Drosophila* experiments mature spermatozoa are irradiated, whereas in mice the dose-rate effect has been observed with irradiation of spermatogonia. When mouse spermatozoa are irradiated, the dose-rate effect is nonexistent, and this fact would indicate that the explanation of the dose-rate dependence in spermatogonia must be found in some characteristic or biologic property of the spermatogonial stage.

Russell does not think that the dose-rate effect in mice is an indication that mutation is induced by a multiple-hit event, for a number of reasons, one of them being that the dose-response relationship fits a straight line and not a concave upward curve. The conclusion seems to be that the dose-rate effect must be due to an intracellular mechanism closely connected with the mutation process itself. Apparently, some repair or prevention of the genetic damage occurs at low dose rates. Since it is unlikely that repair could occur after the mutation process is completed (in fact mutated genes are quite stable), the possibility has been suggested that mutations are induced through a multistage process and that the hypothetical repair mechanism might work by preventing the completion of the mutational process. High dose rates would either leave no time for the repair process to operate or they would damage the repair process itself and thus inactivate it. The repair process may conceivably be more efficient for certain mutations than for others. If so, a mutation for which the repair process is more efficient would be more preventable at low dose rates. The data show that the fraction of preventable mutations is higher for oocytes than for spermatogonia. Why the dose-rate effect is more pronounced in oocytes, less in spermatogonia, and nonexistent in spermatozoa could probably be explained with the possibility that the efficiency of the repair process may be under the influence of naturally occurring biologic variables (metabolic rate, availability of oxygen, etc.) that are known to be different at different stages of gametogenesis. The whole problem deserves further and more detailed study, not only because it may lead to a better understanding of the process of mutation induction itself, but also because of its intimate relations with the problem of the genetic hazard for the human species. It is of extreme importance to know whether or not low dose rate and fractionated radiation (background radiation, fallout, diagnostic X-rays, etc.) are genetically less hazardous than acute radiation is.

The scientific importance of the mammalian studies briefly described above lies in the fact that they have shown that some of the principles of radiation genetics believed to be of general application are actually valid only under certain conditions or only for certain types of germ cells.

STUDIES ON INDUCED MUTATIONS IN MAN
Sources of information

Unfortunately, the species for which we are most eager to obtain reliable information about genetic radiation hazards is not a favorable material for the study of radiation-induced mutations. There is no doubt that like so many other organisms that have been adequately investigated, man is also susceptible to genetic damage from radiation. However, we would like to obtain direct information about dose-response relationships, dose-rate effect, influence of LET, doubling dose, etc., as we have obtained from other organisms. The study of radiation mutagenesis in man is beset with extraordinary difficulties. Even assuming that a large irradiated sample of the human population is available, one almost always encounters a dosimetric problem; in most cases, whether the irradiation was planned or accidental, determination of the radiation dose to which the sample was exposed is difficult or impossible. At best, only rough estimates can be obtained. Secondly, in contrast with organisms of short life cycle used by the experimental geneticist, the human species has a long life-span, which obviously requires several years of observations, even when the study is limited to the first generation only. Finally, irradiated human organisms cannot be arbitrarily mated to any preselected strain, as it is done with mice and Drosophila and cannot be arbitrarily confined while under observation, as it is done with laboratory organisms.

In spite of all these formidable difficulties, several attempts have been made to study the possible induction of mutations by radiation in particular groups of human subjects who were exposed during their lifetime to radiation doses well above the levels to which the general population is exposed. The search for mutations has been carried out in the progeny of the irradiated individuals using the progeny of a suitable unirradiated sample as control.

Among the groups of irradiated subjects that have been most thoroughly studied, we may mention the following:

1. Early American radiologists who by nature of their profession were exposed to significant doses over several years, in part because of the lack of knowledge of the deleterious biologic effects of X-rays; this survey was made by S. H. Macht and P. S. Lawrence and published in 1955 (Amer. J. Roentgenol. **73:**442).

2. Canadian women who received radiation treatment for sterility.

3. A group of about 13,000 patients who received radiation therapy for ankylosing spondylitis (an inflammation of the vertebrae with tendency to fusion) in England between 1935 and 1954; the treatment consisted of the irradiation of the spine with a significant fraction of the dose delivered to the gonads.

4. Populations of villages located in areas (India and Brazil) with a high background radiation level due to considerable amounts of radioactive monazite present in the rocks (see p. 246).

5. Survivors of the atomic bomb explosions in Hiroshima and Nagasaki; the survey of the progeny of these survivors for possible genetic damage was part of a project entrusted to the Atomic Bomb Casualty Commission (ABCC), which was a joint undertaking of the U. S. National Academy of Sciences and the Japanese Institute of Health and financed mainly by the Atomic Energy Commission. Results of the survey have been published on several occasions by

J. V. Neel and W. J. Schull, members of the commission. The ABCC began its work in Japan in 1947.

Unfortunately, in none of these projects were precise dosimetric data available. In spite of this and other difficulties, these studies have made some contribution, however limited, to the knowledge of the genetic effects of radiation in man.

Methods and results

The method used in the studies mentioned above consisted basically of a statistical comparison of two populations of children supposedly similar in all respects except that one population was born from irradiated parents and the other from unirradiated parents. In the study of the two populations, a survey was made of possible indicators of mutations induced in the parental gametes. A high frequency of stillbirths, neonatal deaths, or deaths during the first 9 months of life in the progeny of the irradiated parents as compared with the control progeny might be a possible indication of lethals induced in the gametes. Congenital malformations may be indicators of some types of nonlethal mutations. The induction of sex-linked lethals in the gametes of the parents would be expected to result in significant deviations of the sex ratio of the progeny from its normal value.

The last indicator deserves a few words of explanation. The sex ratio (SR) can be expressed as the number of males born for every hundred births. In the human species, the SR is not exactly 50% (or 0.5) as one would expect on the basis of the mechanism of sex determination, but slightly greater; on the average, for every 1000 females, 1050 males are born. (SR = 0.512). Apparently, the sperm cells carrying the Y chromosome have a slightly better chance to fertilize the egg than do the sperms carrying the X chromosome. The reasons for the difference between the two types of sperms are still unknown.

The induction of lethal mutations in the parental gametes will affect the sex ratio of the offspring in opposite ways, depending on whether the mother or the father is exposed to radiation (Fig. 13-4).

Since the female is the homogametic sex, all egg cells are the bearers of one X chromosome, and the offspring of both sexes receive an X from the mother. When only the mother is irradiated, a lethal mutation is possibly induced on the X chromosome in a few of her egg cells. The offspring of either sex may receive the X bearing the mutation; a daughter receiving the lethal will be unviable only if the mutation is dominant, whereas a son receiving the lethal will never be viable, regardless of whether the mutation is dominant or recessive, because any gene on the X chromosome of males is hemizygous and will express itself in the phenotype. Thus the potential male offspring of an irradiated mother has a greater chance of dying during its embryonic development than the potential female offspring. The result is that the SR is lower than normal, and this decrease, of course, can be observed by surveying the offspring born from a large number of irradiated mothers.

The situation is different when only the father is irradiated. Since the male is the heterogametic sex, 50% of the sperm cells carry an X chromosome and 50% the Y chromosome. The paternal X is transmitted only to the daughters. Because the Y chromosome is relatively inert from a genetic point of view, one may safely assume that a sex-linked mutation is much more likely to be induced

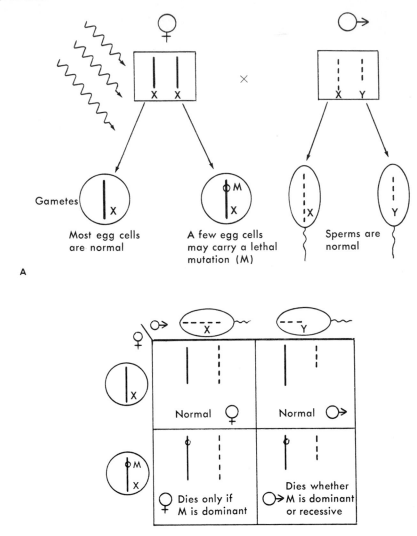

Fig. 13-4. Diagrams showing how irradiation may influence the sex ratio of the offspring when the mother, **A,** or the father, **B,** is exposed. **M,** Radiation-induced lethal mutation.

in the sperms carrying the X chromosome. A few X-bearing sperms of an irradiated father may carry a lethal mutation, which can be transmitted only to a daughter and not to any of the sons. Any potential daughter receiving the mutated gene will not be viable if the mutation is dominant. Thus the potential female offspring of an irradiated father has a greater chance of dying during its embryonic development than the potential male offspring. The result is that the SR is higher than normal.

Actually the above considerations suffer some oversimplification because the possibility of mutations being induced in the autosomes has not been taken into account. The existence of a sort of genetic antagonism between the autosomal genome and the genome carried by the sex chromosomes is well known. Furthermore, the fact is also known that the genetic equilibrium between the two genomes is different in the two sexes. Whether both sexes would be equally susceptible to the effects of detrimental or lethal mutations

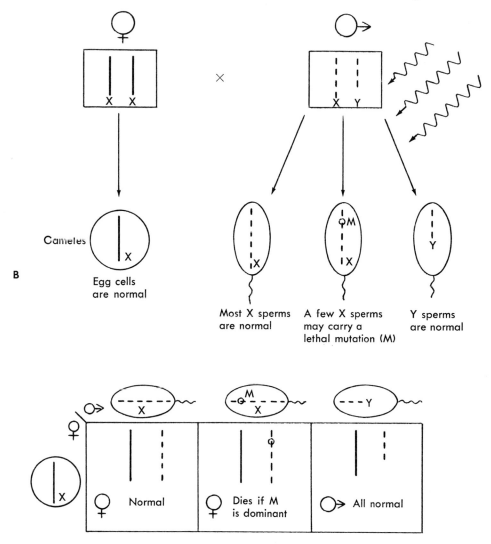

Fig. 13-4, cont'd. For legend see opposite page.

induced in the autosomes is a disputed question. But, even without going into details about this intricate problem, one is safe in believing that although the induction of autosomal mutations might complicate the effects on the sex ratio this interference would deviate the SR toward the same direction as that which can be predicted on the assumption of lethals induced only on the X chromosome; that is, when the mother only is irradiated, the SR of the offspring should be lower than normal, whereas when the father only is irradiated, the SR should be higher than normal.

The authors of the studies mentioned above recognize that the data on the sex ratio as indicator of genetic damage should be interpreted with caution, because the SR may be influenced by factors other than irradiation of one parent, such as parental age, birth order, race, urban or rural residence, and social conditions of the parents. These complications can be minimized with an appropriate choice of the control samples. To a lesser extent, the same can be said for the other indicators of genetic damage (stillbirths, neonatal deaths, etc.), whose mechanism of inheritance is not always straightforward; some of them might even be caused by nongenetic factors.

In view of all these drawbacks, it is not surprising that the findings of the surveys have not met the preliminary expectations and have not contributed much to the knowledge of the genetic effects of ionizing radiations in man. The data of these surveys have been published, tabulated, and summarized in many reports.* Some of them show no statistically significant differences between the irradiated samples and the controls; others are inconsistent or contradictory. The data on SR, and especially those from the Japanese studies, are more significant than those on other indicators and consistent with the theoretical expectations. In the case of maternal irradiation the SR was considerably below the SR of the controls (for example, 0.447 versus 0.546 in the controls, or 0.480 versus 0.533 in the controls). In the case of paternal irradiation, the SR was higher than the SR of the controls (for example, 0.539 versus 0.497 in the controls, or 0.533 versus 0.509 in the controls).

In conclusion, radiation mutagenesis in man has been poorly demonstrated so far by studies of samples of exposed population. Of course, this negative conclusion does not imply by any means that radiation is not mutagenic in man. There is no logical reason why it should not be mutagenic, if it is so in all other organisms studied. Probably the studies on the human species prove that man, when compared with experimental organisms, is not exceptionally sensitive to the mutagenic effects of radiations, although this conclusion should be adopted with caution until further studies can be done with more appropriate methods. In the meantime, the only reliable appraisal of the genetic hazard of radiation for man must be done by extrapolation of data obtained with experiments on other mammals.

SOME GENERAL CONCLUSIONS

A basic point that has been only incidentally mentioned in other parts of this textbook is emphasized here. A complex organism that reproduces sexually can be viewed as being made of two components: the *soma*, consisting of all the cells, tissues, and organs that are not involved in reproduction and will die with the death of the individual, and the *germ cells*, consisting of the gametes (mature or immature), which, if involved in fertilization, will survive in the offspring after the death of the individual producing them. Accordingly, all the biologic effects of ionizing radiation in an organism may be classified into two distinct groups. The somatic effects (cancer, leukemia, radiation burns, radiation sickness, somatic mutations, etc.) are limited to the soma of the irradiated individual; they will be extinguished with the death of the individual. The genetic effects (gametic point mutations and chromosomal aberrations) have the gametes as their targets, and as such they may be perpetuated through the following generations long after the death of the irradiated individual. The uniqueness of the radiation-induced genetic damage stems, therefore, from the fact that it is perpetuated in time long after irradiation and is spread rapidly from one or a few irradiated individuals to large segments of the unexposed population through sexual reproduction. This spreading is made possible by the irrevers-

---

*See, for instance, Neel, and Neel and Schull, both works cited at the end of this chapter.

ibility of mutations and by the self-replicating power of genes and chromosomes. Obviously these considerations are applicable also to the human species.

We speak of genetic "damage" because, in fact, most mutations are harmful to the species, either by killing the individuals who carry them, or by weakening their fitness for survival. In wildlife, new (spontaneous) mutations are continuously introduced into the common genetic pool of a population, yet the genetic load (the sum total of deleterious mutant genes of all kinds) is kept constant by the natural selection pressure, which gradually eliminates the unfavorable genes. A dynamic equilibrium is thus established between the input of spontaneous mutations and their elimination by selection. If additional mutations, such as those induced by radiation, are introduced into the pool, natural selection will still tend to reduce the total number of mutants so that the genetic load remains constant, although eventually a new equilibrium will be reached with regard to the nature of mutants present in the pool. All this may be true in wildlife, wherein natural selection is left free to operate.

In the human species, the situation is different. Man interferes with the selection pressure by greatly reducing it. The unfavorable mutants (carriers of malformations, inherited diseases, mental deficiency, etc.) are not allowed to be wiped out by the blind force of natural selection; rather they are kept in hospitals, asylums, and other appropriate institutions, which are obviously a financial burden for a civilized society. With the natural selection greatly reduced or made inoperative, the introduction of new mutations into the human genetic pool increases the genetic load. For the human population this means a higher percentage of unfit individuals.

When the radiation-induced genetic damage is taken into account, the fact is evident that a human individual is not only responsible to himself when he handles radiation carelessly, but also to the future generations. A recessive lethal mutation induced in an irradiated individual today may be 50% responsible for the death of another individual several generations from now. We are the custodians of the germ plasma of the future generations. The present generation may be responsible for a worsening of the genetic load of the generation living 200 years later. Our concern for the genetic hazards of radiation is increased by the fact (firmly established by a wealth of experimental evidence) that the genetic effects, both intragenic and intergenic, can be obtained with radiation doses much lower than those needed to obtain other types of biologic effects. As a matter of fact, probably no threshold exists for the induction of point mutations and of some types of chromosomal aberrations; this fact does not seem to be true for other biologic effects.

Laymen often believe that geneticists disagree among themselves concerning the real genetic hazard deriving from ionizing radiations. That radiation is mutagenic in man, as it is for so many other organisms, and that human genes are no less sensitive to radiation than those of mice are firmly established facts accepted by everyone. One of the controversial points concerns the possibility that recessive lethal or detrimental alleles might not be completely harmless when heterozygous. This problem is closely related to another general one: does complete dominance (and therefore complete recessiveness) really exist? When we say of two alleles that A shows "complete dominance" over a, are we really sure that the homozygotes AA are phenotypically identical to the heterozygotes Aa? Recently, several experiments and observations have shown that probably the classical mendelian complete dominance is of rarer occurrence than incomplete or intermediate dominance. If so, one may logically suspect that many recessive lethals and detrimentals are not completely harmless in heterozygosity.

Another problem is concerned with the doubling dose value for the human species. The estimate for mice varies between 30 and 100 R, with 40 R accepted by Russell as the most probable. If human genes are at least equally sensitive to radiation mutagenesis as those of mice, then the exposure of all the individuals to 40 R during their entire lifetime would lead to a doubling of the mutation rate at each generation. This doubling would certainly be deleterious for the human species, if not disastrous. The maximum permissible dose for the general population should be well below 40 R during the period of sexual maturity. The present recommended value is set at 5 rems for 30 years (see p. 251).

In the past investigators suspected that spontaneous mutations might be induced by background radiation, but this possibility has been ruled out at least for *Drosophila*. Calculations have shown that the frequency of sex-linked lethals induced by background radiation would amount to only 0.00015%, whereas the actual spontaneous frequency for those mutations is 0.2%. Furthermore, the rate of radiation-induced mutations is independent of temperature, whereas temperature does affect the rate of spontaneous mutations. Apparently then only a small fraction of spontaneous mutations may be caused by background radiation; others may be caused either by external factors hitherto unknown or by a random error in the replication of the genetic material.

Induced mutations may be of useful value in certain domesticated species. Although most mutations are harmful, a few favorable ones do occur; their frequency can be increased with radiation. This method has been successfully applied in agriculture. Large numbers of seeds are irradiated with high doses; some of them do not survive, whereas others germinate and may show mutations. The few favorable or desirable mutants (perhaps 2 or 3 among thousands of seeds irradiated) are selected and perpetuated, whereas the others are discarded. In Sweden, some superior varieties of barley have been obtained with this method. Some of them give a better yield of grain; others a better yield of straw. In other species, mutants that are more resistant to the attack of parasites have been obtained and selected. In another field, *Penicillium* mutants have been developed with an improved penicillin yield. We hope that a similar success will be achieved in animal breeding in the not too distant future.

REFERENCES

GENERAL

Gustafson, A. 1961. The induction of mutation as a method in plant breeding. In Errera, M., and A. Forssberg, editors. Mechanisms in radiobiology. Academic Press, Inc., New York. Vol. I.
King, R. C. 1965. Genetics. Oxford University Press, Inc., New York.
Neel, J. V. 1963. Changing perspectives on the genetic effects of radiation. Charles C Thomas, Publisher, Springfield, Ill.
Purdom, C. E. 1963. Genetic effects of radiations. Academic Press, Inc., New York.
Srb, A. M., R. D. Owen, and R. S. Edgar. 1965. General genetics. W. H. Freeman & Co., Publishers, San Francisco.
Wallace, B., and T. G. Dobzhansky. 1959. Radiation, genes and man. Holt, Rinehart & Winston, Inc., New York.
World Health Organization. 1957. Effects of radiation on human heredity. Report of a study group. Geneva.
International Atomic Energy Agency. 1966. Genetical aspects of radiosensitivity. Vienna.

LITERATURE CITED

Lea, D. E. 1962. Actions of radiations on living cells. Cambridge University Press, Cambridge, England. Chapter 5.

Muller, H. J. 1927. Artificial transmutation of the gene. Science **66**:84.

Neel, J. V., and W. J. Schull. 1956. The effects of exposure to the atomic bombs on pregnancy termination in Hiroshima and Nagasaki. National Academy of Sciences and National Research Council, Washington, D.C.

Patterson, J. T. 1929. X-rays and somatic mutations. J. Hered. **20**:261.

Russell, W. L. 1965. Studies in mammalian radiation genetics. Nucleonics **23**:53.

Stadler, L. J. 1928. Mutations in barley induced by X-rays and radium. Science **68**:186.

Stadler, L. J. 1928. Genetic effects of X-rays in maize. Proc. Nat. Acad. Sci. USA **14**:69.

CHAPTER 14

Effects on embryonic and larval development

RADIOSENSITIVITY AND ORGANISMAL DEVELOPMENT

For Metazoa, ontogeny is the morphologic evolution of an individual organism, leading from the unicellular zygotic stage to the definitive pluricellular structure characteristic of the species. The end of ontogeny coincides with the achievement of sexual maturity.

The complexity of the ontogenic process is a function of the level of morphologic organization attained by a species in the course of its evolution. The most radical transformations are usually observed during the early stages of development. Fertilization is immediately and rapidly followed by cleavage, gastrulation, and formation of the primitive germ layers (ectoderm, endoderm, and mesoderm). During these developmental stages, the cells undergo rapid divisions and at the same time become more and more differentiated. Through the process of differentiation, cells of different embryonic territories are directed along different pathways toward their final morphologic and physiologic specialization.

Organs and organ systems begin their development as organ rudiments or primordia from one or more of the embryonic germ layers. The process of organ formation and development is referred to as organogenesis, and it is an important phase of embryogenesis.

If we confine our considerations only to the two major taxa of the animal kingdom, vertebrates and insects, in which we find the highest degree of morphologic complexity, we notice a variety of modalities and patterns in the ontogenic development. Eggs of insects are laid after they are fertilized. Embryogenesis and most of organogenesis take place before their hatching. In holometabolous insects, hatching is followed by a relatively long metamorphosis, with a number of larval stages (three in *Drosophila*) and a pupal stage. The imago is the final phase in the ontogeny of insects. Although the essential morphologic organization is completed during the embryonic development in the egg, formation of some new organs does occur during metamorphosis, while some larval structures disappear.

Essentially the same pattern can be seen in some lower vertebrates, such as amphibians, in which a metamorphosis follows the embryonic development. Again, the hatching of the eggs is the event that separates the two processes. In other oviparous vertebrates (reptiles, birds, and most fishes) there is no metamorphosis; the young animal emerging from the hatched egg is already organized morphologically like the adult because organogenesis has been completed during the embryonic development. In these cases the ontogenic process continues with the simple growth of the organism. In most mammals the embryonic development takes place entirely within the maternal uterus. Organogenesis is completed at some time long before birth. Between the end of organogenesis and birth, the embryo is more properly called a fetus.

The immature forms of an animal organism (embryos, fetuses, larvae, etc.) are considerably more radiosensitive than its adult stages, in perfect agreement with the law of Bergonié and Tribondeau (see p. 330). This is because cells of immature forms are characterized by high reproductive activity and low degree of differentiation. In general, the earlier the stage in the ontogenic development, the higher is the degree of radiosensitivity. The zygote is known to be extremely radiosensitive, even more than the unfertilized egg. At later stages, neuroblasts (the precursors of neurons) are also very radiosensitive, and perhaps more so than other types of embryonic cells of comparable development stages. This radiosensitivity is rather surprising, when we realize that the neuron in the adult is one of the most radioresistant types of cells (see p. 335). In fact, the neuron can tolerate up to 10,000 R without apparent damage, whereas the neuroblast from which it derives may be damaged with doses as low as 25 to 40 R. Furthermore, the neuroblast seems to be even more radiosensitive than the more primitive neurectoderm from which it develops. As we shall see below, the extreme radiosensitivity of the neuroblast may explain why so many malformations induced by radiation in mammalian embryos involve organs of the central nervous system.

One consequence of the irradiation of immature organisms may be death. Their radiosensitivity in relation to death as end point can be quantitatively measured in terms of median lethal doses (see p. 248). The LD_{50} values for embryos, fetuses, and larvae are considerably lower than the values for the adults of the same species, although they increase with the time of irradiation after fertilization, as expected. In mammals, depending on the developmental stage irradiated, the organism may die before or immediately after birth (prenatal or neonatal death).

If the radiation dose is not sufficient to kill the embryo or larva, it may induce congenital morphologic malformations or abnormalities (also referred to as terata—singular, teras). Normally a malformation is the consequence of the death of some embryonic cells in the course of organogenesis. If the killed cells are parts of an organ primordium, a malformation will be visible in the organ that develops from that primordium. The mammalian embryo possesses enough phagocytes to eliminate the dead cells. In many cases, the space left vacant by the destroyed cells is occupied by the neighboring surviving cells, whose differentiation may be slightly shifted in direction because of the new territory they are forced to occupy. The end result may be an organ that is anatomically quite normal, but reduced in size. Malformations like microcephaly and microphthalmia can be interpreted on the basis of these considerations.

Since congenital malformations are the result of derangement in the formation of organs, they can be induced only if the organism is exposed to radiation in the course of organogenesis or at earlier developmental stages. They can never be induced in adults, not even with high dose levels, such as supralethal doses.

The teratogenic action is not a monopoly or peculiarity of ionizing radiations. It is shared with other agents, and especially with certain drugs. The case of thalidomide of the early 1960s is too notorious to be mentioned in detail.* But because ionizing

---

*Thalidomide was a mild and supposedly safe sedative, used as an active ingredient in tranquilizers and sleeping pills in West Germany during the late 1950s and early 1960s. It had been given to adults and children for some time with no ill effects. When pregnant mothers started using the drug to relieve the symptoms associated with early pregnancy, they gave birth to thousands of infants affected by phocomelia, a malformation consisting in the deformity of all four limbs.

radiations affect the body of an organism with a mechanism that is less specific and selective than that of most drugs, a greater variety of malformations are induced by penetrating radiation than by chemical agents. In fact, the thalidomide syndrome was characterized practically by only one or two types of anomalies (especially phocomelia). An embryo is a mosaic of several primordial and differentiating tissues, and all of them are more or less equally radiosensitive.

Finally, one should remember that gonads start developing quite early in the course of the embryonic development. The first spermatogonia or oogonia can be traced back to very early stages. Consequently, exposure of embryos to ionizing radiation can result not only in malformations of the embryo but also in genetic damage, which of course, will be transmitted to the following generations. As seen in the preceding chapter, mutations induced in spermatogonia and oogonia present a greater genetic hazard for the species than those induced at later stages of gametogenesis.

In the following sections some studies of the effects of ionizing radiations on immature organisms are described and discussed.

STUDIES ON INSECTS

Insects in the imago stage are known to be very radioresistant when compared with other organisms. The LD_{50} for several common species (*Drosophila*, wasps, and bees) can be well above 10^5 R. But the lethal dose for the immature stages is only a small fraction of this value; the earlier the ontogenetic stage, the greater the radiosensitivity. C. Packard has investigated the relationship between age and radiosensitivity in *Drosophila*. This relationship is represented graphically in Fig. 14-1. One can see that the median lethal dose increases sharply during the

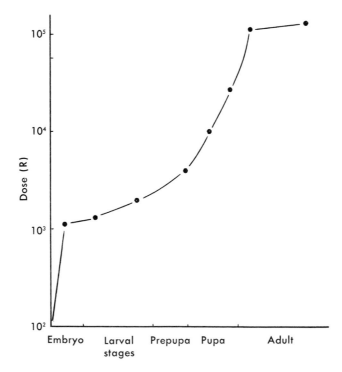

Fig. 14-1. Variation of LD_{50} in *Drosophila* at different stages of development.

embryonic development from 100 to 1000 R. The less sharp increase beyond the embryonic development may be explained with a special aspect of insect ontogeny: after the hatching of the egg, very little cell division occurs during the larval, pupal, and imago stages, so that organismal growth is accomplished mainly by cellular growth rather than by an increase in cell number. Of course, some bursts of mitotic activity occur in coincidence with molting and at the end of pupation. In the adult, with the exception of the germ cells, no mitotic activity exists and the body cells have reached their final stage of differentiation. In *Locusta migratoria* (locust) the LD_{50} for X-rays increases from 304 R in 1-hour-old eggs to 1170 R within 2 hours. Maximum sensitivity is found during cleavage and at the blastula stage, during which the mitotic activity is at its peak, with a decline during gastrulation.

Abnormalities and monstrosities were observed and studied in irradiated *Drosophila* larvae and pupae by C. H. Waddington and C. A. Villee. Larvae, prepupae, and pupae were X-irradiated with doses ranging from 1096 to 7000 R at different dose rates. Except for the lowest doses, a significant retardation of pupation was observed. A dose-rate effect was also found. Flies irradiated as larvae 70 to 100 hours old showed overgrowth, duplication of organs, and malformations such as leglike antennae, formation of palps in the eye, and palps or wings attached to the prothorax. Flies irradiated as late larvae and early prepupae showed deficiency of macrochaetae, small eyes, crippled legs, and abnormality in the anatomy of the mesothorax, scutellum, and abdomen. Practically no malformations were observed in flies irradiated as pupae.

Irradiation may also interfere with molting. In locusts, exposure of the insect immediately before molting does not block that particular molt, but successive molts are prevented. However, if exposure occurs long before a particular molt, that molt is blocked. Apparently, a minimum of time is required for the full development of whatever lesion is responsible to prevent molting.

STUDIES ON RODENTS IRRADIATED IN UTERO

The effects of ionizing radiations on the embryonic development of mammals have been investigated thoroughly in mice and rats by a number of investigators, notably among others R. Rugh, L. B. Russell, and W. L. Russell.

The gestation time in the mouse is about 19 days and in the rat about 22 days. The general method consists of mating sexually mature females in estrus with males. One of several tests can be used to determine whether fertilization has occurred. Pregnant females are exposed to penetrating X-rays (of at least 250 kV) or to gamma rays at preselected gestation times with sublethal (for the mother) doses. Embryos or fetuses are thus irradiated in utero. The effects of radiation on the offspring (neonatal deaths or malformations) are observed, analyzed, and scored after birth. Sometimes survivors are followed until adulthood, because certain types of malformations become evident long after birth. Prenatal deaths and resorptions of embryos are observed and scored after dissection of the maternal uterus.

Mice

The most exhaustive studies on mice have been carried out by the Russels at the Oak Ridge National Laboratory.

In the mouse, as in many other mammals, fertilization normally occurs in

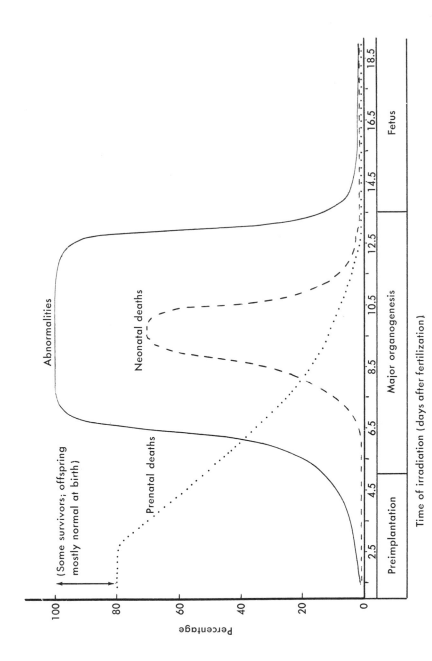

Fig. 14-2. Incidence of prenatal and neonatal death and of abnormal individuals at term after irradiation of mice in utero with 200 R at various intervals after fertilization. (From Russell, L. B., and Russell, W. L. 1954. J. Cell Comp. Physiol. **43**(suppl. 1): 103-149.)

the oviducts, and cleavage of the zygote begins immediately thereafter, while the fertilized egg slowly travels toward the uterus. Implantation in the uterine wall takes place about 5 days after fertilization at the blastocyst stage. Thus the time between fertilization and implantation (5 days) may be referred to as the preimplantation period. The postimplantation period begins with the formation of the primitive germ layers followed by organogenesis, which continues until about the thirteenth day of gestation. By this time all major organs and organ systems are formed; from this time until birth (6 days) the developing mouse is a fetus.

The results of mouse irradiation in utero are presented synthetically and in graphic form in Fig. 14-2, in which the percentages of three end effects obtained (prenatal and neonatal deaths and abnormalities) are plotted as functions of the time of gestation when an acute X-ray dose of 200 R was delivered to the pregnant females. The general procedure was to irradiate a group of pregnant females 1 day after fertilization, another group 2 days after fertilization, etc., and observe the offspring at birth and thereafter. Percentages of neonatal deaths and of offspring with malformations could thus be calculated for each group and therefore for each gestation time when the embryo or fetus was exposed to radiation. Prenatal deaths within each group were observed at autopsy of the females either before or after delivery.

Mice irradiated in the preimplantation stages show a high percentage (up to 80%) of prenatal deaths and very low percentages of malformations and neonatal deaths. This fact indicates that the LD_{50} for 1- or 2-day-old mouse embryos must be well below 200 R and, at any rate, much lower than that of the adult mouse (640 R). The few embryos that escape death after early irradiation have a good probability to develop normally and to survive after birth. The effect of radiation on very young embryos seems to be a sort of all-or-none effect: radiation either kills the embryo outright, or if it does not, the embryo develops normally like the unirradiated controls. In most cases, if any damage is done to the cleaving zygote, it is not compatible with life.

Later irradiation, in the course of organogenesis, brings about a decrease in prenatal deaths but a sharp increase in the production of abnormalities. With 200 R, as many as 100% malformed mice are born, if embryos are irradiated between the seventh and the twelfth day of gestation. This phase of the embryonic development is the most susceptible to malformations, as one would expect, because it coincides with organogenesis. Some of the gross abnormalities produced are not compatible with life, as evidenced by the fact that many of the malformed offspring die shortly after birth.* Neonatal death may reach a peak of 70% of the malformed offspring, if the embryos are irradiated around the tenth day of gestation. The possible reason of this fact may be found in some other experimental results mentioned below.

Irradiation with 200 R after the end of organogenesis seems to be almost ineffective in causing prenatal and neonatal deaths or malformations. Apparently, the fetus is considerably less radiosensitive than the embryo (in terms of LD_{50}), so that 200 R can kill only a very small percentage of fetuses. On the

---

*Care should be taken in distinguishing between natural neonatal death and violent death of these malformed offspring, since we know that the mother mouse frequently eats her abnormal young at birth.

other hand, the fetus is not susceptible to malformations because the basic anatomic structures have already been laid down during the earlier organo-genesis. However, these conclusions must not be construed as meaning that irradiation of fetuses with 200 R is entirely harmless. So far, we have been con-sidering only prenatal and neonatal deaths and gross malformations as end effects of irradiation in utero. The fact that the offspring lives after birth with-out visible gross abnormalities does not mean that it is entirely normal. At least stunted growth and reduction of the life-span have been observed in animals irradiated during the fetal stage.

Less drastic effects can be induced in mice with doses lower than 200 R. However, some types of abnormalities can be produced with 200 R, as well as with 100 R. With lower doses, the incidence will be less, and some types of mal-formations are never observed with less than 50 R. Since the zygote is the most sensitive stage in the embryonic development, the fact is not surprising that only 25 R delivered a few hours after fertilization will cause about 30% of prenatal deaths. Note that with this dose no visible symptom appears in the mother, and even a lymphocyte count (see Chapter 15) might not show that an exposure to radiation has occurred. One type of malformation, exencephaly (see below), has been induced with as little as 15 R at the two-blastomere stage.

What kinds of malformations can be induced with irradiation in utero? Studies on mammals show that a wide range of abnormalities, involving a large number of organs (central nervous system, bones, sensory organs, etc.), may be induced, and the same abnormalities have been observed in rodents and in man. Only some of the abnormalities can be listed, as follows:

Anencephaly—complete absence of the brain
Exencephaly—herniated brain; the brain protrudes through the dorsal cranial bones
Microcephaly—small head
Anophthalmia—complete absence of the eyes
Microphthalmia—small eyes
Hydrocephalus—an abnormal increase in the amount of cerebrospinal fluid within the cranial cavity, with expansion of the cerebral ventricles, enlargement of the skull, and atrophy of the brain
Coloboma—fissure in the iris and/or retina of the developed eye caused by persis-tence of the choroid fissure in the embryonic optic cups
Spina bifida—a vertebral column characterized by double spinous processes in some or all of the vertebrae; the vertebral arch may show a gap between the two spinous processes; the abnormality is caused by failure of the sclerotomes (left and right) to fuse dorsally to the spinal cord
Limb reduction
Rib fusion
Vaulted cranium
Deformed tail
Limb overgrowth

The mouse experiments have clearly shown that although the incidence of all abnormalities is very high only when the embryo is irradiated during organo-genesis, certain specific abnormalities are more likely to be induced at certain stages of organogenesis than at others. In other words, the sensitivity of the embryo with regard to a specific malformation is different at different develop-mental stages. Conceivably, the probability of producing a certain malforma-tion is very high if the embryo is irradiated at a stage during which the first rudiment or anlage of the organ affected is formed. In many cases, this assumption has been confirmed by the embryologic information available on normal development. In other cases, irradiation experiments of the type described in this section have contributed significantly to the identification of the time when certain organ rudiments appear in organogenesis. This is another interesting example of how radiation biology can be used as a research tool at the service of other biologic sciences.

According to data published by the Russells and by Rugh, the sensitivity of the mouse embryo to the production of some malformations reaches a maximum when the embryo is irradiated at the time of gestation, in days, indicated below:

Rib fusion—7 to 8
Microphthalmia—7 to 8, 9 to 10
Coloboma—8 to 10
Narrow iris—11 to 13
Vaulted cranium—9 to 10
Spina bifida—9 to 10
Deformed tail—10 to 13
Reduction of hind feet—10 to 13

Exencephaly occurs most frequently after 50 R on day $5\frac{1}{2}$ and after 200 R on day $8\frac{1}{2}$. This malformation, however, can be produced by X-rays at any time up to 9 days of gestation. Although irradiation during the fetal stage does not result in any gross malformation involving the central nervous system, exposure of the fetus to 100 R causes, after birth, certain behavior reactions that are indicative at least of functional damage to the brain.

The fact that the percentage of neonatal deaths reaches a peak when the embryo is irradiated around the tenth day of gestation (Fig. 14-2) seems to indicate that the number and nature of malformations that can be produced at that time are not compatible with the viability of the animal after birth.

Rats

Since mice and rats have essentially the same gestation period and exhibit more or less the same patterns in their embryonic development, one would expect that the interference of ionizing radiation with the embryonic development should not show significant differences between these two species of rodents. To a certain extent this assumption is correct, so that results obtained with one species can be safely extrapolated to the other species and vice versa. However, the two species have been studied quite often by different investigators, with different aims and also with different experimental methods. Consequently, the findings from the two rodent groups are complementary.

Much attention has been concentrated on the radiosensitivity of the embryonic nervous system in rats. Exposure to less than 100 R results in extensive destruction of the embryonic nervous tissues, with no significant damage to other organ systems. Necrotic cells have been found as early as 2 hours after exposure, with phagocytosis of the detritus occurring after 6 hours. Evidently neuroblasts are considerably more radiosensitive than other types of embryonic tissues.

Like the mouse, rat embryos irradiated during the first week of gestation (preimplantation period) suffer a high percentage of prenatal mortality, but the survivors appear to be practically free from malformations. Irradiation after 1 week brings about several types of malformations, such as anencephaly and severe brain deficiencies on the ninth day, abnormalities of the eyes (anophthalmia and microphthalmia) on the tenth day, and hydrocephalus and spinal deformities on the eleventh day. Irradiation during the following days until the end of organogenesis induces anomalies in specific regions of the brain (cerebral vesicles, hippocampus, corpus callosum, etc.). Interestingly enough, malformations involving the cerebellum can be induced with a high percentage at early or late fetal stages, and even with irradiation of the newborn animal. This late radiosensitivity of the cerebellum should be correlated with the late presence of neuroblasts in the organ. In fact, the cerebellum seems to be one of the last portions of the cerebrospinal axis to differentiate.

More recently, a number of experiments have been conducted to study the effects of prenatal irradiation on the postnatal physiologic and pathologic condition of the nervous system of the survivors, on their behavior, and on their reproductive fitness, growth, weight of organs, and life-span. In one such study (Sikov et al., 1969), rats were irradiated in utero after 10 days of gestation with 20 R or 100 R, or after 15 days (end of organogenesis) with 50 R or 185 R of X-rays. A reduction of weight at birth was observed in all groups, except for the rats exposed to 20 R on the tenth day. Growth was depressed only in the rats irradiated on the fifteenth day. Prenatal irradiation was also responsible for a reduction of the LD_{50} in the adults. When the LD_{50} was determined for 100-day-old survivors, it was significantly lower than the normal value (that is, about 700 R), at least for those animals exposed in utero to a dose of 185 R. This effect of prenatal irradiation had already been previously observed by others (such as Rugh) with different doses delivered at different times of the embryonic development.

In all the experiments described above, embryos were treated with single acute exposures delivered at specific stages of development. A dose-rate effect with regard to the response of the embryo to radiation has been reported several times. In fact, a chronic radiation exposure to 200 R spread over the first 2 weeks after fertilization induced no malformations in the offspring; the only observable effect was a shortening of the breeding cycle. No anomalies were observed either, when the embryos were exposed to as much as 20 R daily throughout the gestation period. With higher chronic exposures (50 R daily), the result was sterility in both male and female adult offspring and a decreased growth rate in males. However, some recent experiments (Coppenger and Brown, 1967) with the use of the same level of chronic exposure seem to indicate that although irradiation at this level has no effect on the total number of implantations a significant increase of prenatal deaths occurs at about the twelfth day of gestation. A retardation of prenatal growth was also observed. Fetuses examined shortly before birth did show some abnormalities, including microencephaly, anophthalmia, sternal malformations, and edema. Nevertheless, the difference in the teratogenic action between acute and chronic exposure is quite evident, especially when one considers that one acute exposure to only 15 R of X-rays during the preimplantation period results in severe brain defects like exencephaly. From all these investigations the fact is clear that the rat embryo can tolerate fairly well large total radiation doses (up to 1000 R) when delivered at relatively low rates throughout its development.

STUDIES ON MAN IRRADIATED IN UTERO

The importance of the results obtained from experiments of prenatal irradiation of rodents resides in the possibility of predicting the response of the human organism irradiated in utero. One can rightly assume that prenatal and neonatal death and malformations induced by radiation in the human embryo could follow the same general pattern as in other mammalian species. Keep in mind that because gestation time in man (about 265 days) is much longer than in mice and rats, the duration of specific developmental stages is greatly extended.

After fertilization, which normally seems to occur in the fallopian tubes, 11 days pass before the embryo is implanted in the uterine wall. The primitive streak, neural groove, and notochordal plate begin appearing at about day 20, which may be taken as the beginning of organogenesis. The organogenesis is practically completed at about day 38, when the embryo has a length of 12 mm. The developing organism is considered a fetus from about day 47 until birth.

According to Rugh, the graphic data presented by the Russells for the mouse, as shown in Fig. 14-2, could be extrapolated to man. Similar curves could be drawn by use of a different scale on the axis of the abscissas (from 0 to 265 days). Consequently one could expect a high percentage of prenatal deaths with irradiation during the first 11 days of gestation (preimplantation period). The highest percentage of malformations would be induced between day 20 and day 38. After this time higher doses would be needed to obtain the same effect. The incidence of neonatal death would reach a peak at about day 28, which happens to be approximately in the middle of the organogenetic stage. One can hardly predict the actual percentages of these responses at any specific irradiation time without a large number of observations. It would also be arbitrary to predict that the same radiation doses used with rodents would elicit the same response in the human embryo.

Some data on the effects of irradiation on the human embryonic development are available. One project of the Atomic Bomb Casualty Commission was the examination of the children born from mothers who were exposed to the nuclear explosion at Hiroshima and Nagasaki during their pregnancy. The most common anomaly observed in these children was small head circumference (microcephaly) associated with mental retardation. In Hiroshima, 33 children showed this malformation; 24 of them had been exposed between 7 and 15 weeks of embryonic age. Children exposed in utero during the latter part of gestation did not show these effects. A crude dose-response relationship has been found for several groups of prenatally irradiated children. The closer the pregnant mother was to ground zero, the more likely she was to bear a malformed child. A high incidence of small head circumference and mental retardation was observed in children irradiated in utero within 1200 meters from ground zero. Estimates of the actual doses to which these children were exposed are not available; however, about 50% of their mothers reported that after the explosion they had experienced the typical symptoms of acute radiation sickness. Besides microcephaly, mental retardation, mongolism, and other similar defects, no other malformations of the type experimentally induced in rodents have been observed at Hiroshima and Nagasaki.

Fetal and neonatal deaths after exposure of the pregnant mothers were evaluated only 6 years after the explosion. Among the women who were within 2000 meters from ground zero at Nagasaki and who reported symptoms of radiation sickness, 43% reported stillbirths or infant deaths as compared with 9% who were within the same distance from ground zero and who had not experienced radiation sickness. These data seem to indicate that prenatal and neonatal deaths were more frequent with higher doses.

Long before the disasters of Hiroshima and Nagasaki, a large body of information was already available on the effects of human intrauterine irradiation.

The information was gathered from the observation of subjects who were born from women exposed to diagnostic or therapeutic X-rays during pregnancy. When the fact that the embryo is extremely radiosensitive was not known, a common practice among gynecologists was to examine pregnant women by means of X-rays (roentgen pelvimetry). Examination was done either radiographically or, worse, fluoroscopically. In the latter case, the dose delivered to the pelvis, and therefore to the embryo, was as high as 30 to 40 R. Furthermore, several women were X-rayed more than once during one single pregnancy. In other instances, embryos were irradiated accidentally, because they were carried by women who were unaware of their pregnancy. These women were exposed to diagnostic or therapeutic X-rays during the first 2 to 3 months of gestation.

A variety of malformations have been observed in children born from women irradiated during gestation, and some of them are identical to those experimentally induced in rodents. A woman who was X-rayed at 4, 5, and 6 months of her pregnancy gave birth to a child with microphthalmia, microcornea, syndactyly (fused fingers), brachydactyly (short fingers), strabismus, hyperopia, and other anomalies!* Other reports lead to the conclusion that the damage to the embryo is greatest during the first 2 months of gestation and gradually decreases at later stages. Irradiation during the first days of development usually causes abortion; in fact, X-rays have been frequently considered as a tool for therapeutic abortion in early stages. Other effects of intrauterine irradiation are mongolian idiocy and hydrocephalus. Statistical studies are underway to ascertain whether irradiation in utero increases the incidence of leukemia and other forms of cancer.

Evidently, the information available from human experience is scanty and scattered. The data at hand are not statistically sufficient to allow rigorous conclusions about dose-response relationships for prenatal and neonatal deaths and congenital malformations or for threshold doses below which no effect is induced. This kind of information is badly needed, because in many situations the exposure of pregnant women to diagnostic or therapeutic X-rays or gamma radiation is a medical necessity. Nonetheless, the few data on the human experience, if correlated with the experimental results obtained in rodents, leave no doubts about certain general principles concerning the effects of intrauterine irradiation of humans. There is no question that the human embryo and fetus, as in other animal species, are markedly more radiosensitive than adults and that the radiosensitivity is highest during the first few weeks of pregnancy, at least with regard to prenatal death. The human embryo, like the mouse and rat embryos, is very sensitive to the production of malformations during organogenesis, that is, between the twentieth and thirty-eighth day of pregnancy. The fetal stages are somewhat more radioresistant than are earlier stages. Threshold doses have not been determined, but many feel that diagnostic exposures of the early embryo to less than 20 R may be teratogenic.

The practical corollary stemming from the above considerations is that the X-irradiation of pregnant women may be a seriously hazardous procedure. Therapeutic irradiation of these women (for example, for cancer treatment), or even a routine diagnostic exposure, should be postponed until after delivery.

---

*Case reported in Stewart, A., J. Webb, D. Giles, and D. Hewitt. 1956. Malignant disease in childhood and diagnostic irradiation in utero. Lancet **2**:447, 1956.

If postponement cannot be done, and the exposure involves areas or organs other than the pelvis and abdomen, these areas still should be well shielded during exposure to protect them from any possible stray radiation. If irradiation of the abdomen is essential and cannot be postponed, therapeutic abortion should be seriously considered to prevent the possible birth of malformed children. What doses could be safely used in a similar situation is a difficult question to answer; all that can be said is that the magnitude of a safe dose depends on the age of the embryo. But whether a 5 to 10 R dose is safe during the early stages is unknown, and if it is unknown, no chances should be taken.

Furthermore, the pelvic irradiation of any woman of childbearing age, for whom a possibility of pregnancy exists (that is, of fertile women who engage in sexual relations without using contraceptive methods) should also be handled with extreme caution. In view of the possibility that such woman might unknowingly be in a very early stage of pregnancy when irradiated, it is recommended that the procedure be limited to the first 8 to 10 days after the onset of menstruation. During this time, pregnancy seems to be highly improbable. Once more the purpose of this recommendation is to prevent the possible birth of malformed children. However, although irradiation in very early stages is more likely to result in prenatal death or resorption, it may also result in the production of malformations affecting viable embryos.

In this chapter we have seen another example of biologic effects of ionizing radiations, which are not limited to the individual directly exposed. As it happens for the genetic damage (see Chapter 13), irradiation of a pregnant woman may result in serious damage to her offspring, with little or no damage to herself. Of course, the congenital malformations induced in the offspring are not inherited by the following generations, although intrauterine irradiation is capable of inducing gametic mutations in the embryo when it is administered after the differentiation of the embryo's gonads. Once more, we come to the conclusion that when we handle ionizing radiations carelessly we may be responsible not only for the damage done to ourselves but also for the damage done to other individuals of the species.

REFERENCES

GENERAL

Miller, R. W. 1969. Delayed radiation effects in atomic-bomb survivors. Science **166**:569.

Rugh, R. 1959. Ionizing radiations: their possible relation to the etiology of some congenital anomalies and human disorders. Milit. Med **124**:401.

Rugh, R 1960. General biology: gametes, the developing embryo, and cellular differentiation. In Errera, M., and A. Forssberg, editors. Mechanisms in radiobiology. Academic Press, Inc., New York, Vol. 2, pp. 36-94.

LITERATURE CITED

Coppenger, C. J., and S. O. Brown. 1967. The gross manifestations of continuous gamma irradiation on the prenatal rat. Radiat. Res. **31**:230.

Packard, C. 1935. The relationship between age and radiosensitivity of Drosophila eggs. Radiology **25**:223.

Rugh, R. 1962. Low levels of X-irradiation and the early mammalian embryo. Amer. J. Roentgen. **87**:559.

Russell, L. B., and W. L. Russell. 1954. An analysis of the changing radiation response of the developing mouse embryo. J. Cell. Comp. Physiol. **43**(suppl. 1):103.

Russell, L. B., and C. S. Montgomery. 1966. Radiation-sensitivity differences within cell-division cycles during mouse cleavage. Int. J. Radiat. Biol. **10**:151.

Sikov, M. R., C. F. Resta, and J. E. Lofstrom. 1969. The effects of prenatal X-irradiation of the rat on postnatal growth and mortality. Radiat. Res. **40**:133.

Villee, C. A. 1946. Some effects of X-rays on development in Drosophila. J. Exp. Zool. **101**:261.

Waddington, C. H. 1942. Some developmental effects of X-rays in Drosophila. J. Exp. Biol. **19**:101.

CHAPTER 15

Radiation accidents and the acute radiation syndrome

In the preceding chapter we have discussed the damage caused by acute or chronic radiation doses to organisms exposed during their embryonic or larval development, with special emphasis on mammals.

When a mammal is exposed during an adult or near-adult stage, the general pattern of damage and injuries may be quite different. A striking difference is observed between the effects produced by acute and chronic exposures, or by partial and whole-body exposures. In this chapter we concern ourselves with the injuries produced in humans by acute radiation doses. If the exposure is partial, the injury will usually be localized to the irradiated organ. In the case of acute whole-body exposure with doses above a certain threshold, the injury may involve the whole organism and will manifest itself as a special type of illness called *radiation sickness*, or more properly *acute radiation syndrome*. If untreated, this disease may or may not be fatal, depending on the total dose accumulated.

COMPARATIVE RADIOSENSITIVITY OF DIFFERENT ANIMAL SPECIES

The response of the mammalian organism to an acute whole-body exposure depends mainly on its radiosensitivity. When death is considered as the end point, the radiosensitivity varies among different mammalian species.

As previously explained on p. 248, the radiosensitivity of a mammal is expressed most conveniently with $LD_{50/30}$, which is the minimum acute, whole-body dose sufficient to kill 50% of a population within 30 days.

The $LD_{50/30}$ can be determined experimentally in mammals by plotting a dose-mortality curve. The procedure consists of dividing a large population of animals into groups, which are then irradiated with different doses. The percent mortality in each group within 30 days is observed and then plotted as a function of dose on linear graph paper. The $LD_{50/30}$ is then read off in correspondence to 50% mortality.

As an example, let us assume that we wish to determine the $LD_{50/30}$ for mice. A large colony of animals is divided into groups, and each group is exposed to different acute, whole-body doses of X- or gamma radiation. After irradiation, the animals are returned to their cages and observed periodically for 30 days, and the deaths within each group are scored. The following could be typical data recorded in the experiment:

Dose (R)	Number of mice exposed	% dead within 30 days
0	15	0
300	12	0
500	18	5
550	20	15
600	24	35
650	20	55
700	17	80
800	15	100

The dose-mortality curve plotted with these data is shown in Fig. 15-1. The curve has a characteristic sigmoidal shape. There is a threshold dose below which no animals die within 30 days; it may be referred to as $LD_{0/30}$. Any dose below the threshold is a *sublethal dose*. Above the $LD_{0/30}$, the curve rises slowly at first and then sharply within a few hundred R, until a plateau is formed that corresponds to 100% lethality. The dose necessary to obtain 100% lethality within 30 days is known as $LD_{100/30}$. A *supralethal dose* is any dose above $LD_{100/30}$. The dose range between $LD_{0/30}$ and $LD_{100/30}$ is known as the *lethal range*, which for mice is between 300 and 800 R. The median lethal dose ($LD_{50/30}$) can be read off from the curve; for mice it is about 640 R.

Similar experiments to determine the median lethal doses of several mammals and nonmammalian species have been carried out and reported in the literature. Table 15-1 gives the $LD_{50/30}$ values for some selected vertebrate species. One can see that mammals are in general more radiosensitive than other

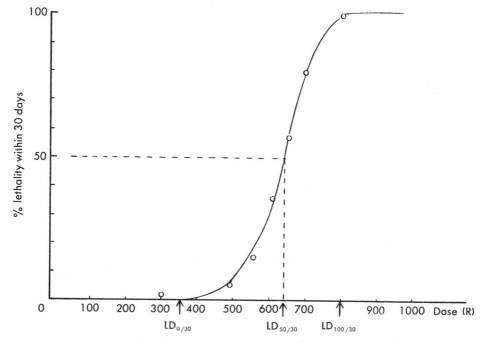

Fig. 15-1. Dose-lethality curve for mice.

Table 15-1. $LD_{50/30}$ of selected vertebrate species

Species	Type of radiation	$LD_{50/30}$
Mammals		
Mouse	200 kVp X-rays	640 rads
Rat	250 kVp X-rays	714 rads
Dog	—	250 rads
Macaca mulatta (monkey)	250 kVp X-rays	600 rads
Rabbit	250 kVp X-rays	750 rads
Guinea pig	250 kVp X-rays	450 rads
Hamster	200–250 kVp X-rays	610–856 rads
Swine	1000 kVp X-rays	250 rads
Goat	200 kVp X-rays	240 rads
Burro	1.1 MeV gamma rays	255 rads
Other vertebrates		
Chicken		600–800 rems
Tortoise		1500 rems
Frog		700 rems
Newt		3000 rems
Goldfish		670 rems

vertebrates. Among mammals, relatively large animals like the burro, the goat, and the swine are more radiosensitive than smaller species. As pointed out in Chapter 9, the $LD_{50/60}$ for man seems to be around 300 to 350 rems.

I should emphasize, however, that the values reported in Table 15-1 are in most cases approximate and obtained from experiments with animals irradiated in normal environmental conditions. Different strains of mice and rats may show markedly different degrees of radiosensitivity. In the case of poikilothermic vertebrates, the lethal dose is temperature dependent. For instance, frogs and goldfish survive several thousand R for more than 30 days if kept at a temperature below 15°C.

The LD_{50} may be a useful indicator of the radiosensitivity of invertebrates and unicellular organisms, although it may not be based on observation periods of 30 days. Therefore LD_{50} values for other organisms are hardly comparable with those reported for vertebrates. The criterion used to determine the percent survival of irradiated protozoa and bacteria is clone formation. The LD_{50} for a bacterial species is understood as the minimum radiation dose needed to prevent 50% of a population of bacterial cells from forming colonies.

In spite of the different criteria adopted for different organisms, the fact is undisputable that many invertebrates and unicellular organisms are much more radioresistant than vertebrates. The LD_{50} for snails is estimated to be around 8000 to 20,000 rems; for many insect species it is 10^4 to 10^5 rems. LD_{50} values for some unicellular organisms are about 100,000 rems for amebas, about 30,000 rems for yeasts, about 5,600 rems for *Escherichia coli*, about 150,000 rems for *Bacillus mesentericus*, and about 300,000 rems for infusoria. The radiosensitivity of some bacterial species has been found to be dependent not only on the strain but also on several environmental factors such as the composition of the culture medium.

Why there should be such enormous difference in radiosensitivity among different species of organisms is one of the unsolved problems in radiation biology. Certainly the law of Bergonié and Tribondeau is insufficient to explain the reasons. After observing that radioresistant organisms are also resistant to cyanides, which are known to poison certain enzyme systems, some investigators are inclined to believe that the degree of radiosensitivity depends on the presence and physiologic role played by cyanide-sensitive enzymes. For instance, although coelenterates are quite radioresistant, hydras are not; in fact, hydras are also sensitive to cyanides, unlike other coelenterates. The degree of radiosensitivity of a species is possibly dependent on the mechanism used to maintain the intracellular osmotic pressure. Among invertebrates, the osmotic pressure is often regulated by means of amino acids, some of which are known to have protective properties against ionizing radiation (see Chapter 16). In vertebrates, the osmotic pressure is maintained by inorganic ions (K^+, Na^+, etc.). The oxygen tension in the tissues of an organism may also affect radiosensitivity. The lower this tension, the greater the resistance to radiation (see the discussion on the oxygen effect on p. 441). Actually, an interplay of several morphologic, physiologic, and biochemical factors is possibly responsible for the degree of radiosensitivity of a species. The whole problem requires further investigation.

SOURCES OF INFORMATION ON THE HUMAN ACUTE RADIATION INJURY

During the past 25 years, our knowledge of the injuries caused by acute whole-body or partial exposure of human individuals to ionizing radiation has made significant progress. Twenty years ago the very diagnosis of radiation sickness was extremely difficult. Today we have sufficient information on diagnosis, as well as prognosis and therapy, to the extent that we can save the lives of individuals who have been exposed to low supralethal doses. The sources of our information on human radiation injury are summarized and briefly discussed below.

Animal experiments

Acute radiation sickness and other injuries can be induced in laboratory mammals with acute whole-body exposures. Different types of radiation (X-rays, gamma rays, and neutrons) have been used with sublethal, lethal, and supralethal doses on dogs, rats, monkeys, and other species. The time taken for the appearance of several symptoms can be easily observed. In these experiments the doses delivered are accurately known, so that dose-response relationships are not difficult to determine. A large body of information has been obtained through animal experiments concerning histopathologic findings (such as effects of different doses on blood cells, on hemopoietic tissues, on the intestinal mucosa, etc.), biochemical changes (for example, levels of some excretion products in the urine), and most of all the relation between dose and type of death. Other symptoms that normally appear in human radiation sickness (nausea, headache, fatigue, etc.) are more difficult to assess in animals. At any rate, some of the findings on mammalian radiation sickness can be cautiously extrapolated to humans.

Patients treated with radiation therapy

Although radiation therapy usually involves partial exposure,* side effects affecting the body as a whole do appear occasionally, and in particular when high

---

*Whole-body radiation therapy has been used several times for the treatment of leukemia patients. Also, whole-body irradiation has been frequently used prior to kidney transplant to depress the recipient's immune response and thus prevent the rejection of the transplanted organ.

doses are used. These unwanted effects are often identical to certain symptoms of radiation sickness such as nausea, vomiting, anorexia (loss of appetite), and fatigue. As in animal experiments, the doses delivered can be accurately measured. However, caution should be used in the interpretation of the results because the irradiated subjects are abnormal individuals, usually cancer patients. One can never be sure as to what extent the response to irradiation of an individual who is already ill is comparable to the response of normal subjects. Nevertheless, these studies do represent controlled exposures, and as such have considerable value for a better understanding of radiation sickness.

Japanese victims of atomic explosions

At Hiroshima and Nagasaki thousands of Japanese were exposed to gamma and neutron radiation. The exposure was practically instantaneous, and a wide spectrum of doses was represented. Unfortunately the doses could only be estimated through indirect inferences. Many individuals heavily exposed, who did not die instantly or within a few minutes or hours, were observed by physicians of the Japanese Army Medical School before their death, which occurred several days after the explosions. These observations have been recorded and studied later by the ABCC. However, the interpretation of the findings is often complicated by a number of difficulties; besides the uncertainties about the doses, many fatalities were caused not only by exposure to ionizing radiation but also by thermal burns, blast, and epidemic infection. Therefore, the precise role of radiation injury in causing the death of some individuals was uncertain.

Victims of radiation accidents
Definition

In a very broad sense, even the spillage of radioactive material with consequent contamination of a laboratory bench or floor may be considered as a radiation accident. But if such an event causes no injury whatsoever to human individuals, it is of no radiobiologic interest. Likewise, if a person inadvertently carries in his pocket a gamma source with the activity of only a few μCi over a period of several days or weeks, he has been accidentally exposed to a low chronic dose. If, as a result of the exposure, chromosomal aberrations and gametic mutations are induced, this incident can also be classified as a radiation accident.

In the context of this chapter, however, a radiation accident is defined as *any unforeseen, unplanned, and unexpected event causing acute whole-body or partial exposure of the human body to some type of ionizing radiation, with a dose large enough to result at least in gross, short-term injuries to the individual(s) exposed.* In an accident so defined the dose may be supralethal, lethal, or even sublethal, but large enough to cause injuries at organ or organismal level and detectable within days or weeks after exposure. Any accident that results only in late effects (such as onset of cancer, cataracts, etc., which are mentioned in Chapter 17) a few or several years after exposure is not considered a radiation accident in the present context.

About two dozen radiation accidents have occurred during the past 25 years, since the beginning of the atomic age. Because these accidents are not anticipated and are often caused by ignorance or carelessness, quite often no dosimetric data are available for an accurate diagnosis and prognosis of the radiation injury. In spite of this shortcoming, radiation accidents usually involve clinically normal people (unlike patients treated with radiation therapy), and as such they are a valuable source of information on short-term radiation effects in man. Moreover, the victims of radiation accidents, especially of the more recent ones, have been kept under close and thorough clinical observation by specialists until their death or recovery, and many times even examined long after recovery.

Possible causes

Some radiation accidents may be classified as reactor accidents; that is, they are caused by a nuclear reactor accidentally going "out of control." In Chapter 8, I have explained that the normal operation of a reactor involves a self-sustaining chain reaction under controlled conditions. The self-sustaining chain reaction is attained when the reaction is critical or slightly supercritical. If the k factor is allowed to reach values much greater than 1, because of the failure of the automatic control system, the energy may be liberated at such a high rate to cause leaks or ruptures in the shielding material. Any person who happens to be in the vicinity of the reactor will be exposed to a mixed neutron and gamma radiation. Other accidents may result in more serious physical damage to the reactor, such as disintegration of the shield and exposure of the core, with falling debris, heat wave, fire, etc.

Criticality accidents may occur even without a reactor. During the processing of fissionable material (enriched uranium, plutonium), a large amount of material may be unintentionally allowed to reach a critical geometric configuration. A self-sustaining chain reaction is thus accidentally triggered, although it may last for only a very short time. Such reaction is often referred to as a *nuclear excursion.* A burst of neutron and gamma radiation is suddenly emitted by the unshielded system; bystanders may be exposed to this radiation and suffer injuries.

Because of the presence of neutron radiation in criticality accidents, radioactivation takes place in the body of the victims. Notable is the radioactivation of the relatively large amounts of sodium present in the body, with production of ^{24}Na, a high-energy gamma emitter. By radioassaying the gamma activity of the victim's body with a whole-body counter, calculation of the neutron radiation dose received by the victim is often possible.

Other radiation accidents have been caused by exposure to X-rays emitted by unshielded Coolidge tubes, Van de Graaff generators, and incidental sources of X-rays (klystrons), or to gamma radiation emitted by unshielded ^{60}Co sources, or to particulate radiation emitted by particle accelerators.

Heavy local fallout was the direct cause of the Bikini accident, which involved a large number of individuals. The exposure was from mixed beta-gamma radiation emitted by the fallout dust ingested, inhaled, or deposited on the body of the victims.

Short-term injuries may also be caused by the accidental ingestion of large amounts of radioactive materials. This is a case of internal radioactive contamination. The radiation dose delivered to the body may be large enough to cause radiation sickness. If the contaminant is a gamma emitter, the whole-body counter may be useful for its identification and the determination of the activity ingested. Occasionally, mild symptoms of radiation sickness have been observed in patients who were given large therapeutic doses of radioisotopes. Accidents caused by dosimetric miscalculations or improper use of teletherapy units have been rare.

Although the use of ionizing radiation and nuclear energy has greatly increased, the frequency of radiation accidents has been surprisingly low when compared with the number of accidents occurring among workers in other industries. This good record can be attributed to the conscientious work of radiation safety experts.

Brief survey of some serious radiation accidents

From 1945 until the time of this writing about 25 radiation accidents with at least slight injuries to humans have occurred in different countries or areas (U.S.A., Soviet Union, Belgium, Argentina, the South Pacific, Mexico, Yugoslavia). The total number of persons exposed was 384, the number of fatalities 14, and the number of seriously injured persons 161. The largest number of exposures (290) and the largest number of injuries (132) occurred in one single accident (the Bikini fallout accident). The following is a brief description, in chronologic order of some of the serious accidents reported in the literature.

1. Pacific Nuclear Testing Grounds, March 1, 1954. The fallout accident occurred after the explosion of a thermonuclear device on the Bikini atoll (see further details on p. 209).

2. Oak Ridge, Tennessee, June 16, 1958. This accident occurred at the Y-12 Plant, where uranium 235 is extracted from natural uranium. A critical geometry was achieved when, by miscalculation, some enriched uranium was allowed to drain into a 55-gallon drum, which was intended to contain only water. At the time of the nuclear burst one of the operators noticed yellow-brown fumes arising from the drum, followed by an odd bluish flash. Five persons were exposed to mixed neutron-gamma radiation with an estimated dose varying from 298 to 461 rems. All five were hospitalized with radiation injuries, from which they later recovered.

3. Vinca, Yugoslavia, October 15, 1958. A heavy-water power reactor, with natural uranium as fuel, became supercritical during an experiment designed to determine the spontaneous fission constant of natural uranium. Six persons were exposed; they were unaware that an accident had occurred until some time later when the dosimeters were checked and when the first symptoms of radiation sickness appeared. However, an odor of ozone was noticed at the time of the accident. The exposure was from gamma and relatively low-energy neutron radiation. All six victims were flown to Paris, where they were treated with bone marrow transplants. The therapy failed to save the life of one victim; the others eventually recovered. There is no agreement on the radiation doses to which the victims were exposed. Probably the highest dose was either below or not much above 1000 rems.

4. Los Alamos Scientific Laboratory, December 30, 1958. (See the details about this accident in the Appendix on p. 436.)

5. Lockport, New York, March 8, 1960. In this accident, injuries were caused by exposure to radiation emitted by an incidental source of X-rays (see p. 73). In the course of a troubleshooting procedure, one klystron tube of a radar station was operated with its shield removed. Two persons suffered injuries from pulsed X-rays emitted by the anodic plate of the tube; one of them received an estimated dose of 1500 R to the head, 300 R to the trunk, and a smaller dose to the legs.

6. Mexico City, April to July 1962. This accident was caused by the exposure of an entire family of 5 to high doses of gamma radiation over a period of several weeks. Although the exposure could not be classified as acute, there were four fatalities. A cobalt-60 source was found in a field by the children of the family, who brought it home and treated it as a play toy. The source was the property of a construction engineer, who used it for radiographic work. When it was found, it was not properly locked. The four persons who died received a total accumulated dose ranging from 3000 to 4700 R. The survivor (the father of the family, who was home only on weekends) was exposed to 1200 R.

7. Wood River Junction, Rhode Island, July 24, 1964. One victim was exposed to a radiation dose, which is probably the highest ever recorded in a radiation accident. A 38-year-old operator was pouring a mixture containing uranium 235 from a plastic

cylinder into a tank containing sodium carbonate. A supercritical volume was attained with the new geometry and a nuclear excursion took place. The operator saw a flash of light and immediately was hurled backward and stunned. Investigators estimated that he absorbed 2200 rads of neutron radiation and 6600 rads of gamma radiation. He died 49 hours after the accident. High levels of radiosodium were measured in his body, and of radiophosphorus in his feces. Two other persons who tried to help the victim were also exposed to low radiation doses but did not suffer any injuries.

8. Pittsburgh, Pennsylvania, October 14, 1967. Three technicians employed with the Gulf Research Development Company were accidentally exposed to a 3 MeV electron beam, while attempting to repair the cooling system of a Van De Graaff generator. Due to failure of the safety system, the machine was on without the knowledge of the operators. One hour after the exposure the three men suffered nausea, which they interpreted as a symptom of influenza. But soon it was evident that they were victims of radiation sickness. One of them had received an estimated dose of 6000 R on his hands and 600 R on the rest of the body. His life was saved with a marrow transplant from his identical twin brother. However a serious gangrene developed in his hands, and seven of his fingers had to be amputated. The other two men received a whole-body dose of 300 R and 100 R, respectively. They suffered only mild radiation sickness and went back to work a few weeks after the accident.

The Medical Division of the Oak Ridge Associated Universities (ORAU) is an important center for the storage of information and for research on the clinical aspects and therapy of radiation accidents.

Dosimetric problems

As we shall see later in this chapter, the prognosis and therapy of radiation injury are largely dependent on the dose received by the patient. In many radiation accidents the dose is not immediately known, either because the victim was not wearing his film badge or pocket dosimeter at the time of the accident or because dosimetric devices were simply not available. Sometimes a rough estimate of the dose can be obtained by reenacting the accident on location (as in the case of the Lockport accident).

With the possible absence of reliable dosimetric data in radiation accidents, efforts have been made to see whether any responses elicited in the victim by irradiation are themselves dose dependent and, as such, whether they can be used as biologic dosimeters.

Several biologic dosimeters have been identified through animal experiments, and their validity has been confirmed for victims of radiation accidents. The linear dose dependence of some of them is still under investigation.

One method is the measurement of the level of β-aminoisobutyric acid (BAIBA) in the patient's urine. BAIBA is an amino acid that has been known for the past 20 years. It is excreted by the kidney as a waste product of the catabolism of thymine, thymidine, and DNA. The excretion of BAIBA is increased after substantial exposure to ionizing radiation and reaches a maximum a few days thereafter. Studies on the victims of the Oak Ridge accident showed that the excretion of BAIBA was roughly dose dependent. Further studies have indicated that this dosimeter may be useful in the 50 to 200 R dose range. The cause of the extra excretion of BAIBA after irradiation is not very clear. It possibly originates from radiation-induced breakdown of the DNA of destroyed cell nuclei, but the explanation is also conceivable that it derives from DNA thymine-containing precursors that are not utilized because of the radiation-induced inhibition of DNA synthesis. Since increased excretion of BAIBA is not specific for radiation injury, this response in victims of radiation accidents should be used as a biologic dosimeter with caution.

More recently, investigators have found that irradiated rats, dogs, and men excrete increased amounts of the nucleoside deoxycytidine in their urine 1 day after exposure to ionizing radiation and that in rats the level of deoxycytidine in the urine is a linear function of dose up to 600 R. This response can also be used as a biologic dosimeter.

Concerning the origin of the extra deoxycytidine excreted in the urine, we are confronted with the same problem as for BAIBA. Some recent experiments utilizing the radiotracer method seem to indicate that the excess of nucleoside originates from the radiation-induced degradation of DNA.

Another procedure that can be used for dosimetric purposes is the study of the variation of the mitotic index (see p. 359) in the victim's bone marrow. The normal mitotic index in human bone marrow is approximately 9/1000 between 10 AM and 1 PM. In the more heavily exposed individuals of the Oak Ridge accident the mitotic index was practically reduced to zero by the fourth day. Observations on the other victims suggested that the depression of the mitotic index is dose dependent. If the victim's bone marrow is examined periodically at least during the first week after the accident, the method may be useful to estimate exposures in the 50 to 200 rad dose range. E. P. Cronkite believes that a mitotic index that approaches zero by the fourth day after an accident is an indication of a dose of 200 rads or more and that serious consequences could be anticipated.

Other biologic dosimeters have also been suggested, but they are much less reliable than the ones mentioned above, because they are less objective and often reflect the bias of the observer.

THE ACUTE RADIATION SYNDROME

In 1897, shortly after the discovery of X-rays, D. Walsh was first to point out that a whole-body exposure of man to radiation may be followed by a "radiation sickness."[*] Since then, the disease has been clinically investigated, especially during the past 20 years.

The acute radiation syndrome is an organismal response quite distinct from isolated local injuries (burns, erythema, epilation, etc.) that may be induced by partial body exposure. The onset of the syndrome is observed only after an acute, whole-body exposure to some type of penetrating radiation with doses above a certain threshold, although occasionally partial exposures involving large portions of the body may also be responsible for the same response. The severity and course of the disease are largely dependent on the total accumulated dose and on whether the exposure is total or partial.

If the symptoms and lesions associated with the disease are taken separately, one must acknowledge that almost none of them is unique and typical of the radiation injury; in fact, some of them mimic other quite unrelated pathologic processes. Only by the array of all the nonunique symptoms can the radiation syndrome be identified as a distinct nosologic entity.

As already mentioned above, a large body of information is available at the present time, concerning the pathology, symptomatology, diagnosis, and treatment of the acute radiation syndrome.

---

*Walsh, D. 1897. Deep tissue traumatism from roentgen ray exposure. Brit. Med. J. **2**:272.

The relationship between dose and survival time

The primary factor that determines the severity of the damage to the organism, its possible lethality, and the time after exposure when death occurs is the dose received. This relationship can be conveniently studied in laboratory mammals. When different groups of animals are exposed to different radiation doses, the mean survival time within the groups decreases with increasing dose. The mean survival time (MST) is the average time elapsed between exposure and death. If the data from experiments with different mammalian species are pooled together, a generalized curve can be obtained by plotting MST as a function of dose on log-log graph paper (Fig. 15-2).

An analysis of the curve shows that for absorbed doses of 100 rads or slightly higher, the MST declines sharply with increasing dose. A short plateau exists in correspondence to the median lethal doses of most mammalian species. Along this plateau, the MST is dose independent to a limited extent, and its value is about 13 to 14 days. As the dose approaches 1000 rads, the MST becomes again dose dependent and reaches a value of about 3.5 days with 1000 rads. This value remains constant over a wide dose range (between 1000 and 10,000 rads); another much longer plateau is thus obtained. Beyond 10,000 rads, the MST falls again with increasing dose, from a few days to a few hours. With very high doses, it may be reduced to zero (death under the beam).

With this type of experiment, the observation has also been made that the im-

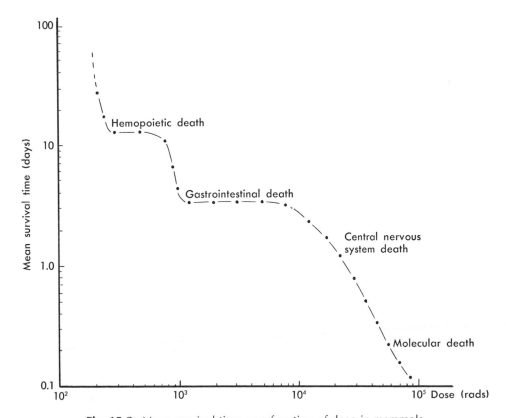

Fig. 15-2. Mean survival time as a function of dose in mammals.

mediate cause of death is not the same for all doses. With doses below 1000 rads, the animal dies because of the destruction of certain cellular elements of the blood, which in turn is caused by irreparable damage to the hemopoietic tissues. This type of death is referred to as *hemopoietic death,* or *bone marrow death,* and the syndrome that precedes death is known as *hemopoietic* or *bone marrow syndrome.* This name should not be construed as meaning that at doses below 1000 rads no injury is induced in other organs or systems. Other injuries and lesions do occur, but they are not severe enough to be directly responsible for the death of the animal. In the 1000 to 10,000 rad range, the injuries to the gastrointestinal system appear shortly after exposure and are serious enough to be the immediate cause of death within an average of 3 to 4 days *(gastrointestinal death).* Although the hemopoietic damage is also very severe, its effects on blood cells would be lethal only several days later, if death had not been quickly caused by the gastrointestinal failure. The gastrointestinal death is preceded by a *gastrointestinal syndrome* of short duration.

With doses above 10,000 rads, death is attributed to irreparable damage to the more radioresistant, but very vital, central nervous system *(CNS death).* For doses about 10^5 rads and higher, the death, which follows within a few minutes or hours after exposure, is conceivably directly caused by massive and extensive inactivation or destruction of molecules of vital biologic importance (enzymes, proteins, etc.). This type of death is often referred to as *molecular death.*

For reasons of convenience, the dose-MST curve can be divided into three components. The first corresponds to the 100 to 1000 rads dose range and is associated with the hemopoietic death; the second corresponds to the 1000 to 10,000 rad range and is associated with the gastrointestinal death; the third, corresponding to doses above 10,000 rads, is associated with CNS death and molecular death. The long plateau of the second component may be interpreted as an indication of the high degree of radioresistance of the central nervous system. In other words, the dose required to induce lethal damage to the CNS is much above the minimum dose necessary to induce lethal damage to the gastrointestinal system.

Although the distinction of the four radiation syndromes discussed above is very convenient and has a sound scientific basis, it should be used with some limitations and without misunderstandings. The distinction of the four syndromes may turn out to be an oversimplification of the clinical picture in acute radiation injuries. The clinical course after accidental nonuniform exposures may not be likely to fit a pattern that one would predict from the study of uniform whole-body exposures. The victim of the Lockport accident is a good example. Although the total body dose he received was not very high, certain symptoms had to be clearly attributed to CNS injury.

The fact has already been stressed that for each syndrome the damage does not involve only one system, but practically the entire organism. But in the different syndromes, death is caused, more or less rapidly, by the failure of different systems. In the transition zones between two syndromes, death may be directly caused by lethal damage to two systems, or to one system in some animals and to the other system in other animals receiving the same dose.

The general symptoms are essentially the same in any radiation syndrome, except for a few symptoms that might be encountered in the gastrointestinal syndrome and even more in the CNS syndrome but not in the hemopoietic syndrome. This is, of course, as much as we can infer from the study of the relatively few human individuals who have been victims of radiation sickness after exposure to very large doses. Another difference between syndromes is that those symptoms common to all syndromes are more severe and appear more rapidly after exposure to high doses than to low doses, although exceptions to this rule have been noted in the victims of some radiation accidents.

The following exposition mainly revolves around the hemopoietic syndrome; only occasional and incidental comments are made in relation to the gastrointestinal and CNS syndromes.

General survey of the symptoms in relation to dose and time after exposure

The earliest symptoms of radiation sickness are the consequence of gastrointestinal distress; they manifest themselves as nausea, vomiting, diarrhea, and anorexia. Walsh also described a severe prostration very similar to that resulting from a sunstroke, which he attributed to radiation-induced "deep tissue traumatism."

The severity of the early symptoms and the time of their onset depend mainly on the dose received and on the sensitivity of the individual exposed, although with high dose levels the individual sensitivity has no effect.

Depending on the dose, the early symptoms may appear within minutes or, at most, within hours after exposure. They are like the prelude to the ensuing complex of pathologic processes that develop some time later. For this reason, according to a recent terminology, the early symptoms are collectively referred to as *prodromal syndrome, premonitory phase,* or *Strahlenkater* (German for "radiation hangover"). The severity of the prodromal syndrome and the time of its onset after exposure can be of significant prognostic value concerning the forthcoming acute radiation syndrome, especially when precise dosimetric data are not available immediately.

The mechanisms by which the prodromal responses are evoked are not clear. We do not know whether they are direct or indirect effects, although we know that they can be induced also by partial irradiation of the abdomen, thorax, or head. Probably, the autonomic nervous system is intimately involved in the appearance of the prodromal symptoms.

After sublethal doses, the prodromal syndrome is usually followed by a latent period, lasting several days, during which no definite symptoms are evident. After the latent period, the damage to the hemopoietic system and to other organs begins to express itself with the symptoms of the acute radiation syndrome, which are general malaise, fatigue, drowsiness, and listlessness (apathy), while vomiting and diarrhea may return at irregular intervals. A mild erythema (reddening of the skin) may become visible, especially on those regions of the body that were directly facing the radiation source. Partial epilation is one of the last symptoms to appear. Gradually all symptoms disappear; recovery is usually complete a few weeks after exposure.

After exposure to lethal doses, the prodromal syndrome is followed by a shorter latent period with no definite symptoms. During the second week, the symptoms of radiation sickness are in part the same as those observed with lower doses. Additional symptoms, indicative of a more severe injury, include loss of weight, bacteremia (bacterial infection of the blood), fever, abdominal pains, and internal bleeding. Stomatitis (inflammation of the oral mucosa) has been observed in a few patients (for instance, in the victim of the Lockport accident). Erythema and epilation are usually more severe than with sublethal doses. Female patients have also noted irregularities in their menstrual cycle. Without treatment, patients exposed to lethal doses have only a 50% chance of recovery.

After exposure to supralethal doses, the latent period is practically nonexistent. With doses much above 1000 rems some of the symptoms typical of lower doses fail to appear simply because the course of the gastrointestinal syndrome is too rapid to allow them to develop. For instance, the pathologic mechanisms leading to bacteremia or epilation take a long time to produce their effects; in the meantime the catastrophe is being precipitated by more direct injuries to other organs. In the Rhode Island and 1958 Los Alamos accidents, the victims, who had been exposed to the highest doses ever recorded for any accident (9000 and 4500 rads respectively), suffered also shock and unconsciousness immediately or within minutes after the accident. Probably these symptoms are typical of the prodromal syndrome that follows exposure to very high doses. For both accidents, curiously enough, it was reported that the patients felt comfortable at least for a few hours during their short survival time, although the clinical tests were clearly showing that the end was only a few hours away. This deceptive and ephemeral state of well-being may be difficult to explain. Low blood pressure (arterial hypotension), insomnia, and restlessness have been observed after exposure to supralethal doses. Serious radiation burns and blisters were seen in patients exposed to nonuniform doses (for example, in one victim of the Pittsburgh accident). Without treatment, death is quite probable or certain. After doses of 2000 to 3000 rads or more, the patient's life cannot be saved with the therapeutic measures currently available.

A synopsis of the clinical symptoms of acute radiation sickness in relation to postexposure time and dose is presented in Table 15-2. The data presented contain several limitations and approximations and therefore should be viewed with a certain degree of caution. They have been gathered from the clinical reports about the few human victims who could be thoroughly studied by competent investigators. The symptoms shown do not necessarily appear in all patients receiving the same dose or at least not at the same time, because of the variations of individual sensitivity. Perhaps they should be viewed in terms of probability of their occurrence. No attempt has been made to time the symptoms and other events accurately, simply because the time of their onset is also affected by individual variability.

Effects on some organ systems in the acute radiation syndrome
The response of the hemopoietic systems and blood

The hemopoietic syndrome is characterized primarily by reversible or irreversible failure of the hemopoietic systems, that is, of the blood cell–forming tissues, which, in the adult, are essentially the bone marrow and the spleen (myeloid and lymphoid tissues, respectively). The failure is caused by hypoplasia or aplasia (partial or total destruction) of the active tissue. The consequence of the damage is a decrease in the number of mature blood cells in circulation.

The hemopoietic tissue is made of a loose stroma, which resembles a meshwork. Within the meshes, several precursors of blood cells can be identified. The spleen is the

Table 15-2. Clinical symptoms of acute radiation sickness in relation to postexposure time and dose

Time after exposure	100 to 250 rems (sublethal dose)		350 to 450 rems (lethal dose)		650 rems or more (supralethal dose)
First week	Nausea and occasional vomiting within hours		Nausea, vomiting, pallor within a few hours		Nausea, vomiting, pallor within a few hours or minutes
	No definite symptoms	(Latent period)	No definite symptoms	(Latent period)	Shock, unconsciousness Diarrhea, abdominal pains and cramps Arterial hypotension Fever, severe erythema, burns or blisters Insomnia, restlessness
Second week			Loss of weight, general malaise, fatigue, stomatitis		Death certain within a few hours, or a few days, depending on dose
			Bacteremia, fever, anorexia, abdominal pains, severe erythema		
Third week	General malaise, anorexia, mild erythema, diarrhea, fatigue, drowsiness		Epilation		
			Internal bleeding and petechiae		
	Epilation				
Fourth week, and later			(Menstrual irregularities)		
			50% chance of death		
	Recovery probable				

site of production of monocytes and lymphocytes. The bone marrow produces granulocytes, erythrocytes, and platelets.

The common precursor of all blood cells is a type of stem cell called hemocytoblast, which is found in both the spleen and the bone marrow (Fig. 15-3). In the spleen hemocytoblasts give origin to monocytes or lymphocytes, by means of a series of transformations, which involve cell divisions and progressive differentiation. Some intermediate stages between the primitive hemocytoblast and the lymphocytes are the lymphoblast and the prolymphocyte. Monocytes are white blood cells with a kidney-shaped nucleus; lymphocytes are also white cells with a relatively large spherical nucleus. The hemocytoblasts of the bone marrow may give origin to granulocytes, through the leukoblast and myelocyte stages; or to erythrocytes through several erythroblast stages; or to platelets (thrombocytes) whose immediate precursors are large megakaryocytes. Granulocytes are white blood cells with bilobed or multilobed nuclei and with typical granules in their cytoplasm, which can be stained with acid, basic, or neutral dyes. Unlike the erythrocytes, the immature erythroblasts are nucleated cells that already contain hemoglobin but are

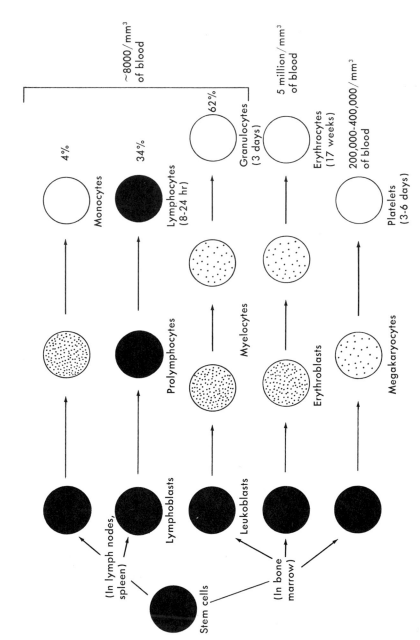

Fig. 15-3. Genesis of the blood cells in the hemopoietic organs. Black spheres represent very radiosensitive stages; dotted spheres indicate less radiosensitive cells; white spheres are radioresistant cells. The lifetimes of mature blood cells are also shown.

still unable to function as oxygen carriers (as in erythroblastosis fetalis). The platelets are believed to be cellular fragments of the megakaryocytes.

In the bloodstream, the lifetime of the mature blood cells varies. Lymphocytes have the shortest life (probably not more than 24 hours). The longest lifetime is that of the erythrocytes (about 17 weeks), whereas granulocytes and platelets live for only a few days. The lifetime of the monocytes is unknown.

The stroma of the hemopoietic tissues is radioresistant, whereas the hemocytoblast is very radiosensitive. In agreement with the law of Bergonié and Tribondeau (see p. 330), the radiosensitivity decreases through the successive hemopoietic stages, so that the mature blood cells are radioresistant, or at least much less sensitive than their precursors. Lymphocytes are an unexplained exception; they are just as radiosensitive as their precursors. As a matter of fact, they must be classified among the most radiosensitive cells of the mammalian body.

The responses of the blood cells after acute whole-body exposures are graphically shown in Fig. 15-4 for doses of 200, 300, and 450 rads. The curves were plotted by averaging the results of data from radiation accidents and a small number of therapeutically irradiated individuals.

The first blood cells to be affected are the lymphocytes. Their number drops sharply within minutes or hours after exposure (lymphopenia) and may remain low throughout the course of the disease. The return to normal levels is a very slow process. With very high doses, their number may be practically reduced to zero. A significant and measurable decrease in the number of lymphocytes is observed with doses as low as 10 to 20 rads, that is, when any other symptom of radiation sickness is completely absent. The magnitude of the reduction of the lymphocyte count may be used for a preliminary estimation of the dose to which the patient has been exposed. Because of the high radiosensitivity of these cells and because their response is so fast, there is no doubt that the destructive effect of radiation on them is a direct one.

The first response of the granulocytes is a sudden increase of their number (granulocytosis), followed by a decrease, which is at first rapid and then slow (granulocytopenia). The curves show that granulocytosis occurs almost immediately after exposure and that the peak value reached by the granulocyte count is a direct function of dose. The victim of the 1958 Los Alamos accident showed a granulocyte count of 28,000/mm^3 of blood 14 hours after exposure (recall that the normal count for all white cells is only 8,000/mm^3 of blood). The mechanism of this radiation-induced granulocytosis is not well known, although it must be related to an accelerated maturation in the bone marrow, which could be interpreted as a sort of mobilization phenomenon in response to widespread injury. According to G. A. Andrews, granulocytopenia reaches its nadir about 30 days after exposure, regardless of the dose. After this time, the count rises slowly if the patient recovers.

Granulocytopenia is not the result of direct destruction of granulocytes by radiation. As already mentioned above, these cells are quite radioresistant. The reduction of their number is caused by the fact that the cells that die of natural death are not promptly replaced by other cells from the bone marrow, due to radiation damage to the more radiosensitive myelocytes and leukoblasts (the precursors of the granulocytes).

The depletion of platelets (thrombocytopenia) begins rather late after expo-

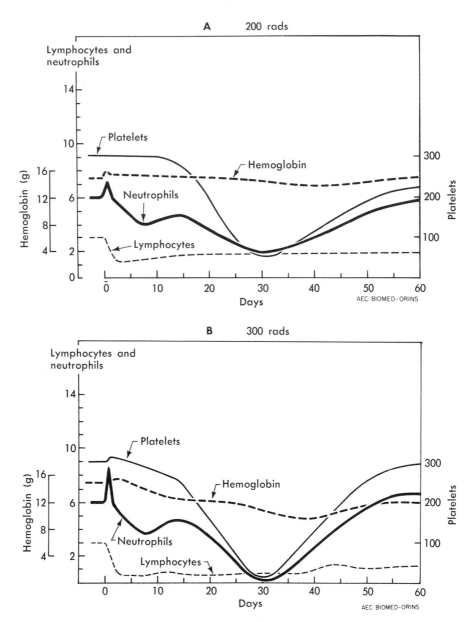

Fig. 15-4. Typical hematologic response of the human body to radiation doses of 200 rads, **A;** 300 rads, **B;** and 450 rads, **C.** Lymphocyte, neutrophil, and platelet values should be multiplied by 1000. Hemoglobin values are in grams per 100 ml of blood. The curves represent the best information available at the present time. They may not be entirely correct; but their improvement has not been possible yet. The extent and duration of the initial leukocytosis are not clear. A different picture is obtained at exceedingly high supralethal doses, as in the last Los Alamos accident and in the Rhode Island accident. (From Andrews, G. A., 1967. Radiat. Res. suppl. 7, pp. 391-392.)

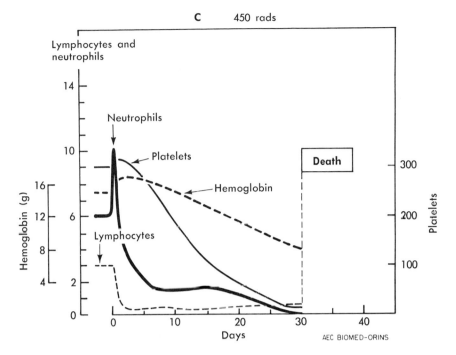

C   450 rads

Fig. 15-4, cont'd. For legend see opposite page.

sure to low doses and earlier after high doses. Thrombocytopenia and granulo-cytopenia reach their lowest points at about the same time. The reduction of the platelet count is also an indirect effect of radiation; it is explained by the fact that the radioresistant platelets that die of natural death are not replaced by others, due to the radiation damage suffered by their precursors, the more radiosensitive megakaryocytes.

The depression of erythrocytes is less severe than that of other blood cells, probably because of their much longer lifetime. Although their precursors (erythroblasts) are definitely damaged or destroyed, this damage is not immediately reflected on the red cell count. The drop in the red cell count can be appraised by measuring the hemoglobin level in grams per 100 ml of blood.

The depletion of cellular blood elements results in secondary consequences, which may be responsible for death from the bone marrow syndrome. Infection develops, in part, as a result of lymphopenia and granulocytopenia. Since white blood cells have the ability to phagocytize bacteria, their disappearance from the blood means the breakdown of one physiologic defense mechanism of the body. Thrombocytopenia is the cause of the frequent hemorrhages in acute radiation syndrome, because of the important role played by the platelets in the process of blood coagulation. Red cell depletion leads to anemia, which may be aggravated by hemorrhages.

Clearly, the depression of the blood cell count in the peripheral blood must be largely attributed to direct radiation damage to the hemopoietic tissues. The damage consists of the killing of the most radiosensitive blood cell precursors, or in a complete inhibition of mitoses, or simply in mitotic delay. The depression of the mitotic index in the bone marrow has already been mentioned in a previous section of this chapter. Several types of morphologic abnormalities have been reported in the irradiated bone marrow cells. Most of them are connected with irregular mitosis and mitotic block. Giant cells, karyomeres (nuclear fragments), binucleated myelocytes, tripolar mitoses, and internuclear and chromosomal bridges are readily visible in the bone marrow shortly after exposure. Such abnormal immature cells, and in particular myelocytes, may be prematurely released by the marrow and can be followed in the peripheral blood.

The gravity of the hemopoietic injury is a function of dose, and the chance of recovery depends largely on the number of stem cells and other precursors that have been left unharmed. If their number is not too small, they may have the remarkable power of regenerating, in due time, the blood cell population and bring it back to normal levels. The fact is significant that the shielding during exposure of even one bone containing active marrow (for example, the femur) can bring about the recovery of the marrow lethally irradiated in the other bones (see more details on p. 434).

The response of the gastrointestinal system

The mucosa is the most radiosensitive of all the tissues that make up the gastrointestinal system. Furthermore, the mucosa lining the small intestine (and especially the duodenum) is more sensitive than the gastric mucosa or the mucosa of the colon. The radiation injury to the mucosa consists of the destruction of its cells, which may be more or less extensive, depending on the dose. Considerable, but repairable, damage of the muscoa is observed even in the hemopoietic syndrome; with the higher doses responsible for the gastrointestinal syndrome, the injury to the mucosa is so severe and irreversible that it may be considered as the direct cause of the gastrointestinal death.

The lining of the small intestine has a velvety appearance, because of the presence of millions of fingerlike projections of the mucosa, called "villi" (Fig. 15-5). The epithelium

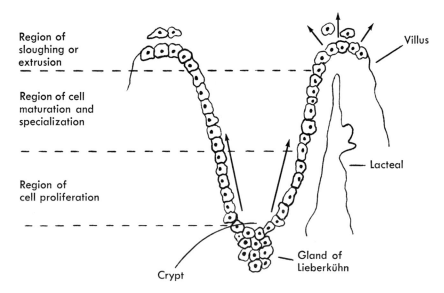

Fig. 15-5. Microscopic morphology and cytokinetics of the intestinal epithelium. The capillary network surrounding the lacteal is not shown.

lining each villus is of the columnar type. One of its functions is the selective absorption of the end products of digestion (amino acids, monosaccharides, etc.). The core of a villus is made of a lacteal (a lymphatic capillary) surrounded by a network of blood capillary vessels. The pits at the base of the villi are known as crypts, which are intimately associated with the microscopic glands of Lieberkühn.

The mucosa cells lining the tips of the villi are continuously sloughed off, as it happens for the most superficial cells of the epidermis. The shedding of the cells into the lumen of the intestine is accelerated by the friction with the moving materials filling the lumen (for instance, chyme). The sloughed cells are replaced by younger mucosa cells, which slowly move upward from the lateral lining of the villi. These cells are, in turn, generated by the proliferative activity of the crypt cells, which constitute the pool for the cell renewal mechanism of the intestinal epithelium. Therefore, in this epithelium one can distinguish at least three regions, which are a region of cell proliferation corresponding to the crypts, a region of maturation and differentiation on the lateral surface of the villi, and a region of sloughing or extrusion at the tips of the villi. With tracer experiments using tritiated thymidine and radioautography, researchers have found that the time taken for the cells to migrate from the first to the third region is about 3 to 4 days (transit time).

Radiation interferes with proliferative activity of the crypt by inhibiting the mitosis of its very sensitive stem cells or, worse, by killing them. With cessation of cell division, there is a complete loss of cell renewal for the other areas of the epithelium of the villi. Typical pyknosis is detectable in the crypt cells. Despite the lack of regeneration, the cells on the tips continue to be sloughed off. The attempt of the cells from the maturation region to fill the voids results in a stretching of the epithelial lining of the villus, which changes from columnar to cuboidal and then to squamous. With a decrease of the overall surface area, the villi show a tendency to become flat and the spaces between them become wider and shallower. Because of the short transit time, these events occur within a few days after exposure. If, in the meantime, no recovery has taken place in the crypts, the villi may become completely denuded and the more delicate underlying tissues are thus exposed. Loss of plasma, petechiae (scattered minute hemorrhagic areas), extensive hemorrhages, and ulcerations are among the serious consequences of the damage to the intestinal mucosa. Since another defensive barrier has broken down, the patient's body is open to bacterial infection from microorganisms that habitually live as harmless commensals in the intestine as well as from exogenous pathogens. *Escherichia coli, Pseudomonas, Klebsiella, Staphylococcus*, and other forms have been identified in therapeutically irradiated patients with late stages of leukemia and malignancy, although one may presume that not all those microorganisms entered the bloodstream through the injured gastrointestinal tract. Germ-free animals can survive larger doses, or at least show a longer mean survival time than other animals exposed to the same dose. This finding shows that bacterial infection is an important factor in the fatal outcome of the gastrointestinal syndrome. The effort made by the organism to combat bacterial infection is manifested with fever, which often is one of the symptoms of radiation sickness after lethal or supralethal doses.

Although the other tissues of the gastrointestinal system are much less radiosensitive than the epithelial lining, their functions are more or less severely impaired by irradiation. The muscular contractions (rhythmic segmentation, peristalsis) are definitely affected and become abnormal; antiperistalsis is the cause of vomiting. Other anomalies of the intestinal motility and altered glandular secretions account for such common symptoms as anorexia, nausea, and diarrhea (often bloody). Animal experiments and the autopsy of victims of fatal radiation accidents show that the food ingested remains in the stomach for a much longer time. The results of recurrent diarrhea are severe dehydration of the body, loss of weight, and emaciation.

For man, the threshold dose beyond which the recovery of the gastrointestinal injury is impossible seems to be much above 1000 rems, in sharp contrast with the much lower threshold observed for the irreversibility of the hemopoietic damage. This is another indication of the greater radiosensitivity of the hemopoietic systems.

Effects on the vascular system

Since the heart, large arteries, and large veins are quite radioresistant, morphologic damage to these organs may be observed only with very high doses. Classic examples are the victim of the 1958 Los Alamos accident, with hemorrhages in the myocardium (see Appendix on p. 436), and the victim of the Rhode Island accident, who showed interstitial myocarditis and acute pericarditis involving chiefly the anterior right atrium and the adjacent upper right ventricle. A victim of another earlier Los Alamos accident showed fibrinous pericarditis, although the dose he received was only 840 rems of soft radiation and 500 R of gamma radiation. In this last case the cardiac response was probably in some way related to a congenital heart defect, which the patient was known to have.

Capillary blood vessels are considerably less radioresistant than other organs of the circulatory system. Radiation may cause the swelling of their endothelial cells, which will result in occlusion of the lumen and disturbance in the supply of blood to the neighboring tissues. Rupture may also occur, with more or less massive hemorrhages or petechiae. Increased permeability of the capillaries has also been observed after moderate doses.

Effects on the skin and hair

Injuries to the skin and its derivatives were among the first lesions observed soon after the discovery of X-rays and radioactivity. Erythema was reported as a common damage suffered by the early radiologists.

The most radiosensitive structures of the skin are the malpighian layer (epithelium germinativum), the sebaceous glands, and the hair follicles. The malpighian layer, between the dermis and the epidermis, is a tissue with a high degree of proliferative activity; it continuously supplies new cells to the abovelying epidermis, as the most superficial epidermal cells are gradually shed.

Radiation-induced erythema is caused by a dilation of the dermal capillary vessels; it can be observed even with moderate doses. More massive doses are required to produce true burns and blisters.

Exposure to radiation inhibits, as expected, cell division in the malpighian layer. In the meantime, the natural sloughing of the more radioresistant superficial layers of the epidermis continues. The lost cells are not replaced by new ones; the consequence is a desquamation of the skin.

Desquamation, burns, and blisters represent a serious damage to the structural integrity of the skin; a skin affected by this type of damage is another open door to exogenous pathogenic bacteria. This type of damage adds to the patient's susceptibility to infection during certain phases of the acute radiation syndrome.

Epilation, which is usually one of the late symptoms of radiation sickness, is the consequence of destruction of the hair follicles. These structures are located deep in the dermis, although they are, in part, of epidermal origin. Their cells, like those of the malpighian layer, undergo multiplication to supply new cells for hair growth.

Concerning epilation, there are considerable differences in sensitivity among mammalian species. Man is particularly sensitive to this kind of skin damage. Epilation in man may be induced even with sublethal doses, whereas doses in excess of 2000 R are needed to produce the same effect in most laboratory mammals.

Depending on the dose, epilation may be temporary or may result in permanent baldness or alopecia. When the hair regrows, it may have a coarse and dry appearance, because of the possible irreversible destruction of the sebaceous glands. Destruction of melanoblasts and melanocytes explains why the new hair may be white or gray.

A unique characteristic of the radiation-induced skin damage is that, since the integumental system is the most superficial of the body, even an external source of nonpenetrating ionizing radiation (like soft X-rays, beta radiation) can induce the same type of injuries described above. These injuries have been induced in animal experiments with the use of 100 kV X-rays and with the application on the skin of plaques containing beta-emitting radionuclides, such as ^{35}S, ^{90}Sr, and ^{32}P. The skin injuries suffered by several victims of the Bikini fallout accident were probably caused by the beta radiation emitted by the fallout dust deposited on their skin. Epilation and beta burns were, in fact, observed in some of those victims.

Effects on immune mechanisms

From the preceding discussion it is already evident that exposure to radiation weakens or destroys several defensive barriers of the body; injuries to the phagocytic white blood cells, to the intestinal mucosa, and to the skin render the organism more sensitive to the attack of pathogens. But radiation interferes with another very effective physiologic mechanism of defense—the antigen-antibody reaction.

An antigen must be considered as any foreign substance (protein, polysaccharide) or organism (bacterium, virus) or cell that elicits in the body a reaction leading to its destruction, rejection, or neutralization. The agents involved in producing these effects on the antigens are specific proteins called antibodies. The presence of an antibody makes the organism immune against a specific antigen. Thus the antigen-antibody reaction is also referred to as the "immune reaction."

The human body is innately immune against several antigens, such as the microorganisms of the normal intestinal flora. An individual of blood type O is innately immune against agglutinogens A and B. This immunity is called *natural* or *innate*. Apparently, the antibodies specific for those antigens are present in the organism since birth. Other times it is the presence of an antigen that stimulates the organism to produce the specific antibody needed to counteract the antigen. The resulting immunity is known as *acquired* immunity, because the antibody did not exist before the antigen entered the body. In this case, the production of the antibody is an *immune response*.

Experimental evidence from laboratory animals and irradiated humans clearly indicates that radiation seldom enhances, but more often inhibits the immune response (immunosuppressive effect).

Enhancement of immunity has been noted in experiments on animals infected with pathogens and irradiated with low doses applied to small areas of the body. Radiation proved to be beneficial to the animals in overcoming the infection. The explanation of these results is rather difficult. We must rule out that the low doses used have a direct germicidal effect, because they are much below the doses required to kill the same microorganisms in vitro. Low doses may stimulate an increase in the number of some cellular elements that are involved in the immune mechanisms, such as macrophages and other elements of the reticuloendothelial system.

There is no question about the more frequent inhibitory action of ionizing radiation on the immune response at least for doses above the sublethal range. The first observations were made in 1908, when Benjamin and Sluka reported that rabbits X-irradiated before the injection of bovine serum albumin, formed the specific antibody (precipitin) at a lower level than the unirradiated controls. They also reported that no depression of the antibody level occurs when the animals are irradiated 4 days after the injection of the antigen.

Recently, W. H. Taliaferro et al. have reported the results of more accurate experiments on the inhibitory effect of radiation on the immune response of rabbits against a sheep antigen. If sheep red cells (containing an antigen) are injected into a rabbit, they elicit an immune response in the recipient's body; the latter produces a specific antibody, a hemolysin, that lyses the foreign red cells. However, the hemolysin is not detectable in the rabbit's blood until a few days after the injection of the antigen. Apparently, there is a latent period between the injection and the immune response. After the latent period, the

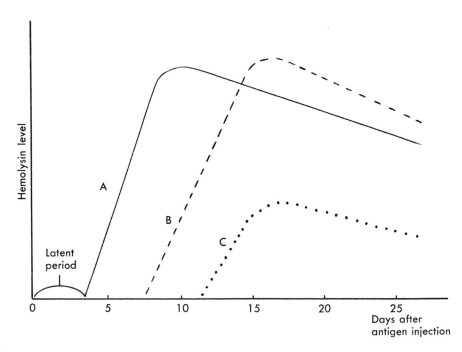

Fig. 15-6. Primary immune response in unirradiated and irradiated rabbits against injection of sheep red blood cells. **A,** Not irradiated (controls). **B,** Irradiated with 500 R *after* injection of the antigen. **C,** Irradiated with 500 R *before* injection of the antigen.

level of hemolysin climbs sharply, reaches a peak, and slowly declines (Fig. 15-6, *A*). This immune response following an initial injection of the antigen is known as the primary antibody response. A second injection of the same antigen elicits a secondary (or anamnestic) response, which, in terms of variation of hemolysin level with time, resembles the primary response, except that the latent period is considerably shorter and the hemolysin peak level is higher than the peak reached in the primary response.

Irradiation of the animal affects its primary immune response differently, depending on whether the injection of the antigen precedes or follows irradiation. If a rabbit is X-irradiated with a dose of 500 R *after* the injection of the antigen, the latent period is lengthened, but the peak level reached by hemolysin is not lower than that observed in unirradiated animals (Fig. 15-6, *B*). The lengthening of the latent period is (roughly) inversely proportional to the time elapsing between antigen injection and exposure to radiation; if this time is short, the lengthening of the latent period is more pronounced, and vice versa.

But when the animal is X-irradiated with a dose of 500 R *before* the injection of the antigen, there is not only a lengthening of the latent period but also a depression of the peak level reached by hemolysin (Fig. 15-6, *C*). This time, the lengthening of the latent period as well as peak depression are (roughly) in direct proportion with the time elapsing between irradiation and antigen injection, at least up to 24 hours.

These findings indicate that the type and magnitude of the interference of radiation with the primary immune response are affected not only by the dose (as some other experiments had already shown) but also by the time correlation between exposure and injection of the antigen. The maximum damage to the immune response is caused when the animal is irradiated 24 hours before the administration of the antigen, although a considerable damage can be observed also when the antigen is injected much later.

Although the physiologic mechanism of antibody formation is not perfectly understood at the present time, it is believed that radiation delays and depresses the antibody response by injuring the antibody-forming cells, and especially the most efficient ones, such as lymphocytes, monocytes, macrophages, and plasma cells that derive from lymphocytes when certain antigens are present in the body.*

The radiation-induced damage to the immune response, so accurately investigated with animal experiments, has been confirmed for irradiated human subjects. In the course of the acute radiation syndrome, the immune response is depressed, or even completely suppressed for a considerable time after exposure, if the dose is in the high lethal or supralethal range. The immunosuppressive effect contributes, with other injured defense mechanisms or structures, to the patient's increased susceptibility to infections from exogenous pathogens. However, there is evidence that radiation causes damage also to the body's natural immunity. In fact, the blood of irradiated animals and humans is invaded by bacteria normally present in the gastrointestinal tract (such as *E. coli*).

The damage to the immune mechanisms has another important consequence. The graft (or transplantation) of an organ or tissue to a normal animal or

---

*The formation of plasma cells in human blood is quite evident during an infection of German measles.

human is usually unsuccessful unless the donor is closely related, genetically, to the recipient. After a certain time, the grafted organ or tissue is rejected. The rejection is the consequence of an antigen-antibody reaction. The cells of the graft produce and release certain proteins that act as antigens. The host, therefore, produces specific antibodies that either destroy or isolate the graft. Consequently, the graft does not take. The depression or suppression of the immune response in irradiated animals makes a graft less likely to be rejected. This fact has been exploited for several years for kidney transplants in humans and more recently for heart transplants. Before the transplant, the patient is given a whole-body dose of X- or gamma radiation just sufficient to produce a temporary depression of his immune mechanisms. The technique has been more successful for kidney transplants than for heart transplants. More about these problems appears in the discussion on bone marrow transplants on p. 433.

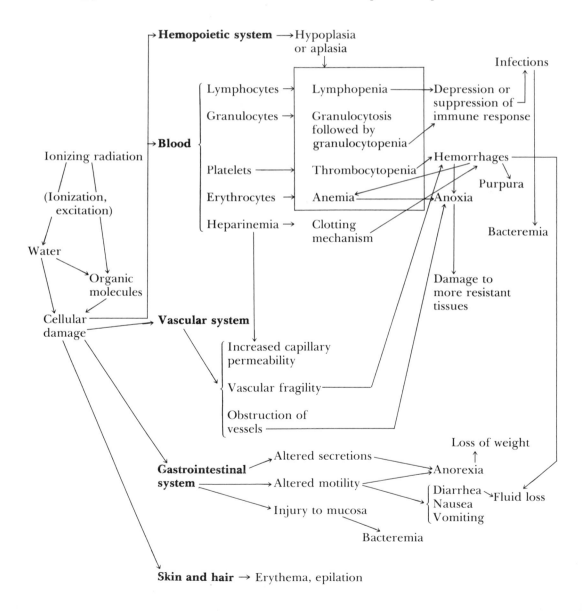

Summary of interrelationships

Even from a cursory analysis of the symptomatology and physiopathology of the acute radiation syndrome the fact is quite apparent that this disease is a complex nosologic entity with many cause-effect relationships. Furthermore, the same symptom may be effected by several causes, and many pathologic processes that cause the symptoms are causally connected. Intricate interactions are observable among different injured mechanisms or systems. The gravity and rapidity of appearance of the symptoms (and therefore the prognosis) reflects the gravity of the injury to critical organ systems.

The diagram shown on p. 430, gives a summary of the most important relationships between injuries and symptoms in the acute radiation syndrome, with special emphasis on the hemopoietic syndrome. Most of the summary should be intelligible and self-explanatory, if the previous discussion has been carefully followed. Only a few additional words of comment are in order.

As it happens for other biologic effects of ionizing radiations at organ and organismal level, also in radiation sickness the damage begins at molecular and cellular level. By means of ionizations and excitations, by indirect or direct action, protoplasmic molecules are injured, and this results in cellular damage, which affects the functions of the organs and systems indicated in the summary.

In animal experiments irradiation causes release of heparin from the mast cells of the liver into the bloodstream. The resulting heparinemia is believed to be the cause of the increased permeability of the capillary vessels; it is probably also a contributing factor to hemorrhages. In fact, heparin is known to interfere with the blood-clotting mechanism (it is an anticoagulant).

The anoxic condition caused by anemia and hemorrhages is harmful to certain tissues and organs that are quite radioresistant, but very sensitive to even a transient lack of oxygen (such as brain tissue).

THERAPY OF THE ACUTE RADIATION SYNDROME

With the knowledge gained from the human experience of the past 25 years much can be done to save the life of patients who are exposed to acute whole-body lethal, or even supralethal, doses of radiation. Unfortunately, no effective therapeutic measure is known yet to save the lives of victims who receive doses large enough (about 2000 rems or higher) to show a severe gastrointestinal or CNS syndrome. All that can be done in these cases has a palliative value and serves only the purpose of alleviating the suffering of the patient during his short survival time.

On the other hand, in cases of mild hemopoietic syndrome (after doses of 200 rems or less) no treatment may be necessary; all that may be needed is hospitalization with a restful environment and avoidance of stress or exposure to infection. The moderate radiation injury does not prevent the body from recovering spontaneously by means of the normal physiologic processes.

A well-planned therapeutic program is necessary for victims of exposure doses in the high sublethal, lethal, or low supralethal ranges. The treatment is designed to carry the patient through the critical period that accompanies the depression and slow recovery of the hemopoietic organs and of the gastrointestinal mucosa.

From the proceedings of symposia and from other literature on the therapy of the acute radiation syndrome, a slight disagreement concerning the general methodology to follow in the therapy would appear to exist among the investigators interested in this problem. One point of view, authoritatively advocated by E. P. Cronkite, is a conservative one: nothing should be done unless there are clear-cut signs and indications that call for specific therapeutic measures. The management of a case of radiation sickness should not be influenced by the knowledge of the dose to which the patient was exposed, even when dosimetric data are available. The patient's course is what should inform the physician as to the appropriate therapy, and the decisions should be made at the bedside. Dosimetry is not essential for the treatment of the patient; it may only be useful for scientific purposes and for the prognosis of possible late effects. However, the patient should be closely watched. If exposure to neutron radiation is involved, it is urgent to obtain a blood sample to estimate the dose from sodium activation. Other blood tests to follow the behavior of the blood cells should be carried out as often as every day. Fluid and electrolyte balance should also be watched closely and corrected as necessary with appropriate solutions. Although the management should not be influenced by dosimetric considerations, attempts should be made to obtain some estimates of the dose from the health physicist or from biologic dosimeters. No drugs should be given to the patient unless they are indicated clinically.

Other investigators are in favor of a less conservative and more aggressive type of management of radiation sickness. Their point of view is that although the treatment should be generally symptomatic some prophylactic measures should also be used to prevent the development of some pathologic processes. To this end, dosimetric data should be sought, so that a well-coordinated therapeutic program can be planned in advance. With a knowledge of the dose, one can predict what the course of the disease is likely to be.

Regardless of these differences of opinions, experts now agree that the most specific therapeutic measures of considerable value in the treatment of the acute radiation syndrome fall into three groups, which are anti-infectious measures, platelet transfusions, and bone marrow transplants.

Anti-infectious measures

I have repeatedly stressed above that a patient suffering from radiation sickness is particularly susceptible to infections from endogenous and exogenous microorganisms. A sterile environment must be created around the patient to protect him at least from contact with exogenous pathogens. Plastic-tent isolators are useful to create an environment as germ-free as possible.

Antibiotics are necessary to bring under control infections already present in the patient's body, especially when associated with fever. The type of antibiotic to administer is dictated by the nature of the infecting bacterium, which can be identified by means of bacterial cultures from blood, sputum, urine, etc. It is advisable to use one antibiotic at a time, although doses should be three times larger than normally used. Until a few years ago, the administration of antibiotics for prophylactic reasons was discouraged. The availability of controlled-environment procedures to prevent exogenous infection and the proved effectiveness of some of the new antibiotics have recently changed this point of view. We now believe that patients with severe marrow depression, who are in a sterile environment facility, should be given antimicrobial drug therapy before there are clinical signs of infection. Broad-spectrum antibiotics should be avoided, because they have the tendency to encourage fungal infections, which, on the other hand, can be prevented with the help of some oral fungicidal drugs. Gamma globulins may be of some value as anti-infectious agents, although not much information is available in this regard.

If possible, transfusions of leukocytes from a suitable donor may be very helpful to a patient with severe leukopenia. The problem is that not enough cells may be available for each transfusion. Patients affected by chronic granulocytic leukemia would be the best sources of leukocytes. However, such sources should be used with great caution because of the possibility of immunologic complications in the recipient, and most of all because of the remote but serious risk of transplanting leukemia from one individual to another. Nevertheless, a leukocyte transfusion may be particularly effective in combating *Pseudomonas* infection.

Platelet transfusions

Frequent transfusions of whole blood to overcome the anemia that develops in one of the late stages of the syndrome seem to be of little value. In experiments with dogs, such transfusions have little effect on mortality. Especially when there is massive bleeding, the lost erythrocytes cannot be easily replaced by the transfused ones. To control bleeding and therefore the loss of erythrocytes, restoration of the integrity of the clotting mechanism is more useful; it may be accomplished with transfusions of platelets. Either fresh blood (not more than 4 hours old) or platelet-rich plasma can be used for this purpose; these transfusions can stop hemorrhages at least temporarily. Fresh blood is preferable to plasma when the patient's hematocrit is low.

In the absence of bleeding, one has difficulty knowing when to start platelet transfusion. Certainly the transfusion should be done when blood tests indicate a severe thrombocytopenia. Patients with high platelet counts suffering from polycythemia vera are convenient sources of platelets for transfusions.

Bone marrow transplants

The transplantation of bone marrow cells from a suitable donor is the most radical and drastic therapeutic measure that can be used in severe cases of radiation sickness and should be considered when all other forms of treatment have failed to bring about significant improvement in the conditions of the patient. The theoretical principle behind this therapy is quite simple. Since most of the pathologic symptoms of the hemopoietic syndrome are caused by a depression or temporary suppression of the blood-forming system, a significant improvement should be expected if the functions of the disabled system are taken over by healthy hemopoietic tissue. Of course, one hopes that the transplanted tissue will not be rejected by the patient's body and that it will not cause any unwanted effects that might make the transplantation harmful rather than beneficial.

The point of the matter is that a bone marrow graft may be extremely useful, useless, or even dangerous, depending on a variety of factors, such as dose to which the patient was exposed, source of the transplanted marrow, time of transplantation after irradiation, etc. Furthermore, certain types of interactions between host and graft are not yet entirely clear. Because of these difficulties, the bone marrow transplantation therapy should be handled judiciously and after consultation with specialists.

With regard to the donor-recipient relationship, the following are four types of bone marrow transplants (the same classification can be used for the transplants of any other organ or tissue):

1. *Autologous.* The donor and recipient are the same individual. A transplant of this kind is possible and practical (but not necessarily useful) when exposure of an individual to radiation is expected or deliberately planned, as in the case of therapeutic irradiation of malignancies. Before exposure, some bone marrow is removed from the subject, stored in a suitable manner, and then reinjected into his body after exposure, if radiation sickness develops and its severity is such as to warrant marrow transplant.

2. *Isologous (syngeneic).* The donor and recipient are genetically similar or identical, as in the case for laboratory animals of the same inbred strain, or for human identical twins.

3. *Homologous (allogeneic).* The donor and recipient are not genetically identical or similar, but simply belong to the same species.

4. *Heterologous (xenogeneic).* The donor and recipient belong to different species, such as the case in the transplantation of bone marrow from rats to mice or vice versa.

The enthusiastic interest in bone marrow transplantation as therapy for patients with acute radiation syndrome began with some early experiments on mice carried out by L. O. Jacobson and his colleagues. In 1949 Jacobson discovered that the chance of survival of irradiated mice can be strikingly improved, if the animals are irradiated with their spleen exteriorized and shielded with lead. While 700 R of X-rays killed all the controls (their spleen was exteriorized but not shielded), the survival of the animals in the experimental group was 96%! Further experiments showed that if the shielded spleen is removed from the animals within a few minutes after irradiation, there are no survivors; whereas if it is removed a few hours after irradiation, there is a significant percentage of survivors. The fact seems clear, then, that the shielded spleen does not protect the animal during irradiation but during the postirradiation time and that the longer the spleen is left in place the more protection it provides. Apparently, something that is responsible for the survival of the animals is transferred from the spleen to the rest of the exposed body after irradiation.

In other experiments, other organs were exteriorized and/or shielded. Exteriorization and shielding of kidneys did not provide any protection. Shielding of one hind leg with thigh was effective in increasing survival, but shielding of a leg from which the femur had been removed was ineffective. Therefore, it is the bone, and not any other organ of tissue of the leg, that provides protection when shielded. Later, it was discovered that the bone marrow, and not the bony tissue, is the protective factor, because the shielding of bones that are known to contain inactive or no bone marrow (such as the bones of the tail) is ineffective in increasing survival.

What is transferred, during the postirradiation time, from the shielded spleen or bone marrow to the rest of the body? What is responsible for the increased chance of survival of the irradiated animal? Jacobson and others suggested a humoral hypothesis: the unirradiated spleen or marrow contains a humoral factor capable of stimulating the regeneration of the irradiated blood-forming tissues. The humoral factor was envisaged as a chemical substance (perhaps a protein) released by the shielded organs. The humoral hypothesis seemed to be confirmed by the fact that mouse spleen homogenates from normal mice are active in increasing the chance of survival of mice irradiated without shielding their spleen. More refined studies brought up some experimental evidence against the humoral hypothesis, especially when it was discovered that the homogenates of the experiments mentioned above were not entirely cell-free. Investigators therefore believed that in Jacobson's and other similar experiments intact hemopoietic cells move from the unirradiated spleen or marrow, via the bloodstream, to the injured hemopoietic tissues and repopulate them (cellular hypothesis). The cellular mechanism has been definitely confirmed mainly through the work of J. F. Loutit. There is no doubt at the present time that stem cells migrate from a shielded spleen or bone marrow to other areas of the irradiated body. In recent studies (see for instance, Fujioka et al.) the recovery of hemopoiesis in the irradiated, partially shielded animals has been followed with the examination of ^{59}Fe uptake.

In the experiments described above, the migration of hemopoietic tissue from shielded to unshielded areas of the body is comparable to an autologous transplant. In fact, if bone marrow is removed from the animals before they are irradiated with a lethal dose and then reinjected after irradiation, a therapeutic effect is also observed. But what results are obtained when the irradiated animal is treated with isologous, homologous, or heterologous marrow transplants? Are these grafts accepted by the recipient? If so, do they have therapeutic properties?

As already mentioned on p. 430, normal, unirradiated animals reject homologous or heterologous grafts of bone marrow or of any other organ or tissue. The explanation for the rejection is to be found in an immune response. Since the host and graft are genetically different, the graft behaves as an antigen and thus elicits the production of specific antibodies in the host, which will result in the rejection of the graft. On the contrary, isologous, like autologous, grafts are not rejected; apparently no immune response is elicited in this case, because of the close genetic relationship between donor and recipient.

Irradiation may depress or suppress entirely the immunologic mechanisms of the body. This fact should make the body of an irradiated subject unable to reject a homologous or even a heterologous graft. As a matter of fact, the survival rate of mice exposed to supralethal doses has been considerably improved with bone marrow transplants from mice of other strains, or even from rats. Homologous bone marrow was also given to some of the victims of the Vinca accident. Whether the transplant was responsible for the recovery of the survivors is a much disputed question. Probably it produced a functional but transient graft. The patient who died showed a pronounced rise in the granulocyte and platelet counts after the transplant, and this was taken as evidence of the success of the marrow graft. But the fact that the rise was noted too soon after transplantation casts doubts on its cause.

An isologous transplant of bone marrow to the patient is always successful (at least in the sense that it is not rejected). Isologous bone marrow (donated from an identical twin) was given to the most seriously injured victim of the Pittsburgh accident. It is believed that this transplant was the decisive factor in saving his life.

Unfortunately, isologous bone marrow cannot be used for all patients with radiation sickness, because they very seldom have identical twin brothers or sisters. Homologous bone marrow can be more readily obtained in all cases;

but because of the possible immunologic complications, homologous transplants should be considered with great care.

These transplants are definitely hazardous and contraindicated if the patient has been exposed to sublethal doses. In these cases the immune response is not depressed enough to prevent the rejection of the graft. The graft is in fact rejected, and, in addition, the stress associated with the immune response may aggravate the radiation syndrome. Homologous transplants have killed sublethally irradiated animals, which would otherwise have recovered.

As to the success of a homologous transplant following lethal doses, no general rule can be given. Again, the likelihood that the graft will be accepted by the host, depends on how much the immune response has been depressed; and this cannot always be ascertained. Several investigators have found that a homologous transplant is useless and sometimes even harmful after irradiations in the midlethal dose range.

Better success is obtained with transplants after exposures to low supralethal doses; in these cases, the immune response is so much depressed, or even temporarily suppressed altogether, that rejection of the grafted bone marrow is unlikely. After very high doses, transplants are ineffective, because, although the marrow graft will not be rejected and may start repopulating the damaged hemopoietic tissue, the patient will die from a gastrointestinal syndrome.

One can see, therefore, that homologous bone marrow transplants can be successful and useful only after exposure to doses that fall within a rather narrow range (from midlethal to low supralethal doses).

From the above brief discussion of the problems connected with bone marrow transplantation one may now understand why this form of therapy for radiation sickness should be used very cautiously, at least until some hitherto obscure immunologic aspects are further clarified. To circumvent the problem, a great deal of research has been done to see if the regeneration of a radiation-damaged hemopoietic system can be stimulated in some way other than by bone marrow transplant. Several drugs (some vitamins, for instance) are known to enhance hemopoiesis in unirradiated subjects. These agents have been tested in the postirradiation bone marrow syndrome, but, so far, the results have not been very encouraging. The marrow transplant, if feasible, must still be considered as a hopeful therapeutic measure to save the life of an irradiated patient, when all other forms of therapy have failed.

APPENDIX: A case history

An example of a radiation accident is presented here in detail to illustrate the circumstances that may lead to an accident, the radiation injuries caused, and the clinical course of radiation sickness. The following description consists of partial quotations* from a lengthier report published by the investigators who were directly in charge of the treatment of the victim:

The Los Alamos accident was a criticality accident and resulted in one fatality. The victim was exposed to one of the largest doses ever reported for any accident. He died of direct damage to his heart, although he must have also suffered from a severe gastrointestinal syndrome. Death followed within less than 2 days after the accident. With the high supralethal dose involved, nothing could be done to save his life. This accident,

---

*From Shipman, T. L. 1961. A radiation fatality resulting from massive overexposure to neutrons and gamma rays. In "Diagnosis and treatment of acute radiation injury." International Documents Service, New York. Pp. 113-133.

like many others, was thoroughly studied by experts; their studies have been a significant contribution to the knowledge of human radiation injury. The following are the details of the 1958 Los Alamos accident:

On the 30th of December 1958, an accident occurred at the Los Alamos Scientific Laboratory which was of particular interest because of the extremely high dose of radiation delivered to the principal victim. . . . The accident occurred in a complex of buildings known as DP West, situated some thousands of feet from any housing areas or from any other concentration of work. DP West is primarily concerned with the chemical and metallurgical processing of plutonium. The procedure being carried out when the accident occurred was the recovery of plutonium from liquid wastes, a lengthy process involving many steps. Not more than a few hundred grams of plutonium were normally processed at any one time. At the time of the accident, the operation was nearing the final step with plutonium in a tank containing water and a solvent, tributyl phosphate. The solvent and aqueous phases were in separate layers but were to be mixed by stirring. . . .

This was an extremely complicated accident resulting from the fortuitous conjunction of several quite unrelated factors. In the first place, the process was not the normal routine one that had been carried out many times over a three-year period; it was the end-of-the-year cleanup in preparation for the annual inventory of plutonium . . . . It is difficult to imagine that the simple action of a stirring device in a tank could draw a subcritical configuration of fissionable material into a truly critical geometry . . . . The system had been in operation for a number of years and many batches of plutonium had been processed. With each batch a little more went in than came out, and it was assumed that the deficit was irrecoverable loss which had gone down the drain. What was not realized was that over the years the system had actually retained, bit by bit, a total of almost 3.0 kg of plutonium . . . .

The chemical operator K was a man of no great technical education but with many years of practical experience in this and related operations. He was repeating a process he had carried out many times before. It is possible that over the years he had introduced a few short-cuts in the process without the knowledge of his supervisors . . . . On the afternoon of the accident, K was standing on a short stepladder, looking through a viewing port into the tank [where plutonium was] . . . . Within seconds after . . . the stirrer was started there was a muffled boom and K fell backwards off the stepladder . . . . The blades of the stirrer drew material down in the centre and forced it up the outer sides of the tank, and for an unfortunate instant the geometry in the solvent layer brought the material together in a critical configuration.

There was only a single critical excursion without subsequent oscillations such as occurred at Oak Ridge. Later calculations showed that there had been a burst of 1.5×10^{17} fissions. Fortunately K was the only man in the room, but . . . there were two men in the adjacent room . . . . There were a great number of tanks of various sizes which fortunately shielded the other two men, D and R. Both these men heard the boom of the critical excursion . . . . In a matter of seconds D had left his work station to see what had happened in the next room . . . . By the time he got there, K had already picked himself up off the floor and had gone to and opened the outside doors. When D reached him, K was standing outside in the snow . . . . D found K ataxic and disoriented. He needed support to remain erect, and all he could say was: "I'm burning up, I'm burning up" . . . . K's face appeared flushed even at this early time . . . . [Thinking that K had been the victim of alpha contamination,] D guided and supported K back into the room, where they were met by R, and the three continued . . . on to an emergency shower . . . . D and R stripped off his outer clothes and held him under the shower, . . . because he could not stand unaided . . . . Perhaps 5 minutes after the accident, he was virtually unconscious. While R called for assistance, D returned to the room of the accident . . . . He certainly passed within a few feet of the tank at least two or more times.

The plant nurse arrived on the scene approximately 10 minutes after the accident and was puzzled to find a patient obviously in shock and unconscious, but with nice, rosy-pink cheeks; she did not realize that his color was due to radiation-induced erythema. The patient was nearly pulseless . . . . The man was admitted to the emergency room of the Los Alamos Medical Center . . . 25 minutes after the accident . . . .

The patient was a powerfully-built man of 38; he weighed approximately 170 lbs and was 71 inches tall. By the time he arrived at the hospital . . . he was semiconscious, but disoriented. He was moving around restlessly on the stretcher and all visible skin areas were of a dusky purplish color. He seemed to be in severe pain, apparently abdominal. His conjunctivae were markedly reddened, but his excessive restlessness made careful examination difficult. He retched frequently but vomited only small amounts of watery fluid. About 10 minutes after admission he had an episode of explosive watery diarrhoea. Some of this faecal fluid was radioassayed and showed a significant content of ^{24}Na, indicating a copious passage of fluids into the gastrointestinal tract.

His blood pressure was found to be 80/40 mm Hg with a pulse rate of 160 per minute. He had repeated mild shaking chills, and his restlessness was so great that he had to be restrained . . . . An indication that the dose had been massive was the fact that a portable gamma survey instrument held to the surface of the body gave a reading of 15 mR/hour . . . .

The patient . . . was placed in an oxygen tent. His hypotension . . . and his rapid pulse . . . still persisted and his rectal temperature was found to be 103° F . . . . Physical examination did not reveal impressive findings. His optic fundi were normal, but the conjunctivae were intensely injected. His eyes looked as though they should have been painful . . . but the patient denied any discomfort . . . . There was definite erythema over the anterior surface of the body down to the level of the knees . . . .

About 5 hours after the accident the patient appeared to be in a satisfactory condition. He was rational, comfortable, and emotionally at ease. By this time it was also apparent from the dosimetric studies that his radiation exposure had unquestionably been supralethal and of greater magnitude than in any of the cases previously reported. The changes in his white cell counts reflected this very definitely. The total white cell count rose steadily to a peak of 28,000 per mm³, but the lymphocytes had virtually disappeared from the circulating blood in less than 6 hours. This we regarded as a very grave prognostic sign.

A very dramatic finding was the marked degree of urinary retention . . . . There was a total urinary output of less than 600 ml with a total fluid intake of approximately 14 litres! . . .

On the second evening, more than 30 hours after the accident, the patient's condition deteriorated rather abruptly. He developed increasing abdominal cramps and fairly heavy sedation failed to control his restlessness . . . . Despite administration of oxygen by mask he showed increasing cyanosis. Sedation was given and he lapsed into a coma from which he never roused. Death supervened from cardiac arrest 34¾ hours after the accident, his heart having been the target of nearly 12,000 rad of ionizing radiation . . . .

The neutron dose was determined by measurement of induced ^{24}Na activity in the blood, in selected body tissues, and in the whole body, as well as from induced activity in other materials such as brass overall buttons and nearby chemicals . . . . It now appears that the combined neutron and gamma dose delivered to K's anterior chest wall, and thus to the right side of the heart and the anterior wall of the stomach, was approximately 12,000 rads. The total dose to the face and to the front of the skull was less, but still in excess of 10,000 rads. The dose to the lower legs was probably less than 1000 rads . . . .

At autopsy . . . the most striking finding . . . was the edematous, water-logged appearance of practically all tissues except the lungs. The general picture was quite characteristic of acute right heart failure resulting from right-sided myocarditis complicated by excessive fluid intake . . . . The first loop of the jejunum, the gastric pyloric bulb and the surface of the left lobe of the liver contained numerous petechial hemorrhages . . . . The spleen was wrinkled and flabby . . . . The right side of the heart was dilated and filled with blood, while the left heart was in systole. Externally the right auricle and the anterior portion of the right atrium also showed hemorrhages similar to those in the pericardium . . . .

This man had received more than enough radiation to his bone marrow to kill him in 3 or 4 weeks, if he had had no other injuries. The injury to his gastrointestinal tract would have killed him in 1 or 2 weeks had not a more vital insult killed him first . . . . In our case the man received at the same time another and quite distinct injury to his heart, one which, physiologically, was quite overwhelming. It seems clear that the injury to the heart muscle in this case must be regarded as the primary cause of death.

REFERENCES

GENERAL

Andrews, G. A. 1967. Radiation accidents and their management. Radiat. Res. **7**:390.

Berdjis, C. C. 1970. Pathology of irradiation. The Williams & Wilkins Co. Baltimore.

Bond, V. P., T. M. Fliedner, and J. O. Archambeau. 1965. Mammalian radiation lethality. Academic Press, Inc., New York.

Congdon, C. C. 1971. Bone marrow transplantation, Science **171**:1116.

Cronkite, E. P. 1964. The diagnosis, treatment and prognosis of human radiation injury from whole body exposure. Ann. N. Y. Acad. Sci. **114**:341.

Cronkite, E. P. 1965. Radiation injury in man. In Schwartz, E. E., editor. The biological basis of radiation therapy. J. B. Lippincott Co., Philadelphia. Chapter 5.

Cronkite, E. P., and V. P. Bond. 1960. Radiation injury in man. Charles C Thomas, Publisher, Springfield. Ill.

Ingram, M., J. W. Howland, and C. H. Hansen. 1964. Sequential manifestations of acute radiation injury vs. "acute radiation syndrome" stereotype. Ann. N. Y. Acad. Sci. **114**:356.

International Documents Service. 1961. Diagnosis and treatment of acute radiation injury. New York.

International Atomic Energy Agency. 1965. Personnel dosimetry for radiation accidents. Vienna.

International Atomic Energy Agency. 1969. Bone-marrow conservation, culture and transplantation. Vienna.

International Atomic Energy Agency. 1969. Handling of radiation accidents. Vienna.

Krayevsku, N. A. 1965. Studies in the pathology of radiation disease. Pergamon Press, Inc., New York.

Lanzl, L. H. 1965. Radiation accidents and emergencies in medicine, research and industry. Charles C Thomas, Publisher, Springfield, Ill.

Lushbaugh, C. C. 1969. Reflections on some recent progress in human radiobiology. In Augenstein, L. G., et al., editors. Advances in radiation biology. Academic Press, Inc., New York. Vol. 3, pp. 277-310.

Lushbaugh, C. C., F. Comas, and R. Hofstra. 1967. Clinical studies of radiation effects in man. Radiat. Res. suppl. **7**:398.

Lushbaugh, C. C., and G. Castañeda. Radiation accident review. (To be published.)

Mathé, G., J. L. Amiel, and L. Schwarzenberg. 1964. Treatment of acute total-body irradiation injury in man. Ann. N. Y. Acad. Sci. **114**:368.

Micklem, H. S., and J. F. Loutit. 1966. Tissue grafting and radiation. Academic Press, Inc., New York.

Quastler, H. 1956. The nature of intestinal radiation death. Radiat. Res. **4**:303.

Saenger, E. L. 1963. Medical aspects of radiation accidents. USAEC Handbook. Government Printing Office, Washington, D. C.

Taliaferro, W. H., L. G. Taliaferro, and B. N. Jaroslow. 1964. Radiation and immune mechanisms. Academic Press, Inc., New York.

Van Bekkum, D. W. 1961. Recovery and therapy of the irradiated organism. In Errera, M., and A. Forssberg, editors. Mechanisms in radiobiology. Academic Press, Inc., New York. Vol. 2, Chapter 5.

Van Bekkum, D. W., and M. J. de Vries. 1967. Radiation chimaeras. Academic Press, Inc., New York.

Warren, S. 1961. The pathology of ionizing radiation. Charles C Thomas, Publisher, Springfield, Ill.

LITERATURE CITED

Benjamin, E., and E. Sluka. 1908. Antikörper-Bildung nach experimenteller Schädigung des hämatopoietischen Systems durch Röntgenstrahlen [Antibody formation after experimental irradiation of the hemopoietic system with X-rays]. Wein. Klin. Wschr. **21**:311.

Fujioka, S., K. Hirashima, T. Kumatori, F. Takaku, and K. Nakao. 1967. Mechanism of hematopoietic recovery in the X-irradiated mouse with spleen or one leg shielded. Radiat. Res. **31**:826.

Jacobson, L. O., E. K. Marks, E. O. Gaston, M. Robson, and R. E. Zirkle. 1949. The role of the spleen in radiation injury. Proc. Soc. Exp. Biol. Med. **70**:740.

Kurnick, N. B., and N. Nokai. 1968. Hematological response to isologous and homologous bone marrow transplantation: mechanism of homologous failure. Radiat. Res. **36**:31.

The modification of radiation response

In the preceding chapters I have pointed out several times that the response given by a biologic system to a specific absorbed radiation dose may vary with the variation of a number of factors. These factors may affect both the nature and the magnitude of the biologic effects induced by an absorbed dose. This statement is true whether the biologic system is a bacterium, a chromosome, a fungus, a culture of mammalian cells, an embryo, or a system as complex as the whole body of a mammal.

The factors in question are therefore capable of *modifying* the radiation response. With regard to the magnitude of the response, some factors modify the response by enhancing it, others by diminishing it. In the first case, the result is an increased radiosensitivity of the irradiated system; commonly this is referred to as *sensitization,* and the responsible factors are classified as sensitizing factors. In the second case, the result is increased radioresistance; commonly this is referred to as *protection,* and the responsible factors are classified as protective factors. Sensitization and protection are, therefore, the two opposite directions that may be taken by the modification of the magnitude of the radiation response.

Some of the modifying factors are not related to the biologic system being irradiated, but rather to the physical properties of the radiation used or to some irradiation parameters. Typical examples of such factors are the type of ionizing radiation (X- or gamma rays, or alpha, beta, or neutron particles), its specific ionization, LET, penetrating power, and rate at which radiation is delivered. How these factors affect the quality or magnitude of a variety of biologic effects has been shown several times in the preceding chapters. Radiation sickness in mammals may result from the absorption of 400 rads of X-rays, but only if this dose is absorbed in a relatively short time (with high dose rate and acute exposure). Chronic exposure to 400 rads of X-rays does not result in radiation sickness but in other qualitatively different biologic effects. On the other hand, 400 rads of neutron radiation absorbed at high dose rate produce in the same animal a more severe type of radiation sickness than 400 rads of X-rays, because several physical properties (LET, ionizing power, etc.) of neutron radiation are different from those of X-rays. Neutron radiation is more efficient (that is, has a higher RBE) than X-rays in inducing a wide variety of biologic effects.

Dose rate does not seem to affect the yield of mutations in *Drosophila,* as it does in mammals (see Chapter 13). Dose rate and fractionation of the total dose affect the dose-response curve for two-break aberrations. The frequency of chromosomal aberrations is also influenced by the LET (see p. 346). The influence of dose rate on the onset and duration of mitotic arrest in *Tradescantia* pollen grains has been discussed on p. 361.

Other modifying factors are not related to the physical properties of the radiation used, nor to the method of irradiation, but rather to the biologic system itself that is the target of irradiation. Some of these factors are related to physical or chemical properties of the external or internal environment of the biologic system, such as temperature, presence or absence of oxygen, and presence or absence of certain chemical compounds within the irradiated system. Others are related to certain biologic states of the irradiated organism, such as age, sex, health, hibernation, etc. All these factors may modify the radiation response quite independently of the physical properties of the radiation used. The study of the influence of these factors on the radiation response is the purpose of this chapter.

The modification of radiation response to be discussed here is brought about by factors operating *during* irradiation. This modification should not be confused with one that may be brought about by factors operating *after* irradiation. For instance, the term "protection" is used only when a chemical compound diminishes the magnitude of a biologic effect while present in an organism during irradiation. The diminution of the response resulting from any type of postirradiation treatment is better referred to as *restoration.* The various therapeutic measures used to diminish the severity of radiation sickness are examples of restoration, because they are applied after the radiation exposure. Restoration does not really modify the response but simply helps the organism to recover from a radiation damage that has already been produced. Generally speaking, restoration is possible only when the radiation damage is reversible.

The practical importance of the problems discussed in this chapter should be evident. The knowledge of how the modifying factors protect or sensitize a biologic system can be used for several applications. It would be desirable to sensitize tumor cells to make them more vulnerable to radiation therapy. On the other hand, it would be desirable to protect human subjects who are likely to be exposed to radiation, as may be the case for radiation workers, for space travellers, or in the event of a nuclear war. Specific examples of these practical applications are mentioned in more detail in the following sections.

THE OXYGEN EFFECT
Definition and nature

A biologic system is more radiosensitive when irradiated in the presence of oxygen (aerobic condition) than when irradiated in its absence (anaerobic condition or anoxia). This enhancement of the radiosensitivity by oxygen is known as the *oxygen effect.*

The oxygen present during exposure acts as a "dose-multiplying factor," in the sense that the magnitude of a given effect produced in its presence by a given absorbed dose is identical to that obtained with a higher dose absorbed in its absence. Consequently, if the production of a given effect requires dose D_a in anaerobic conditions, the dose D_0 required in the presence of oxygen will be

$$D_0 = D_a/r$$

and therefore

$$D_a = D_0 r$$

or

$$r = D_a/D_0$$

where r is a number greater than 1, which may be called the *oxygen enhancement ratio*. This ratio gives a quantitative appraisal of the magnitude of the oxygen effect. For instance, if the pattern of response given by mammalian cells in tissue culture to 800 rads in air is similar to the pattern of response given by the same cells to 2000 rads under anoxia, the enhancement ratio is 2000/800 = 2.5. With other biologic systems, the enhancement ratio can be as high as 3 or 4.

The fact should be stressed that for observation of an oxygen effect, the oxygen must be present during irradiation. In experiments using very sophisticated apparatus, it was possible to supply oxygen to dysentery bacilli 20 msec after they had been irradiated with a burst of high energy electrons in anaerobic conditions. No oxygen effect was observed, as it was when oxygen was made available to the bacilli during irradiation. The important results of these experiments may throw light on the possible mechanism of the action of oxygen, as discussed in one of the following sections.

The magnitude of the oxygen affect (and therefore the enhancement ratio) is influenced by the LET and specific ionization of the radiation used. The oxygen effect is more pronounced with radiation of low LET (X- or gamma rays) than with radiation of high LET (alpha or neutron radiation). With this latter type of radiation, the effect may be insignificant or absent altogether. The possible reasons for the influence of LET are analyzed on p. 444.

Examples

Probably the existence of an oxygen effect was first noticed by the early radiologists. Some of them reported that human skin in a state of ischemia (decreased blood supply) was less radiosensitive than normal skin. Ischemia could be induced by simply applying pressure during irradiation. With this expedient the blood circulation in the skin could be reduced, and this reduction would result in hypoxia. In 1921, Holthusen reported that anoxic *Ascaris* eggs are more resistant to the killing action of X-rays than eggs with a normal oxygen supply.

Since those early times, research has shown that the oxygen effect is of universal occurrence, regardless of the biologic system or the end point under investigation; death of an organism, anatomic lesions, biochemical injuries, gene mutations, chromosomal aberrations, etc., are all radiation-induced biologic effects that are influenced in their magnitude by the presence or absence of oxygen during irradiation. Even in vitro systems of biologically important molecules seem to be affected. Generally, the presence of oxygen enhances enzymatic inactivation during X- or gamma irradiation. Trypsin is inactivated more rapidly in the presence of oxygen, with an enhancement ratio of about 2. An oxygen effect has also been noticed with dry ribonuclease and egg-white lysozyme (Stevens et al.).

Hollaender et al. studied the oxygen effect on *E. coli* cultures irradiated in oxygen-saturated and nitrogen-saturated buffers. The survival curve for the cultures irradiated in the presence of oxygen had a steeper slope than the curve for the cultures irradiated in the presence of nitrogen. For any dose, the oxygen enhancement ratio was about 3. A surviving fraction of 1% was obtained with only about 28 kR in presence of oxygen, whereas in its absence an exposure dose of about 85 kR was required. Likewise, a dose of 40 kR reduced the surviving fraction to 0.1% in presence of oxygen, whereas 120 kR were required to obtain the same effect in its absence.

The effect of oxygen on the lethal damage produced by radiation to dry spores of microorganisms has also been reported. Spores of *Aspergillus terreus* are more sensitive in presence of oxygen. The enhancement ratio is about 1.5. Dry spores of *Bacillus subtilis* are considerably more radiosensitive when irradiated in air than in vacuum. With some bacterial spores (*B. megaterium*) the oxygen effect seems to be reduced and even suppressed at very low temperatures (for example, −120° C).

An oxygen effect on the induction of chromosomal aberrations in *V. faba* root tips was discovered by J. M. Thoday and J. M. Read in 1947. With X-rays, the enhancement ratio was about 3; no oxygen effect was observed with alpha rays (high LET). The influence of oxygen on the frequency of chromosomal aberrations has also been noted in Ehrlich ascites tumor cells, *Tradescantia* microspores, and *Drosophila* gametes. Kimball has observed an oxygen effect in the induction of mutations in micronuclei of *Paramecium*.

In a recent study, G. A. Legrys and E. J. Hall have observed an oxygen effect on the X-ray sensitivity of synchronously dividing cultures of Chinese hamster cells. The X-ray sensitivity was measured under aerobic and hypoxic conditions at several stages of the generation cycle. The enhancement ratio was found to be the same (about 2.6) throughout the cycle; an indication that the radiosensitivity of these cells is influenced by the presence or absence of oxygen to the same degree throughout the generation cycle.

Mice in utero become more radioresistant when their mothers breathe a mixture of 5% oxygen and 95% helium. Under these conditions, the doses required to induce congenital anomalies are increased 100% to 160%.

The $LD_{50/30}$ of adult rats is increased to 1400 R if they are allowed to breathe air containing only 5% oxygen during irradiation.

In other experiments, 32 of 42 mice exposed to 7% oxygen survived an acute, whole-body exposure dose of 800 R, whereas none survived the same dose when exposed to 10% oxygen. These experiments indicate that hypoxia in mammals has a marked protective effect against radiation damage.

Possible mechanisms of action

Several hypotheses have been suggested to explain the mechanism of action of the oxygen effect. However, some of them are not in agreement with the important experimental finding that oxygen must be present during irradiation in order to enhance the magnitude of the radiation effects and that no effect is noted when oxygen is supplied to the irradiated system as early as a few milliseconds after exposure (see above). This fact would suggest that the responsible mechanism must be operative during the physicochemical and chemical stages of the sequence of events leading to biologic damage (see p. 300). These stages are of extremely short duration.

It is commonly believed that oxygen interacts with some free radicals produced by radiation. Among the products of radiolysis of water, the hydroperoxy radical ($HO_2^{\bullet}$) results from the interaction of the $H^{\bullet}$ radical with a molecule of oxygen (see p. 306). The hydroperoxy radical may be more harmful to some biologic molecules.

The presence of oxygen may interfere with the restoration of a damaged organic molecule. In fact, certain free radicals deriving from the hydrolysis of water may remove hydrogen from organic molecules. What is left is an organic radical. For instance:

$$RH + OH^{\bullet} \longrightarrow R^{\bullet} + H_2O$$

In absence of oxygen, the organic radical $R^{\bullet}$ possibly combines with a hydrogen radical; thus the original organic molecule is restored:

$$R^{\bullet} + H^{\bullet} \longrightarrow RH$$

If oxygen is present, it will compete with the hydrogen radical by combining with the organic radical, so that the restoration of the original organic molecule is prevented:

$$R^{\bullet} + O_2 \longrightarrow RO_2$$

In summary, oxygen would enhance the magnitude of the biologic effects by acting during the very early phases of the process that leads to the final biologic damage and specifically by increasing the number of harmful radicals and/or by blocking the restoration of damaged organic molecules.

The reduction or absence of the oxygen effect with radiation of high LET may be explained with the fact that these radiations have a high specific ionization density. Ion pairs are produced so close together that the occurrence of several ionizations within the same macromolecule is not unlikely. With such amount of molecular damage, there is no need for oxygen to enhance a lesion that is already irreparable.

Unfortunately, experiments with biologically active molecules in vitro have not yet added much light to the precise mechanism of the oxygen effect. More data in this field are urgently needed.

Applications to radiation therapy

Radiologists have been aware for a long time that some types of tumors are more radioresistant than the surrounding normal tissues, or, more interestingly, that the cells of the same tumor mass may show different degrees of radiosensitivity, in spite of their histologic similarities. The radioresistance of these tumors is a consequence of their poor vascularization and therefore of their hypoxic or anoxic state.

In normal tissues the intercapillary distance is on the average 20 μm, which is about the diameter of most cells, so that almost every cell is near or adjacent to a capillary vessel. In many neoplastic tissues, on the contrary, cell proliferation may be so rapid that some of the cellular elements may happen to be at a great distance from the nearest capillary vessel. Experiments have shown that cells farther than 150 μm from a capillary vessel are definitely anoxic and therefore less radiosensitive than the cells closer to the sources of oxygen supply. If a specific dose of therapeutic radiation is delivered to such a mass of tumor cells, it may kill those having a good supply of oxygen available, but the anoxic cells may survive the treatment. The survivors may be responsible for the recurrence of tumors that is so frequently observed after radiation therapy. After the death of the well-oxygenated cells, the distance between the survivors and the nearest capillary vessel is decreased. The oxygen supply to these cells is improved; this improvement brings about resumption of their rapid proliferative activity and thus recurrence of the tumor.

A definite correlation between radiosensitivity and oxygen supply has been established with a series of experiments for several types of mouse tumors. Ehrlich ascites tumor, sarcoma, mammary carcinoma, lymphosarcoma L_1, and others, all show an oxygen effect. Any technique or procedure that is capable of improving the oxygen supply to the hypoxic or anoxic cells of a tumor, should also increase their radiosensitivity.

To achieve this purpose, the local injection of vasodilators has been tried. With the dilation of the capillary vessels surrounding a tumor mass, the rate of blood (and therefore of oxygen) supply should conceivably be increased. In many instances this technique has given only doubtful results. Other avenues investigated include hypertransfusion of washed red blood cells before the radiotherapeutic treatment, to increase hemoglobin concentration and oxygen-carrying capacity to maximal levels.

A more drastic solution to the problem would be to increase the amount of oxygen in the inspired air during the radiotherapeutic treatment. Although the hemoglobin of arterious blood is practically saturated with oxygen when natural atmospheric air is breathed, any extra oxygen available in the inspired air could be dissolved in the blood plasma. In normal conditions the plasma of arterious blood is not saturated with oxygen.

For an increase of oxygen tension in neoplastic tissues, patients may be irradiated while breathing pure oxygen either at normal atmospheric pressure or at 3 to 4 atmospheres (hyperbaric condition). Although oxygen at 4 atmospheres of pressure may be toxic to the central nervous system, the toxicity may be prevented with the administration of drugs, like barbiturates and chlorpromazine, or with general anesthesia. With a different technique the patient may be allowed to breathe normal air (20% oxygen) at 3 to 4 atmospheres of pressure.

Hyperbaric conditions are realized with special compression chambers. The patient is positioned in one of these chambers; air or pure oxygen is allowed to fill the chamber with pressure gradually rising to 3 or 4 atmospheres, and an X-ray or gamma beam from a source outside the compression chamber is directed through a window against the target region of the patient's body. Some of the patients treated in hyperbaric conditions have been followed for years after the treatment; according to a few reports, these patients have shown a better recovery ratio than similar patients not treated with hyperbaric techniques. Also the histopathologic examination of tumors removed after radiotherapeutic treatment in hyperbaric conditions showed that these tumors had suffered greater radiation damage than similar tumors treated in normal conditions.

An alternative technique, which also takes advantage of the oxygen effect, is aimed at reducing the oxygen tension in the whole tumor as well as in the surrounding normal tissues, so that a general condition of hypoxia prevails in the whole area to be irradiated. In this way the radiosensitivity of the normal tissues is lowered, while the sensitivity of the normally anoxic tumor cells is not affected. In these conditions a larger radiation dose can be delivered to the tumor without too much risk of damage to the normal tissues. This goal could be attained by temporarily occluding local blood circulation. Although the principle is theoretically sound, the practical procedure is still in an experimental phase, because several technical difficulties must be overcome to occlude local blood circulation without producing unwanted side effects.

CHEMICAL PROTECTION
Definition

Modification of radiation response is also obtained by means of chemical substances that can significantly decrease the magnitude of the response when present in a biologic system during irradiation. This type of modification may be classified, therefore, as *chemical protection*, and the substances responsible for it may be termed "chemical protectors" or "chemical radioprotectors." The administration of these substances after irradiation is without protective effect.

Chemical protection was first observed in pure chemical systems containing two solutes; investigators noticed that one of the solutes may protect the other from radiochemical changes (see p. 311). The protection was interpreted as a consequence of the competition between the solutes for the free radicals produced in the radiolysis of water.

In vivo chemical protection was not discovered until 1949, when H. M. Patt at the Argonne National Laboratory reported that the amino acid cysteine reduces radiation mortality in rats if injected before irradiation. No protective activity was observed if the chemical was administered as soon as 5 minutes after irradiation.

The discovery of chemical protection for organisms obviously stimulated considerable interest among radiation biologists, especially in view of its possible practical applications. Thousands of chemicals were screened in several laboratories for possible radioprotective activity. Most of the substances screened were found to be inactive; but a few showed significant protective effects for a variety of biologic systems, from bacteria to mammals. According to Z. Bacq (1965), research on chemical protection occupies about 10% of the time and of the publications in symposia, meetings, and journals devoted to radiation research.

Quantitative appraisal of protective activity

The degree of protection provided by chemical protectors varies from one organism to another and for different end points. It is also affected by factors like radiation dose and radiation type.

Chemical protection can be quantitatively measured in a number of ways, depending on the system and end point under investigation. One method that may be employed in a large number of cases consists of the measurement of the *dose reduction factor* (DRF), which may be defined as the ratio between equi-effective doses in the presence and in the absence of the radioprotector used, or

$$DRF = \frac{\text{Radiation dose to induce a given effect in presence of protector}}{\text{Radiation dose to induce the same effect in absence of protector}}$$

Since the presence of the radioprotector actually increases the dose necessary to induce a given effect, the DRF is always a number greater than 1.

If the end point is mortality in mammals, the DRF may be expressed as the ratio between the $LD_{50/30}$ of protected animals and the $LD_{50/30}$ of unprotected ones:

$$DRF = \frac{LD_{50/30} \text{ of protected animals}}{LD_{50/30} \text{ of controls}}$$

Chemical protection of biologic macromolecules irradiated in vitro is more conveniently measured in terms of *percent protection*, which is defined as follows:

$$\frac{I_c - I_p}{I_c} \times 100$$

where I_c is the fraction of target molecules destroyed or inactivated by a given dose of radiation in the absence of a protector, and I_p is the fraction of the same target molecules destroyed or inactivated by the same dose in the presence of the protector.

For example, suppose that the protective activity of a substance is being investigated on the radiation-induced inactivation of an enzyme in solution. If 21.4% of the enzyme molecules are inactivated by a given dose in the absence of protector and 8.6% are inactivated by the same dose in the presence of protector, the percent protection is expressed as follows:

$$\frac{21.4 - 8.6}{21.4} \times 100 = 60\%$$

Survey of chemical protectors

Only a limited number of substances that have been found to possess considerable radioprotective activity are classified and described in this section. They have been selected either because of their widespread use in radiobiologic research or because of the light they may throw on the possible mechanisms of chemical protection to be discussed later. A more complete list may be found in Bacq's review (1965).

Sulfur-containing compounds

Aminothiols. Aminothiols are organic compounds containing a sulfhydryl radical (—SH) and an amino radical (−NH$_2$) in their molecules.

Cysteine (an amino acid, whose formula appears in Fig. 16-1) was the first compound discovered with radioprotective activity for mammals (see p. 446). The natural L-cysteine is not more active than the D-form, which is not utilized for protein synthesis.

Cystine, an amino acid characterized by two amino radicals and a disulfide bridge, has no protective activity.

Cysteamine, which derives from cysteine by decarboxylation (see Fig. 16-1), is a potent radioprotector. It is also known with the name β-mercaptoethylamine (MEA).

Cystamine, an —S—S— derivative of MEA, is as active as MEA, in spite of the absence of —SH radicals. Its radioprotective properties may be due to the fact that it is rapidly reduced in the body (cystamine + H$_2$ → MEA). Its reduction is possibly an essential condition for radioprotective activity. In fact, certain in vitro systems, in which there is no reason to believe that reduction occurs, are not protected by cystamine.

Glutathione (GSH) is a tripeptide consisting of cysteine, glutamic acid, and glycine. It has shown moderate radioprotective properties in rats and mice and is naturally present in large amounts inside of the cells.

AET (S-[2-aminoethyl]isothiuronium) is a derivative of MEA. Its remarkable radioprotective activity was discovered by D. G. Doherty and his associates

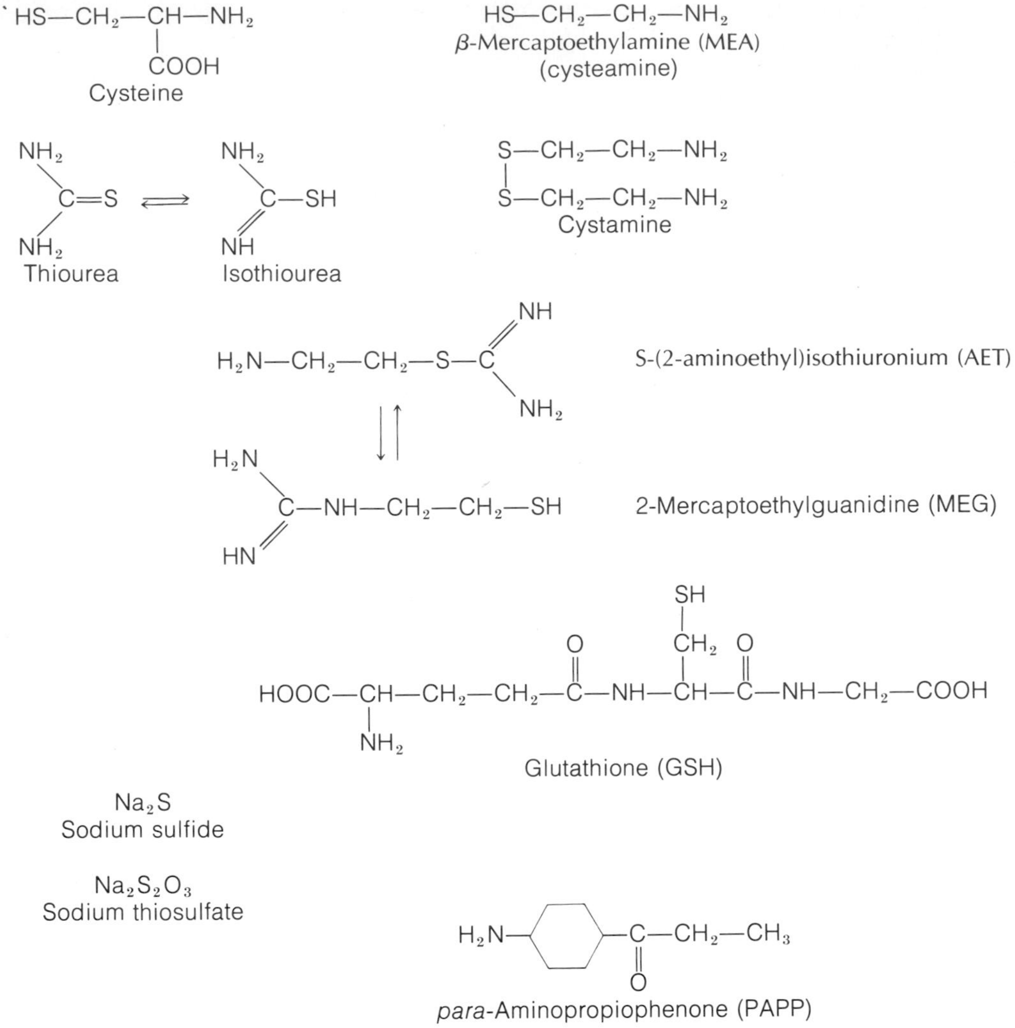

HS—CH$_2$—CH—NH$_2$
|
COOH
Cysteine

HS—CH$_2$—CH$_2$—NH$_2$
β-Mercaptoethylamine (MEA)
(cysteamine)

NH$_2$
|
C=S ⇌
|
NH$_2$
Thiourea

NH$_2$
|
C—SH
‖
NH
Isothiourea

S—CH$_2$—CH$_2$—NH$_2$
|
S—CH$_2$—CH$_2$—NH$_2$
Cystamine

H$_2$N—CH$_2$—CH$_2$—S—C
     NH
     ‖
     \
     NH$_2$
S-(2-aminoethyl)isothiuronium (AET)

H$_2$N
\
C—NH—CH$_2$—CH$_2$—SH
/
HN
2-Mercaptoethylguanidine (MEG)

                              SH
                              |
                        O   CH$_2$ O
                        ‖ | ‖
HOOC—CH—CH$_2$—CH$_2$—C—NH—CH—C—NH—CH$_2$—COOH
     |
     NH$_2$
Glutathione (GSH)

Na$_2$S
Sodium sulfide

Na$_2$S$_2$O$_3$
Sodium thiosulfate

H$_2$N—⬡—C—CH$_2$—CH$_3$
        ‖
        O
para-Aminopropiophenone (PAPP)

Fig. 16-1. Structural formulas of some chemical protectors.

at Oak Ridge. In alkaline solution, AET undergoes a tautomeric molecular rearrangement (transguanylation) leading to the formation of 2-mercapto-ethylguanidine (MEG), which contains an amino radical and a sulfhydryl radical. The radioprotective property of AET is attributed to MEG. In mammals, AET is generally more active than MEA, which is, in turn, five to eight times more efficient than cysteine.

Since not all aminothiols have radioprotective activity, the presence of —SH and —NH$_2$ radicals is not a sufficient prerequisite for such protection. The protective aminothiols and their derivatives have a strong basic function (—NH$_2$ or guanidine) separated from the sulfhydryl radical by no more than three carbon atoms.

Thiourea and derivatives. Thiourea shows moderate radioprotective action for some organisms. In solution, it undergoes tautomeric transformation lead-ing to the formation of isothiourea (Fig. 16-1), which has one amino radical and one sulfhydryl radical. Some derivatives of thiourea (phenylthiourea, guanylthiourea, and others) are also moderately good protectors.

Inorganic sulfur compounds. Sodium sulfide reduces the mutagenic effect of X-rays in *Drosophila,* although it is inactive in mammals. Sodium thiosulfate is a good protector of macromolecules in vitro. In vivo, it protects extracellular molecules or structures, like the mucopolysaccharides of the intercellular substance of connective tissues.

Selenium compounds

Some theoretical considerations (concerning ionization potential, bond strength, and electropositivity) have suggested that compounds in which selenium replaces sulfur should also be good radioprotectors. Several attempts made in the past to test this assumption were not met with success. However, in a recent study (Breccia et al.) with rats, organoselenium compounds have shown radioprotective activity at least equal to that of similar sulfur compounds.

Compounds interfering with oxygen transport

These are drugs that interfere with the oxygen-binding ability of hemoglobin in mammals. Thus protection would seem to be correlated with the hypoxic conditions of the tissues (see the discussion on oxygen effect on p. 441). Carbon monoxide provides chemical protection by transforming part of the hemoglobin to carboxyhemoglobin, which is unable to transport oxygen. Para-aminopropiophenone (PAPP, see Fig. 16-1) probably protects by oxidizing hemoglobin to methemoglobin, which is also unable to transport oxygen. The DRF values of these radioprotectors for mice vary between 1.4 and 1.6.

Compounds interfering with blood circulation

The drugs interfering with blood circulation are vasoactive substances that constrict or dilate the blood vessels. In both cases, hypoxia of the tissues may be the result. Epinephrine, histamine, metacholine (a derivative of acetylcholine), melatonin, and dopamine belong to this group. Melatonin, for instance, in a dose of 75 mg/kg of body weight resulted in 85% survival of albino rats after rectal administration 30 minutes prior to whole-body irradiation with 900 rads of gamma radiation. The proposed mechanism of protection was that melatonin causes spasm of the blood vessels in the radiosensitive organs, with hypoxia as a consequence.

Recent studies (see Prasad and Van Woert) seem to suggest that the radioprotective mechanism of dopamine in vivo may be very complex and may involve more than simple interference with blood circulation.

Respiratory depressants

The drugs that depress the nervous respiratory centers are moderately good radioprotectors in mammals. Some examples are morphine, ethanol, reserpine, amphetamine, chlorpromazine, serotonin (5-hydroxytryptamine = 5-HT), and pentobarbital.

Hormones

Some hormones like ACTH, deoxycorticosterone, and estradiol have been reported to have radioprotective properties for mice and rats. With regard to estradiol, researchers have observed that when this hormone is administered to mice during or after X-irradiation, it increases mortality markedly by interfering with hemopoietic recovery. On the contrary, the same hormone confers radioprotection when administered about 10 days prior to irradiation. The possible mechanism of estradiol-induced radioprotection is discussed by J. S. Thompson et al.

Toxicity

Many radioprotectors (and especially the sulfur compounds) are known to be toxic when administered in high doses. The toxicity of the same compound may vary for different organisms. When these substances are employed as chemical protectors against ionizing radiation, it is important to know the lethal dose as related to the organism under investigation, so that doses considerably below the threshold of toxicity should be used.

Mice tolerate doses of MEA as high as 150 mg/kg of body weight, if injected intraperitoneally. Cystamine is used with rats in doses of 600 mg/kg by oral route and with mice in doses of 150 mg/kg intraperitoneally. Cysteine is somewhat less toxic; it can be used in doses of 500 to 1000 mg/kg.

The maximal sublethal dose of AET for mammals is between 200 and 300 mg/kg when injected intraperitoneally. When used intravenously, it is less.

The degree of radioprotection provided is often a function of the dose administered. For some compounds, protection is not demonstrable unless they are used in almost toxic amounts. The route of administration may also influence radioprotection. Cysteine is more efficient when administered intravenously than when given intraperitoneally. AET may be given orally, whereas cysteine is almost useless when administered orally.

Sublethal doses of certain chemical protectors may produce temporary minor symptoms. AET induces nausea and vomiting in humans.

Manifestations of chemical protection

Chemical protection seems to be almost universal, in the sense that any irradiated system, from in vitro macromolecules to mammals, can be protected against practically any radiation-induced effect. Viruses, bacteria, plant organs, seeds, spores, embryos, insects, cells in tissue cultures, and the whole mammalian body can all be chemically protected. Chemical protectors decrease the magnitude of such diverse responses like inactivation of enzymes, chromosomal aberrations, gene mutations, congenital malformations caused by irradiation in utero, acute radiation syndrome in mammals, erythema, epilation, delayed effects (cataracts, sterility, carcinogenesis), etc. However, these statements should not be construed as meaning that every single protector shows activity in any organism tested or protects against any type of damage. Some protectors are active in some organisms, but not in others. For instance, cysteamine, which is so active in mammals, is completely useless in chickens. Why a chemical protector should be active in one system and inactive in another is often explained by its metabolic fate, that is, whether it is incorporated in the body cells or quickly excreted, or else whether it is retained in its radioprotective chemical form or quickly catabolized or reduced to a form that has lost its protective properties.

In any event, with the wide spectrum of radioprotectors available, one can almost always find at least one of them that will protect any organism against a given biologic damage.

A large body of information about the chemical protection of in vitro systems has been obtained from studies on synthetic polymers. One of the most thoroughly investigated is polymethacrylate, which results from the union of several basic monomers of this type:

$$-CH_2-\overset{\overset{\displaystyle CH_3}{|}}{\underset{\underset{\displaystyle COO^-}{|}}{C}}-$$

Irradiation of polymethacrylate in aqueous solution leads to its degradation. However, the extent of degradation can be diminished considerably by adding to the solution thiourea, cystamine, or a few other chemical protectors.

Biologic macromolecules in vitro may be protected in solution as well as in the dry state. GSH increases the inactivation dose of certain enzymes (catalase, invertase). Cysteamine protects the protein zein. The radiation-induced conversion of oxyhemoglobin to methemoglobin (see p. 319) is significantly reduced by the addition of cystamine, cysteine, cysteamine, AET, and sodium thiosulfate.* Phages and tobacco mosaic virus RNA are protected against inactivation by sulfhydryl compounds.

The slope of the survival curve for many bacterial species is drastically decreased if the organisms are chemically protected during irradiation. Recently, D. M. Ginsberg and H. K. Webster reported that MEA protects E. coli against DNA strand breakage in gamma-irradiated stationary-phase cultures.

Chemical protection can also be obtained against intragenic and intergenic damage. Pretreatment of spores of Aspergillus niger with MEA decreased the frequency of radiation-induced mutations. Also E. coli, Paramecium, and conidia of Neurospora crassa are protected by MEA against mutagenic effects. MEA protects mouse sperm cells against the induction of dominant lethals; AET is less effective than MEA against the same effect. In insects, protection against genetic damage is difficult to demonstrate, at least with cysteine and MEA. What happens to these substances when they are introduced into the insect body is not known.

Protection against chromosomal aberrations with cysteine, MEA, cystamine, and AET has been observed in onion root tips, in Tradescantia and Vicia roots, in bone marrow cells of mice, and in human cells in tissue culture.

Cystamine and MEA, injected into pregnant mice, have been tested for their protective value against radiation-induced prenatal and neonatal death, as well as against congenital anomalies. Significant chemical protection has been observed.

Chemical protection of mammalian cells in tissue cultures against lethal X-ray damage has been intensively investigated by W. K. Sinclair and co-workers at the Argonne National Laboratory. Synchronized Chinese hamsters cells in cultures are differentially protected by cysteamine at different stages of the cell cycle. As already mentioned in a previous chapter (p 357), these cells are most radiosensitive to interphase death in mitosis and in G_2, less in G_1, and even less in the latter part of the S phase. The efficiency of protection is greatest for the cells in the most sensitive stages, so that in cultures protected by cysteamine the differences of radiosensitivity at different stages are less pronounced. DRFs of about 5 were observed for sensitive mitotic and G_2 cells, about 4.2 for G_1 and early S cells, and 3.3 for the resistant late S cells. The variation of DRF with stages of cell cycle is in contrast with the lack of variation of the oxygen enhancement ratio. This finding seems to suggest that cysteamine does not protect these cells by inducing hypoxia but by some other mechanism (see p. 452). Cystamine has no protective effect on Chinese hamster cells.

The in vivo cell-cloning technique of Till and McCulloch (see p. 329) for hemopoietic cells was employed to test the radioprotective effect of AET and

*V. Arena, preliminary unpublished observations.

anoxia on marrow cells irradiated in vivo (Cole and Davis). Mice were injected before irradiation with AET, or subjected to anoxia. After irradiation, their bone marrow cells were injected into lethally irradiated mice to assay their colony-forming ability in the recipients' spleens. For both, AET and anoxia, a DRF of about 2 was observed.

Different mammalian species are differently protected by sulfhydryl drugs against radiation lethality. Generally AET is an efficient protector for most mammals. Surprisingly enough, it is effective when given orally to mice, but not to rats. The toxicity of AET is also different for different species. For instance, it is somewhat less toxic for monkeys than for dogs. Equally effective for all mammals are cysteamine and cysteine.

Chemical protectors are effective in mammals against some delayed effects, but not against others. Cataracts and life-shortening could not be prevented in sublethally irradiated mice by AET. Likewise, the incidence of leukemia and other forms of neoplasms does not seem to be reduced in mice and rats. On the other hand, cysteine, cysteamine, and glutathione administered intravenously have protected rabbits against certain eye-lens disorders.

Local application of an aminothiol results sometimes in local protection. If newborn albino rats (hairless) are injected subcutaneously with a small amount of MEA and then irradiated with soft X-rays, hair will grow only on the area of the skin where the drug was injected. An interesting fact to note is that a subcutaneous injection of histamine protects the whole skin and not only the area of injection. This proves that MEA and histamine protect with two different mechanisms: histamine is known to protect by reducing the oxygen tension in the tissues (it is a vasodilator, see p. 449), whereas MEA does not act through an oxygen effect.

Mechanism of action

The problem of how chemical protectors operate in diminishing the effects of irradiation has been the object of much (and sometimes heated) discussion during the last 15 years. Suggestions and theories are still controversial and debatable at the present time.

In spite of the astonishing diversity of chemical structure of all the radioprotective drugs, the idea is still advanced from some quarters that all of them may protect with the same mechanism. Anyone who attempts to develop a hypothesis of a general mechanism valid for all chemical protectors should take into account the following experimental facts:

1. The universal action of radioprotectors, which can be effective on a variety of systems, including chemical systems in vitro, against a variety of effects
2. The possibility of obtaining local protection with certain drugs but not with others
3. The protector must be present at the time of irradiation; if administered afterwards it is entirely without effect

This last finding indicates that whatever mechanism suggested as being responsible for protection must be operative during the short-lived physicochemical and chemical stages of the sequence of events leading to radiation-induced biologic damage.

On the other hand, there is no valid reason to believe that all radioprotectors must act with the same mechanism. Furthermore, the same substance may quite possibly protect with different mechanisms in different systems.

The fact seems to be definitely established that the substances interfering with oxygen transport or blood circulation and respiratory depressants (see p. 449) protect by inducing hypoxia or anoxia in tissues. A few among nonsulfur-containing compounds may protect with special mechanisms. The problem arises with the sulfur compounds, mainly because they can be effective in a variety of systems, from in vitro solutions of macromolecules to bacteria and mammals. For these compounds a number of hypotheses have been suggested. Only those backed by some experimental evidence are surveyed here. Once again, I should emphasize that the same substance may protect with more than one mechanism.

Free-radical scavengers

The free radical–scavenging mechanism has been strongly advocated by Bacq and Alexander. In an aqueous system (as an in vitro solution or protoplasm) aminothiols are oxidized by the free radicals produced by the radiolysis of water, so that the molecules of other solutes (such as molecules of biologic importance) are prevented from being attacked by the same radicals. This is tantamount to a removal of free radicals from the solution; the radioprotector scavenges free radicals by competing for them with other solutes. Aminothiols and their derivatives have a marked affinity for the free radicals $HO^\cdot$ and $HO_2^\cdot$ because of their particular molecular configuration. Electron spin resonance (ESR) studies have shown that, in fact, the protective aminothiols do remove free radicals from an irradiated solution. The same has been verified in bacteria, spores, yeast cells, viruses, fish sperms, and seeds.

Energy transfer

When radiation energy is absorbed by a certain carbon atom of an organic molecule, a molecular lesion may not necessarily occur at the level of that carbon. The energy may be transferred from that carbon to other points of the same molecule in which the lesion may occur (intramolecular energy transfer—Fig. 16-2), or it may be transferred to another neighboring molecule of a different species (intermolecular energy transfer). If the different species is an aminothiol, the probability that energy will be transferred to its molecule from other

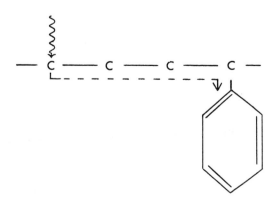

Fig. 16-2. Example of intramolecular energy transfer. A benzene ring may protect the main carbon chain by this mechanism.

molecules directly hit is high, as observed in irradiated films of polymethacrylate mixed with MEA. This radioprotective mechanism is effective against direct action and therefore also in dry systems, whereas radical scavenging would protect only against indirect action.

Immediate repair of damaged molecules

A biologic molecule (RH) may lose an atom of H as a consequence of direct or indirect action:

$$RH \longrightarrow R^{\bullet} + H$$

or

$$RH + HO^{\bullet} \longrightarrow R^{\bullet} + H_2O$$

Without a radioprotector, the free radical R may react with another identical radical or with oxygen; in either case the original molecule has been permanently changed.

In the presence of a sulfhydryl protector (PH) the original molecule may be restored through donation of a hydrogen atom by the protector:

$$R^{\bullet} + PH \longrightarrow RH + P^{\bullet}$$

Formation of mixed disulfides

The formation of mixed disulfides as a mechanism has been suggested by Eldjarn and Pihl, who have also accumulated an impressive wealth of evidence in its favor.

Protein molecules would be protected by aminothiols with this mechanism. The —SH group of the aminothiol interacts with an —SH group of the protein forming a disulfide bridge, and therefore a mixed disulfide compound resulting from the union of the protein molecule and the aminothiol molecule. The protein is thus protected from the indirect action of ionizing radiation (that is, from the free radicals produced by radiolysis of water) and from direct hits. How this protection is envisaged is illustrated in Fig. 16-3.

Despite the efforts made by the authors of the mixed disulfide hypothesis to prove the general occurrence of this mechanism, many arguments may be advanced against it. DNA does not contain any sulfur, and yet it is well protected by aminothiols. Thiourea also protects proteins and yet it does not form mixed disulfides with them. MEA protects bovine serum albumin very well without formation of mixed disulfide. On the other hand, although some nonprotective aminothiols do not form mixed disulfides, cystine does, and yet it is inactive as a protector.

Anoxia

It has been suggested that since many protective aminothiols (cysteine, MEA, glutathione, MEG) are readily oxidized in neutral or slightly alkaline solutions to S-S compounds, their protective properties are related to their consumption of oxygen. This would bring about a state of hypoxia or anoxia in the tissues, which would result in protection.

Cysteamine does indeed consume oxygen when it is oxidized to cystamine. If air is bubbled in a suspension of *E. coli* bacteria containing MEA, the protective activity of the drug is decreased or abolished. On the other hand, cystamine, which does not consume oxygen (as a matter of fact it is an oxidant), shows some protection. The phenomenon of local protection obtainable with MEA is also

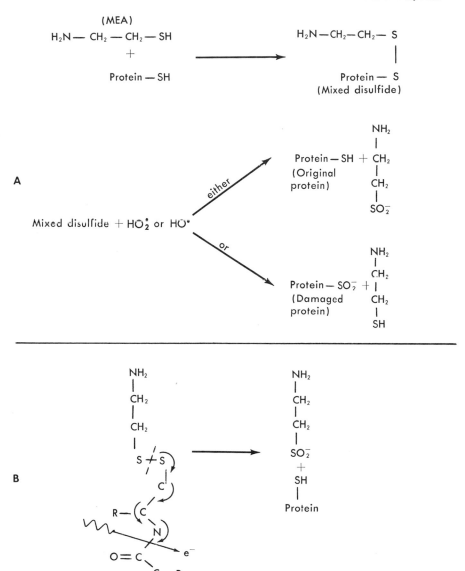

Fig. 16-3. Mixed disulfide mechanism of chemical protection. The protector is MEA. **A,** Mechanism of protection against the indirect action of ionizing radiation. MEA combines with the protein molecule as a result of the oxidation of the two —SH groups. A mixed disulfide is thus formed. A free radical from radiolysis of water may attack one or the other of the two sulfur atoms of the disulfide bridge. The consequence is either the oxidation of the MEA moiety of the complex and the restoration of the original protein, or the oxidation of the protein molecule and the restoration of the original MEA molecule. The final result is that less protein molecules are attacked by the free radicals in the presence of the protector than in its absence. **B,** Mechanism of protection against a direct hit. The ionizing particle or photon hits the protein moiety of the mixed disulfide directly. An electron is ejected, but its vacancy is filled by movements of electrons along the peptide chain. This "bucket brigade" of electrons begins at the disulfide bridge. One of the electrons of the covalent S—S bond (S:S) is thus lost; the bond is consequently ruptured, with the restoration of the —SH of the protein and the formation of an oxidized MEA molecule. Again, MEA has protected the protein.

a strong argument against the anoxia mechanism. In fact, local protection can be observed even when MEA is introduced into cavities like the rectum or vagina; no significant absorption takes place in these conditions, and therefore the induction of anoxia seems to be out of the question. Furthermore a large number of easily oxidizable sulfhydryl compounds show no protective activity.

In spite of the conflicting evidence, we believe that certain aminothiols in certain conditions may protect with the anoxia mechanism.

The possibility that different chemical protectors may act with different mechanisms in the same system has prompted some investigators to study the degree of radioprotection offered by mixtures of several protective drugs ("radioprotective cocktails"). This technique is also necessitated by the problem of the toxicity of the single drugs. As mentioned above, when these chemicals are used singly at high concentrations, their protective value is limited by their toxicity. Hopefully, a mixture of chemical protectors, each at a low concentration, could provide at least the same degree of protection provided by single protectors when they are used at subtoxic concentrations. Several experiments with radioprotective cocktails have been successful; some of the most recent ones are described here.

In one experiment, a radioprotective mixture was used in rats (R. Wang and A. T. Hasegawa), since in these rodents protection with single chemicals has not been as successful as in mice. The mixture consisted of 5-HT (5 μmoles), AET (100 μmoles), and MEA (150 μmoles) and was injected intraperitoneally in 0.4 ml of solution 10 minutes before X-irradiation. The $LD_{50/30}$ for protected rats was 1580 R versus 860 R for the controls. The DRF was therefore 1.83. The protective activity of the mixture was found to be greater than that of any of the components.

H. H. Vogel et al. have tested the effectiveness of a combination treatment in protecting and restoring mice irradiated with X-rays or fission neutrons. The combination treatment consisted of the preirradiation injection of a protective mixture (AET, MEA, and 5-HT), postirradiation administration of homologous bone marrow cells, and daily subcutaneous injections of streptomycin for 10 days after irradiation. The animals were exposed to supralethal doses of X-rays or neutron radiation. With the $LD_{50/30}$ as end point, the DRF for X-irradiated mice that received the multiple treatment was 2.5, whereas for neutron-exposed animals it was less than 1.3. In other experiments, animals were treated exclusively with the preirradiation radioprotective mixture. The DRF for X-irradiated mice was 1.8, whereas for neutron-exposed animals it was less than 1.1. These studies show not only that the radiation damage is further minimized when protective and restorative measures are combined, but also that protection and restoration are less effective in neutron irradiation (high LET) than in X-irradiation (low LET).

Possible practical applications

The information currently available on chemical protection opens the door to interesting applications in humans. Chemical protectors could be useful in radiation therapy, nuclear warfare, and space travel and for radiation workers.

Before any such practical application is considered, one needs to ascertain how the human body tolerates the radioprotective drugs that are so widely used in animal experiments. AET and MEA have been tested in human volunteers. Researchers have observed that doses of AET as low as 10 mg/kg of body weight are not well tolerated, whether they are given orally or intravenously; the administration of the drug is followed by undesirable symptoms like nausea, vomiting, skin rash, and tachycardia.

With regard to MEA, adults seem to be able to tolerate 300 mg daily without ill effects. It is possible that even higher doses can be tolerated. Certainly MEA is less hazardous to humans than AET.

Chemical protectors would be very useful in radiation therapy, if they could protect the normal tissue, but not the malignant tissues that are the target of irradiation. In this case, the hazard of radiation damage to the normal tissues would be minimized, and one could afford to deliver larger radiation doses to the cancerous tissues. Unfortunately, the experimental evidence is conflicting; attempts to demonstrate a differential radioprotection of normal tissues over tumors in vivo have been successful in certain instances and unsuccessful in others. The degree of protection provided to a tumor depends on whether the tumor concentrates the injected drug in its cells or not. The fact is not surprising, if one considers the great structural and biochemical variability of tumors, that the intracellular concentration of a chemical protector may occur in some of them but not in others. Because, in practice, one knows only with difficulty whether or not there is differential radioprotection for a given tumor and with a given drug, the routine use of chemical protectors in radiotherapy is not acceptable yet. However, a much simpler and less hazardous procedure that takes advantage of local temporary protection can be used. One case is rather typical. Radiation therapy of cancer of the uterus is often handicapped by severe secondary troubles arising from the irradiation of neighboring tissues and organs, like the bladder and rectum. When these complications appear, the therapy must be discontinued or replanned. These problems could be circumvented by protecting the mucosa of the neighboring organs, during irradiation only, with tampons impregnated with MEA. The temporary application of the drug protects those organs; but the drug does not have time to diffuse into the tumor and the rest of the body.

Radiation workers, in general, would need no chemical protection, if the nature of their work and the safety measures they use are such that they receive not more than the maximum permissible doses recommended by the NCRP (see p. 250). However, scientists and technicians who might use techniques and procedures (certain experiments with nuclear reactors, manipulation of large amounts of fissionable material) wherein a radiation accident is not a remote possibility could safeguard themselves against such an occurrence with chemical protectors. Chemical protectors would be useful for the general population in the event of a nuclear war. They would certainly protect at least against the radiation damage from heavy local fallout.

Some benefits from chemical protectors should also be expected in space flight. The environmental radiation surrounding man in space is not only quantitatively but also qualitatively different from the environmental radiation we are accustomed to, on the surface of the earth. In space, primary cosmic rays (see p. 110), X-rays from the sun and other stars, bursts of protons, and atomic nuclei emitted with solar flares and solar winds are the predominant components; most of them are high-LET particles. How organisms respond to these unusual types of ionizing radiations is known only in part. Only recently living material has been irradiated in the laboratory with high-energy and high-LET particles (heavy ions, protons, pions, deuterons). The biologic effects are still being investigated. The space-flight experience of the past decade seems to show that astronauts are reasonably well protected by their spaceships against space radiation. The doses to which they are exposed while in flight are not large enough to produce short-term biologic damage. In the future, when space missions of much longer duration will be undertaken and when astronauts will spend more time outside of their spaceships, exposures to higher doses should be expected. Chemical protection, combined with better physical protection (more adequate space suits), may be one answer to the problem.

CHEMICAL SENSITIZATION

It is possible to enhance the magnitude of the biologic effects induced by ionizing radiations by means of certain organic chemicals, which therefore may be referred to as *chemical sensitizers*. Like protective agents, chemical sensitizers must be present in a biologic system during irradiation in order to be effective. Most of them are toxic by their nature, if they are used in concentrations above a certain threshold. Subtoxic doses are harmless when used alone but sensitize the organism to radiation. A few chemical sensitizers are illustrated here; their structural formulas are presented in Fig. 16-4.

The sensitizing properties of *N*-ethylmaleimide (NEM) were discovered by Bridges in 1960 with *E. coli*. NEM sensitized *E. coli* irradiated in anoxic or aerated conditions. In both cases, the dose-survival curve was steeper in the presence of NEM, although the degree of sensitization was higher with anoxia. Other bacteria (*Micrococcus radiodurans, Bacillus subtilis, B. cereus*) cannot be sensitized with NEM; but some bacteriophages can.

Radiation-induced mutagenesis in *E. coli* is enhanced by NEM. Enhancement was also observed in the frequency of X-ray–induced autosomal recessive lethal mutations in *Drosophila*. Sensitization has been obtained, more recently, in HeLa

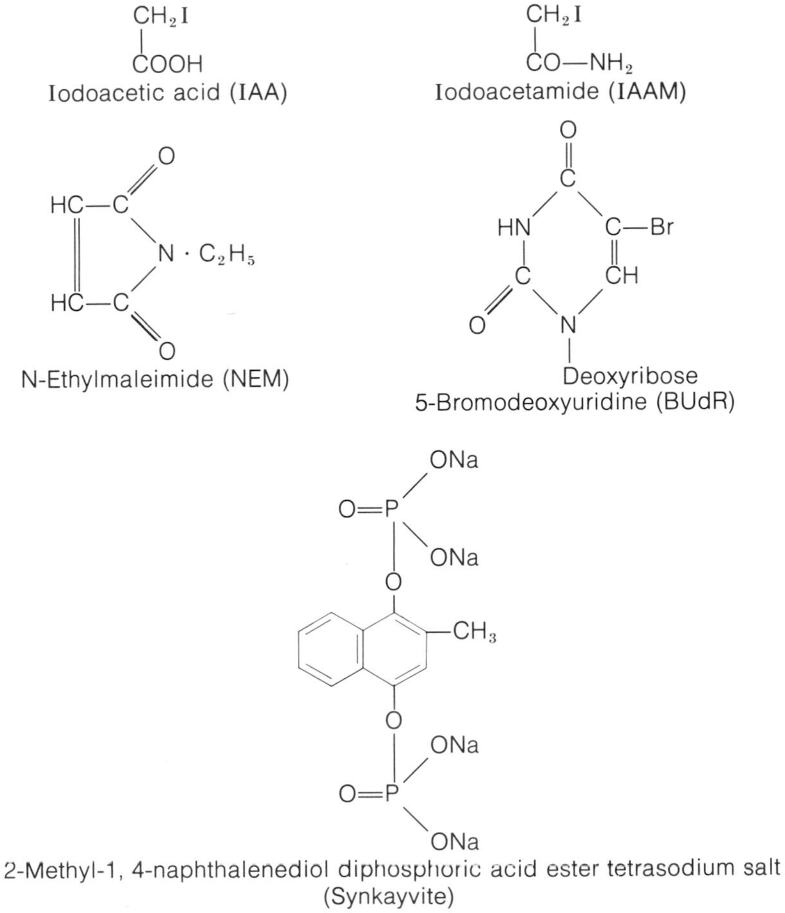

Fig. 16-4. Examples of chemical sensitizers.

cells and Chinese hamster cells in tissue culture; but since mammalian cells cannot tolerate NEM as well as bacteria and phages, very low concentrations had to be used.

Iodoacetic acid (IAA) and iodoacetamide (IAAM) are very effective in sensitizing bacteria, and in particular the most radioresistant species. *Pseudomonas*, *M. radiodurans*, and even budding cells (radioresistant) of the yeast *Saccharomyces cerevisiae* are easily sensitized. IAAM reduces the D_{37} of the very radioresistant *M. radiodurans* from 263 krads to 3 krads, with a dose enhancement factor of about 90!

Radiosensitization by IAA of mouse lymphoma cells in tissue culture has also been obtained. Rat erythrocytes irradiated in the presence of IAAM are more susceptible to hemolysis by snake venoms. This sensitization seems to be due to a radiation-induced reaction between the sensitizer and certain chemical constituents of the plasma membrane (Myers and Slade).

More recently, neoarsphenamine, an arsenic compound that has been used for years in the treatment of syphilis, has shown radiosensitizing activity. It sensitizes anoxic cultures of *Micrococcus sodonensis* and *E. coli* (Kligerman and Schuloff). A much less toxic drug, methylglyoxal, has been found effective in sensitizing *Serratia marcescens* and Chinese hamster cells in tissue cultures.

The radiosensitizers mentioned up to this point (NEM, IAA, IAAM, neo-arsphenamine, methylglyoxal) are known to be sulfhydryl-binding agents. An example of this binding is illustrated by the reaction of NEM with a thiol:

Therefore one should not be surprised that some investigators attribute the sensitizing properties of these compounds to their ability to neutralize sulfhydryl-containing substances, which are naturally present in the intracellular environment and which would function as natural radioprotectors. In fact, the protective activity of some chemical protectors, like cysteine and MEA, is sharply reduced by the presence of NEM during irradiation (as in Chinese hamster cells). However, the fact has also been established that the radiosensitizing properties of IAA and IAAM should be attributed to short-lived products of radiolysis (of unknown nature) that have an acute damaging effect for the cell membrane. The results of several experiments indicate that other mechanisms of radiosensitization for the sulfhydryl-binding agents cannot be ruled out.

The radiosensitizing properties of Synkayvite (which is chemically related to vitamin K) were discovered back in 1946 by J. S. Mitchell and co-workers at Cambridge. They found that chick fibroblasts pretreated with this drug were sensitized to the action of X-rays. The end point investigated was mitotic inhibition. Whether Synkayvite sensitizes also mammals is still controversial. Attempts to sensitize microorganisms like bacteria and yeasts with this drug have not been successful.

One group of radiosensitizers deserves special mention—the halogenated pyrimidine analogs, of which bromodeoxyuridine (BUdR) is the most important and most intensively investigated.

BUdR (see structural formula in Fig. 16-4) is a thymidine analog. This nucleoside differs from thymidine in one point: the methyl radical attached to carbon in position 5 in thymidine is substituted with a bromine atom in BUdR. In spite of this difference, if BUdR is made available to cells, they may utilize it for the synthesis of DNA in lieu of thymidine. However, it has also been observed that DNA-incorporated BUdR has a radiosensitizing effect. BUdR can therefore be classified as a chemical radiosensitizer. To a lesser extent, the same can be said for other halogenated thymidine analogs (like IUdR).

Radiosensitization by BUdR has been demonstrated in *E. coli*, for which the slope of the dose-survival curve is significantly increased. Enhancement of the yield of chromosomal aberrations in *Vicia faba* roots is observed, if the roots are immersed in a solution of BUdR before exposure to 50 R. Cancer cells in culture, viruses, and isolated DNA are also sensitized.

The experiments with BUdR and other halogenated thymidine analogs have a considerable scientific importance because they give an insight into the mechanism of radiation damage. The sensitizing effect of BUdR, when it is incorporated in DNA, is an important piece of evidence in favor of the hypothesis that DNA is indeed the critical site of radiation damage and that DNA damage is the chief contributor to cellular lethality. BUdR-substituted DNA is evidently more radiosensitive than normal DNA. The reason for the greater sensitivity is an interesting, but unsolved problem. Possibly DNA-incorporated halogenated thymidine analogs may magnify the radiochemical damage to DNA by increasing the yield of nonrepairable double-strand breaks, or they may interfere with postirradiation repair mechanisms. Further discussion of this problem may be found in the review of Szybalski (1967).

The practical applications envisioned for chemical sensitizers are correlated, of course, with radiation therapy. A chemical sensitizer would be invaluable if with its use the therapeutic radiation dose necessary to kill a tumor could be reduced. This dose reduction would be possible if tumor cells could be sensitized differentially with respect to the surrounding normal tissues. Another aspect of the problem is the fact already mentioned above (see p. 444) that certain tumors are radioresistant because of the presence of small foci of anoxic cells in their depth. NEM, which sensitizes anoxic cells more efficiently than oxygenated ones, should be very useful for the sensitization of these tumors. Experiments designed to verify these assumptions are underway in many laboratories. In one experiment (Moroson et al.), mice carrying an Ehrlich carcinoma were injected with NEM and other chemical sensitizers and then given whole-body irradiation with sublethal doses of X-rays. Enhanced tumor cell killing was observed; differential radiosensitization of the tumor was strongly suspected on the basis of the experimental data.

HIBERNATION

In wildlife, hibernation has a considerable survival value for certain sedentary mammalian species; by hibernating, these animals can successfully survive through the cold wintery temperatures. The animal falls in a long deep state of sleep resembling coma. Metabolic activities are reduced to a minimum; the body temperature also drops considerably. The energy needed for such essential functions like circulation and breathing is derived from the combustion of the reserve materials stored as fats and carbohydrates during summer. In the laboratory, hibernation of these mammals can be induced by placing them in a controlled environment at a temperature of a few degrees above freezing.

Irradiation of hibernating animals has produced interesting results. The state of hibernation turns out to be a physiologic factor that modifies the radiation response. In earlier studies, the experimental procedure was to irradiate two groups of animals, one in the active state and the other in hibernation, with the same lethal dose. The hibernating animals were aroused several days after irradiation; their percentage of survival as a function of postirradiation time was compared with that of the irradiated active animals. In one of those experiments (Künkel et al.), hibernating European dormice (*Glis glis*) were exposed to 700 R of X-rays, kept in hibernation after irradiation for 3 weeks, and then aroused. None of these animals died within 30 days, whereas dormice exposed to the same dose in the active state showed symptoms of radiation sickness within a few days, and their mortality pattern was consistent with the median lethal dose of the species. After arousal, the animals irradiated in hibernation began showing symptoms of radiation sickness and began dying within a few days with a pattern similar to that of the irradiated controls. These findings were interpreted in the sense that hibernation simply delays the radiation response until posthibernation time, without altering the median lethal dose (taking time of arousal as time 0) or reducing the magnitude of the radiation-induced biologic damage.

In other studies, blood cell analysis has revealed that hibernating marmots exposed to 650 R show only those blood changes characteristic of the state of hibernation. The hematologic response typical of radiation sickness (see pp. 418 to 424) develops only after the animals are aroused.

Recent work with the American ground squirrel (*Citellus tridecemlineatus*) has considerably expanded our knowledge of the influence of hibernation on radiation response. In addition, the dose-response relationship has been investigated.

Musacchia and Barr irradiated active and hibernating ground squirrels with gamma-ray doses ranging from 900 to 20,000 rads; the hibernating animals were aroused immediately after irradiation. The results showed that the animals irradiated in hibernation were more radioresistant than those irradiated in the active state. The difference in radiosensitivity was assessed in terms of both mean survival time (Fig. 16-5) and percentage of survival. In Musacchia and Barr's opinion, although the mechanism of radioprotection provided by hibernation is not exactly known, one may reasonably suspect that the reduced mitotic activity and the tissue hypoxia associated with the state of hibernation could afford some protection.

In another experiment, Jaroslow et al. investigated the correlation between increased survival of gamma-irradiated hibernating ground squirrels and postirradiation hibernation. The $LD_{50/90}$ was 1127 rads for animals irradiated in hibernation and kept in hibernation for 3 weeks after irradiation, and 1024 rads for the controls. Survival was higher in animals aroused 1 day after exposure to 1160 rads than in animals aroused 2 weeks after exposure to the same dose. The results of this and other experiments with fractionated doses seem to indicate, in those investigators' opinion, that the initial radiation damage is the same in irradiated hibernating and nonhibernating animals; the difference in lethal dose and percentage of survival could be explained with a difference in the recovery process. In both groups, considerable intracellular repair occurs

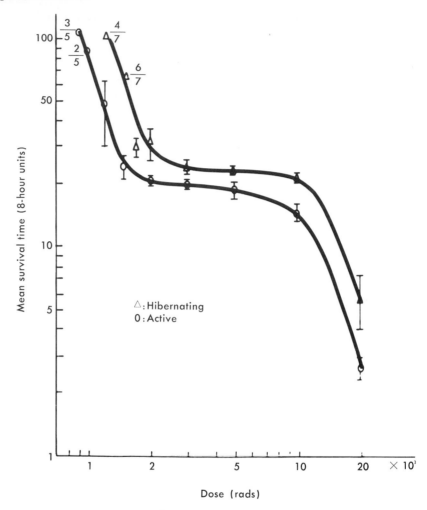

Fig. 16-5. Dose-response curves showing mean survival time of whole body–irradiated hibernating and active ground squirrels. Data points with standard error of the mean are given for groups in which all animals died. Data-point ratios show number of decedents over total number of animals in a specific group. (From Musacchia, X. J., and Barr, R. E. 1968. Survival of whole body–irradiated hibernating and active ground squirrels, Citellus tridecemlineatus. Radiat. Res. **33:**348-356.)

during the first postirradiation day. This repair process is probably interfered with by the large metabolic requirements existing in nonhibernating animals; the small metabolic requirements associated with the state of hibernation could facilitate the postirradiation recovery process. Therefore, survival is improved if the return to a state of higher metabolic activity is delayed to allow maximum intracellular repair.

Künkel's experiments mentioned above were designed to study the influence of both hibernation and cysteine on the survival of irradiated dormice. He observed that cysteine protects irradiated hibernating dormice even when it is administered as late as 21 days *after* irradiation, immediately after arousal.* This finding is in contrast with what we

---

*In later experiments Künkel found that cysteine protects even when injected 10 minutes after irradiation, while the animals are still in hibernation.

know about chemical protection from all other experiments, described previously in this chapter, showing that chemical protectors are inactive if administered after irradiation. Künkel's findings are difficult to explain in the light of the current opinions about the protective mechanisms of sulfhydryl compounds. D. E. Smith duplicated Künkel's experiments with the North American ground squirrel but failed to confirm the results obtained with the dormouse. Unfortunately, he also failed to show whether or not cysteine is active when administered *before* irradiation in hibernation. (Preirradiation treatment with cysteine is effective in protecting nonhibernating ground squirrels).

More experiments similar to those performed by Künkel urgently need to be undertaken with other hibernating species, in order to clarify the problems posed by the dormouse.

OTHER MODIFYING FACTORS

Age

The variation of radiosensitivity with age has been particularly investigated in mammals. The higher radiosensitivity of the unborn animal has already been discussed before (see Chapter 14). After birth the radiosensitivity decreases progressively with age and then levels off during adulthood. A slight increase may be observed in senescence. In certain strains of mice and rats a short period of relative radioresistance may be observed immediately after birth. For SAS/4 mice, the $LD_{50/30}$ is about 850 R at birth, drops to 700 R within 4 weeks, climbs to above 900 R at about 40 weeks of age, and then declines to 700 R in senescence.

The relative radioresistance of newborn mice is attributed by some to the transmission of nutritional or hormonal elements present in the maternal milk to the offspring. These elements might be helpful to some postirradiation recovery processes. The increase of radiosensitivity with senescence is probably caused by the reduced ability of the organism to repair cells or tissues injured by radiation.

Sex

In some mammalian species, females are slightly more radioresistant than males of the same age, although the difference is not always statistically significant. The slight difference, whenever real, may be attributed to the high levels of female sex hormones; some estrogens are known to have radioprotective properties (for instance, estradiol).

Health

The general state of health of an organism at the time of irradiation is, of course, a factor determining the magnitude of the radiation damage. This statement is true for any injurious agent that attacks the body. We know that individuals in good general health are better equipped to overcome infectious processes caused by pathogenic viruses or bacteria than are individuals in poor health or in a state of malnutrition at the time of the attack. The same can be said for the ability of recovery from chemical poisoning, burns, etc. Radiation is no exception; like so many agents it threatens the organism. The ability of the organism to defend itself and to repair the damage depends largely on how functional the many defensive mechanisms are at the time of exposure to radiation.

REFERENCES

GENERAL

Bacq, Z. M. 1965. Chemical protection against ionizing radiation. Charles C Thomas, Publisher, Springfield, Ill.

Bacq, Z. M., and P. Alexander. 1961. Fundamentals of radiobiology. Pergamon Press, Inc., New York. Chapters 11 and 19.

Hamilton, L. D., editor. 1964. Physical factors and modification of radiation injury. Ann. N. Y. Acad. Sci., Vol. 114.

Hollaender, A., editor. 1960. Radiation protection and recovery. Pergamon Press, Inc., New York.

Moroson, H., and M. Quintiliani, editors. 1970. Radiation protection and sensitization. Proceedings of an international symposium. Barnes & Noble, Inc., New York.

Schwartz, E. E. 1966. The modification of radiation response. In Schwartz, E. E., editor. The biological basis of radiation therapy. J. B. Lippincott Co., Philadelphia. Chapter 7.

Thomson. J. F. 1962. Radiation protection in mammals. Reinhold Publishing Corp., New York.

LITERATURE CITED

Breccia, A., R. Badiello, A. Trenta, and M. Mattei. 1969. On the chemical radioprotection by organic selenium compounds in vivo. Radiat. Res. **38:**483.

Bridges, B. A. 1969. Sensitization of organisms to radiation by sulfhydryl-binding agents. In Augenstein, L. G., R. Mason, and M. Zelle, editors. Advances in radiation biology. Academic Press, Inc., New York. Vol. 3, pp. 123-176.

Cole L. J., and W. E. Davis. 1968. Chemical radioprotection of hemopoietic colony-forming cells: comparative effects of AET, anoxia and urethan. Radiat. Res. **36:**555.

Eldjarn, L., and A. Pihl. 1960. Mechanisms of protective and sensitizing action. In Errera, M., and A. Forssberg, editors. Mechanisms in radiobiology. Academic Press, Inc., New York. Vol 2, Chapter 4.

Ginsberg, M., and H. K. Webster. 1969. Chemical protection against single-strand breaks in DNA of gamma-irradiated E. coli. Radiat. Res. **39:**421.

Hollaender, A., W. K. Baker, and E. H. Anderson. 1951. Effect of oxygen tension and certain chemicals on the X-ray sensitivity of mutation production and survival. Cold Spring Harbor Symposia on Quantitative Biology **16:**315.

Holthusen, H. 1921. Beiträge zur Biologie der Strahlenwirkung. Untersuchungen an Askarideiern. [Contributions to the biology of the effect of radiation. Investigations on ascarid eggs]. Arch. Ges. Physiol. Pflügers **187:**1.

Jaroslow, B. N., D. E. Smith, M. Williams, and S. A. Tyler. 1969. Survival of hibernating ground squirrels after single and fractionated doses of cobalt-60 gamma radiation. Radiat. Res. **38:**379.

Kimball, R. F. 1955. The role of oxygen and peroxide in the production of radiation damage to Paramecium. Ann. N. Y. Acad. Sci. **59:**638.

Kligerman, M. M., and L. Schulhof. 1969. Sensitization of bacteria to X-rays by neoarsphenamine. Radiat. Res. **39:**571.

Künkel, H. A. 1961. Strahlenbiologische Untersuchungen an hibernisierten Siebenschläfer [Radiobiological investigations on hibernating dormice]. Chemotherapia **3:**200.

Künkel, H. A., G. Höhne, and H. Maass. 1957. Der Einfluss von Cystein und Winterschlaf auf die Überlebensrate röntgenbestrahlter Siebenschläfer (Glis glis) [The influence of cysteine and hibernation on the survival rate of irradiated dormice]. Z. Naturforsch. **12b:**144.

Legrys, G. A., and E. J. Hall. 1969. The oxygen effect and X-ray sensitivity in synchronously dividing cultures of Chinese hamster cells. Radiat. Res. **37:**161.

Moroson, H., M. Schmid, and M. Furlan. 1968. Enhancement of murine ascites tumor cell killing with X-irradiation by thiol binding agents. Radiat. Res. **36:**571.

Musacchia, X. J., and R. E. Barr. 1968. Survival of whole-body–irradiated hibernating and active ground squirrels, Citellus tridecemlineatus. Radiat. Res. **33:**348.

Myers, D. K., and D. E. Slade. 1967. Radiosensitization of mammalian cells by iodoacetamide and related compounds. Radiat. Res. **30:**186.

Myers, D. K., T. A. Tribe, and R. Mortimer. 1969. On the radiation-induced reaction of iodoacetamide with albumin and with the erythrocyte membrane. Radiat. Res. **40:**580.

Patt, H. M., E. B. Tyree, R. L. Straube, and D. E. Smith. 1949. Cysteine protection against X-irradiation. Science **110:**213.

Prasad, K. N., and M. H. Van Woert. 1969. Radioprotective action of dihydroxyphenylethylamine (dopamine) on whole-body X-irradiated rats. Radiat. Res. **37:**305.

Sinclair, W. K. 1969. Protection by cysteamine against lethal X-ray damage during the cell cycle of Chinese hamster cells. Radiat. Res. **39:**135.

Smith, D. E. 1960. Failure of cysteine given postirradiation to protect the hibernating ground squirrel. Radiat. Res. **12:**79.

Stevens, C. O., B. M. Tolbert, and G. R. Bergstrom. 1970. Effects of oxygen on irradiated solid egg-white lysozyme. Radiat. Res. **42:**232.

Szybalski, W. 1967. Molecular events resulting in radiation injury, repair and sensitization of DNA. Radiat. Res. **7:**147.

Szybalski, W., and Z. Opara-Kubinska. 1965. Radiobiological and physicochemical properties of 5-bromodeoxyuridine–labeled transforming DNA as related to the nature of the critical radiosensitive structures. In Cellular radiation biology. The Williams & Wilkins Co., Baltimore. Pp. 223-240.

Thoday, J. M., and J. M. Read. 1947. Effect of oxygen on the frequency of chromosome aberrations produced by X-rays. Nature **160:**608.

Thoday, J. M., and J. M. Read. 1949. Effect of oxygen on the frequency of chromosome aberrations produced by alpha-rays. Nature **163:**133.

Thompson, J. S., E. L. Simmons, M. K. Crawford, and C. D. Severson. 1969. Studies on the mechanism of estradiol-induced radioprotection. Radiat. Res. **40:**70.

Vogel, H. H., A. T. Hasewaga, and R. I. H. Wang. 1969. Comparative protection by a combination treatment in mice irradiated with fission neutrons or X-rays. Radiat. Res. **39:**57.

Wang, R. H., and A. T. Hasewaga. 1969. Radioprotective effect of a chemical mixture in rats. Radiat. Res. **40:**310.

CHAPTER 17

Delayed somatic effects

NATURE OF THE DELAYED EFFECTS

Most of the radiation-induced biologic effects discussed in some of the preceding chapters are detectable within a relatively short time after exposure to ionizing radiation. Examples of these short-term effects at different levels of organization are chromosomal aberrations, effects on the cell cycle and on cell division, cell death, congenital anomalies resulting from irradiation in utero, epilation, burns, radiation sickness, and the death of an organism exposed to lethal or supralethal doses. But one peculiar characteristic of radiation injury is that certain damaging effects are detectable only a long time (often several years) after exposure and after a long latent period during which no damage of any kind is apparent. These effects are commonly referred to as *delayed effects*.

In a certain sense, genetic effects could be classified as delayed effects because they manifest themselves phenotypically only among the offspring of the irradiated individuals. However, one could argue that a radiation-induced gene mutation that is the cause of the later damage is not a delayed effect, because it is produced at the time of irradiation. Other delayed effects are restricted to the soma of the exposed individuals, and therefore will be detectable during their lifetime. The delayed somatic effects about which a great deal of information is available are malignant diseases, including leukemia, shortening of life-span, impairment of fertility, and cataracts. The discussion of these effects is the object of this chapter.

One may logically assume that the effects mentioned above are delayed because, by their very nature, they are the end products of long and slow multistage biologic processes, whose mechanisms are often unknown. Exposure to radiation initiates these processes. Obviously, then, these effects will be detectable only when death does not follow shortly after irradiation. Therefore, no cataracts, or leukemia, or any form of cancer has ever been observed in victims of radiation accidents who died within days or weeks after exposure.

Delayed somatic effects, unlike several other radiation-induced injuries, may follow after acute or chronic, whole-body or partial exposures. However, the type of exposure may be a determining factor for the induction of specific effects. For instance, leukemia is more likely to follow after a whole-body exposure than after a partial exposure, whereas bone cancer usually follows after irradiation of bones. Whether there is a threshold dose for delayed effects is still a debatable question. Actually, the problem is quite complex, because the threshold value (if it exists) may depend on several radiation parameters, and in particular on dose rate.

The knowledge of the fact that exposure to ionizing radiation may result in some injuries many years later has been widely disseminated among the lay public. No matter how accurate this knowledge is, there is the danger that the statistical aspects of the delayed effects are not properly appreciated, and they may engender exaggerated fears in individuals who have recovered from some type of short-term radiation injury. To dispel unwarranted fears, we should stress that delayed effects do not necessarily follow every single case of radiation exposure. What is true is simply that an individual who has received a radiation dose above a certain threshold is statistically more likely to be a victim of cancer, or leukemia, or cataracts, etc., than unexposed individuals.

RADIATION CARCINOGENESIS

Ionizing radiation increases the incidence of cancer; as such, it must be classified as a carcinogenic agent.

Nature of cancer

In the United States, as well as in most other countries, cancer is the second leading cause of death. (The highest number of deaths is caused by cardiovascular diseases.) According to statistics published by the American Cancer Society, in 1967, 870,000 Americans were under medical care for cancer, 570,000 discovered they had cancer for the first time, and 311,000 died of the disease.

What is commonly known as cancer is not a single disease, but a broad spectrum of different (although closely related) nosologic entities, which should be more properly called *neoplasms*. Tumors are neoplasms that may grow to moderate or enormous size. Like communicable diseases, neoplasms may be caused by a variety of agents and their effects on the patients range from minor inconveniences to life-threatening pathologic processes.

Since neoplasms are so different from one another, one has difficulty finding a satisfactory definition or description that could be valid for all of them. As an approximation, it may be stated that a neoplasm begins as a local affection consisting of atypical tissue proliferation, characterized by progressive growth without a definite final developmental stage and by a "rebellion" against the laws that regulate the orderly growth of the organisms and the orderly cohabitation of its tissues and organs. In other words, a neoplasm originates as a rebellion of a group of cells to the rules of interdependent life of all the cells of the body. The neoplastic cells form a tissue of their own, they live a life of their own, and, by multiplying unlimitedly at the expense of the host organism, they harm it or destroy it.

No tissue or organ of the body seems to be immune from neoplasia. The following partial classification of neoplasms is based on the tissues from which they originate:
1. Neoplasms of tissues of mesodermal origin, generally referred to as sarcomas — myomas are neoplasms of muscular tissues, lipomas of adipose tissues, chondromas of cartilaginous tissues, and osteomas of bony tissue
2. Neoplasms of nervous tissues — neuromas and gliomas
3. Neoplasms of epithelial tissues — carcinomas

Another classification well known to the lay public distinguishes benign from malignant tumors. A benign tumor does not show an unlimited growth and keeps, so to speak, good neighborhood relations with the normal adjacent tissues, without invading them or subtracting nutrients from them. Its cells are not as atypical as those of a malignant tumor, and, most of all, do not show tendency to metastasize, that is, to be transported by the blood or lymph to other points of the body and start there other foci of neoplasms, as it happens so frequently for malignant tumors. However, because of their anatomic location, benign tumors may not be entirely harmless. A neuroma of the brain may exert pressure on normal vital tissues and organs and interfere with their functions.

Furthermore, a benign tumor, in due time, may quite possibly assume characteristics of malignancy; the trivial melanomas of the skin (black moles), which are nothing but benign tumors, may become foci of malignant carcinomas. The term "cancer" is commonly applied to any malignant neoplasm.

As mentioned above, the neoplastic process begins with the rebellion of one or a few cells against certain laws that govern the growth and development of the body. To illustrate this concept, let us take a neuroma and a carcinoma as examples.

Normal neurons have lost their proliferative activity since birth or some late stage of the embryonic development. As explained in Chapter 14, the loss of proliferative activity is accompanied by a process of specialization and differentiation. Indeed, neurons are the most specialized cells of the human body. When a neuroma develops in the brain of an adult, one or a few neurons suddenly start behaving like the embryonic neuroblasts from which they derived. The neoplastic process is accompanied by dedifferentiation (anaplasia) and by reacquisition of the embryonic proliferative activity. The neoplastic neurons start losing the morphologic and physiologic characteristics of normal neurons; even their metabolic processes become atypical and may result in the production of abnormal waste products. From one or a very few initial cells, through rapid divisions, hundreds and thousands of cells originate; the mass of cancerous cells invades the surrounding normal tissues, interfering with their functions, competing with them for blood supply and nutrients, and living and growing as a parasite of the body.

The cells of a normal epithelial tissue, although fairly differentiated, have not completely lost their original proliferative activity. Some of them multiply in an orderly fashion to replace the more superficial cells, which die of normal death or are sloughed off through normal wear and tear. When a carcinoma develops, one or more epithelial cells begin multiplying wildly, purposelessly, and independently of the rules of a good neighborhood.

The mystery of cancer is related to the unsolved basic problem of the causes of the neoplastic process. What causes the rebellion of the cancerous cells against the laws that regulate the orderly cohabitation of the normal cells? Why does a cell that had reached its final stage of differentiation become suddenly the ancestor of a mass of cells that multiply wildly and autonomously? Many years ago investigators understood that if the problem of cancer defies a solution this difficulty is due, at least in part, to the fact that the laws against which cancerous cells rebel are not yet clear. What is responsible for the slowdown of the proliferative activity of the cells destined to become neurons or muscular fibers, or what accounts for the loss of this activity for some cells and not for others, is not entirely clear. The problem of cancer will be much closer to its solution when all these problems concerning normal cells are better clarified.

Nevertheless, many hypotheses and theories have been suggested during the last 50 years to explain why a neoplasm originates. Commendable efforts have been made to seek evidence in favor of one theory or another; none of the cancer theories is overwhelmingly convincing. An old theory (Cohnheim-Durante) suggests that a tumor originates from a supernumerary embryonic cell that has never reached its final stage of differentiation to begin with. Its ability to multiply remains dormant for a long time until it is resumed with the development of a tumor as a consequence. Other investigators have seen the cause of neoplasms in the very morphologic anomalies (atypical mitoses, chromosomal aberrations, missing or supernumerary chromosomes, etc.) so frequently shown by the cancerous cells. According to Warburg's old theory, which has been recently revived and tied to modern concepts, cancer originates as a result of irreversible injury to a normal cell's respiration. This injury is then compensated for by a shift to the anaerobic state and increased glycolysis. Perhaps, a theory that enjoys a great deal of favor at the present time is the somatic mutation theory. Cancer cells could originate from normal cells by somatic mutation followed by clonal selection. The mutation must either give the mutated cell an advantage in growth rate or must bring it within one step of such advantage. Since we know that some types of neoplasms are definitely induced by viruses, virus particles can cause a change equivalent to that caused by the mutation by adding genetic information to the cell, possibly in a manner similar to bacterial transformation or transduction. After its initiation, the evolution of the cancer cell clone is a random process of selection similar to Darwin's concept of the evolution of the species.

Regardless of the possible causes of the carcinogenic process itself, the process can be initiated by certain external agents. Some types of neoplasms may be initiated by viruses (such as Rous sarcoma virus), and others by chemical substances of the aromatic series, such as those found in coal tar, lubricating oils, aniline dyes, and cigarette smoke. Some animal parasites (such as the cysticercus larva of the tapeworm) can initiate a neoplasm through their irritating action on a normal tissue. Prolonged exposure to ultraviolet radiation may cause carcinoma of the skin. Ionizing radiations must be added to the large family of carcinogenic agents. According to some current opinions, confirmed by studies of experimental carcinogenesis in animals, latent tumor cells are produced by one process and brought to a malignant state by an external agent referred to as a "cocarcinogen." The neoplastic process can be envisaged as consisting of two phases—an "initiating" phase and a "promoting" phase.

That ionizing radiations may be carcinogenic and yet be able to cure cancer (see Chapter 10) may sound like an unbelievable paradox. Very appropriately P. Alexander considers radiation as a "two-edged sword." The apparent contradiction is clarified by the complex effects of ionizing radiations on the cells. The therapeutic use of radiation depends on its immediate lethal effects, whereas when radiation induces cancer, it does so by inducing whatever modification (somatic mutation?) is necessary to convert a normal cell into a cancerous cell.

The human experience

The carcinogenic properties of ionizing radiation have been repeatedly demonstrated by a number of surveys of groups of human individuals who had been acutely or chronically exposed to radiation from external sources or internal emitters. These individuals have been observed for several years after exposure and the incidence of cancer among them has been compared with that of appropriate unexposed groups used as controls.

The following is a summary of the most representative studies of radiation carcinogenesis in humans:

> Radiation carcinogenesis
>   Chronic exposure
>   From external sources—radiologists
>   From internal emitters
>   Uranium ore miners
>   Watch dial painters
>   Patients treated with radium
>   Users of Thorotrast
>   Acute exposure—atomic bomb survivors

The first cases of cancer among radiologists were reported within a decade after the discovery of X-rays. At that time, the biologic effects of X-rays were poorly known, and no shielding precautions were taken by those early radiation workers. We believe that the early radiologists were chronically exposed to more than a thousand R over a period of several years. Carcinoma of the skin (mainly of the squamous cells) developed in many of them as a complication of radiation dermatitis.

A high incidence of lung cancer was observed among the miners of Joachimsthal (Bohemia). Uranium ore had been extracted from the Joachimsthal mines for years, long before radioactivity was discovered; uranium compounds were used mainly to stain glasses. Pitchblende, one of the minerals particularly abundant in those mines, contains uranium as well as radium and other elements of the natural radioactive series. It was used by Mme. Curie as the starting material from which she discovered and isolated radium (see Chapter 5). At the beginning of the nineteenth century it was already known that a mysterious disease ("mountain disease") was responsible for a high rate of death among the Bohemian miners. In 1879 the disease was recognized as lung cancer. Today we know that the responsible agent was the radioactive gas radon, which is the first decay product of radium. Radon was inhaled by those miners with the air for several years. The concentrations of radon in the atmosphere of those mines have been estimated to be about 10^{-3} μCi/liter of air. The dose to the lungs from this concentration has been calculated at 500 to 1700 rads/year. The exposure was from alpha particles (radon is an alpha emitter, like radium). Recent statistical studies have shown that in the 1930s the mortality from lung cancer among the Joachimsthal miners was 50 times higher than that of the general population of Vienna, and that the average period of time before the onset of the disease was 17 years.

Radium as an internal emitter has been identified as a cause of cancer of the bones. Major studies in this field are in progress at the Argonne National Laboratory, the Argonne Cancer Research Hospital, the Massachusetts Institute of Technology, and the New Jersey State Department of Health. These studies are concerned with people who acquired body burdens of radium some 30 to 45 years ago. The largest group consists of women who were employed in factories of New Jersey, Connecticut, and Illinois to paint watch dials with a luminous paint containing radium and mesothorium. The accepted practice for those young women was to bring the brush to a sharp point by shaping it on their lips. Thus, significant quantities of the radioactive material were ingested. Unfortunately, the hazards associated with this practice were not clearly recognized until after a significant number of individuals had ingested substantial quantities. By 1926, however, the practice of tipping the brushes was discontinued.

The radium body burden of those women has been studied with the whole-body counter and by radon breath analysis. The latter procedure takes advantage of the fact that the decay product of radium, which is radon, is released from the skeleton and exhaled at a predictable rate so that a quantitative relationship can be determined between the level of radon in the breath and the level of radium in the skeleton. An attempt has also been made to measure the terminal burden of radium by autopsy of the fatalities and to determine the nature of any disease present at the time of death.

The studies have shown that a relationship exists between radium content in bones and the incidence of tumors. The changes in bone range from a slight coarsening of the trabeculae, spotty areas of rarefaction, and isolated patches of osteosclerosis to more severe changes like aseptic necrosis, pathologic fractures, and malignant tumors. The last have been mainly of two types — sarcomas of bone and carcinomas of the mastoid air cells and of the paranasal sinuses. Thirteen malignant tumors have been associated with current or terminal radium body burdens greater than 0.5 μCi. Ten malignant epithelial tumors of the mastoids and sinuses have been found in subjects with body burdens of 0.17 to 4.7 μCi. The average latent period before the onset of these tumors was between 15 and 25 years.

Some of the dial painters who ingested large amounts of radioactive material died within a few years of anemia, hemorrhages, and infections. It is believed that death was caused by bone marrow damage.

In the 1920s, radium was used orally or parenterally as an accepted therapeutic drug for a wide variety of ailments, from arthritis to mental disorders. Many patients were advised by their physicians to drink radioactive mineral waters. More than 15 years later a number of these patients developed bone tumors, and significant concentrations of radium were detected in their skeletons.

Between 1930 and 1945 several radiologists used a substance called Thorotrast as a contrast medium in diagnostic radiography. Thorotrast is a colloidal suspension of thorium dioxide, which is also an alpha emitter, but unlike radium it is mainly concentrated in the reticuloendothelial system. Carcinomas and sarcomas of the liver and of the biliary ducts were found 13 to 19 years later in patients who had been injected with Thorotrast. Other malignant tumors have been found at the sites of injection. After the discovery of its carcinogenic effects, the drug was quickly discontinued by radiologists.

High incidence of cancer induced by acute exposure to radiation has been reported among the Japanese atomic bomb survivors by the ABCC. The evidence strongly suggests that the risk of lung cancer among the survivors who received 90 rads or more is higher than among the unexposed Japanese population. Six cases of breast cancers were observed among women who were exposed to 90 rads or more, as compared with 1.53 cases expected. The atomic bomb survivors have also shown an incidence of thyroid cancer higher than expected.

Although all the studies conducted by the ABCC have been handicapped by serious difficulties and biases, a statistical analysis of the data shows that the thyroid cancers found among the survivors are certainly radiation induced, whereas the evidence concerning cancers of the breast and lungs is still very much in doubt.

Radiation carcinogenesis in animals

The carcinogenic nature of ionizing radiation has been repeatedly demonstrated with animal experiments. Several types of malignant tumors may be induced in laboratory animals, although some organs are more prone to neoplasia than others. A variety of external sources has been used for whole-body as well as partial irradiation. Current research is concentrated on the carcinogenic effects of internal emitters, in view of the fact that some radionuclides associated with nuclear energy operations, if ingested or inhaled, could pose a serious hazard to man.

Most of the animal experiments have shown that there is a latent period which may be a significant fraction of the life-span and which seems to increase as the dose is lowered. They have also shown that induced tumors may continue to develop throughout the animal's life and that the spectrum of neoplasms induced by radiation is much broader than that induced by other carcinogenic agents. The relation between incidence of tumors and dose cannot, however, be specified precisely for any neoplasm. The available data indicate that this relation may vary, depending on the type of neoplasm and other variables, such as species, sex, physiologic state, environmental factors, and radiation parameters.

Radiation-induced ovarian tumors have been reported especially in mice. Animals exposed to single or repeated doses of X-rays showed a fifteenfold increase in incidence of these tumors as compared with that of the controls. Mice were also exposed to radiation from a nuclear explosion; ovarian neoplasms were observed with doses as low as 30 R. At higher doses the incidence was lower, probably because of the shortening of the life-span, since the latent period for these tumors is about 12 to 15 months. Chronic exposure to gamma radiation (0.4 to 8.8 R daily) resulted in 60% to 100% incidence versus 12% incidence in unirradiated controls. The incidence can be reduced if the irradiated animals are injected with estradiol benzoate; this and other findings show that a relation exists between the induction of ovarian tumors and female sex hormones. An influence of the pituitary gonadotropic hormones on the induction has also been suggested by some experiments.

Cutaneous tumors have been induced in animals by X-rays and by beta radiation. Glass pellets containing ^{90}Sr have been implanted subcutaneously and intraperitoneally in rats. The tumor incidence seems to be a function of the activity of the source. No tumors have been observed with an activity of only 1.5 μCi. Rats irradiated with electrons from a Van de Graaff generator have shown the highest incidence of epitheliomas after doses of about 1200 rads, with up to 5 or 6 tumors per rat. Higher doses (2000 rads or more) caused less tumors.

Cancer of the mammary glands has been investigated especially in female Sprague-Dawley rats, because the latent period for these tumors is only about 3 months and the difference between the incidence of spontaneous tumors and the incidence of radiation-induced ones is large. Some experiments indicate that a linear dose-response relationship exists for doses in the 25 to 400 R range. Above this range, the incidence remains constant or decreases. Ovariectomy greatly reduces the incidence of radiation-induced mammary tumors; an indication of some relationship between ovarian functions and these tumors.

Lung cancer has been induced in rodents by external radiation and by inhaled radioactive particulate matter (^{144}Ce, ^{239}Pu, ^{90}Sr, and ^{35}S). Recently, the possible carcinogenic properties of the alpha-emitter polonium 210 have been the object of intensive studies, especially in view of the fact that the high incidence of lung cancer in cigarette smokers seems to be attributable in part to this radionuclide, which is found in significant quantities in tobacco smoke. In one series of experiments (Yuile et al.), rats were allowed to inhale aerosols with small amounts of polonium added. Besides shortening of life-span in the highest dose group, 41 lung tumors were found in 288 exposed rats versus none in the control animals. The activities delivered to the lungs in these experiments were between 0.15 and 0.02 μCi. Similar results have been obtained with hamsters injected intratracheally with polonium.

Experiments show that with regard to the induction of bone tumors external irradiation is effective only with single whole-body doses of 600 to 700 R or higher. Chronic irradiation from an external source does not seem to cause this type of tumor in mice, guinea pigs, and rats. A large number of data are available on the induction of bone tumors by bone-seeking internal emitters; extensive work in this field has been done at the Argonne National Laboratory. Different bone-seekers are differently effective in the induction of bone tumors. The effectiveness decreases in the following order: plutonium, radium, strontium 90, strontium 89, calcium 45. Fifty-four percent of rats injected with as little as 1.89×10^{-3} μCi of plutonium per gram of body weight exhibited bone tumors. Recent studies (Howard et al.) with miniature swine have shown that animals fed or injected daily with rather large doses of ^{90}Sr develop bone tumors, mostly in the skull. Fourteen tumors, classified as either osteosarcomas or giant cell tumors, were observed in nine animals.

At the Argonne National Laboratory, plutonium has been injected under the skin of the abdominal wall in rats and mice to study the late effects of alpha radiation. The fate of plutonium after injection depends on its chemical form. When an absorbable form is injected, most of the radionuclide is carried to the skeleton and bone tumors result after several months. When a particulate form is used, about half of it is carried to the skeleton and the rest remains at or near the site of injection. With large concentrations of the particulate form, a few sarcomas have developed at local sites and many more bone tumors have resulted from the fraction that goes to the skeleton. With lower doses, bone tumors do not appear, but the local sarcomas still occur. It has been reported that castrated Sprague-Dawley female rats have a much higher incidence of osteogenic sarcomas than noncastrated animals, when injected subcutaneously with a similar dose of particulate plutonium. Studies are continuing to determine whether this difference is due to the production of a higher plutonium body burden in castrated animals.

Some investigators have noticed that for the induction of tumors in a given organ, irradiation of the same organ is not always an essential condition. Partially irradiated rats show tumors in both irradiated and shielded areas, although the incidence in the latter is much less (Carsten and Innes). Kaplan found that thymic lymphoma may develop in a normal thymus reimplanted into an irradiated thymectomized animal. These findings suggest some type of *abscopal effect*,* that is, an effect due to interactions between irradiated and nonirradiated tissues. The mechanism of this abscopal effect in Kaplan's results is difficult to explain, especially in the light of subsequent investigations which showed that the lymphoma arose from cells that had never been irradiated and not as a result of metastasis of cancerous cells migrating into the thymus from irradiated organs.

RADIATION LEUKEMOGENESIS

Leukemia is a neoplasm of the hemopoietic tissues and is characterized by the disorderly proliferation of the precursors of leukocytes and by production of abnormal cells, which may infiltrate most of the tissues and organs of the body. The disease is accompanied by characteristic disturbances of the normal elements of the circulating blood.

Depending on the particular type of hemopoietic tissue affected, leukemias are classified as myelocytic (or granulocytic) if the neoplasia involves the precursors of granulocytes (neutrophils), lymphocytic if it involves the precursors of lymphocytes, and monocytic if it involves the precursors of monocytes. Leukemias may be acute or chronic. The acute form has a rapid and fulminating course leading to death within a few weeks or months. It strikes children more frequently than adults, and the therapeutic measures currently available do not have a high probability of success, although they may prolong the survival time of the patients. Chronic leukemias may be brought under control much more easily with a variety of treatments, and sometimes are compatible with an almost normal life.

Proliferation of the neoplastic leukemic cells causes hyperplasia of the bone marrow and enlargement of the spleen, liver, and lymph nodes. The disease may or may not be associated with a high white cell count in the circulating blood, depending on the rate

---

*Literally, effect watched at a distance.

of proliferation of the neoplastic tissue and on the survival time of the abnormal leuko-cytes. One consequence of the abnormal proliferation of the leukemic cells in the hemo-poietic tissues is the crowding-out of the normal blood-forming elements, which leads to anemia and thrombocytopenia. Leukemia is also associated with an impairment of the immune mechanisms of the body, thus producing increased susceptibility to infec-tion. In the course of the disease, leukemic cells may infiltrate other systems such as the intestinal tract, CNS, kidneys, lungs, and joints.

Since leukemia is a form of cancer, its origin is surrounded by the same veil of mystery as the origin of other forms of cancer; the many theories mentioned in the preceding section of this chapter may be extended to the origin of leukemia. Leukemia is not one disease with many expressions, but rather a group of disorders that may be originated by different causes with different mechanisms. However, although leukemia may origi-nate spontaneously without any apparent external cause, several external agents may also be responsible for initiating the disease; some of these agents are identical to those mentioned above for other forms of cancer.

In 1969, 15,000 Americans died of leukemia and 19,000 new cases of the disease were diagnosed; 50% to 60% of the cases were of the acute form.

Studies on humans and experiments on animals have definitely shown that ionizing radiation is leukemogenic. In general, leukemia is most often induced by external irradiation of the whole body, or of large portions of it, with sub-lethal doses; only a few clear-cut cases of leukemia induction by internal emit-ters have been reported from animal experiments. Furthermore, the data show that radiation, may induce acute as well as chronic leukemia; however, no induc-tion of chronic lymphocytic leukemia has been demonstrated yet in humans.

The leukemogenic effect of radiation in man has been firmly established for American radiologists, Japanese atomic bomb survivors, patients who received radiation therapy for ankylosing spondylitis, and patients who received large amounts of radioiodine in the therapy of carcinoma of the thyroid gland.

A survey study of American radiologists showed that between 1929 and 1943 the mortality from leukemia among radiologists was 10 times greater than that among other physicians who do not have much contact with ionizing radiation. In the 1944 to 1948 period the mortality ratio was 7, and in the 1952 to 1955 period it was 4. Obviously, the fall of the ratio must be attributed to the improved shielding and safety measures in the operation of X-ray equipment. The expo-sure responsible for the induction of leukemia in radiologists was a chronic fractionated dose to the whole body or to significant portions of the bone mar-row, spread over a period of many years.

As a consequence of the knowledge obtained from the studies on radiolo-gists, the ABCC expected that a higher incidence of leukemia should be found among the Hiroshima and Nagasaki survivors. In fact, they found that acute exposure of the survivors to the neutron-gamma radiation from the explosions had indeed increased the frequency of leukemia among the survivors (Fig. 17-1). A significant difference in the frequency between exposed and unexposed indi-viduals was detected as early as 2 years after the explosions; the peak incidence was in the years 1951 and 1952. Both, acute and chronic granulocytic leukemias were observed but not the chronic lymphocytic form. Acute leukemia was more common among children. In all age groups acute leukemia continued to occur at higher than usual rates through 1966, while by that time, chronic granulocytic leukemia had dropped to near-normal values.

The data show that the latent period for leukemia is considerably shorter than for other forms of cancer.

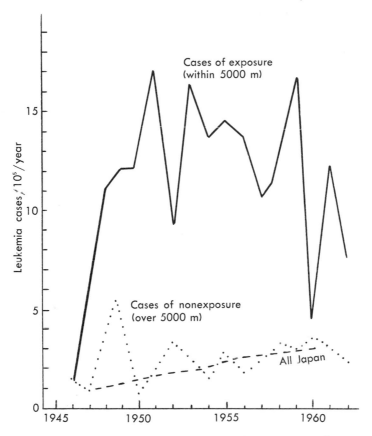

Fig. 17-1. The incidence of leukemia among the Hiroshima survivors between 1946 and 1962. (From U. N. Scientific Committee on the Effects of Atomic Radiation—Nineteenth Session. 1964. Suppl. 14 (A/5814)

A dose-response relationship was also found among the Hiroshima survivors. The following data show how the incidence of leukemia decreased with increasing distance of the exposed individuals from ground zero:

Distance from ground zero (meters)	Incidence of leukemia (per 10,000 persons)
Less than 1000	128
1000–1500	28
1500–2000	4
2000–3000	2
3000 or more	1.6
Unexposed Japanese population	1.5

The British spondylitic patients (see p. 386) received partial body exposure; they showed a high incidence of chronic granulocytic leukemia, but not the lymphocytic form. Among these patients a total of 52 cases of the disease were observed versus 5.48 expected on the basis of the national mortality for England. The peak incidence was reached about 5 years after the first exposure.

The patients who showed a higher than normal incidence of leukemia after being treated with radioiodine had received doses of ^{131}I in excess of 1 Ci.

Other groups of individuals studied are children who were exposed in utero to diagnostic pelvic X-rays, children who received radiation therapy for cancer of the thymus, patients with polycythemia vera treated with X-rays or ^{32}P (or both), and children born from parents who had a history of diagnostic radiation exposure before their children were conceived. An accurate statistical analysis of the data from these studies has failed to prove, beyond any reasonable doubt, a causal relationship between radiation and the cases of leukemia observed. Probably those data need a more critical evaluation.*

At the present time one cannot attribute to radiation exposure the increased incidence in leukemia that has been reported in several countries, including the United States (it more than doubled between 1925 and 1940). One reason is the fact that a substantial portion of the increased incidence is due to chronic lymphocytic leukemia, the occurrence of which has not yet been shown to be influenced by ionizing radiations. Another reason is that in the modern environment there are many chemical carcinogens that have not been evaluated adequately with respect to their influence on leukemogenesis even though some of them are carcinogenic in laboratory animals.

Leukemia has been induced in animals with whole-body exposure; some strains (for example, of mice) seem to be much more sensitive than others. In some strains, total accumulated doses as low as 50 R may cause an incidence significantly higher than in the controls. Chronic irradiation with as little as 4 R per day induces leukemia.

Experiments performed by Upton et al. with mice of the so-called RF strain have shown that the incidence of granulocytic leukemia is increased in animals who have undergone acute whole-body exposure within a dose range from 50 to 400 rads of X-rays. The yield of leukemias declines as the dose is increased above an optimal level. Chronic exposure to X- or gamma radiation is less leukemogenic than acute exposure. Fast neutrons show the same general leukemogenic effectiveness as X-rays. For the same total accumulated dose, partial exposure is less effective than whole-body exposure.

Recent experiments at the Oak Ridge National Laboratory indicate that the relatively high susceptibility of male mice of the RF strain to radiation-induced granulocytic leukemia, as compared with that of females of the same strain, is paralleled by a similar sex difference in vulnerability to viral transmission of the disease. In both cases, the female's higher resistance may be related to the inhibitory effects of the female hormone estrogen, as shown from experiments in which estrogen was administered to RF mice of both sexes and found to inhibit viral induction of the disease. As mentioned above, a protective effect of female sex hormones had been observed also against the induction of other forms of cancer.

Use of germ-free, caesarian-derived mice that are also free of conventional mouse viruses (when appropriately isolated from contamination) has aided the study of the viral aspects of radiation leukemia at the Oak Ridge National Laboratory. A low-level irradiation experiment began in 1962; mice of the RFM/Unf strain have been monitored continuously as one means of checking the effectiveness of barrier techniques designed to prevent contamination. All tests indicate that the mouse colony has remained free of conventional mouse

---

*For further details see R. Miller. 1969. Delayed radiation effects in atomic-bomb survivors. Science **166:**569.

viruses. However, despite germ-free ancestry and barrier protection against conventional viruses, such mice have still developed leukemia in response to irradiation, and viruslike particles have been observed in their leukemic tissues. These findings imply that the mice possess the leukemia virus from birth and need not acquire it from their environment and that irradiation activates the virus. Possible mechanisms of activation are under study.

With regard to the leukemogenic effects of internal emitters, 0.3 to 0.5 μCi of ^{32}P per gram of body weight, given to mice twice a week 10 times, induced leukemia. Induction of leukemia was also observed with 0.3 μCi of ^{144}Ce per gram given 10 times, twice a week. Strangely enough, with some radionuclides (^{32}P, ^{89}Sr), leukemia appears after the administration of low doses, but not of high doses; the latter induce only osteosarcomas.

• • •

For many years, considerable discussion has been centered around the extremely important question of whether a threshold dose exists for radiation carcinogenesis in man and whether there is a straight-line relationship between radiation dose and incidence of cancer or leukemia. Actually the problem exists not only for radiation but also for all other known carcinogens. Opinions are divided in this very controversial area. Although a few animal experiments would indicate an absence of threshold, an extrapolation of the data to man seems premature. Many radiation biologists are inclined to believe in a non-linear dose-response relationship and in the existence of a threshold. A. M. Brues, for example, presents his arguments against the linear theory in an article which those interested may wish to consult.*

EFFECTS ON LIFE-SPAN

Animal experiments have clearly shown that one of the delayed effects of exposure to ionizing radiation is a shortening of the life-span. This effect is quite independent of the life shortening caused by deaths from other seriously damaging delayed effects, like malignancies. In fact, even in the absence of cancer or leukemia, irradiated animals show a reduction of life expectancy when their mean survival time is compared with that of unirradiated animals. Investigators believe that the shortening of the life-span is due to accelerated aging as well as to degenerative processes occurring after irradiation, which render the animals more susceptible to certain diseases.

A proportionality seems to exist between percent of life shortening and the total accumulated dose, although it is difficult to prove whether there is a threshold dose. For rats and mice, according to some data, life is shortened 2% to 5% for every 100 R delivered as a single sublethal whole-body dose of X-rays. However, the percentage of life shortening is influenced by the dose rate; with the same total accumulated dose, acute exposure is more effective than chronic or fractionated exposure. A threshold dose rate exists; mice exposed to 0.1 R/day throughout their life did not show any life shortening, although they experienced more neoplasms than the controls.

Radiation-induced life shortening has also been investigated in mammals with a life-span much longer than in rodents. In a recent study (Andersen and

---

*Brues, A. M. 1958. Critique of the linear theory of carcinogenesis. Science **128**:693.

Rosenblatt), female beagles were given single or fractionated X-ray doses of 100 or 300 R at 10 to 12 months of age. They were kept under observation for many years until death, and their mean survival time was compared with that of other beagles used as controls. The animals exposed to 100 R showed 9.5% life short-ening and those exposed to 300 R, 20.7%. On a linear scale this represented a loss of 6.7% per 100 R. In the 300 R group, differences were observed be-tween animals irradiated with a single dose and those irradiated with a frac-tionated dose; in this latter case the life shortening was reduced when the time interval between the first and the second fraction was increased. Causes of death among the irradiated animals were infectious disease (canine distemper, hepa-titis), heart failure, uremia, endocrine dysfunction, mastitis, and neoplasms of the mammary glands and genital organs.

Attempts to demonstrate life shortening in groups of exposed humans have proved to be extremely difficult. The data are open to controversy. The statis-tical analysis of the data for American radiologists suggests a slight shortening of the life-span, whereas the same effect has not been observed for that of British radiologists. Whether this difference lies in faulty elaboration of the data or because the practice of radiology in the two countries was different is an open question. The evidence for the Japanese atomic bomb survivors is somewhat more clear cut. The most recent data published by the ABCC con-cern deaths in the two bombed cities occurring between 1950 and 1960 among 99,393 survivors of all ages. Even excluding leukemia as a cause of death, the mortality for survivors who had been within 1200 meters from ground zero was 15% higher than that found in a comparable group of unexposed individuals. This increase was found to be statistically significant.

At the present time there is a great deal of speculation concerning the exact mechanism by which radiation induces a shortening of the life-span. As men-tioned above, the experimental evidence is that radiation accelerates aging and increases susceptibility to a number of diseases. How radiation accelerates aging is not clear, in part because the very process of aging is obscure, although much research is being done today to clarify the problem. One effect of sublethal doses is possibly a complex of many minute lesions involving several tissues and or-gans and especially the finest capillary vessels; or a complex of many slight in-juries to functions. The organism is thus debilitated and therefore less capable of overcoming the effects of the multiple environmental stresses. Another theory is based on the assumption that aging is caused by the accumulation of deleterious somatic mutations. Exposure to radiation would be responsible for an increase of these mutations and therefore for the acceleration of the aging process.

Interestingly enough, exposure to radiation in certain conditions may lengthen the life-span, rather than shorten it. This was at least suspected by R. H. Mole* in the course of some experiments with mice and guinea pigs irradiated chronically with a dose rate of less than 1 rem/week. Further experiments with rats exposed to 0.8 R/day gamma radiation suggested the same conclusion. Perhaps the best-documented experiments that prove an extension of the life-span are those performed by Cork with flour beetles.† Although exposure of these animals to single doses above 11,000 R definitely

---

*See Mole, R. H. 1957. Nature **180:**456.

†Actually a radiation-induced increased life-span in insects was reported for the first time by W. P. Davey in 1919.

caused a shortening of their life-span, animals exposed to 3000 R at a rate of 10 R/min. showed a longer survival time than the unirradiated controls. For instance, 240 days after irradiation, 90% of the irradiated beetles were still alive, whereas only 70% of the controls were surviving. The difference in survival rate was consistently in favor of the irradiated animals throughout the postirradiation time.

Although some reports on life lengthening cannot be blindly accepted because of erroneous experimental conditions or faulty statistical treatment of the data, the fact is undeniable that low radiation doses delivered at low dose rate certainly can prolong the life-span. The case in question proves that the effects of ionizing radiations are not always deleterious. Some investigators have speculated that radiation may have a sort of therapeutic effect against diseases already existing in the irradiated animals, or may produce harmless physiologic conditions of a undetermined nature that discourage the infection by pathogenic agents. Of course, since the optimal doses and dose rates necessary to obtain lengthening of the life-span are not known, it is extremely premature and hazardous to think of practical applications to man.

STERILITY

Radiation damage to the gonads was reported shortly after the discovery of X-rays, when H. E. Albers-Schönberg described the deleterious effects of radiation in the testes of rabbits and guinea pigs (1903). Later, a massive destruction of several types of germ cells was observed in X-rayed testes, and this destruction was thought to be the cause of radiation-induced permanent or temporary sterility.

It may be of historical interest to recall here that at the beginning of this century the origin of the sex hormones responsible for the male sexual secondary characteristics was a subject of much heated discussion. Some investigators believed that the hormones were secreted by the cells of the germinal line; others believed that they were secreted by the interstitial cells. The strongest argument in favor of the second hypothesis was that X-irradiation of the testes does not interfere with the sexual secondary characteristics, and yet it destroys the germ cells thus causing sterility. The high radioresistance of the interstitial cells was implicitly recognized in those experiments.

The effects of radiation on the gonads and the consequent induced sterility follow different patterns in the two sexes, at least insofar as mammals are concerned. The male gonad is basically made of two histologically and physiologically different components—seminiferous tubules and interstitial tissues. Spermatogenesis takes place within the tubules, starting from spermatogonia, which multiply quite rapidly to give origin to primary spermatocytes, followed by the secondary spermatocytes, spermatids, and mature sperm cells. At least three types of spermatogonia may be recognized. Type A spermatogonia are the primordial germ cells; each of them is believed to undergo three divisions, thus producing eight cells. One of them will remain for some time in a dormant stage without divisions; the other seven will become spermatogonia of the intermediate type and then will divide once to form 14 type B spermatogonia. Through one more cell division, each type B spermatogonium produces two primary spermatocytes. The existence of a dormant stage for some spermatogonia explains why in the mammalian male the production of mature sperm cells is a continuous process.

Sertoli cells are found within the seminiferous tubule, but they are not members of the spermatogenic line. They are nutrient cells attached to the germinal epithelium of the tubule and mixed with the spermatogonia. The interstitial tissue is located between the tubules like a sort of loose connective tissue. After the onset of puberty, it functions as an endocrine gland; its cells secrete male sex hormones.

The relative radiosensitivity of the different elements of the spermatogenic series has been the object of considerable discussion in the past. At the present time there seems to be a general agreement, with few exceptions about certain details that need further clarification. The radiosensitivity of the male sex cells is in accordance with the predictions of the law of Bergonié and Tribondeau. Of all the cellular components of the mammalian testis, the most radiosensitive are the spermatogonia, and the most radioresistant are the Sertoli and interstitial cells. The radiosensitivity gradually decreases from the spermatogonial to the mature sperm stage. Significant destruction of spermatogonia may be observed with doses as low as 20 to 30 rads, whereas doses as high as 1500 rads are not sufficient to kill mature sperms or to impair their motility and ability to fertilize the egg. Whether spermatids and sperms are equally radioresistant has not been firmly established yet. Type B spermatogonia are believed to be as radiosensitive as active type A spermatogonia, whereas there is some evidence that dormant type A spermatogonia are slightly less sensitive (Fig. 17-2, *A*).

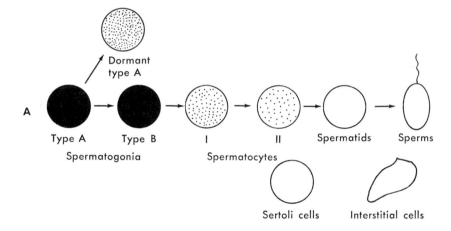

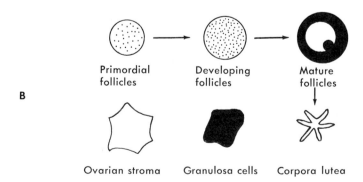

Fig. 17-2. Relative radiosensitivity of the cellular elements in the mammalian testis, **A,** and ovary, **B.** Elements shown in black are the most radiosensitive. Elements shown in white are the least radiosensitive. Dotted circles represent stages with intermediate degrees of sensitivity.

Depending on the dose received, an irradiated male may experience either temporary or permanent sterility. In either case, sterilization is never immediate, regardless of the level of exposure; this justifies the classification of radiation-induced sterility as a delayed effect. The animal will remain fertile for some time after exposure because of the survival of the radioresistant mature sperms and spermatids. The later sterility is a consequence of the failure of production of more sperms; this failure is caused by the more or less extensive destruction of spermatogonia at the time of irradiation. A reacquisition of fertility may follow the sterile period if a few spermatogonia did survive irradiation to repopulate the seminiferous tubules with mature sperms. Such will be the case if the dose was not too high. Sterility may become permanent if the dose was so high that none of the spermatogonia survived. The sterility period is not necessarily associated with a complete absence of mature sperm cells in the testis. As a matter of fact, in mammals, total male sterility may be caused not only by an aspermic semen but also by an insufficient concentration of sperms. After irradiation, whether the sperms are absent or in insufficient number in the semen during the sterile period depends very much on the number of surviving spermatogonia and therefore on the dose.

Male sterility is not associated with impotency or impaired libido, even after recovery from acute radiation syndrome. The secretion of male sex hormones (which are responsible for potency and libido) is not suppressed or depressed by radiation because of the high radioresistance of the interstitial cells.

In general, the ovary of the mammalian female is more radiosensitive than the testis of the same species, despite the fact that ovarian hormones give the female a degree of radioresistance not found in the male. In most species, the radiosensitivity increases from the primordial follicle to the mature follicle, although there is still some uncertainty as to whether the mature follicle is really the most radiosensitive stage of the series. Granulosa cells (the cells lining the mature follicle) are also very sensitive; corpora lutea and the ovarian stroma are radioresistant (Fig. 17-2 *B*). The primordial follicles of mice show a peculiar radiosensitivity, which might explain why the mouse ovary responds to radiation somewhat differently from the ovaries of other mammals.

Despite a few contradicting reports, the evidence indicates that, unlike what happens in the male, radiation-induced female sterility is an "all-or-none" response. There appears to be no level of exposure that renders the female only temporarily sterile. A radiation dose either leaves the animal fertile or induces permanent sterility. In this latter case, sterility begins much earlier than in males. In rats at least 600 R are required to cause marked degeneration in the ovaries within 2 days and a disappearance of ova within 4 months. For the observation of similar effects in mice, much lower doses are sufficient, although there are differences among strains. Studies at the Argonne National Laboratory have shown that a single dose of 80 rads of fission neutrons causes complete and permanent sterility in the female mouse within several weeks after exposure.

Although many believe that irradiation never induces temporary female sterility, it may be responsible for reduced fertility. The two effects should not be confused. Reduced fertility (which is manifested by small litter size) is caused by doses that are not large enough to kill all the potential sex cells. Those killed will not be replaced, in contrast to what happens in the testis. Those which survive will eventually reach maturity and become fertilizable. On the other hand, one should remember that in the ovary of an adult mammal not all the stages of the oogenic series are found, but only primary and secondary oocytes. If all the oocytes are destroyed, they cannot be replaced by the proliferation and transformation of germ cells in earlier oogenic stages; the result is permanent sterility. Likewise, there is no evidence suggesting that radiation-damaged oocytes may recover in the sense that the mature eggs developing from them after ovulation may undergo normal development if fertilized.

For sterility in either sex to occur, the gonads themselves must be exposed. Therefore sterility may follow only after whole-body or partial exposure including the gonads. Whole-body exposure with the gonads shielded does not induce sterility, no matter how high the dose is. No abscopal effects have been observed with regard to sterility.

Data on radiation-induced sterility in humans are obviously scarce. Data on Japanese survivors are not available. A few studies have been carried out on surviving victims of radiation accidents and on women X-rayed for therapeutic purposes. A dose of 600 R applied to the testes has been reported to induce permanent sterility; temporary sterility has been observed after doses of 250 R. A victim of one of the Los Alamos radiation accidents was kept under observation for some time after exposure and provided valuable information on radiation-induced sterility (see Hempelmann et al.). The victim, a 34-year-old man, had been exposed to an estimated dose of 416 R; he suffered acute radiation sickness from which he recovered. Seven months after the accident, his semen was aspermic. After 10 months, biopsy showed normal Sertoli cells, but no spermatogenesis. Signs of recovery appeared later; the sperm count began increasing and reached almost normal values 58 months after the accident, when this man fathered a child.

We suspect that doses in the 320 to 625 R range may induce permanent sterilization in human females. A single exposure of 625 R caused permanent sterility in a group of 72 women. Radiation-induced sterility in women has been often found associated with disturbances of the menstrual cycle (including amenorrhea) and diminished libido. Apparently, unlike what happens in males, irradiation at levels high enough to sterilize upsets the secretion and balance of the female sex hormones. Although the production of progesterone should not be affected because of the high resistance of the corpora lutea, it is possible that the secretion of estrogens (which are mainly responsible for libido) is interfered with because of the high radiosensitivity of the mature follicle, which produces some of them.

It has been reported that low-level irradiation of the human ovary may stimulate ovulation through mechanisms that are not entirely clear. Although this type of irradiation has been used in the past for practical purposes, it has been discouraged and discontinued in more recent times for fear of mutagenic effects.

Radiation-induced sterilization has been exploited as an aid in biologic pest control. The most spectacular success has been obtained with the eradication of the screw-worm fly.* The screw-worm fly (*Callitroga hominivorax*) is a deadly parasite of livestock. The female lays hundreds of eggs in the wounds of warm-blooded animals. The maggots hatching from the eggs feed upon the flesh of their host and cause bleeding and other wounds, which attract more flies. The infested animal dies if left untreated. The maggots

*See Knipling, E. F. Oct. 1960. The eradication of the screw-worm fly. Sci. Amer. **203:**54.

go through the pupal stage in the ground; the adults emerge from the pupal cases after 8 days and mate about 3 days later.

Since females mate only once during their lifetime (as it happens for many other insect species), some investigators thought that by releasing in the field a large number of sterilized males, these males could compete with other males in mating with the females. Females mating with sterilized males would lay unfertilized (and therefore sterile) eggs. The number of individuals should decrease in a few generations and hopefully could be also reduced to zero. The most practical method to sterilize the males was found to be gamma irradiation. Laboratory experiments showed that a dose of 2500 R was sufficient for sterilization, without causing side effects and without impairing mating ability.

The knowledge gained in the laboratory was first applied in large scale in the early 1950s on the island of Curaçao in the Netherlands Antilles, where the screw-worm fly was causing serious losses among goats. Four hundred sterilized males per square mile were released every week on the island from airplanes. The pest was completely eradicated after only four generations. The same method was later used for the eradication of the same fly from the entire southeastern United States. Obviously, because of the much larger area, millions of sterilized males had to be released, but the eradication of the pest was accomplished as on the island of Curaçao. This method of biologic pest control has been considered and tried with other species of parasitic flies with considerable success.

Experiments with weevils and other beetles, however, have been unsuccessful, apparently because of their radiosensitivity to acute lethal effects, making their survival of massive radiation doses unlikely. Since no somatic cell division occurs in adult Diptera, these insects can tolerate huge X-ray doses without lethal effects, whereas adult beetles show cell renewal in the midgut epithelium and die as a result of intestinal damage after much lower radiation doses.

The sterile male–release method presents several advantages over other conventional methods of pest control. It is highly selective, because it involves only the target species. There is no possibility that females could become immune to sterile matings. Chemical insecticides lose their effectiveness after some time because of the survival of a few insecticide-resistant mutants, which with their reproduction bring the population back to its normal levels. Furthermore, most insecticides (as discovered recently) are not entirely harmless to other animal and plant species. None of these disadvantages seems to be associated with the sterile-male method.

CATARACTS

A cataract is a disorder consisting of the opacity of the eye lens with possible loss of vision. It may be congenital or induced in the course of life by senile degenerative changes, trauma, diabetes, ingestion of certain toxic substances, diets with high content of galactose, and exposure to ionizing radiation.

The development of cataracts after exposure to ionizing radiation has been demonstrated in both experimental animals and man, as a result of acute doses. Cataracts have not been reported yet as a result of low-level chronic exposure, but probably further studies are needed before this possibility is unequivocally ruled out. No abscopal effects are involved in radiation cataracts; the eye lens must be directly exposed.

With doses lower than certain values, true cataracts fail to appear but some opacity of the lens can still be induced. Permanent damage to the lens can be induced already with 250 R; in mice opacities appear after doses as low as 15 R. In man, the threshold for the induction of true cataracts appears to be about 600 R. Neutron radiation is particularly effective; its RBE is between 5 and 10. Neutron doses of only 1 rad produce definite opacification in mice (Bateman and Bond).

Cataract must be classified as a delayed effect, because it develops after a minimum latent period that is proportionate to the length of the life-span of the species. The latent period is a few months in rodents and at least 3 to 4 years in man.

The ABCC has reported about 10 cases of severe radiation cataracts among Japanese atomic bomb survivors. In 1963 and 1964, 1627 residents of Hiroshima and 841 residents of Nagasaki were examined; 40% of them had received doses above 200 rads at the time of the explosions. In the high-dose group there were more lens opacities than in groups who received lower doses. Impaired visual acuity was also found in children who had been exposed in utero. One surviving victim of a Los Alamos radiation accident showed cataracts in both eyes 3 years later. Some physicists working with cyclotrons have also experienced cataracts from accidental neutron irradiation of their eyes.

REFERENCES

GENERAL
Blum, H. F. 1959. Environmental radiation and cancer. Science **130**:1545.
Casarett, G. W. 1965. Experimental radiation carcinogenesis. Progr. Exp. Tumor Res. **7**:49.
Glucksmann, A., L. F. Lamberton, and W. V. Mayneord. 1958. Cancer. Butterworth & Co. (Publishers) Ltd., London.
Hamilton, L. D. 1964. The Hiroshima and Nagasaki data and radiation carcinogenesis. Ann. N.Y. Acad, Sci. **114**:241.
Harris, R. J. C., editor. 1963. Cellular basis and aetiology of late somatic effects of ionizing radiation. Academic Press, Inc., New York.
Mandl, R. 1964. The radiosensitivity of germ cells. Biol. Rev. **39**:288.
Miller, R. 1969. Delayed radiation effects in atomic-bomb survivors. Science **166**:569.
Rugh, R. 1960. General biology: gametes, the developing embryo and cellular differentiation. In Errera, M., and A. Forssberg, editors. Mechanisms in radiobiology. Academic Press, Inc., New York. Vol. 2, Chapter 1.
U. N. Scientific Committee on the Effects of Atomic Radiation. Nineteenth session. 1964. Suppl. 14 (A/5814).
Upton, A. C., V. K. Jenkins, and J. W. Conklin. 1964. Myeloid leukemia in the mouse. Ann. N. Y. Acad. Sci. **114**:189.
Van Cleave, C. D. 1968. Late somatic effects of ionizing radiation. AEC-TID-24310. Clearing House for Scientific and Technical Information. U. S. Department of Commerce, Springfield, Va.
Zeldis, L. J., S. Jablon, and M. Ishida. 1964. Current status of ABCC-NIH studies of carcinogenesis in Hiroshima and Nagasaki. Ann. N. Y. Acad. Sci. **114**:225.

LITERATURE CITED
Andersen, A. C., and L. S. Rosenblatt.1969. The effect of whole-body X-irradiation on the median life span of female dogs (beagles). Radiat. Res. **39**:177.
Bateman, J. L. and V. P. Bond. 1967. Lens opacification in mice exposed to fast neutrons. Radiat. Res. **7**:239.
Brues, A. M. 1958. Critique of the linear theory of carcinogenesis. Science **128**:693.
Carsten, A. L. and J. R. M. Innes. 1964. The effects of partial-body irradiation on delayed responses. Ann. N. Y. Acad. Sci. **114**:316.
Cork, J. M. 1957. Gamma radiation and longevity of the flour beetle. Radiat. Res. **7**:551.
Hempelmann, L. H., H. Lisco, and J. G. Hoffman. 1952. The acute radiation syndrome: a study of nine cases and a review of the problem. Ann. Intern. Med. **36**:279.
Howard, E. B., W. J. Clarke, M. T. Karagianes, and R. F. Palmer. 1969. Strontium-90–induced bone tumors in miniature swine. Radiat. Res. **39**:594.
Kaplan, H. S., and M. B. Brown. 1952. A quantitative dose-response study of lymphoid tumor development in irradiated C57 black mice. J. Nat. Cancer Inst. **13**:185.
Knipling, E. F. Oct. 1960. The eradication of the screw-worm fly. Sci. Amer. **203**:54.
Yuile, C. L., H. L. Berke, and T. Hull. 1967. Lung cancer following polonium-210 inhalation in rats. Radiat. Res. **31**:760.

Epilogue

When ionizing radiation is absorbed by matter, interactions occur according to one or more of the several mechanisms discussed in Chapter 6. Most of these interactions are peculiar to ionizing radiations (as opposed to other types of radiations) because of the very high energies associated with them. The absorbed energy is responsible for molecular changes, which in solutions, may be induced by direct or indirect action.

When ionizing radiation is absorbed by living matter, the same types of interactions take place as in nonliving materials. Chemical changes may be induced in any kind of protoplasmic compounds, but those affecting certain macromolecules of vital importance (nucleic acids, proteins, enzymes, etc.) are particularly damaging. The metabolic activities of the protoplasm amplify the initial biochemical damage because the lesioned molecules are links of intricate and interwoven chains of chemical reactions.

The biochemical damage is the direct cause of morphologic alterations, which in cells, are observable at submicroscopic and microscopic level. The actual response given by a cell when it absorbs ionizing radiation energy depends on several factors; in the worst case the response may even be death. However, cellular structures are not equally vulnerable; the evidence shows that certain structures and molecules are more sensitive and critical than others. The critical sites of cellular radiation damage have been identified with different cellular constituents by different authors; however, there is mounting evidence in favor of the hypothesis that DNA damage is the chief contributing factor to cellular lethality.

Why different types of cells show different radiosensitivity is not entirely clear. The law of Bergonié and Tribondeau can provide only part of the answer.

In multicellular organisms, because of the interdependence of the various organs and of their functions, the radiation-induced cellular damage is the beginning of other lesions visible at organ or organismal level, either shortly after exposure to radiation or after a long latent period. The damage affecting the genome of the reproductive cells may be transmitted to the following generations (genetic damage).

Fortunately, radiation damage is not always irreversible. The mysterious self-repairing power that is a unique characteristic of all organisms is operative also against radiation damage. Whether repair mechanisms succeed in bringing about restoration depends mostly (but not exclusively) on the total accumulated dose and on the dose rate.

485

Although ionizing radiation is almost always harmful and interferes adversely with the welfare of the species, science and technology have found useful applications of ionizing radiations and of their sources. The radiotracer method, radiation therapy, the diagnostic use of X-rays and of radionuclides, the use of radiation in biologic pest control are only a few examples. However, most of these applications entail a possible increased exposure of the human population to radiation sources; man must be well equipped with the necessary knowledge required to provide better and efficient protection against such exposure.

We should have every reason to be happy for the magnificent progress made by radiation biology in the last 20 years. However, one must frankly admit that there are still many misty and nebulous areas in our understanding of the biologic effects of radiation. The number of radiobiologic problems that are still awaiting satisfactory solutions is unfortunately large. The patient reader can find evidence of this throughout this book. More information is urgently needed concerning the long-term biologic effects of low-level chronic exposure, the safety of the currently recommended maximum permissible doses, the consequences of radioactive fallout, the biologic hazards of certain internal emitters, radiation carcinogenesis, the medical management of the victims of radiation accidents, chemical protection, and radiation hazards in space flights. Since many brilliant minds with sound ideas are working on these problems, we hope that their solutions are not too distant in the future.

We also hope that radiobiologic knowledge will be a strong deterrent for man against unwise use of nuclear energy and against the temptation of deploying nuclear weapons for the "solution" of international conflicts. With the experience of Hiroshima and Nagasaki, we like to believe that man will have enough wisdom in the future to avoid the destruction of his own species in a nuclear holocaust.

General references

BOOKS, BROCHURES, AND PAMPHLETS
Atomic and nuclear physics

Atomic Power. 1955. Simon & Schuster, Inc., New York.

This is a collection of articles published in Scientific American, *in paperback form. Among others, the fol lowing titles should be mentioned:* Reactors, Fission products, The Atomic Energy Act, The lethal effects of radiation, Tracers, *and* The hydrogen bomb.

Barnes, D., R. Batchelor, and A. Maddock. 1962. Concise Encyclopedia of nuclear energy. Interscience Publishers, Inc., New York.

Frisch, O. R. 1958. The nuclear handbook. D. Van Nostrand Co., Inc., Princeton, N. J.

Glasstone, S. 1958. Sourcebook on atomic energy. Ed. 2. D. Van Nostrand Co., Inc., Princeton, N. J.

This is an outstanding concise encyclopedia of information in atomic, nuclear, and radiation physics. The style is scientific, and yet clear and readable, even for the student without any background in the field. Complex mathematical treatments are avoided. Essential as a reference for the radiation biology student.

Hogerton, J. F., editor. 1963. Atomic energy deskbook. Reinhold Publishing Corp., New York.

Lapp, R. E., and H. L. Andrews. 1954. Nuclear radiation physics. Prentice-Hall, Inc., Englewood Cliffs, N. J.

Another useful reference. Special feature in this work is the wealth of illustrative examples and problems.

Romer, A. 1960. The restless atom. Doubleday & Co., Inc., Garden City, N. Y.

This is an easy-to-read paperback, written for the general public.

Semat, H., and H. White. 1959. Atomic age physics. Holt, Reinhart & Winston, New York.

Discusses a variety of problems of modern physics (such as relativity and electronic energy levels) in a rather elementary manner.

Radionuclide methodology and nuclear instrumentation

Chase, G. D., and J. L. Rabinowitz. 1967. Principles of radioisotope methodology. Ed. 3. Burgess Publishing Co., Minneapolis, Minn.

A classical and successful laboratory manual. Each experiment is preceded by theory in details. Very useful diagrams and summaries, not frequently found in other sources. Essential for every chemist and biologist interested in radiotracer methods.

Lederer, C. M., J. M. Hollander, and I. Perlman. 1967. Table of isotopes. John Wiley & Sons, Inc., New York.

An up-to-date, self-contained, and readily usable compilation of nuclear data. Some of these (conversion electrons, decay schemes, etc.) are not readily available in similar works.

Overman, R. T., and H. M. Clark. 1960. Radioisotope techniques. McGraw-Hill Book Co., New York.

Price, W. J. 1955. Nuclear radiation detection. McGraw-Hill Book Co., New York.

This is perhaps the most comprehensive and accurate review of nuclear instrumentation. Useful to the student who wants to know more about characteristics and uses of the variety of instruments available today for radiation detection.

Wang, Y., editor. 1969. CRC handbook of radioactive nuclides. The Chemical Rubber Co., Cleveland, Ohio.

Biologic and medical uses of radionuclides

Abbot Laboratories. Technical Bulletins.

These brochures describe the techniques for the use of the radioactive chemicals and drugs for diagnostic and therapeutic purposes. They provide valuable information for the biologist who wishes to use the same chemicals in tracer experiments. The booklets can be obtained free by writing to Abbott Laboratories, North Chicago, Ill.

Beierwaltes, W. H., W. H. Johnson, and P. C. Solari. 1957. Clinical use of radioisotopes. W. B. Saunders Co., Philadelphia, Pa.

Blahd, W., F. Bauer, and R. Cassen. 1958. The practice of nuclear medicine. Charles C Thomas, Publisher, Springfield, Ill.

Broda, E. 1960. Radioactive isotopes in biochemistry. D. Van Nostrand Co., Inc., Princeton, N. J.

Comar, C. L. 1955. Radioisotopes in biology and agriculture. McGraw-Hill Book Co., New York.

Besides general techniques, the author discusses the type of research that can be done with each of several radionuclides and the special methods involved.

Delario, A. J. 1953. Roentgen, radium, and radioisotope therapy. Lea & Febiger, Philadelphia, Pa.

Fried, M., editor. 1962-1963. The use of radioisotopes in animal biology and the medical sciences. Academic Press, Inc., New York.

Kamen, M. 1958. Isotopic tracers in biology (an introduction to tracer methodology). Academic Press, Inc., New York.

A classical work written by one of the pioneers of the tracer method.

Owen, C. 1959. Diagnostic radioisotopes. Charles C Thomas, Publisher, Springfield, Ill.

Quimby, E. H., and S. Feitelberg. 1963. Radioactive isotopes in medicine and biology—physics and instrumentation. Lea & Febiger, Philadelphia, Pa.

Silver, S. 1962. Radioactive isotopes in medicine and biology—medicine. Lea & Febiger, Philadelphia, Pa.

Wagner, H. N., editor. 1968. Principles of nuclear medicine. W. B. Saunders Co., Philadelphia, Pa.

An up-to-date treatise of nuclear medicine.

Wang, C. H., and D. L. Willis. 1965. Radiotracer methodology in biological science. Prentice-Hall, Inc., Englewood Cliffs, N. J.

Many advanced biologic tracer experiments are described in this book with step-by-step procedures. Also the essentials on radionuclide theory, instrumentation, and general methods are covered.

Wolf, G. 1964. Isotopes in biology. Academic Press, Inc., New York.

Biologic effects of ionizing radiations (radiation biology)

Alexander, P. 1957. Atomic radiation and life. Penguin Books, Inc., Baltimore, Md.

A paperback written for the general public.

Andrews, H. L. 1961. Radiation biophysics. Prentice-Hall, Inc., Englewood Cliffs, N. J.

Augenstein, L. G., R. Mason, H. Quastler, and M. Zelle, editors. Advances in radiation biology. Academic Press, Inc., New York.

This is one of the many scholarly "Advances" published by Academic Press. In this series, three volumes have been published so far (1964, 1966, and 1969). As the title implies, reviews about current research and trends in radiobiology are written by leading investigators in the field.

Bacq, Z., and P. Alexander. 1961. Fundamentals of radiobiology Pergamon Press, Inc., New York.

A rather advanced, well-written textbook, useful to the graduate student.

Casarett, A. P. 1968. Radiation biology. Prentice-Hall, Inc., Englewood Cliffs, N. J.

Claus, W., editor. 1958. Radiation biology and medicine. Addison-Wesley Publishing Co., Inc., Reading, Mass.

Ellinger, F. 1957. Medical radiation biology. Charles C Thomas, Publisher, Springfield, Ill.

Errera, M., and A. Forssberg, editors. 1960–1961. Mechanisms in radiobiology. Academic Press, Inc., New York. Vol. I: General principles; Vol. II: Multicellular organisms.

This work may be considered as an advanced textbook of radiation biology.

Grosch, D. S. 1965. Biological effects of radiation. Blaisdell Publishing Co., New York.

A paperback designed for undergraduate and graduate students in biology.

Hollaender, A. 1954. Radiation biology. McGraw-Hill Book Co., New York.

This is one of the first works (in three volumes) published with this comprehensive title. It covers the biologic effects of ionizing radiation, as well as of visible and ultraviolet light (therefore it includes what today we would call more properly "photobiology"). Although written several years ago, it is still interesting to read for gaining an insight into the research problems and trends of the immediate postwar period.

Lea, D. E. 1955. Actions of radiations on living cells. Ed. 2, Cambridge University Press. Cambridge, England.

Douglas E. Lea is the radiobiologist who developed the target theory in the 1930s, when he also wrote this book for the first time. Many chapters of this classical work are devoted to the discussion of the theory. The book is still read and quoted by investigators today.

Pizzarello, D., and R. Witcofski. 1967. Basic radiation biology. Lea & Febiger, Philadelphia, Pa.

Zirkle, R. E., editor. 1954. Biological effects of X- and gamma radiation. McGraw-Hill Book Co., New York.

Miscellanea

Eisenbud, M. 1963. Environmental radioactivity. McGraw-Hill Book Co., New York.

Although the author concerns himself with the special problems of environmental radioactivity, I strongly recommend this book because it contains valuable information on dose problems, internal emitters, background radiation, and radioactive fallout.

Putman, J. L. 1960. Isotopes. Penguin Books, Inc., Baltimore, Md.

A paperback written for the general public, presenting information on radionuclides in general and on their applications in biology, medicine, agriculture, and industry.

Radiological health handbook. 1960. U. S. Department of Health, Education and Welfare, Division of Radiological Health. Washington, D. C.

This reference book contains data, tables, and graphs useful to the user of radionuclides and to the student of radiation biology. Decay schemes for all radionuclides are also included.

Shaw, E. 1965. Laboratory experiments in radiation biology. Ed. 2. U. S. Atomic Energy Commission, Division of Nuclear Education and Training. Washington, D. C.

Selected experiments on the properties of radionuclides; there are a few irradiation and tracer experiments.

Shilling, C. W., editor. 1964. Atomic energy encyclopedia in the life sciences. W. B. Saunders Co., Philadelphia, Pa.

A concise encyclopedia, useful for obtaining information quickly on topics of atomic and nuclear physics, radionuclides, radiation biology, etc.

"Understanding the Atom." U. S. Atomic Energy Commission, Division of Technical Information. Washington, D. C.

This is a series of booklets on a variety of topics related to atomic energy. Most of them are written in an informal style for the general educated public. They may be obtained free from the A. E. C.

PERIODICALS

Although radiation biology, radionuclide technology, and allied sciences are relatively new fields, the number of specialized journals that publish research papers in these sciences is already over two dozen. They are mainly published in the United States, Soviet Union, England, Germany, Japan, Switzerland, and Italy. The oldest of them were originally devoted to radiology, but at the present time they have expanded their coverage to include papers in other related fields. I am giving a list of most of them, starting with those that, in my opinion, are the most important and likely to be found in average university libraries. The year of publication of the first volume is also given.

American journal of roentgenology, radiotherapy, and nuclear medicine. T. Leucutia, editor. The American Roentgen Ray Society and the American Radium Society. Charles C Thomas, Publisher, Springfield, Ill. 1906. Monthly.

Health physics, the official journal of the American Health Physics Society. Pergamon Press, Inc., New York. 1958. Monthly.

International journal of applied radiation and isotopes. Pergamon Press, Inc., New York. 1956. Monthly.

International journal of radiation biology. With related studies in physics, chemistry, and medicine. W. M. Dale, editor. Taylor & Francis. London. 1959. Monthly.

Journal of nuclear medicine (Society of Nuclear Medicine). G. E. Thoma, editor. Samuel N. Turiel & Associates, Inc. Chicago. 1960. Monthly.

Nucleonics. McGraw-Hill Book Co. 1947. Monthly.

Radiation botany. A. H. Sparrow, editor. Pergamon Press, Inc. 1961. Quarterly.

Radiation research (official organ of the Radiation Research Society). Academic Press, Inc., New York. 1954. Monthly.

Other journals

British journal of radiology. British Institute of Radiology. London. 1896. Monthly.

Isotopes and radiation technology. U. S. Atomic Energy Commission, Washington, D. C. 1963. A quarterly technical progress review.

Journal of labeled compounds. J. Sirchis, editor. Presses académiques européenes, Brussels. 1965. Quarterly.

Journal of nuclear biology and medicine. Società Italiana di Biologia e Medicina Nucleare. L. Donato, editor. Minerva Medica, Turin. 1957. Quarterly.

Journal of radiation research. Official organ of the Japan Radiation Research Society. Tokyo. 1960. Quarterly.

Minerva fisico-nucleare (Giornale di fisica sanitaria e protezione contro le radiazioni). Minerva Medica, Turin. 1957. Quarterly.

Nuclear instruments and methods (a journal on accelerators, instrumentation and techniques in nuclear physics). K. Siegbahn, editor. North Holland Publishing Co. Amsterdam. 1956. Monthly.

Nuclear-Medizin (Isotope in Medizin und Biologie). Organ der Gesellschaft für Nuclear-Medizin. Stuttgart. 1959. Quarterly.

Nuclear safety. U. S. Atomic Energy Commission. Washington, D. C. 1959. A bimonthly technical progress review.

Polish review of radiology and nuclear medicine. Polish Radiological Association. Warsaw. 1936. Bimonthly.

Radiobiologia-Radiotherapia. Berlin. 1960. Bimonthly.

Radiobiologiya. Akademiya Nauk SSR. Moscow. 1961. Bimonthly.

Radioisotopes. Japan Radioisotope Association. Tokyo. 1952. Bimonthly.

Radiologia clinica et biologica (Internationl Radiological Review). Organ of the Swiss Association of Radiology, Nuclear Medicine, and Radiobiology. Basel. 1932. Bimonthly.

Radiology. A monthly journal devoted to clinical radiology and allied sciences. Radiological Society of North America. Easton, Pa. 1915.

Seminars in nuclear medicine. Grune & Stratton, Inc. New York. 1971. Quarterly.

At the present time, no periodical exists specifically devoted to the educational aspects and problems connected with the teaching of radiation biology and radionuclide methodology. Educators in these fields earnestly hope that this vacuum will be filled soon. The only periodical publication that somehow meets these needs is a four-page newsletter: *The Radiation Biology Institute Letter*, published by the American Institute of Biological Sciences, Washington, D. C. (1960). It contains news items, announcements, and short notes and articles of interest especially to teachers. The publication has been discontinued as of 1969.

Abstracts of articles and other publications in radiation biology and tracer methods applied to biologic research are found in *Biological Abstracts* in the appropriate sections. However, a more complete abstracting service in these and related fields is found in the following periodical:

Nuclear science abstracts. U. S. Atomic Energy Commission, Division of Technical Information. Washington, D. C. 1948. Bimonthly.

Each issue is divided into the following sections: chemistry, earth sciences, engineering, instrumentation (radiation detection instruments, radiation dosimeters, radiometric instruments, miscellaneous instruments and components, and radiation effects on instruments and instrument components), life sciences (biochemistry, physiology, and molecular biology, ecology, genetics and cytogenetics, health physics and safety, medicine, radiation effects on animals, radiation effects on plants, radiation effects on microorganisms, radiosterilization, and radiopreservation), metals, ceramics and other materials, physics (general, high-energy, and nuclear), reactor technology, and general science.

Each issue has an author index (corporate and personal) and a subject index. Quarterly, semiannual, and annual indexes are also issued.

Some useful abbreviations and symbols

A	mass number
ABCC	Atomic Bomb Casualty Commission
AC	alternating current
AEC	Atomic Energy Commission
AET	S-(2-aminoethyl)isothiuronium
amu	atomic mass unit, dalton
BG (or BKG)	background radiation
BUdR	bromodeoxyuridine
CF	carrier-free
cgs	centimeter-gram-second
Ci	curie
cpm	counts per minute
cps	counts per second
DC	direct current
DF	discrimination factor
dpm	disintegrations per minute
dps	disintegrations per second
DRF	dose reduction factor
E_b	binding energy
EC	electron capture
E_{max}	maximum energy (of beta particles)
ESR	electron spin resonance
esu	electrostatic unit
eV	electron volt
GeV	giga electron volt, billion electron volt
GSH	reduced glutathione
5-HT	5-hydroxytryptamine (serotonin)
HVL	half-value layer
IAA	iodoacetic acid
IAAM	iodoacetamide
IAEA	International Atomic Energy Agency
ICRP	International Commission on Radiological Protection
ICRU	International Commission on Radiological Units
ip	ion pair
IT	isomeric transition
KE	kinetic energy

keV	kilo electron volt
kV	kilovolt
kw	kilowatt
LD	lethal dose
LD_{50}	median lethal dose
LET	linear energy transfer
MEA	β-mercaptoethylamine (cysteamine)
MEG	2-mercaptoethylguanidine
MeV	mega electron volt, million electron volt
mks	meter-kilogram-second
MPB	maximum permissible body burden
MPC	maximum permissible concentration
MPE	maximum permissible exposure
MST	mean survival time
N	neutron number
NBS	National Bureau of Standards
NCRP	National Council for Radiation Protection
NEM	N-ethylmaleimide
PAPP	*para*-aminopropiophenone
POPOP	4-bis-2(5-phenyloxazolyl)benzene
PPO	2,5-diphenyloxazole
R	roentgen
rad	radiation absorbed dose
RBE	relative biologic effectiveness
rem	roentgen-equivalent mammal (or man)
SA	specific activity
T (or $T_{1/2}$)	half-life
T_b	biologic half-life
T_e	effective half-life
T_p	physical half-life
TLD	thermoluminescent dosimetry
TOD	target-to-object distance
USP	United States Pharmacopeia
WBE	whole-body exposure
Z	atomic number

Main characteristics of radionuclides of biologic importance

The main characteristics of, and further information and data about, a limited number of radionuclides that are frequently encountered in radiation biology and biologic radiotracer methodology are presented in this appendix. The radionuclides are arranged in alphabetical order of their chemical symbols. The chemical symbol, with *A*, *N*, and *Z* numbers, is shown in the upper right corner. Atomic mass values are rounded to the fourth decimal digit. They have been taken from the *Radiological Health Handbook* (1960. U. S. Department of Health, Education, and Welfare, Division of Radiological Health, Washington, D. C.).

The physical half-lives are those reported in the *Chart of the Nuclides* (1966) published by Knolls Atomic Power Laboratory (operated by the General Electric Company). The biologic and effective half-lives have been taken from the "Report of Committee II on permissible dose for internal radiation," Table 12, published in *Health Physics* **3**:154–230, 1960. The organ of reference is shown in parentheses. The decay schemes have been simplified and adapted from those published in the *Radiological Health Handbook* and in the *Table of Isotopes* (Lederer, C. M., J. M. Hollander, and I. Perlman. 1967. John Wiley & Sons, Inc. New York). Refer to Chapter 5 for their correct interpretation. The energy values are in MeV.

For gamma emitters, gamma scintillation spectra are also shown. They have been redrawn, in a simplified form, from *Scintillation Spectrometry and Gamma-ray Spectrum Catalogue* (Heath, R. L. 1964. Ed. 2. U. S. Department of Commerce, Washington, D. C.). The spectra represent the responses of a $3'' \times 3''$ NaI(Tl) detector to sources placed at a standard distance of 10 cm, with a beryllium absorber between source and crystal. The sources were prepared on card mounts. Energy (in MeV or keV), rather than pulse height, has been plotted as abscissa in the spectra presented here, in order to facilitate the identification of the energy values corresponding to the different peaks of a spectrum. Counts per channel are plotted as ordinates on a log scale; no specific values are given because they depend, of course, on the activity of the source. The intensity of the high-energy portion of some spectra has been increased by a factor of 10 or 100 for a better display of important spectral characteristics. Photopeaks are labeled with their energies (in MeV). Many satellite peaks superimposed on the Compton smear are also labeled with reference to their origin.

When an artificial radionuclide is produced in a reactor or in a cyclotron, the nuclear reaction used for its preparation is shown with a short-form nuclear equation (see the appendix to Chapter 6).

The information on the chemical form in which the radionuclides are available has been drawn mostly from the catalogues of leading commercial suppliers of radioactive materials.

For the definition of critical organ, see p. 253. The information on critical organs presented in the tables has been taken from the NBS Handbook 69, *Maximum permissible body burdens and maximum permissible concentrations of radionuclides in air and in water for occupational exposures* (1959. U. S. Department of Commerce, Washington, D. C.). The

MPB (maximum permissible body burden) values reported in the tables are those published in the same NBS handbook. The MPB is the amount of radionuclide that, when deposited in the total body, produces the maximum permissible exposure (MPE) to the organ shown in parentheses after the MPB value. This organ is usually the critical organ.

For each radionuclide, a summary of its biologic and medical importance is also given, whenever pertinent. Radiobiologic aspects, uses in radiotracer methodology, and diagnostic and therapeutic applications are listed. More details are found in the appropriate chapters of the text.

Gold
Atomic mass: 198.0305
Physical half-life: 64.8 hours
Biologic half-life: 120 days (total body)
Effective half-life: 2.6 days (total body)

$$^{198}_{79}\text{Au}_{119}$$

Decay scheme

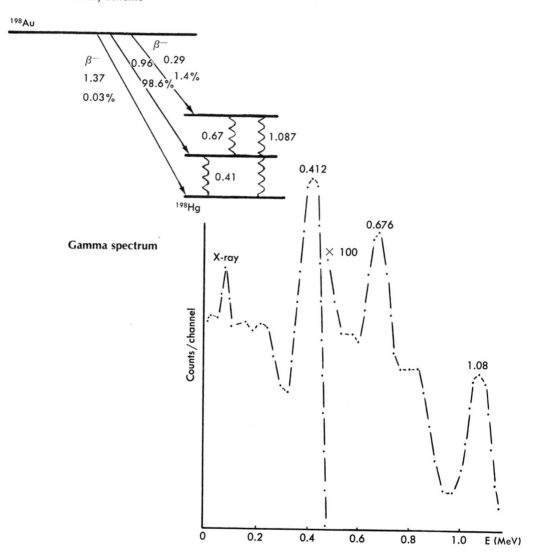

Gamma spectrum

Source: ^{197}Au (n,γ) or ^{198}Pt (p,n)

Available form: chloride; colloidal suspension; metal in the form of needles, seeds, wires, etc.

Critical organ: gastrointestinal tract

MPB: 30 μCi (total body)

Medical importance:
In diagnosis, bone marrow scanning
In therapy, in colloidal form (Aurcolloid) for pleural and peritoneal malignant effusions, carcinoma of the prostate, and cervix uteri; as seeds or wires for interstitial implantation in tumors

Carbon

Atomic mass: 14.0077
Physical half-life: 5730 years
Biologic half-life: 10 days (total body)
Effective half-life: 10 days (total body)

$$^{14}_{6}C_{8}$$

Decay scheme

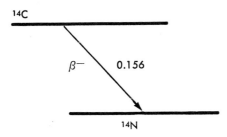

Source: ^{14}N (n,p); can be obtained with a maximum specific activity of 2 Ci/g of C;
also naturally occurring

Available form: soluble or insoluble carbonate (for example, Na_2CO_3 or $BaCO_3$) from
which $^{14}CO_2$ can be easily obtained by treatment with an acid

Critical organ: fat

MPB: 300 μCi (fat)

Biologic importance:

Carbon 14 is one of the natural radionuclides normally present in every living organism. It originates in the upper atmosphere through a nuclear reaction between the cosmic radiation neutrons and nitrogen. More has been produced in atom bomb testing by
radioactivation of the atmospheric nitrogen by fission neutrons. It exists in the environment mainly as carbon dioxide; in this form it is utilized by plants and thus becomes
incorporated in all organisms through the various food chains. The normal body burden
for an adult man is about 0.1 μCi. It is used to label, by chemical synthesis or biosynthesis,
a very large variety of organic molecules for a wide range of radiotracer experiments
(in physiology, biochemistry, cytology, molecular biology, ecology, etc.)

Calcium
Atomic mass: 44.9703
Physical half-life: 163 days
Biologic half-life: 1.64×10^4 days (total body and bone)
Effective half-life: 162 days (total body and bone)

$_{20}^{45}\text{Ca}_{25}$

Decay scheme

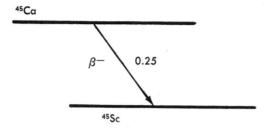

^{45}Ca

$\beta-$ 0.25

^{45}Sc

Source: ^{44}Ca (n,γ)

Available form: chloride or carbonate

Critical organ: bone

MPB: 30 μCi (bone)

Biologic importance:
   Used as a tracer in studies on calcium metabolism in bones and teeth, on differential intestinal absorption and utilization in bones of calcium and strontium, and for the determination of endogenous calcium as affected by diet, hormones, exercise, etc.

Cerium
Atomic mass: 143.9584
Physical half-life: 285 days
Biologic half-life: 563 days (total body)
                    1500 days (bone)
Effective half-life: 191 days (total body)
                     243 days (bone)

$^{144}_{58}Ce_{86}$

Decay scheme

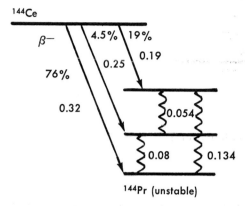

^{144}Ce

β⁻

4.5%   19%

0.25   0.19

76%

0.32

0.054

0.08   0.134

^{144}Pr (unstable)

Gamma spectrum of ^{144}Ce-^{144}Pr

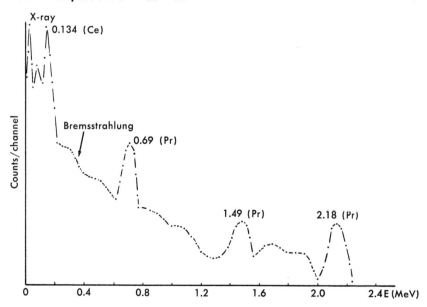

Source: fission product

Available form: chloride

Critical organs: bone, gastrointestinal tract, liver

MPB: 5 μCi (bone)

Biologic importance:
    Used as a tracer in studies of the toxicity of rare earths. It has also been used to tag insects because it appears to bind to tissues and is therefore retained well in the body.

Chlorine
Atomic mass: 35.9797
Physical half-life: 3×10^5 years
Biologic half-life: 29 days (total body)
Effective half-life: 29 days (total body)

$$^{36}_{17}\text{Cl}_{19}$$

Decay scheme

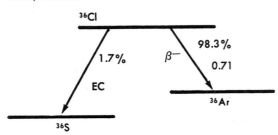

Source: $^{35}\text{Cl}\,(n,\gamma)$

Available form: hydrochloric acid or sodium chloride

Critical organ: total body

MPB: 80 μCi (total body)

Biologic importance:
  Used as a tracer for the determination of chloride in biologic fluids (serum, spinal fluid, etc.) and for the measurement of chloride "space" and total exchanging chloride in man.

Cobalt
Atomic mass: 56.9540
Physical half-life: 272 days
Biologic half-life: 9.5 days (total body)
Effective half-life: 9.2 days (total body)

$$_{27}^{57}\text{Co}_{30}$$

Decay scheme

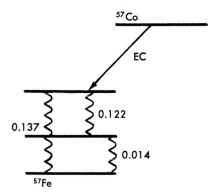

Gamma spectrum

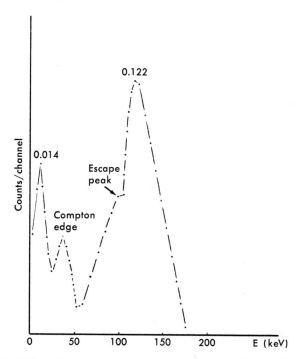

Source: ^{58}Ni (γ,p) ^{56}Fe (p,γ) ^{56}Fe (d,n) ^{55}Mn $(\alpha,2n)$

Available form: chloride

Critical organ: gastrointestinal tract

MPB: 200 μCi (total body)

Biologic and medical importance:
Used to label vitamin B_{12} in physiologic studies of this vitamin. The labeled vitamin is also used for the diagnosis of pernicious anemia (Schilling test).

Cobalt
Atomic mass: 59.9529
Physical half-life: 5.24 years
Biologic half-life: 9.5 days (total body)
Effective half-life: 9.5 days (total body)

$$_{27}^{60}\text{Co}_{33}$$

Decay scheme

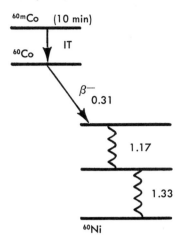

Gamma spectrum

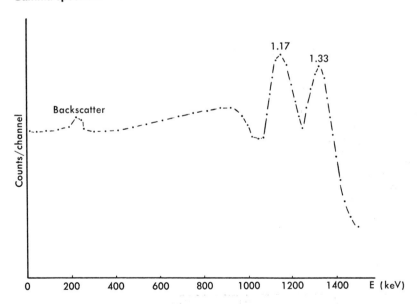

Source: ^{59}Co (n,γ)

Available form: chloride, acetate, or as metal (for teletherapy units)

Critical organs: gastrointestinal tract, total body

MPB: 10 μCi (total body)

Biologic and medical importance:
   Used to label vitamin B_{12} in physiologic studies of this vitamin. The labeled vitamin is also used for the diagnosis of pernicious anemia.
   In therapy, cobalt 60 is used in teletherapy units as a source of high-energy gamma radiation (with activities of hundreds or thousands of curies). In the form of wires or needles (50 to 75 mCi), it is used for intracavitary and interstitial therapy of tumors.

Chromium
Atomic mass: 50.9608
Physical half-life: 27.8 days
Biologic half-life: 616 days (total body)
Effective half-life: 26.6 days (total body)

$$^{51}_{24}\text{Cr}_{27}$$

Decay scheme

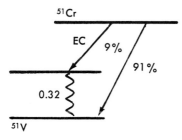

Gamma spectrum

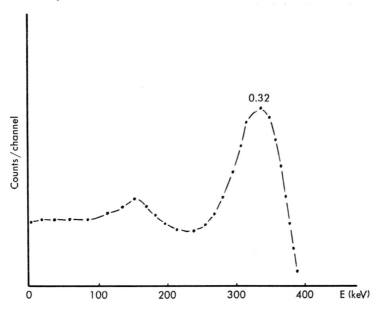

Source: ^{50}Cr (n, γ)

Available form: sodium chromate, chromium chloride

Critical organs: gastrointestinal tract, total body

MPB: 800 μCi (total body)

Medical importance
  In diagnosis:
    As sodium chromate, used to tag red cells for the determination of red cell mass and for studies of red cell survival time; also used in spleen-scanning tests.
    As chromium chloride, used to tag plasma proteins for the determination of blood plasma volume; also used for the determination of cardiac output.

Cesium
Atomic mass: 136.9509
Physical half-life: 30 years
Biologic half-life: 70 days (total body)
                   140 days (muscle)
Effective half-life: 70 days (total body)
                     138 days (muscle)

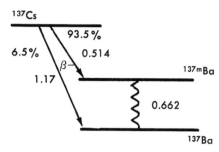

$^{137}_{55}\text{Cs}_{82}$

Decay scheme

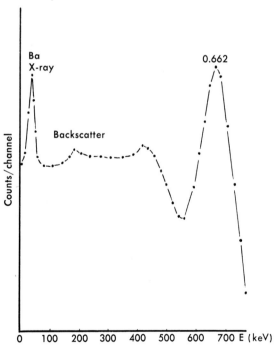

Gamma spectrum

Source: fission product

Available form: chloride, nitrate

Critical organs: total body, liver, spleen, muscle

MPB: 30 μCi (total body)

Biologic and medical importance:
    One of the long-lived fission products and therefore present in local and global fallout. Normally present in the human body since the beginning of the atomic age. Introduced mainly with milk and meat. Easily detected by whole-body counters. It is one of the two significant internal gamma emitters (the other is ^{40}K).
    In therapy, used (less frequently than cobalt 60) in teletherapy units as a source of moderately penetrating gamma radiation for the treatment of shallow tumors.

Iron
Atomic mass: 54.9556
Physical half-life: 2.4 years
Biologic half-life: 800 days (total body)
Effective half-life: 437 days (total body)

$$_{26}^{55}\text{Fe}_{29}$$

Decay scheme

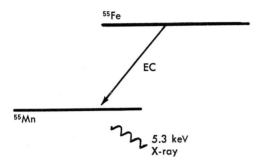

Source: ^{54}Fe (n, γ)

Available form: ferrous sulphate or citrate, and ferric chloride

Critical organ: spleen

MPB: 1000 μCi (spleen)

Biologic importance:
    Used as tracer in iron metabolism and hemopoietic studies. Its use is limited by its very low energy X-ray, which is the only radiation emitted in the decay process.

Iron

Atomic mass: 58.9536
Physical half-life: 45.3 days
Biologic half-life: 800 days (total body)
Effective half-life: 42.7 days (total body)

$$^{59}_{26}\text{Fe}_{33}$$

Decay scheme

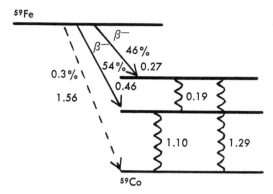

Gamma spectrum

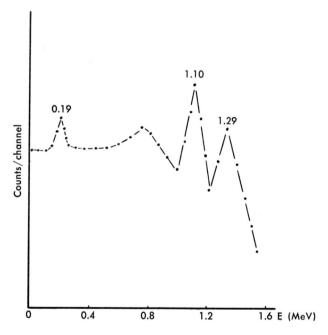

Source: ^{58}Fe (n, γ)

Available form: ferrous sulphate or citrate, and ferric chloride

Critical organs: gastrointestinal tract, spleen

MPB: 20 μCi (spleen)

Biologic and medical importance:

Used as a tracer in studies of plant metabolism, chlorosis, etc. and for the measurement of absorption of iron by the gastrointestinal tract. Diagnostic tests include hemopoietic studies, determination of plasma iron turnover and disappearance, and erythrocyte iron turnover.

Hydrogen
Atomic mass: 3.0170
Physical half-life: 12.26 years
Biologic half-life: 12 days (total body)
Effective half-life: 12 days (total body)

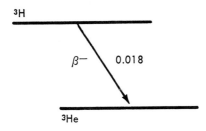

Decay scheme

Source: Li (n,α)

Available form: free hydrogen gas, tritiated water, and as a label for a large number of organic compounds

Critical organ: total body

MPB: 2000 μCi (total body)

Biologic importance:
Traces of tritium are always naturally present in hydrogen. This nuclide originates in the atmosphere through several interactions between cosmic rays and nitrogen. It is also produced and released into the atmosphere by thermonuclear explosions. It is one of the internal emitters normally present in the human body, although the body burden is very small.

Tritium is very extensively used in biologic research as a label for organic molecules, especially when the detecting methods used are liquid scintillation counting and radioautography.

In diagnosis, tritiated water is used for the determination of the total body water.

Mercury
Atomic mass: 203.0365
Physical half-life: 46.6 days
Biologic half-life: 10 days (total body)
Effective half-life: 8.2 days (total body)

$$^{203}_{80}Hg_{123}$$

Decay scheme

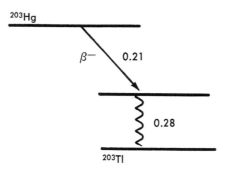

Gamma spectrum

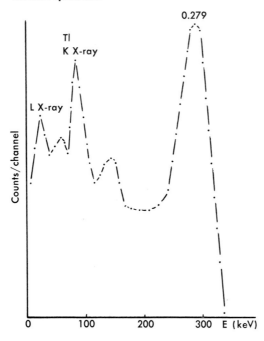

Source: $^{202}Hg\ (n,\gamma)$

Available form: mercuric chloride, nitrate, and acetate

Critical organ: kidney

MPB: 4 μCi (kidney)

Medical importance:

Used at times for brain scanning (localization of tumors). ^{203}Hg-chlormerodrin is used for kidney function studies.

Iodine
Atomic mass: 124.9442
Physical half-life: 60 days
Biologic half-life: 138 days (total body)
Effective half-life: 42 days (total body)

$$125 \;|$$
$$53 \;|\; 72$$

Decay scheme

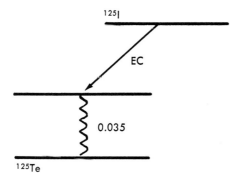

Source: ^{123}Sb $(\alpha, 2n)$

Available form: elemental iodine, sodium iodide, and as a label of some organic compounds

Critical organ: thyroid

MPB: (data not available)

Medical importance:
In diagnosis, it is used for thyroid scanning. This radionuclide is also used as a gamma source in gamma radiography.

Iodine
Atomic mass: 130.9476
Physical half-life: 8.05 days
Biologic half-life: 138 days (total body)
Effective half-life: 7.6 days (total body)

131

53   78

Decay scheme

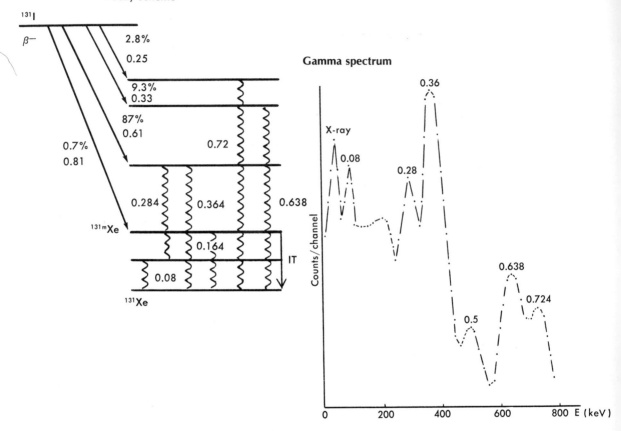

Source: fission product, also produced in cyclotrons with the ^{130}Te (d, n) reaction

Available form: elemental iodine, sodium iodide, and as label for many organic compounds (iodoacetic acid, iodotyrosine, insulin serum albumin, etc.)

Critical organ: thyroid

MPB: 0.7 μCi (thyroid)

Biologic and medical importance:
  Widely used in thyroid physiologic studies
  In diagnosis:
    As sodium iodide for thyroid uptake tests (1 to 50 μCi), for the determination of protein-bound iodine (PBI test, 25 to 100 μCi), for thyroid scintiscanning (100 to 300 μCi)
    As triiodothyronine for the determination of red cell uptake and for in vitro studies of thyroid function
    As iodinated serum albumin for plasma volume determination (3 to 20 μCi), localization of brain tumors (300 to 500 μCi) and determination of cardiac output (10 to 30 μCi)
    As sodium iodohippurate for kidney function studies
    As a label in rose bengal for liver function tests
  In therapy, for the cure of certain forms of hyperthyroidism, of thyroid neoplasms, and of congestive heart failure (used with doses of several mCi)

Potassium

Atomic mass: 39.9766
Physical half-life: 1.3×10^9 years
Biologic half-life: 58 days (total body)
Effective half-life: 58 days (total body)

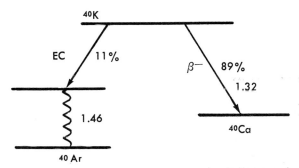

Decay scheme

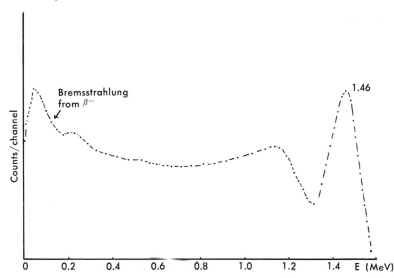

Gamma spectrum

Source: One of the three natural isotopes of potassium; its natural abundance is 0.0118%

Available form: any potassium compound; maximum activity of ^{40}K per gram of compound is found in KOH, KF, and KCl

Biologic importance:

Since potassium is an essential element for practically every living protoplasm and organism, ^{40}K is found in every cell but especially in muscles, and its concentration is higher within the cells than in the extracellular fluids. An adult 70 kg man contains 1.65×10^{-2} g of this radionuclide, corresponding to a body burden of about 0.1 μCi. ^{40}K is one of the internal gamma emitters, and the radiation dose it delivers to the body (about 20 mrems/year) is greater than that delivered by any other single internal emitter. The amount of ^{40}K in the human body is easily detected and measured with whole-body counters. This measurement is often made for the diagnosis of certain muscular diseases.

Potassium
Atomic mass: 41.9758
Physical half-life: 12.42 hours
Biologic half-life: 58 days (total body)
Effective half-life: 0.52 day (total body)

$$^{42}_{19}\text{K}_{23}$$

Decay scheme

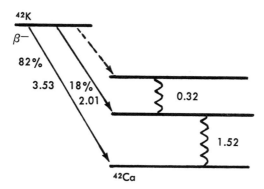

Gamma spectrum

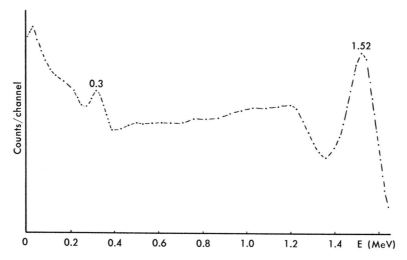

Source: ^{41}K (n,γ)

Available form: potassium carbonate, isotonic potassium chloride solution

Critical organ: gastrointestinal tract

MPB: 10 μCi (total body)

Biologic importance:
   Used as a tracer in physiologic research for the study of potassium excretion, plasma membrane permeability, active transport, muscular physiology, potassium uptake in plants, etc.

Sodium
Atomic mass: 22.0014
Physical half-life: 2.6 years
Biologic half-life: 11 days (total body)
Effective half-life: 11 days (total body)

$$^{22}_{11}Na_{11}$$

Decay scheme

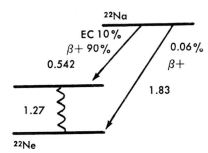

Gamma spectrum

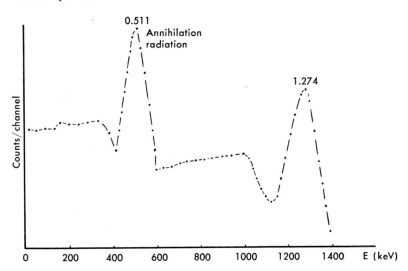

Source: $^{19}F\ (\alpha,n)$ $^{24}Mg\ (d,\alpha)$

Available form: carbonate, bicarbonate, and chloride

Critical organ: total body

MPB: 10 μCi (total body)

Biologic importance:
    Used as a tracer for the study of the metabolism and distribution of sodium, of active transport, etc.

Sodium
Atomic mass: 23.9985
Physical half-life: 15 hours
Biologic half-life: 11 days (total body)
Effective half-life: 0.6 day (total body)

$^{24}_{11}Na_{13}$

Decay scheme

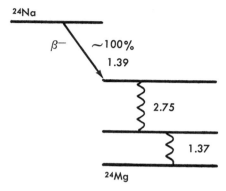

Gamma spectrum

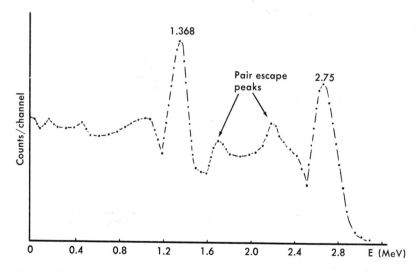

Source: ^{23}Na (n,γ)

Available form: carbonate, bicarbonate, and chloride

Critical organ: gastrointestinal tract

MPB: 7 μCi (total body)

Biologic and medical importance:

Large amounts may be produced in seawater from the underwater detonation of nuclear devices. Also formed in significant amounts in the body of victims of neutron radiation accidents (such as nuclear excursion accidents).

Used as a tracer in plasma-membrane and active-transport studies.

In diagnosis, used to measure blood circulation time and efficiency and for the detection of certain types of vascular disorders.

Phosphorus

Atomic mass: 31.9840

Physical half-life: 14.3 days

Biologic half-life: 257 days (total body)

1155 days (bone)

Effective half-life: 13.5 days (total body)

14.1 days (bone)

$$^{32}_{15}P_{17}$$

Decay scheme

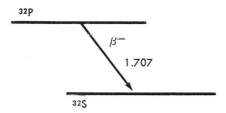

Source: ^{31}P (n,γ) ^{32}S (n,p) ^{34}S (d,α)

Available form: phosphoric acids, sodium or potassium phosphates, phosphorus chloride, and as label in many biologically important organic molecules (such as ATP)

Critical organ: bone

MPB: 6 μCi (bone)

Biologic and medical importance:

Used as tracer in nucleic acid and virus studies, in physiology of bony tissues, and in plant physiology (uptake from soil by plants).

In diagnosis, used for the localization of skin cancers, ocular tumors, and brain tumors exposed by surgery.

In therapy, used as sodium phosphate for the palliative treatment of polycythemia vera, chronic leukemia (lymphatic or myeloid), metastatic bone cancer, primary hemorrhagic thrombocythemia; as colloidal chromic phosphate, used for the treatment of malignant effusions (pleural or peritoneal) and of carcinoma of the prostate.

Polonium
Atomic mass: 210.0485
Physical half-life: 138.4 days
Biologic half-life: 30 days (total body)
                 70 days (kidney)
Effective half-life: 25 days (total body)
                  46 days (kidney)

$$^{210}_{84}\text{Po}_{126}$$

Decay scheme

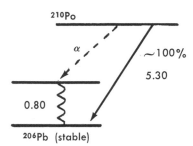

Source: naturally occurring; daughter of ^{210}Bi, which may also be artificially produced
    with the ^{209}Bi (n, γ) reaction

Available form: nitrate

Critical organs: spleen and kidneys

MPB: 0.03 μCi (spleen)

Biologic importance:
    Very hazardous as an internal emitter like most internal alpha emitters. Recently
its presence has been suspected in cigarette smoke; possible contributing cause of lung
cancer induced by cigarette smoking.

Praseodymium
Atomic mass: 143.9581
Physical half-life: 17.27 minutes
Biologic half-life: 750 days (total body)
                    1500 days (bone)
Effective half-life: 17 minutes

$^{144}_{59}Pr_{85}$

Decay scheme

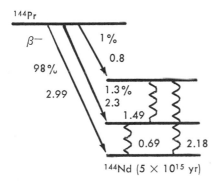

See gamma spectrum of ^{144}Ce

Source: short-lived daughter of ^{144}Ce, which is a fission product. Always found associated with ^{144}Ce. The mixture ^{144}Ce-^{144}Pr accounts for three fourths of the beta activity and one third of the gamma activity of the fission products during the first year after their formation.

Plutonium
Atomic mass: 239.1270
Physical half-life: 24,360 years
Biologic half-life: 6.5×10^4 days (total body)
Effective half-life: 6.4×10^4 days (total body)

$^{239}_{94}\text{Pu}_{145}$

Decay scheme

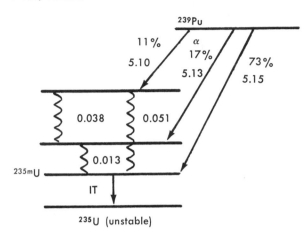

Source: transuranic element, daughter of ^{239}Np; produced in nuclear reactors with the reaction

$$^{238}\text{U (n,}\gamma) \ ^{239}\text{U} \ \xrightarrow{\beta^-} \ ^{239}\text{Np} \ \xrightarrow{\beta^-} \ ^{239}\text{Pu}$$

Available form: plutonium dioxide or in nitric acid solution

Critical organ: bone

MPB: 0.04 μCi (bone)

Biologic importance:
   Extremely hazardous as an internal emitter, because of its alpha emission and because of its very long effective half-life.
   After some accidents of ingestion or inhalation of plutonium by human subjects, studies are underway in several laboratories to learn, with animal experiments, about the distribution and localization of this nuclide in various organs (liver, bone, placenta, etc.) and about the possibility of removing it from the body with chelating agents.

Radium

Atomic mass: 226.0254
Physical half-life: 1620 years
Biologic half-life: 900 days (total body)
                            1.6×10^4 days (bone)
Effective half-life: 900 days (total body)
                            1.6×10^4 days (bone)

$$^{226}_{88}Ra_{138}$$

Decay scheme

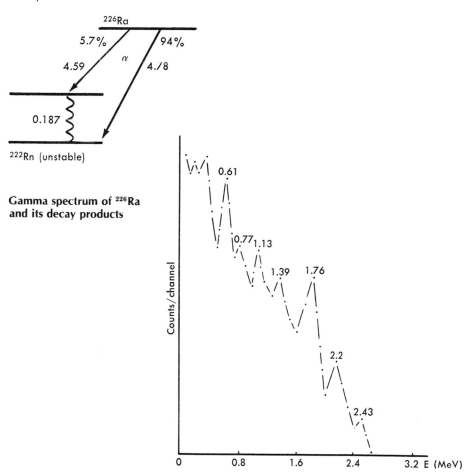

Gamma spectrum of ^{226}Ra and its decay products

Source: naturally occurring radionuclide (daughter of ^{230}Th)

Available form: metal and chloride

Critical organ: bone

MPB: 0.1 μCi (bone)

Biologic and medical importance:

Being chemically similar to calcium, radium follows calcium in the food chain leading to man. As such, it is another internal emitter that has always been part of the total body burden of the human population. It enters the body through water and a variety of foods. All natural waters contain at least small amounts of this element. The average radium body burden for the general adult population is estimated at 1.6×10^{-4} μCi, and it is mostly concentrated in the bones. Accidental ingestion of radium is extremely hazardous because of its long effective half-life in the bones and because of its alpha emission.

In therapy, used for the treatment of certain neoplastic diseases.

Sulfur
Atomic mass: 34.9800
Physical half-life: 86.7 days
Biologic half-life: 90 days (total body)
1530 days (skin)
Effective half-life: 44.3 days (total body)
82.4 days (skin)

$$^{35}_{16}S_{19}$$

Decay scheme

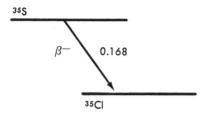

Source: ^{34}S (n,γ) ^{37}Cl (d,α)

Available form: mainly as sulfuric acid and sodium sulfate; also in elemental form; as a label in many organic compounds (including amino acids and certain proteins)

Critical organs: skin, testis

MPB: 90 μCi (testis)

Biologic importance:
Used as a tracer for a variety of physiologic investigations; for example, distribution and excretion of sulphates in rats, metabolism of methionine, utilization of sulfate by laying hens and its incorporation into cystine, metabolism of mucopolysaccharides in cartilage, penicillin studies, and sulfur metabolism of wheat, barley, and corn.

Strontium
Atomic mass: 84.9400
Physical half-life: 65 days
Biologic half-life: 1.8×10^4 days (bone)
Effective half-life: 64.8 days (bone)

$$^{85}_{38}\text{Sr}_{47}$$

Decay scheme

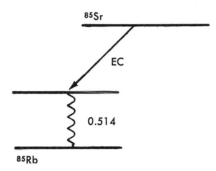

Gamma spectrum

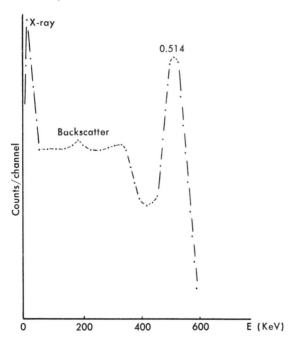

Source: $^{84}\text{Sr} (n,\gamma)$ $^{85}\text{Rb} (d,2n)$ $^{85}\text{Rb} (p,n)$

Available form: chloride

Critical organ: total body

MPB: 60 μCi (total body)

Biologic and medical importance:
    Used as a tracer in studies of bone metabolism and physiology, in experiments to investigate the differential intestinal absorption and deposition of Sr and Ca in bones.
    Also used for the diagnosis of certain bone diseases, including metastatic cancer.

Strontium
Atomic mass: 89.9357
Physical half-life: 28.8 years
Biologic half-life: 1.8×10^4 days (bone)
Effective half-life: 6.4×10^3 days (bone)

$$_{38}^{90}\text{Sr}_{52}$$

Decay scheme

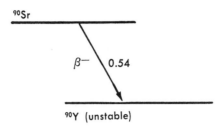

Source: fission product

Available form: nitrate

Critical organ: bone

MPB: 2 μCi (bone)

Biologic and medical importance:

This radionuclide, being a long-lived fission product, is a regular component of local and global radioactive fallout. It is slowly deposited on the surface of the soil and plants over a long time after a nuclear explosion. Hence ^{90}Sr is another internal emitter that has become part of the total body burden of the general population since the beginning of the nuclear age. It reaches the human body primarily through the consumption of dairy products and foods of plant origin. Like radium, strontium is a bone-seeker and, of course, is readily assimilated in the bones and teeth of young individuals. The turnover of ^{90}Sr in those organs is very slow because of its long effective half-life. This justifies the concern about the internal hazard created by this radionuclide.

In therapy, ^{90}Sr is used as a beta source in special applicators for the treatment of certain diseases of the eye, particularly of the cornea and epibulbar regions.

Technetium
Atomic mass: 98.9385
Physical half-life: 6 hours
Biologic half-life: 1 day (total body)
Effective half-life: 0.2 day (total body)

$$^{99m}_{43}Tc_{56}$$

Decay scheme

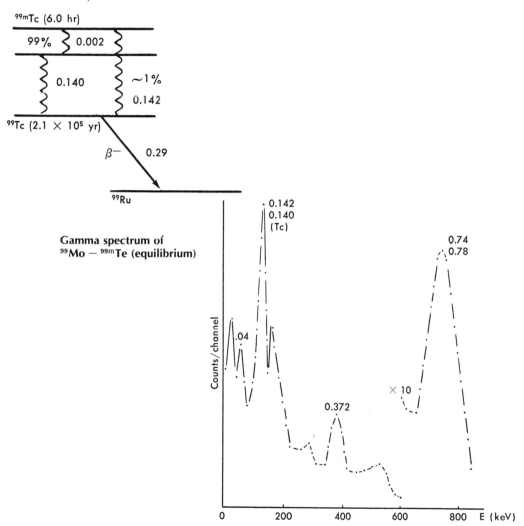

Gamma spectrum of
^{99}**Mo** — ^{99m}**Te (equilibrium)**

Source: daughter of ^{99}Mo, which is a fission product

Available form: pertechnetate; ^{99m}Tc can be obtained by elution of an ion-exchange resin column on which the parent nuclide is adsorbed ("radioactive cow")

Critical organ: gastrointestinal tract

MPB: 200 μCi (total body)

Medical importance:

Recently, technetium 99m has become increasingly popular in diagnosis for scanning tests of several organs. Because of its extremely short half-life and the complete lack of primary particle radiation, the exposure dose to the patient is extremely low. Furthermore, because of the rapid uptake of the radionuclide, the scanning study can be done almost immediately after administration with considerable time saving. The gamma energy (140 keV) seems to be ideal for good resolution on the scintiscans.

Used especially in thyroid and liver studies and for the localization of brain tumors.

Uranium
Atomic mass: 235.1175
Physical half-life: 7.13×10^8 years
Biologic half-life: 100 days (total body)
                    15 days (kidney)
Effective half-life: 100 days (total body)
                     15 days (kidney)

$${}^{235}_{92}U_{143}$$

Decay scheme

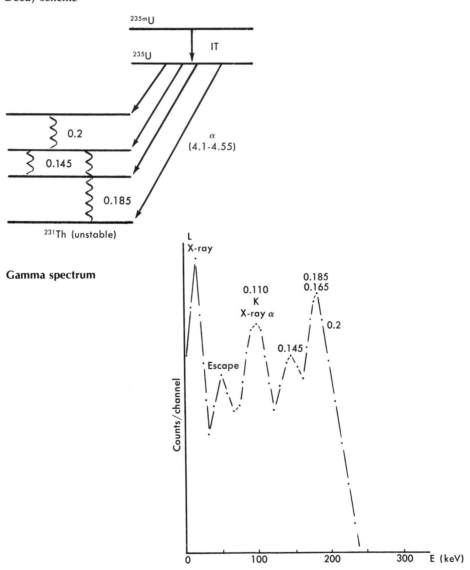

Gamma spectrum

Source: one of the naturally occurring isotopes of uranium (natural abundance is 0.7%); also daughter of ^{239}Pu

Available form: uranium dioxide or metal foils

Critical organs: bone, gastrointestinal tract, kidneys

MPB: 0.03 μCi (kidney)

Importance:
    Uranium 235 is fissionable with thermal as well as high-energy neutrons; used as fissionable material in nuclear reactors and in nuclear explosives.

Uranium

Atomic mass: 238.1252
Physical half-life: 4.51×10^9 years
Biologic half-life: 100 days (total body)
              15 days (kidney)
Effective half-life: 100 days (total body)
              15 days (kidney)

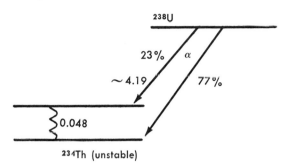

Decay scheme

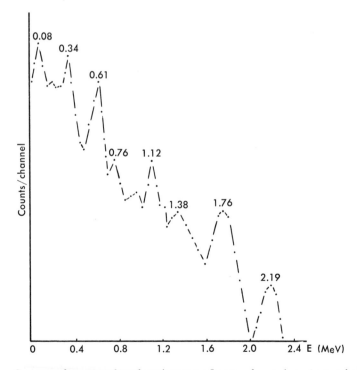

Gamma spectrum of uranium ore

Source: the most abundant isotope of natural uranium (natural abundance is 99.2%)

Available in every uranium compound

Critical organs: gastrointestinal tract, kidney

MPB: 0.005 μCi (kidney)

Importance:
    Fissionable by fast neutrons. Used for the production of plutonium.

Yttrium
Atomic mass: 89.9352
Physical half-life: 64.2 hours
Biologic half-life: 1.4×10^4 days (bone)
Effective half-life: 64 hours (bone)

$^{90}_{39}Y_{51}$

Decay scheme

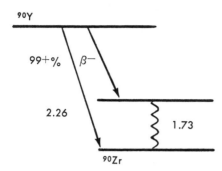

Source: daughter of ^{90}Sr ^{89}Y (n,γ)

Available form: chloride and oxide

Critical organ: gastrointestinal tract

MPB: 3 μCi (bone)

Medical importance:
    In therapy, pellets of yttrium oxide are introduced into bones near the pituitary to destroy the gland.

Zinc

Atomic mass: 64.9498
Physical half-life: 243 days
Biologic half-life: 933 days (total body)
Effective half-life: 194 days (total body)

$$_{30}^{65}\text{Zn}_{35}$$

Decay scheme

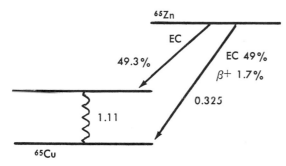

Gamma spectrum

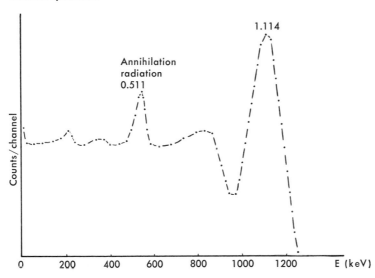

Source: ^{64}Zn (n,γ) ^{65}Cu (d,2n)

Available form: chloride

Critical organs: total body, prostate, liver.

MPB: 60 μCi (total body)

Biologic importance:
   Used as a tracer in a large variety of studies of zinc metabolism in plants and animals.

Author index

Cloutier, R. J. 296
Cole, L. J. 452, 464
Comar, C. L. 296, 488
Comas, F. 439
Compton, A. H. 102, 197
Congdon, C. C. 439
Conklin, J. W. 484
Coolidge, W. D. 6, 49
Coppenger, C. J. 402, 405
Cork, J. M. 478, 484
Corliss, W. 116
Corrigan, K. E. 75
Crafts, A. S. 188
Crawford, M. K. 465
Cronkite, E. P. 414, 432, 439
Crouthamel, C. E. 187
Crowther, J. A. 6, 364
Curie, M. 4, 76, 128
Curie, P. 4, 301
Curran, S. C. 187

D

Dale, W. M. 311, 326
Dalton, J. 25
Daniel, J. W. 369
Daniels, E. W. 337, 369
Daniels, J. 5
Daudin, A. 116
Davey, W. P. 478
Davis, W. E. 452, 464
Dearnaley, G. 187
Deeley, T. J. 296
Deering, R. A. 47
Delario, A. J. 75, 488
Delbrück, M. 6, 364
Democritus 25
Desrosier, N. W. 323, 326
De Vries, M. J. 439
Dobzhansky, T. 392
Doherty, D. G. 447
Doida, Y. 362, 369
Duryee, W. R. 338, 369
Dželepow, B. 93

E

Early, P. J. 286, 289, 296
Edgar, R. S. 392
Edwards, C. L. 296
Einstein, A. 4, 21, 38
Eisenbud, M. 211, 254, 489
Eldjarn, L. 454, 464
Elkind, M. M. 368
Ellinger, F. 489
Errera, M. 326, 368, 369, 439, 484, 489
Evans, E. A. 297
Ewen, H. I. 47

F

Feinendegen, L. E. 297
Feitelberg, S. 297, 488
Fermi, E. 6, 34, 189, 197
Fields, T. 297
Fliedner, T. M. 439
Forssberg, A. 326, 369, 439, 484, 489
Fowler, J. M. 211
Francis, D. S. 359, 369
Francis, G. E. 297
Freund, L. 5
Fricke, H. 5, 302, 311, 318, 326
Fried, M. 488
Frisch, O. R. 189, 487
Fujioka, S. 435, 439
Furchner, J. E. 254
Furlan, M. 464

G

Galilei, G. 12
Gass, A. E. 320, 326
Gaston, E. O. 439
Geiger, H. 6, 144
Giacconi, R. 47
Gibbs, D. W. 188
Giles, D. 404
Ginsberg, D. M. 451, 464
Ginzburg, V. L. 116
Glass, H. B. 379
Glasstone, S. 34, 204, 211, 487
Glucksman, A. 484
Gofman, J. W. 254
Goldfeder, A. 355, 369
Grosch, D. S. 489
Gross, W. 297
Gude, W. 171, 174, 188
Gustafsson, A. 392
Guttes, S. 329, 361, 369

H

Haber, A. H. 332, 333, 369
Hahn, O. 189, 211
Haissinsky, M. 326
Hall, E. J. 357, 369, 443, 464
Hamilton, L. D. 464, 484
Hämmerling, J. 338
Hansen, C. H. 439
Harris, R. J. C. 369, 484
Hasegawa, A. T. 456, 465
Haskill, J. S. 326
Hassid, W. Z. 259, 277
Hayes, F. N. 187
Hayes, R. L. 188
Heath, R. L. 142, 157, 187, 494
Hempelmann, L. H. 482, 484
Henry, H. F. 254
Henshaw, P. S. 359, 369

Subject index

534

535

536